MATHS SKILLS FOR SUCCESS AT UNIVERSITY

MATHS SKILLS FOR SUCCESS AT UNIVERSITY

» KATHY BRADY » TIFFANY WINN

OXFORD
UNIVERSITY PRESS
AUSTRALIA & NEW ZEALAND

OXFORD
UNIVERSITY PRESS

Oxford University Press is a department of the University of Oxford.

It furthers the University's objective of excellence in research, scholarship, and education by publishing worldwide. Oxford is a registered trademark of Oxford University Press in the UK and in certain other countries.

Published in Australia by
Oxford University Press
253 Normanby Road, South Melbourne, Victoria 3205, Australia

First published 2017
Reprinted 2020 (D)

National Library of Australia Cataloguing-in-Publication data

Creator: Brady, Kathleen M. author.
Title: Maths skills for success at university / Kathleen Brady,
      Tiffany Winn.
ISBN 9780190302931 (paperback)
Notes: Includes index.
Target Audience: For undergraduate students.
Subjects: Mathematics—Study and teaching.
Mathematics—Textbooks.
Study skills.
Other Creators/Contributors: Winn, Tiffany author.

Edited and proofread by Karen Jayne
Typeset by Newgen KnowledgeWorks
Indexed by Jon Jermey
Printed and bound in Australia by Ligare Book Printers Pty Ltd

BRIEF CONTENTS

CONTENTS

ABOUT THE AUTHORS

Dr. Kathy Brady

After a career spanning over thirty years in secondary schooling in leadership and management, and as a mathematics teacher, Kathy commenced at Flinders University in 2007 as a Lecturer in Mathematics Education in the School of Education. In the later part of 2012, Kathy was appointed the Head of the Flinders University Student Learning Centre. Kathy has a particular interest in developing students' academic numeracy skills, and she has been instrumental in establishing whole-of-university approaches to do so, including the development and delivery of an innovative modular credit-bearing course that has been adopted across all Faculties at Flinders University. Kathy is a co-author of *Teaching Mathematics: Foundations to Middle Years* (2nd Ed.), Oxford University Press, 2015.

Dr. Tiffany Winn

Tiffany is a Lecturer at Flinders University, working in the Student Learning Centre. She lectures in the Flinders University credit-bearing academic numeracy course, and teaches numeracy skills in context across a range of Faculties, including Nursing and Social Sciences. Tiffany has published several papers in the area of mathematics teaching and learning, focusing on teaching mathematics through patterns, on identifying well-designed software for mathematics teaching, and on the use of metaphors in teaching mathematics. Tiffany's qualifications include a science degree and a PhD in computer science.

ACKNOWLEDGMENTS

An innovative publication such as this does not occur without the authors receiving much support and many contributions from a range of individuals, who rightfully need to be acknowledged.

Firstly, we are indebted to the team at Oxford University Press. In particular, we acknowledge the support that Debra James, Publishing Manager (Higher Education), who has given this publication enthusiastic support, from the first inkling of an idea that Kathy put to her, right through to its fruition. We would also sincerely like to thank Jennifer Butler, Managing Editor (Higher Education), and Karen Jayne for their outstanding editorial services. It has been a privilege to work with the staff at Oxford University Press.

As individuals there we would also like to acknowledge and thank a number of people who have made a contribution to the success of this publication. Kathy is most grateful for the support she received from her husband and family, and her work colleagues, during the development of this publication. She would also like to particularly acknowledge the vision of Professor Andrew Parkin (formerly DVC(A) at Flinders University) for his promotion of academic numeracy being of key importance for success in university study.

Tiffany wishes, firstly, to acknowledge the importance of her parents who brought up three very engaged learners. She also recognises the significance and importance of her children Jesse, Jordan and Soraya, in how much they have taught her. Finally, Tiffany offers a special thanks to Kathy, for being a wonderful boss and mentor, and Adam, for his support throughout the writing and development process.

Kathy Brady and Tiffany Winn

SYMBOLS AND TERMINOLOGY

NUMBER

+	add, addition, plus, total
−	minus, subtract, take away
×	multiply, multiplication, times
÷	divide, division
=	equals, equals to
≠	is not equal to
<	less than
>	greater than
≤	less than or equal to
≥	greater than or equal to
%	per cent, percentage
:	ratio, 'is to'
$\sqrt{}$	square root
$n!$	factorial notation

GEOMETRY

°	degree
∠	angle
∟	right angle or 90°
‖	parallel
π	The Greek letter, Pi

MEASUREMENT

m	metre
cm	centimetre
mm	millimetre
km	kilometre
kg	kilogram
g	gram
mg	milligram
μg	microgram
ML	megalitre
kL	kilolitre
L	litre
mL	millilitre
cm^2	square centimetre
m^2	square metre
km^2	square kilometre
cm^3	cubic centimetre
m^3	cubic metre
km^3	cubic kilometre

STATISTICS

$\bar{x}$ or μ	mean of data
σ	standard deviation (statistics)

1

IT'S NOT AS SCARY AS YOU THINK!

Welcome to *Maths Skills for Success at University*. We have written this book for students just like you – who need to do some maths as part of their course but are concerned about doing so. For some students this is because they may not have done much maths since leaving school (which could have been quite a while ago). For other students maths might have always been a bit of a struggle and so they may be lacking in confidence or even feeling quite anxious.

We would like to start this book by emphasising that maths is not as scary as you think and that by using this book you will be prepared to conquer the maths requirements in your course. In this introductory chapter, we will look at a few strategies you can use on your road to success. We will give you a chance to record how you feel about maths because doing so is a great way to identify what your hurdles might be. We have found that when we encourage students to reflectively record their attitudes towards maths they develop a more positive outlook.

So why might university students think that maths is scary? Well, there are many reasons that may include:

- Only those that are good at 'it' can do maths.
- I haven't done 'it' in years.
- I was never good at 'it' at school.
- I just haven't got a maths brain.
- I can't use a calculator.
- There's always too much information that is taught too fast.
- All of those numbers and symbols are plain confusing.

REFLECTIVE ACTIVITY 1

So … is maths a bit scary for you and what makes it so? Record your thoughts in your notebook.

LEARNING MATHS IS LIKE LEARNING A NEW LANGUAGE

There are some things that are peculiar to learning and using maths that can make it a challenge. To begin with, learning maths is like learning a new language. Maths is full of words that can be foreign, including terms such as:

- quotient
- exponent
- factor
- reciprocal
- hypotenuse
- histogram.

A glossary is included in this book for you to use for maths words that may be unfamiliar.

And maths is not just about new or unfamiliar words and terms; it also involves the use of many symbols and other forms of notation that replace words. Knowing that learning maths is like learning a new language means that some of the strategies that work well when learning a language will also work well when learning maths such as:

- creating you own glossary with new terms and their definitions
- continually reviewing the meaning of new terms
- making a conscious effort to use the language of maths correctly.

REFLECTIVE ACTIVITY 2

Use the index at the back of this book to locate the terms that have been listed above, and then try to write in your own words a basic definition for each term. Alternatively, locate the meaning of these terms in an easy to use web-based resource (such as www.mathsisfun.com). Hopefully, you can see that this process is both easy and useful.

MATHS IS SOMETHING THAT YOU DO, NOT JUST READ OR VIEW

Another aspect that is specific to learning and using maths is that it simply must be done, and not just read or viewed. We have heard students say many times that they understood everything perfectly well in our lectures when watching us complete examples on the board, and then when they sat down to do exercises themselves they went blank. What these students do not realise is that they do not really get 'it' when watching the lecture, even if they think they do. You can only get 'it' by doing 'it' with repeated practice. Think of learning and using maths in the same way as learning and using a sport skill, such as playing golf. If only learning to play golf was as easy as watching the great champions on television. Anyone who has only partially mastered the game of golf will tell you that it only results from constant, repeated practice.

The message here is not to get lulled into a false sense of security, thinking that you understand something new without having done it repeatedly by yourself. You can also optimise your learning by applying these study strategies:

- Do not complete all of your out-of-class maths study in one great big block of time – 3 or 4 hours – on just one day.
- Do a little bit of maths each day – 20 to 30 minutes is all that is required.

This will keep the maths concepts and processes fresh in your mind.

STUDYING MATHS AT UNIVERSITY

Studying anything at university is quite different to being at school, and maths is no exception. For students who are already thinking that maths is a bit scary, these differences can provide further challenges. Some of the differences are:

- At school attendance is mandatory; at university attendance is optional and your lecturer is not going to follow-up if you do not turn up!
- At school teachers constantly monitor student progress and achievements; at university students receive their grades at the end of the semester and need to monitor their own progress.
- At school students see their teacher every day; at university students normally see their lecturer once or twice a week.
- At school teachers prepare students for tests that occur frequently; at university assessment activities occur less frequently and students are required to manage their own preparation.

Given these differences, the key to success at university is to manage your time well and to take responsibility for your progress through good study habits such as:

- doing some maths every day
- attending all your classes, unless it is unavoidable
- ensuring you catch up using resources that your lecturer puts online, if you miss a lecture
- keeping an organised notebook for maths
- finding or organising a study group with other students to support each other.

MATHS ANXIETY

Maths anxiety, which is a feeling of tension, apprehension or fear that interferes with learning or using maths, has been well documented as a key reason for people to have an aversion to studying maths and using even simple maths in their daily lives. Research has

Kathy Brady

found that the brain areas that are active when maths-anxious people prepare to do maths overlap with the same brain areas that register the threat of bodily harm. So the main problem is that fear stops people from using maths, and not a lack of maths skills.

There are a number of factors that cause maths anxiety. The first is that some people have less than positive maths learning experiences at school especially at a young age. These experiences can trigger a fear of not being good at maths, or even not being able to do maths at all. A second factor is that, unfortunately, there exists a general societal view that maths is hard and that only smart people can do maths. This can create unreasonable expectations and pressure, especially during the school years. Once the seeds of maths anxiety have been sown, many negative consequences can arise. For example, maths-anxious people might:

- lose confidence in themselves and in their academic abilities
- trust blindly any bills they receive, because they do not want to engage with figures and numbers
- not be able to help their children with their homework
- avoid enrolling in courses in case they contain maths
- leave courses when they encounter the 'maths part'.

So, what can be done about maths anxiety? As we have already noted, reflectively articulating your attitudes towards maths is one strategy that can develop a more positive outlook. Specifically, we have found that encouraging our students to create a personal maths metaphor is a useful reflective activity that can highlight deep-seated emotions that might exist regarding maths.

PERSONAL MATHS METAPHOR

When creating a personal maths metaphor, you compare maths to specific things or objects as a way of focusing how you feel about learning or using maths. We have found that in creating a personal maths metaphor our students have been helped to recognise the value of mathematics, which has enhanced their chance of success in their studies. So why not try it yourself?

REFLECTIVE ACTIVITY 3

Let's start creating your personal maths metaphor by firstly imagining that you are describing what maths is to someone. In your notebook, write a list of the words or phrases you might use.

REFLECTIVE ACTIVITY 4

Now imagine that you are using maths either at university or in your everyday life. In your notebook, write a list of words or phrases to describe what doing or using maths feels like to you.

REFLECTIVE ACTIVITY 5

Next think about things or objects that reflect what maths is like for you. For example:

- If mathematics was weather, what kind of weather would it be?
- If mathematics was a food, what food would it be?
- If mathematics was a colour, what colour would it be?
- If mathematics was an animal, what animal would it be?
- If mathematics was a type of music, what type of music would it be?

In your notebook, write a list of things or objects that you think mathematics is like.

REFLECTIVE ACTIVITY 6

Read over the list of words and phrases that best describes mathematics for you and the list that describes how you feel about learning or using mathematics. Now from the list of things or objects that you think mathematics is like, select the item on your list that *best* describes what mathematics is like for you. In your notebook, note all the ways that mathematics and this thing or object are alike.

REFLECTIVE ACTIVITY 7

Now you are ready to create your personal maths metaphor. Start your metaphor with the phrase: 'For me, maths is like …' Complete your metaphor by adding a short paragraph to describe the ways the thing or object you have selected and maths are similar. Think in particular about how this metaphor describes how you feel about using or doing mathematics. Here are some metaphors our students have written as an example:

- For me maths is like a roller coaster because you put a lot of courage into getting on it and doing it. And after a crazy ride, you get off the roller coaster that is maths, and you feel pride for having done it.
- For me maths is like a snowstorm. Firstly, it looks cold and terrible but once you have the right warm clothes you can see the beauty of the snow.
- For me maths is like an electrical circuit. When all the components are in place, the light bulb comes on. When a component breaks down, the circuit is cut and you're left in the dark.

In your notebook, create your personal maths metaphor.

So, we hope that we have convinced you that maths is not as scary as you think. If you apply any or all the strategies that we have suggested in this chapter, we hope that you will continue to develop your Maths Skills for Success at University.

Kathy Brady

2

WHOLE NUMBER FUNDAMENTALS

CHAPTER CONTENT

- » Introduction to whole numbers
- » Place value for whole numbers
- » Adding and subtracting whole numbers
- » Multiplying whole numbers
- » Dividing whole numbers
- » Order of operations
- » Rounding whole numbers
- » Significant figures

CHAPTER OBJECTIVES

- » Determine the place value of digits in whole numbers
- » Add and subtract whole numbers
- » Multiply and divide whole numbers
- » Apply the correct order of operations to whole number calculations
- » Round whole numbers to a specified number of decimal places or significant figures

USING WHOLE NUMBERS FOR ESTIMATION IN EVERYDAY LIFE

Whether we realise it or not, we use whole numbers extensively in our everyday life, and estimation is one of the most important skills involving whole numbers. When you shop for groceries, you might need to stick to a budget. So you probably keep a running total of the cost of items in your trolley, discarding the cents and just using whole dollar values—whole numbers. This ensures that you have enough money when you get to the checkout. For example, if you had $25 left until payday and bought milk for $3.50, cheese for $6.99, bread for $3.90, apples for $5.50 and bananas for $2.50, you could add up $4 + $7 + $4 + $6 + $3 = $24 and know that you would come in under budget.

Another shopping situation where you might estimate with whole numbers could be in calculating a discount. If you wanted a pair of jeans that were $59 but they were on sale for 30% off, a quick estimation would tell you that 10% of $60 is $6, three lots of $6 is $18, and $60 − $18 = $42, which would approximately be the cost of the jeans when on sale.

What about when you go out to dinner with friends? Rather than add up exactly who has spent what, there are probably times when you have simply divided the total cost of the bill by the number of people at dinner and obtained a rough, whole number

Making connections Decimal numbers are further discussed in Chapter 6.

estimate of each individual cost. For example, if the cost of dinner for eight people was $241.45, you could divide $241 by 8 to get approximately $30 per person as the individual cost. The reality is that whole numbers are used extensively in many everday life situations for estimation purposes.

2.1 INTRODUCTION TO WHOLE NUMBERS

In ancient times, there were no numbers to represent quantities like two, three and four. Fingers, rocks and sticks came in handy for communicating quantities. Words like 'flock', such as a flock of sheep, or 'swarm', such as a swarm of bees, indicated groups of similar animals (flock) or a large group of insects (swarm); the exact number of animals or insects represented was non-specific. Symbols for numbers came into being with the advent of villages and settlements, where bartering occurred between neighbouring groups.

Early numerical systems used symbols instead of the numbers we use now. For example, the early Egyptians used symbols such as those shown top right and the Chinese used sticks laid on tables to represent numbers, as shown bottom right. Roman numerals are still used in some contexts today; for example, the year that a movie is created is usually given in Roman numerals. A movie made in 2014 would have MMXIV listed for its year, with M being the symbol for 1000, X the symbol for 10, and the I (symbol for 1) being positioned to the left of the V (symbol for 5) together indicating 4 (IV = 5 − 1 = 4). Over many years, counting systems have evolved to give us the numbers from 0 to 9 that we use today.

1	10	100	1000
|	∩	☉	𝍦

| Stroke | Arch | Coiled Rope | Lotus Flower |

10000	100000	1000000
⟋	🐟	🧍

| Pointed Finger | Tadpole | Surprised Man |

|	||	|||	||||	|||||	T	TT	TTT	TTTT
1	2	3	4	5	6	7	8	9

—	=	≡	≣	≣	⊥	⊥	⊥	⊥
10	20	30	40	50	60	70	80	90

Tiffany Winn

2.2   PLACE VALUE FOR WHOLE NUMBERS

Place value refers to the numerical value that a digit has according to its position in a number, as shown below. Each place has a value of 10 times the place to its right. The rightmost digit in a **whole number** is in the 'ones place'—because the rightmost digit has a value of itself multiplied by one. The next digit to the left is in the 'tens place', because it has a value of itself multiplied by ten. Moving to the left digit by digit, each digit has a place value of ten times the previous one. For example, based on the position of the digits in the number 64 905, each digit has the following value:

- 5 means $5 \times 1 = 5$
- 0 means $0 \times 10 = 0$
- 9 means $9 \times 100 = 900$
- 4 means $4 \times 1000 = 4000$
- 6 means $6 \times 10\,000 = 60\,000$

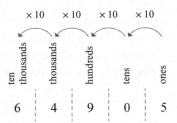

In other words, 64 905 can be written as:

$6 \times 10\,000 + 4 \times 1000 + 9 \times 100 + 0 \times 10 + 5 \times 1$

$= 60\,000 + 4000 + 900 + 0 + 5$

$= 64\,905$

As another example, the digits in the number 4 876 091 have the place value shown below and the number can be written as:

$4 \times 1\,000\,000 + 8 \times 100\,000 + 7 \times 10\,000 + 6 \times 1000 + 0 \times 100 + 9 \times 10 + 1 \times 1$

$= 4\,000\,000 + 800\,000 + 70\,000 + 6000 + 0 + 90 + 1$

$= 4\,876\,091$

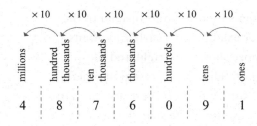

<aside>
Whole number
A positive number with no fraction or decimal part that is comprised of at least one of the digits from 0 to 9
</aside>

EXAMPLE 1

Write in words the place value of the underlined digit in each of the given numbers.

Solutions

a   1236

a   The digit 2 is in the hundreds place in 1236. (Counting left from the rightmost digit: ones, tens, hundreds)

b   5596084

b   The digit 4 is in the ones place in 5596084. (The ones place is the rightmost digit of a whole number.)

c   9214695288

c   The digit 9 is in the billions place in 9214695288. (Counting left from the rightmost digit: ones, tens, hundreds, thousands, ten thousands, hundred thousands, millions, ten millions, hundred millions, billions)

PRACTICE 1

Write in words the place value of the underlined digit in each of the given numbers.

a   412536

b   10559608467

c   97042

EXAMPLE 2

Underline the place value of the digit in the named place in the following numbers.

Solutions

a   ten thousands place in 345903

a   345903

b   ones place in 12384

b   12384

c   hundred millions place in 9760954321

c   9760954321

PRACTICE 2

Underline the place value of the digit in the named place in the following numbers.

a   hundreds place in 345903

b   ten thousands place in 12384

c   billions place in 97609546321

EXAMPLE 3

Write the following numbers in place value form.

Solutions

a   2906

a $2906 = 2 \times 1000 + 9 \times 100 + 0 \times 10 + 6 \times 1$
$= 2000 + 900 + 0 + 6$

b   1 083 920

b $1\,083\,920 = 1 \times 1\,000\,000 + 0 \times 100\,000 + 8 \times 10\,000$
$+ 3 \times 1000 + 9 \times 100 + 2 \times 10 + 0 \times 1$
$= 1\,000\,000 + 0 + 80\,000 + 3000 + 900$
$+ 20 + 0$

c   40 067

c $40\,067 = 4 \times 10\,000 + 0 \times 1000 + 0 \times 100 + 6 \times 10$
$+ 7 \times 1$
$= 40\,000 + 0 + 0 + 60 + 7$

PRACTICE 3

Write the following numbers in place value form.

a   12 086   b   291   c   4 670 068

2.3   ADDING AND SUBTRACTING WHOLE NUMBERS

Adding or subtracting whole numbers involves two steps:

1   lining up two numbers according to their place value

2   adding or subtracting corresponding digits in each number, working from right to left.

Addition

The first step in addition is lining up numbers according to place value. This means lining up the numbers so that digits with the same place value are directly above or below each other, as shown on the left below. When adding 782 and 7, the 2 in 782 and the 7 in 7 are both in the ones place, so these digits need to be directly above and below each other. On the right below, crossed out in green, are two incorrect ways of lining up the two numbers 782 and 7.

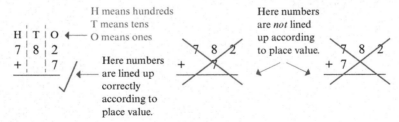

The second step in addition—adding digits in the same column, working from right to left—is shown below for the sum 782 + 7. The three substeps involved in the second addition step are shown as (a), (b) and (c). First (a), the digits in the rightmost column (the ones column) are added (2 + 7), then (b) the digits in the tens column (8 + 0), and (c) finally the digits in the hundreds column (7 + 0). This gives the result 782 + 7 = 789.

(a): 2 + 7 = 9   (b): 8 + 0 = 8   Note that 7 is the same   (c): 7 + 0 = 7
as 07 and 007, so the
blank spaces in front of
the 7 are counted as
zeros.

Th means thousands

Th	H	T	O
	7	8	2
+			7
			9

Th	H	T	O
	7	8	2
+			7
		8	9

Th	H	T	O
	7	8	2
+			7
	7	8	9

The second step in addition is shown below for the sum 433 + 52. The three substeps involved in the second addition step are shown as (a), (b) and (c). First (a), the digits in the rightmost column (the ones column) are added (3 + 2), then (b) the digits in the tens column (3 + 5), and finally (c) the digits in the hundreds column (4 + 0). This gives the result 433 + 52 = 485.

(a): 3 + 2 = 5   (b): 3 + 5 = 8   (c): 4 + 0 = 4

Th means thousands
H means hundreds
T means tens
O means ones

Th	H	T	O
	4	3	3
+		5	2
			5

Th	H	T	O
	4	3	3
+		5	2
		8	5

Th	H	T	O
	4	3	3
+		5	2
	4	8	5

Note that 52 is the same as 052, so the blank space in front of 52 is counted as zero.

EXAMPLE 4

Add 40 672 and 7213.

Solution

Line up the numbers according to place value, and then add individual digits one by one, moving from right to left, as shown below.

(a): Add rightmost column (ones column): 2 + 3 = 5

(b): Add next column to the left (tens column): 7 + 1 = 8

(c): Add next column to the left (hundreds column): 6 + 2 = 8

(d): Add next column to the left (thousands column): 0 + 7 = 7

(e): Add next column to the left (ten thousands column): 4 + 0 = 4

```
  40672        40672        40672        40672        40672
+  7213      +  7213      +  7213      +  7213      +  7213
      5           85          885         7885        47885
```

The result is 40 672 + 7213 = 47 885.

PRACTICE 4

Add the following numbers.

a 451 032 and 64   b 528 and 9431   c 45 274 and 4312

Addition with carrying

When adding two numbers together, if two digits are added and the result is more than 9, then this number needs to be carried and added onto the digits in the next column to the left. An example of addition with carrying is shown below. Note that the numbers have been lined up correctly according to place value before the digit-by-digit addition is completed.

The three substeps involved in the second addition step are shown below as (a), (b) and (c). First (a), the digits in the rightmost column (the ones column) are added (6 + 7). Since 6 + 7 = 13 and this is greater than 9, the 3 from 13 is written as the result in the ones column and the 1 from 13 is carried to the tens column. Then (b), the digits in the tens column are added (5 + 4), and the carry (1) is added to them, giving 5 + 4 + 1 = 10. Since 10 is greater than 9, the 0 from 10 is written as the result in the tens column and the 1 from 10 is carried to the hundreds column. Finally (c), the digits in the hundreds column are added (7 + 0), and the carry (1) is added to them, giving 7 + 0 + 1 = 8. This result is written in the hundreds column. Since there are no more digits, the addition is complete.

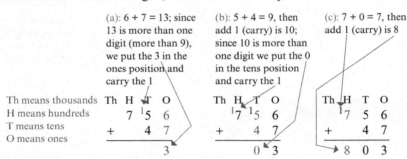

(a): 6 + 7 = 13; since 13 is more than one digit (more than 9), we put the 3 in the ones position and carry the 1

(b): 5 + 4 = 9, then add 1 (carry) is 10; since 10 is more than one digit we put the 0 in the tens position and carry the 1

(c): 7 + 0 = 7, then add 1 (carry) is 8

Th means thousands
H means hundreds
T means tens
O means ones

Th	H	T	O
	7	^{1}5	6
+		4	7
			3

Th	H	T	O
^{1}7		^{1}5	6
+		4	7
		0	3

Th	H	T	O
^{1}7		5	6
+		4	7
8		0	3

EXAMPLE 5

What is the sum of 93 672 and 7253?

Solution

The numbers are lined up according to place value and then added, digit by digit, as shown below.

(a): Add rightmost column (ones column): 2 + 3 = 5. Since 5 is less than 10, the 5 is simply written in the ones column of the result.

(b): Add next column to the left (tens column): 7 + 5 = 12. Since 12 is greater than 9, write the 2 in the result and carry the 1.

(c): Add next column to the left (hundreds column) and add the carry (1): 6 + 2 + 1 = 9. Since 9 is less than 10, there is no carry.

(d): Add next column to the left (thousands column): 3 + 7 = 10. Since 10 is greater than 9, write the 0 in the result and carry the 1.

(e): Add next column to the left (ten thousands column) and add the carry: 9 + 0 + 1 = 10. Since this is the leftmost column, the 10 can be written directly in the result instead of carrying the 1.

```
  93672
+  7253
_____
      5
```

```
  93$^1$672
+  7253
_____
     25
```

```
  93$^1$672
+  7253
_____
    925
```

```
 $^1$93672
+  7253
_____
   0925
```

```
 $^1$93672
+  7253
_____
 100925
```

The sum of 93 672 and 7253 is 100 925.

PRACTICE 5

Determine the sum of the following numbers.

a 761 992 and 99

c 45 658 and 4808

b 528 and 9723 and 10 198

d 0909 and 444

EXAMPLE 6: APPLICATION

Jim's coffee shop sold eight hundred and thirty-three cups of coffee on Monday, six hundred and forty on Tuesday, five hundred and ninety-two on Wednesday, six hundred and twelve on Thursday, and eight hundred and twenty-nine on Friday. How many cups of coffee did they sell for the week?

Solution

The total number of cups of coffee sold is calculated by adding all the cups sold on individual days, as shown below.

$$
\begin{array}{r}
{}^2 8 {}^1 3\ 3 \\
6\ 4\ 0 \\
5\ 9\ 2 \\
6\ 1\ 2 \\
+\ 8\ 2\ 9 \\
\hline
3\ 5\ 0\ 6
\end{array}
$$

The shop sold a total of 3506 cups of coffee during the week.

PRACTICE 6: APPLICATION

Hospital admissions for two hospitals are shown in the table below. Which hospital had more admissions for the week shown?

	SUN	MON	TUE	WED	THU	FRI	SAT
Hospital A	22	33	18	19	21	30	42
Hospital B	31	22	14	23	46	18	30

Subtraction

Subtraction is lined up the same way as addition but the individual digits are subtracted, and instead of carrying, if necessary, the technique of borrowing is used (see next subsection on subtraction with borrowing). The first step in subtraction is lining up the digits according to their place value, as shown overpage. The second and final step is to subtract the corresponding digits in each of the two numbers, going from right to left, and taking the bottom digit away from the top digit—just the same as for addition except that the bottom digit is subtracted from the top digit instead of being added.

An example of subtraction is shown below. The three substeps involved in the second subtraction step are shown as (a), (b) and (c). First (a), the digits in the rightmost column (the ones column) are subtracted $(3 - 2 = 1)$ and the result (1) is written in the ones column. Then (b), the digits in the tens column—the next column to the left—are subtracted $(8 - 5 = 3)$ and the result (3) is written in the tens column. Finally (c), the digits in the hundreds column are subtracted $(4 - 0 = 4)$ and the result (4) is written in the hundreds column. For example:

	(a): $3 - 2 = 1$	(b): $8 - 5 = 3$	(c): $4 - 0 = 4$

	Th H T O	Th H T O	Th H T O
Th thousands H hundreds T tens O ones	4 8 3 − 5 2	4 8 3 − 5 2	4 8 3 − 5 2
	1	3 1	4 3 1

Note that 52 is the same as 052, so the blank space in front of 52 is counted as zero.

EXAMPLE 7

Subtract 8443 from 998 576.

Solution

Line up the numbers vertically according to place value and then proceed as shown below.

(a): Subtract rightmost column (ones column): $6 - 3 = 3$	(b): Subtract next column to the left (tens column): $7 - 4 = 3$	(c): Subtract next column to the left (hundreds column): $5 - 4 = 1$	(d): Subtract next column to the left (thousands column): $8 - 8 = 0$	(e): Subtract next column to the left (ten thousands column): $9 - 0 = 9$	(f): Subtract next column to the left (hundred thousands column): $9 - 0 = 9$
998 576 − 8 4 4 3	998 576 − 8 4 4 3	998 576 − 8 4 4 3	998 576 − 8 4 4 3	998 576 − 8 4 4 3	998 576 − 8 4 4 3
3	3 3	1 3 3	0 1 3 3	9 0 1 3 3	9 90 1 3 3

The result is 990 133.

PRACTICE 7

Subtract the following numbers.

a 2605 from 163 798   b 417 from 9839   c 2027 from 5278

Subtraction with borrowing

When subtracting, if the bottom digit is larger than the corresponding top digit, the bottom digit cannot be subtracted from the top one without getting a negative number as a result. In this case, to avoid a negative number result, the strategy is to borrow from the next column to the left. This is done by reducing the digit in the next column to the left by one, and then adding ten to the current digit. An example of subtraction with borrowing is shown overleaf.

Note that the two numbers have been lined up according to place value before the digit-by-digit subtraction is completed.

The three substeps involved in the second subtraction step are shown below as (a), (b) and (c). First (a), the digits in the rightmost column are considered. Since the top digit (3) is less than the bottom digit (4), ten is borrowed from the next column to the left; the 8 is reduced to 7 and a 1 is placed in front of the 3 in the rightmost column so that it becomes 13. Subtraction for the digits in the rightmost column is then performed ($13 - 4 = 9$) and the result (9) is written in the rightmost column of the answer. The digits in the tens column are then considered (b) and since the top digit (7) is greater than or equal to the bottom digit (5), subtraction is performed ($7 - 5 = 2$) and the result (2) is written in the corresponding column in the answer. Similarly, for the final column (c), the subtraction is performed ($4 - 0 = 4$) and the result (4) is written in the answer. Overall, this subtraction demonstrates that $483 - 54 = 429$.

(a): 4 is larger than 3, so borrow from the tens column: $13 - 4 = 9$	(b): Subtract next column to left (tens column): 7 is larger than 5, so just subtract to get $7 - 5 = 2$	(c): Subtract next column to left (hundreds column): 4 is larger than 0, so just subtract to get $4 - 0 = 4$	

Th H T O
4 $^7\!\!\!\not8$ ^{1}3
+   5 4
———
          9
	Th H T O
	4 $^7\!\!\!\not8$ ^{1}3
	+ 5 4
	———
	      2 9
		Th H T O
		4 $^7\!\!\!\not8$ ^{1}3
		+ 5 4
		———
		   4 2 9
			Note that 54 is the same as 054, so the blank space in front of 54 is counted as zero.

EXAMPLE 8

Subtract 54 from 402.

Solution

The three substeps involved in the second subtraction step are shown overpage as (a), (b) and (c). First, in part (a), the digits in the rightmost column are considered. Since the top digit (2) is less than the bottom digit (4), we look to borrow ten from the next column to the left (the tens column). However, the tens column has a zero in it so cannot be borrowed from—so we look to the next column to the left (the hundreds column) and borrow from the 4 in the hundreds column, making it 3. This means we can add 10 to the tens column, making that now 10 instead of 0. Now we can borrow from the tens column, making that 9, and adding 10 to the ones column, making the 2 now 12. Subtraction for the digits in the rightmost column is then performed ($12 - 4 = 8$) and the result (8) is written in the rightmost column of the answer. The digits in the tens column are then considered and since the top digit (9) is greater than or equal to the bottom digit (5), subtraction is performed ($9 - 5 = 4$) and the result (4) is written in the corresponding column in the answer. Similarly, for the final column ($3 - 0 = 3$) and the result (3) is written in the answer. Overall, this subtraction demonstrates that $402 - 54 = 348$.

(a): 2 is smaller than 4, so look to borrow from tens column. But tens column is zero, so borrow from hundreds column, reducing hundreds column from 4 to 3 and adding ten to tens column (step not shown). Then borrow from tens column, reducing it from 10 to 9 and adding 10 to ones column to give 12. Then perform subtraction: $12 - 4 = 8$

$$\begin{array}{ccccc} \text{Th} & \text{H} & \text{T} & \text{O} \\ & \overset{3}{\cancel{4}} & \overset{9}{\cancel{0}} & \overset{1}{}2 \\ - & & 5 & 4 \\ \hline & & & 8 \end{array}$$

(b): Now subtract digits in the next column to the left (tens column): $9 - 5 = 4$

$$\begin{array}{ccccc} \text{Th} & \text{H} & \text{T} & \text{O} \\ & \overset{3}{\cancel{4}} & \overset{9}{\cancel{0}} & \overset{1}{}2 \\ - & & 5 & 4 \\ \hline & & 4 & 8 \end{array}$$

(c): Now subtract digits in the next column to the left (hundreds column): $3 - 0 = 3$

$$\begin{array}{ccccc} \text{Th} & \text{H} & \text{T} & \text{O} \\ & \overset{3}{\cancel{4}} & \overset{9}{\cancel{0}} & \overset{1}{}2 \\ - & & 5 & 4 \\ \hline & 3 & 4 & 8 \end{array}$$

The result is 348.

PRACTICE 8

Subtract the following numbers.

a   2605 from 163 284

b   487 from 9013

c   36 297 from 43 006

d $804 - 56 - 49$

EXAMPLE 9: APPLICATION

Viv's pay is $1204 but she owes a friend $709. How much does Viv have left to spend?

Solution

The amount left to spend is the result when 709 is subtracted from 1204:

$$\begin{array}{cccc} \overset{11}{\cancel{2}} & \overset{9}{\cancel{0}} & \overset{1}{}4 \\ - & 7 & 0 & 9 \\ \hline & 4 & 9 & 5 \end{array}$$

So, the amount left for Viv to spend is $495.

PRACTICE 9: APPLICATION

When baking recently, Tiffany made 12 dozen cookies for a party. Before they had cooled down, 25 of them had been eaten by the family. How many cookies were left for the party?

2.4 MULTIPLYING WHOLE NUMBERS

Multiplication is the process of adding a number to itself a certain number of times. For example, '5 times 2' means five lots of 2, or '2 plus 2 plus 2 plus 2 plus 2'. In other words, $5 \times 2 = 10$.

When two numbers are multiplied together, each of the numbers being multiplied is called a **factor** and the result of the multiplication is called a **product**.

$$6 \times 2 = 12$$

Factor Factor Product

Factor
A whole number that multiplies with another whole number with the result being the product. A factor can also be defined as a whole number that divides into another whole number without a remainder.

Product
The result of the multiplication of two or more numbers

The multiplication of two, single-digit numbers is usually done simply by using the times tables of those digits. When one or more of the numbers has more than one digit, long multiplication may be useful. The first step in doing long multiplication is lining up numbers according to place value. The second step in doing long multiplication is multiplying the top number by each of the digits in the bottom number, in turn, going from right to left. Each answer is lined up according to the place value of the digit in the bottom row. Note that when you multiply individual digits and the result of the multiplication has more than one digit, you may need to carry.

Consider the example of multiplying 49 by 63:

Step 1: line up numbers according to place value

```
H | T | O
  | 4 | 9
× | 6 | 3
```

Step 2a: multiply 49 by 3

first: 9×3

```
Th  H   T   O
        ²4   9
    ×   6   3
                7
```

$9 \times 3 = 27$
Note that instead of writing 27 as the result, we write the 7 in the ones column and *carry* the 2.

second: 40×3 (think of this as 4×3, put result in tens place), but *you must add the carry to the result*

```
Th  H   T   O
        ²4   9
    ×   6   3
    1   4   7
```

$4 \times 3 = 12$ and $12 + 2 = 14$
We write the 4 in the tens column and since 4 is the leftmost digit of the top number, we have now finished multiplying and can write 14 straight in as the result.

Step 2b: multiply 49 by 60 (think of this as 49×6, put the zero in the ones place, and then start writing the result from the tens place)

first: 9×60 (think of this as 9×6, put 0 in ones place and start result in tens place)

$9 \times 6 = 54$

```
Th  H   T   O
        ⁵4   9
    ×   6   3
    1   4   7
        4   0
```

Note that instead of writing 54 as the result, we put the 4 in the tens place and carry the 5.

second: 40×60 (think of this as 40×6, put result in hundreds place)

```
Th  H   T   O
        ⁵4   9
    ×   6   3
    1   4   7
2   9   4   0
```

Note that our result is 29 ($4 \times 6 + 5$, where 5 is the carry from the first part of Step 2b). Since 4 is the leftmost digit of the top number, we have now finished multiplying and can write 29 straight in as the result.

Tiffany Winn

Step 3: add each answer obtained from the digit-by-digit multiplication

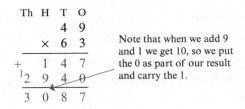

Note that when we add 9 and 1 we get 10, so we put the 0 as part of our result and carry the 1.

Overall, the result of this multiplication is that $49 \times 63 = 3087$.

EXAMPLE 10

Multiply 4903×812.

Solution

The two numbers are lined up according to place value (Step 1). For Step 2, going from right to left, each digit in the bottom number is multiplied by the top number. Finally, for Step 3 the results of each substep of the multiplication are added together. Steps 2 and 3 are shown below.

Step 2a: Multiply 4903 by 2.

Step 2b: Multiply 4903 by 1; since 1 is in the tens column put 0 in the ones column.

Step 2c: Multiply 4903 by 8; since 8 is in the hundreds column put zeros in the ones and tens columns.

Step 3: Add the result of each step of multiplication to get the final answer.

```
  ¹4 9 0 3              4 9 0 3           ⁷4 9²0 3              4 9 0 3
×     8 1 2         ×      8 1 2        ×     8 1 2         ×      8 1 2
  9 8 0 6              9 8 0 6              9 8 0 6           ¹9 8 0 6
                      4 9 0 3 0          4 9 0 3 0          ²4 9 0 3 0
                                       3 9 2 2 4 0 0      + 3 9 2 2 4 0 0
                                                            3 9 8 1 2 3 6
```

The result is 3 981 236.

PRACTICE 10

Multiply the following numbers.

a 493×89 **b** $96\,074 \times 318$ **c** 4152×5934

EXAMPLE 11

Multiply 576 × 804.

Solution

Step 1 is to line up numbers according to place value. Steps 2 and 3 are shown below.

Step 2a: Multiply 576 by 4.

Step 2b: Put 0 in ones place and then multiply 576 by 0. Note that any number multipled by 0 is 0.

Step 2c: Put 0 in ones and tens place. Then multiply 576 by 8.

Step 3: Add up the results of Steps 2a, 2b and 2c.

$$
\begin{array}{r}
{}^{3}{}_{5}{}^{2}76 \\
\times\ 804 \\
\hline
2304 \\
\end{array}
$$

$$
\begin{array}{r}
576 \\
\times\ 804 \\
\hline
2304 \\
0000 \\
\end{array}
$$

$$
\begin{array}{r}
{}^{6}{}_{5}{}^{4}76 \\
\times\ 804 \\
\hline
2304 \\
0000 \\
460800 \\
\end{array}
$$

$$
\begin{array}{r}
576 \\
\times\ 804 \\
\hline
{}^{1}2304 \\
0000 \\
+460800 \\
\hline
463104 \\
\end{array}
$$

NOTE: When two numbers are arranged one above the other for multiplication, according to place value, if one of the digits in the bottom number is zero, then the step that involves that multiplication (Step 2b in the diagram above) can be expedited as shown below.

Step 2a: Multiply 576 by 4.

Step 2b: Put 0 in ones place. The next step is to multiply 576 by 0, but since 576 × 0 will result in a row of zeros, as shown in Step 2b above, it is sufficient to put a zero in the tens place and then continue with Step 2c on the same row, starting from the hundreds place.

Step 2c: Multiply 576 by 8. (Extra zeros are already in place in ones and tens places.)

Step 3: Add up the results of Steps 2a, 2b and 2c.

$$
\begin{array}{r}
{}^{3}{}_{5}{}^{2}76 \\
\times\ 804 \\
\hline
2304 \\
\end{array}
$$

$$
\begin{array}{r}
576 \\
\times\ 804 \\
\hline
2304 \\
00 \\
\end{array}
$$

$$
\begin{array}{r}
{}^{6}{}_{5}{}^{4}76 \\
\times\ 804 \\
\hline
2304 \\
460800 \\
\end{array}
$$

$$
\begin{array}{r}
576 \\
\times\ 804 \\
\hline
{}^{1}2304 \\
+460800 \\
\hline
463104 \\
\end{array}
$$

The result is 463 104.

PRACTICE 11

Multiply the following numbers.

a 9879 × 809   **b** 847 × 900   **c** 1603 × 8042

EXAMPLE 12: APPLICATION

Apples are packed in boxes of 128. A particular truck can carry 236 boxes. How many apples can be taken in one trip on the truck?

Solution

Step 1 is to line up the numbers according to place value. Steps 2 and 3 are shown below.

Step 2a: Multiply 236 by 8.

Step 2b: Put 0 in ones place and then multiply 236 by 2.

Step 2c: Put 0 in ones and tens place. Then multiply 236 by 1.

Step 3: Add up the results of Steps 2a, 2b and 2c.

$$
\begin{array}{r}
{}^{2}{}_{2}{}^{4}3\,6 \\
\times\ 1\,2\,8 \\
\hline
1\,8\,8\,8
\end{array}
$$

$$
\begin{array}{r}
2\,{}^{1}3\,6 \\
\times\ 1\,2\,8 \\
\hline
1\,8\,8\,8 \\
4\,7\,2\,0
\end{array}
$$

$$
\begin{array}{r}
2\,3\,6 \\
\times\ 1\,2\,8 \\
\hline
1\,8\,8\,8 \\
4\,7\,2\,0 \\
2\,3\,6\,0\,0
\end{array}
$$

$$
\begin{array}{r}
2\,3\,6 \\
\times\ 1\,2\,8 \\
\hline
{}^{2}1\,{}^{1}8\,8\,8 \\
4\,7\,2\,0 \\
+\,{}^{1}2\,3\,6\,0\,0 \\
\hline
3\,0\,2\,0\,8
\end{array}
$$

So, 30 208 apples can be taken in one trip on the truck.

PRACTICE 12: APPLICATION

A group of thirty-six people is planning a trip to London together. The adult fare from Sydney to London is $2236. What will be the total airfare cost for the group?

2.5   DIVIDING WHOLE NUMBERS

Dividing one number into another means working out how many times one number fits into another one. For example, when we say 6 divided by 2 is 3, what we mean is that 2 fits into 6, 3 times. This can be written in many ways, as shown below.

$$
6 \div 2 = 3 \quad\Big|\quad \frac{6}{2} = 3 \quad\Big|\quad 6/2 = 3 \quad\Big|\quad 2\overline{)6}^{\;3}
$$

All of these statements are different ways of saying that 2 fits into 6, 3 times.

As shown, the 6 is called the **dividend** (that's the number we are dividing 'into'), the 2 is called the **divisor** (that's the number of 'groups' we have), and 3 is called the **quotient** (that's our answer). The quotient represents how many items are in each of the groups that is formed.

$$6 \div 2 = 12$$

Dividend   Divisor   Quotient

Dividend
The number being divided into

Divisor
The number of groups that the dividend is being split into

Quotient
The result of a division

Division method 1: Short division

Short division involves working out how many times one number fits into each digit of another number, digit by digit, going from left to right. For example, the question 'What is 230 divided by 5?' or 'What is 230 ÷ 5?' is answered using short division below.

Step 1: **Does 5 go into 2?** No it does not. 5 is bigger than 2.

Step 2: **Does 5 go into 23?** Note that since 5 did not go into 2, we now look at 23 rather than just 3.

Step 3: **Does 5 go into 30?** Note that the 30 is made up of the carry (3) and the next digit in the dividend (0).

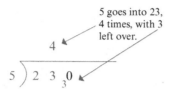

5 goes into 23, 4 times, with 3 left over.

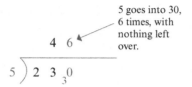

5 goes into 30, 6 times, with nothing left over.

Overall, the result of this division is 230 ÷ 5 = 46.

EXAMPLE 13

What is 356 ÷ 2?

Solution

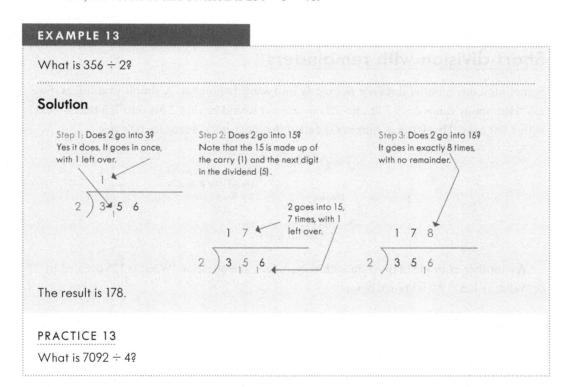

Step 1: Does 2 go into 3? Yes it does. It goes in once, with 1 left over.

Step 2: Does 2 go into 15? Note that the 15 is made up of the carry (1) and the next digit in the dividend (5).

2 goes into 15, 7 times, with 1 left over.

Step 3: Does 2 go into 16? It goes in exactly 8 times, with no remainder.

The result is 178.

PRACTICE 13

What is 7092 ÷ 4?

EXAMPLE 14

What is 93 168 ÷ 3?

Solution

As shown below, the answer (the quotient) when 93 168 is divided by 3 is 31 056. First, 3 goes into 9 three times exactly. Second, 3 goes into 3 one time exactly. Third, 3 does not fit into 1, so the 1 is carried to the fourth step. Fourth, 3 goes into 16 five times with 1 left over. Fifth, three goes into 18 six times exactly.

```
      3  1  0  5  6
   ────────────────
 3 ) 9  3  1  6  8
            1  1
```

The result is 31 056.

PRACTICE 14

What is 23 176 ÷ 8?

Short division with remainders

Sometimes, one number does not fit exactly into another number. A simple example is if we ask 'How many times does 2 fit into 7?' our answer would be that 2 fits into 7, 3 times—but with 1 left over. This leftover number is called the remainder, as shown below.

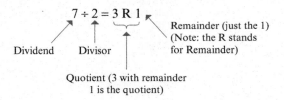

$$7 \div 2 = 3 \text{ R } 1$$

Dividend   Divisor   Remainder (just the 1)
(Note: the R stands for Remainder)

Quotient (3 with remainder 1 is the quotient)

As another example of division with remainders, the question 'What is 125 divided by 2?' or 'What is 125 ÷ 2?' is shown below.

Step 1: Does 2 go into 1? No it does not. 2 is bigger than 1.

```
   ─────────
 2 ) 1  2  5
```

Step 2: Does 2 go into 12? Note that since 2 did not go into 1, we now look at 12 rather than just 2.

```
          6
   ─────────
 2 ) 1  2  5
```

2 goes into 12, 6 times exactly. Note there is nothing left over, so no carry.

Step 3: 2 goes into 5, 2 times, with 1 left over. Since we still have a number left over after dividing into the last digit of 125, we write Remainder 1 (or R1 for short) at the end of our answer.

```
       6  2   Remainder 1
   ─────────
 2 ) 1  2  5
```

Our answer (or quotient) is 62 remainder 1.

EXAMPLE 15

What is 5387 ÷ 6?

Solution

Step 1: Does 6 go into 5?
No it does not — it's too big.
So in the next step, we look
not just at the next digit (3)
but at 53 (the number
comprised of this digit and
the next digit).

Step 2: Does 6 go into 53?
It goes in 8 times (6 × 8 = 48),
with 5 left over. Note the 8 is
written above the 3 in 53.

Step 3: Does 6 go into 58?
It goes in 9 times (6 × 9 = 54),
with 4 left over. Note the
58 is made up of the carry
from the previous step (5)
and the current digit (8).

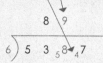

Step 4: Does 6 go into 47? It
goes in 7 times (6 × 7 = 42), with
5 left over. Since we are at the
last digit the 5 left over is our
remainder.

```
      8 9 7 R 5
   _____
6 ) 5 3 8 ₄7
```

The result is 897 remainder 5.

PRACTICE 15

What is 4236 ÷ 9?

EXAMPLE 16

What is 70 216 ÷ 7?

Solution

Step 1: Does 7 go into
7? Yes it does, exactly
1 time (7 × 1 = 7) with
nothing left over.

Step 2: Does 7 go into 0? No it does not
— it's too big. But since the 0 is not the
first (leftmost) digit in the dividend,
we need to write in that 7 does not fit
into 0 (or that 7 fits into 0, zero times),
so we write a 0. Since 7 does not fit
into 0, in the next step we look not
just at the next digit (2) but at 02 (the
number comprised of this digit and the
next digit) — but as it happens, 02 is
the same as 2, so we can just focus on
the 2.

Step 3: Does 7 fit into 2? No it
does not. But since the 2 is not
the first (leftmost) digit in the
dividend, we need to write in
that 7 does not fit into 2 (or
that 7 fits into 2, zero times),
so we write a 0. Since 7 does
not fit into 2, in the next step
we look not just at the next
digit (1) but at 21 (the
number comprised of this
digit and the next digit).

```
        1
   _____
7 ) 7 0 2 1 6
```

```
        1 0
   _____
7 ) 7 0 2 1 6
```

```
        1 0 0
   _____
7 ) 7 0 2 1 6
```

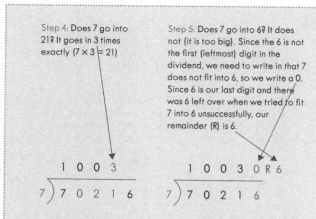

Step 4: Does 7 go into 21? It goes in 3 times exactly (7 × 3 = 21)

Step 5: Does 7 go into 6? It does not (it is too big). Since the 6 is not the first (leftmost) digit in the dividend, we need to write in that 7 does not fit into 6, so we write a 0. Since 6 is our last digit and there was 6 left over when we tried to fit 7 into 6 unsuccessfully, our remainder (R) is 6.

$$7 \overline{)70216} = 1003$$

$$7 \overline{)70216} = 10030 \ R \ 6$$

The result for 70 216 ÷ 7 is 10 030 remainder 6.

PRACTICE 16

What is 48 063 ÷ 6?

EXAMPLE 17: APPLICATION

Grandma has $1300 allocated for buying birthday presents for her grandchildren this year. She likes to spend the same amount on each child and she has fifteen grandchildren. How much money will she be able to spend on each grandchild? Ignore cents in your answer and state how many dollars she will be able to spend on each grandchild.

Solution

The amount grandma has to spend on each child will be the result when the total amount of money ($1300) is divided by the number of grandchildren (15). This division is shown below.

Step 1: Does 15 go into 1? No it does not — it's too big. So in the next step, we look not just at the next digit (3) but at 13 (the number comprised of this digit and the next digit).

$$15 \overline{)1300}$$

Step 2: Does 15 go into 13? No, it's still too big. So in the next step, we look not just at the next digit (5) but at 130 (the number comprised of 13 and the next digit).

$$15 \overline{)1300}$$

Step 3: Does 15 go into 130? Yes. It goes in 8 times (15 × 8 = 120), with 10 left over.

$$15 \overline{)130_{10}0} = 8$$

Step 4: Does 15 go into 100? Yes. It goes in 6 times (15 × 6 = 90), with 10 left over. (Note the 100 is made up of the carry from the previous step (10) and the current digit (0)). Since we are at the last digit, the 10 now left over is our remainder.

$$8\ 6\ R\ 10$$
$$15\,\overline{)\ 1\ 3\ 0\ _{10}0}$$

The result of the division is 86 remainder 10. This means that ignoring cents, Grandma has $86 to spend on each grandchild.

PRACTICE 17: APPLICATION

Mr Collings is buying pizza for his class of 16 students. He is buying 13 pizzas, each of which have 8 pieces. In order to share the pizzas fairly, how many pieces should he advise each student that they can have and still leave enough for everyone?

Making connections
The remainder of ten can be used to work out how many cents Grandma has. The techniques for division with decimals are described in Chapter 6.

Division method 2: Long division

Long division involves the same underlying process as short division: it involves working out how many times one number fits into each digit of another number, digit by digit, going from left to right. However, with long division, there is more writing down of calculations and less making calculations in your head. Either method can be used for division according to personal preference. The question 'What is 230 divided by 5?' or 'What is 230 ÷ 5?' is answered using long division below.

Step 1: Does 5 go into 2? No it does not – it's too big. So in the next step, we look not just at the next digit (3) but at 23 (the number comprised of this digit and the next digit).

Step 2: Does 5 go into 23? Yes, it fits in 4 times (5 × 4 = 20) with 3 left over. In long division, the calculation of what is left over (3 in this case) is shown explicitly as 23 – 20.

Step 3: Bring the next digit (0) down and put it on the end of the remainder (3) from the previous step to get 30. Does 5 go into 30? Yes. It goes in 6 times exactly (5 × 6 = 30), so the remainder is zero (there is no remainder).

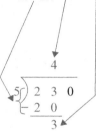

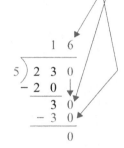

The result when 230 is divided by 5 is 16.

EXAMPLE 18

Use long division to divide 2754 by 17.

Solution

The long division process to solve this problem is shown below.

Step 1: Does 17 go into 2? No it does not — it's too big. So in the next step, we look not just at the next step we look but at 27 (the number comprised of this digit and the next digit).

$$17 \overline{)2\ 7\ 5\ 4}$$

Step 2: Does 17 go into 27? Yes, it fits in once (17 × 1 = 17) with 10 left over. In long division, the calculation of what is left over (10 in this case) is shown explicitly as 27 − 17.

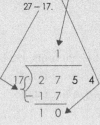

Step 3: Bring the next digit (5) down and put it on the end of the remainder (10) from the previous step to get 105. Does 17 go into 105? Yes. It goes in 6 times (17 × 6 = 102), with 3 left over.

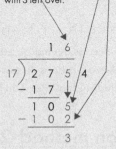

Step 4: Bring the next digit (4) down and put it on the end of the remainder (3) from the previous step to get 34. Does 17 go into 34? Yes. It goes in 2 times exactly (17 × 2 = 34), so the remainder is zero (there is no remainder).

```
          1  6  2
    17 ) 2  7  5  4
       −  1  7
          1  0  5
       −  1  0  2
                3  4
             −  3  4
                   0
```

So, 2754 divided by 17 is 162.

PRACTICE 18

Use long division to calculate 1646 ÷ 18.

Long division with remainders

Remainders when using long division are dealt with in the same way as for short division. The following example uses long division and demonstrates how the remainder is found and recorded as part of the answer.

EXAMPLE 19

Use long division to calculate 32 035 ÷ 13.

Solution

Step 1: Does 13 go into 3? No it does not – it's too big. So in the next step, we look not just at the next digit (2) but at 32 (the number comprised of this digit and the next digit).

$$13 \overline{)\ 3\ 2\ 0\ 3\ 5}$$

Step 2: Does 13 go into 32? Yes, it fits in 2 times (13 × 2 = 26) with 6 left over. In long division, the calculation of what is left over (6 in this case) is shown explicitly as 32 − 26.

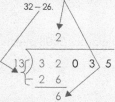

Step 3: Bring the next digit (0) down and put it on the end of the remainder (6) from the previous step to get 60. Does 13 go into 60? Yes. It goes in 4 times (13 × 4 = 52), with 8 left over.

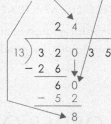

Step 4: Bring the next digit (3) down and put it on the end of the remainder (8) from the previous step to get 83. Does 13 go into 83? Yes. It goes in 6 times (13 × 6 = 78), with 5 left over.

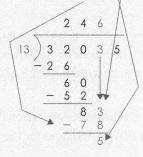

Step 5: Bring the next digit (5) down and put it on the end of the remainder (5) from the previous step to get 55. Does 13 go into 55? Yes. It goes in 4 times (13 × 4 = 52), with 3 left over. Since we are at the last (rightmost) digit, the 3 left over is our remainder.

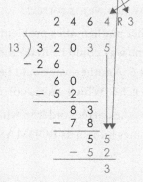

The result (quotient) when 32 035 is divided by 13 is 2464 with a remainder of 3.

PRACTICE 19

Use long division to calculate 8216 ÷ 17.

EXAMPLE 20: APPLICATION

A group of 1520 Adelaide Crows fans are travelling to Melbourne to watch the Crows play in the AFL final. Available buses have 46 seats each. How many buses are needed to transport the entire group?

Solution

The number of buses needed is obtained by dividing the total number of fans (1520) by the number of seats per bus (46), as shown below.

```
           3  3  R 2
      46 ) 1  5  2  0
         - 1  3  8 ↓
             1  4  0
           - 1  3  8
                   2
```

So, all except two of the fans will fit in 33 buses ... but that means we need 34 buses to fit in everyone.

PRACTICE 20: APPLICATION

The dinner for a large conference anticipates 3408 guests. Tables seat 16 people each. How many tables are needed?

2.6   ORDER OF OPERATIONS

Consider the following mathematics problem: $3 + 4 \times 5$. This problem might be tackled in two ways. One way could be to do the $3 + 4$ first, with the result being 7, and then to do 7×5 to get 35. Another way could be to do the 4×5 first, with the result being 20, and then to do $3 + 20$ to get 23. Which way is correct?

Now consider another mathematics problem: $3 \times 10 \div 5$. This problem might also be tackled in two ways. One way could be to do the 3×10 first, with the result being 30, and then to do $30 \div 5$ to get 6. Another way could be to do the $10 \div 5$ first, with the result being 2, and then to do 3×2 to get 6. Which way is correct?

The rule that tells us the correct order in which to carry out mathematics operations is BEDMAS (or sometimes known as BODMAS). The letters in BEDMAS stand for the following:

- B: brackets
- E: exponents (also known as O operators)
- D: division   ⎱ Division and multiplication have *equal* precedence
- M: multiplication ⎰ but must be done in order *from left to right.*
- A: addition   ⎱ Addition and subtraction have *equal* precedence
- S: subtraction   ⎰ but must be done in order *from left to right.*

So in the first example above, the second method—doing the 4×5 first, then $3 + 20$ to get 23—is the correct method and gives the correct answer. In the second example above, the first method—doing the 3×10 first and then $30 \div 5$ to get 6—is the correct method and gives the correct answer, because when two operations have equal precedence (multiplication and division have equal precedence), the order in which they should be calculated is from left to right.

In general, BEDMAS says that calculations within brackets in a mathematics problem need to be done first. Then, any numbers raised to a power (operators/exponents) are worked out next. Division and multiplication then have equal precedence, as do addition and subtraction. However, division and multiplication must be done going from left to right, and the same with addition and subtraction, because in some problems, not going from left to right will give a different answer. The following examples highlight how to use BEDMAS to guide the order in which you tackle individual calculations within a larger mathematics problem. Note that the following examples do not include exponents. Exponents are covered in Chapter 8 and examples of BEDMAS with exponents will be provided there.

EXAMPLE 21

Solve the following problems.

a $30 - 8 \times 2$

b $4 + 7 \times 3 - 6 \div 2$

c $20 \div 4 + 6 - 2 \times 5 + 1$

d $4 + 7 \times (14 - 6) \div 2$

e $20 \div 4 + (40 - 2 \times 5) \div 3 \times 5 + 1$

Solutions

a BEDMAS tells us that multiplication has higher precedence than subtraction, so multiplication is done first in the equation $30 - 8 \times 2$. Calculations are as follows:

$$30 - 8 \times 2 = 30 - 16$$
$$= 14$$

b BEDMAS tells us that multiplication and division are calculated before addition and subtraction, and that multiplication and division have equal precedence and should be carried out from left to right. Calculations are as follows:

$4 + 7 \times 3 - 6 \div 2$	Multiplication is on the left so done before division
$= 4 + 21 - 6 \div 2$	Division is done next
$= 4 + 21 - 3$	Addition is on the left so done before subtraction
$= 25 - 3$	Subtraction is done next
$= 22$	

c   BEDMAS tells us that multiplication and division are calculated before addition and subtraction, and that multiplication and division have equal precedence and should be carried out from left to right. Calculations are as follows:

$20 \div 4 + 6 - 2 \times 5 + 1$	Division is on the left so done before multiplication
$= 5 + 6 - 2 \times 5 + 1$	Multiplication is done next
$= 5 + 6 - 10 + 1$	Addition of 5 and 6 is leftmost so is next
$= 11 - 10 + 1$	Subtraction is leftmost so done next
$= 1 + 1$	Addition is the only remaining operation
$= 2$	

d   BEDMAS tells us that the first calculation to be done is any calculation inside brackets. After brackets, comes exponents, but there are no exponents in this problem. Next, BEDMAS tells us that multiplication and division are calculated before addition and subtraction and that multiplication and division have equal precedence and should be carried out from left to right. Calculations are as follows:

$4 + 7 \times (14 - 6) \div 2$	Brackets first
$= 4 + 7 \times 8 \div 2$	Multiplication is further to the left than division so is done next
$= 4 + 56 \div 2$	Division is done next
$= 4 + 28$	Addition is done next
$= 32$	

e   BEDMAS tells us that the first calculation to be done is any calculation inside brackets. After brackets, comes exponents, but there are no exponents in this problem. Next, BEDMAS tells us that multiplication and division are calculated before addition and subtraction and that multiplication and division have equal precedence and should be carried out from left to right. Calculations are as follows:

$20 \div 4 + (40 - 2 \times 5) \div 3 \times 5 + 1$	Brackets first, within brackets multiplication first
$= 20 \div 4 + (40 - 10) \div 3 \times 5 + 1$	Brackets first, within brackets subtraction next
$= 20 \div 4 + 30 \div 3 \times 5 + 1$	Division of 20 by 4 is leftmost division/multiplication so next
$= 5 + 30 \div 3 \times 5 + 1$	Division of 30 by 3 is further left than multiplication so is done next
$= 5 + 10 \times 5 + 1$	Multiplication is done next
$= 5 + 50 + 1$	Addition of 5 and 50 is leftmost so done next
$= 55 + 1$	Addition of 55 and 1 is all that is left to do
$= 56$	

PRACTICE 21

Solve the following problems.

a $12 + 18 \div 9$

b $24 - 20 \div 5 + 8 \times 4$

c $6 + 20 \times 4 \div 2 - 3 \times 6 + 2$

d $4 + 7 \times ((14 - 6) \div 2)$

e $40 \div (20 - 2 \times 6) \times 4 \div 2 \times 5 + 1$

2.7 ROUNDING WHOLE NUMBERS

Sometimes an approximation of a number is sufficient. For example, if three houses are sold for $540 510, $423 200 and $681 300, respectively, then when comparing the sale prices, hundreds of dollars are unlikely to be relevant. It is sufficient to consider an approximation of each sale price to, say, the nearest ten thousand dollars: $540 000, $420 000 and $680 000.

To **round** a whole number to a given place value, the process is as follows:

1   Underline the digit in the place to which the number is being rounded.
2   Note that the digit to the right of the underlined digit is called the critical digit.
   a   If the critical digit is 5 or more, then add one to the underlined digit.
   b   If the critical digit is less than 5, then the underlined digit remains unchanged.

Rounding
The process of approximating a number to a given place value

EXAMPLE 22

Round the number 123 405 to the following places.

a   thousands   b   tens   c   hundred thousands

Solutions

a   Underline the digit in the thousands place: 12<u>3</u>405. The digit to the right of the thousands place is a 4. Since this is less than 5, the 3 remains unchanged and the rounded number is 123 000.

b   Underline the digit in the tens place: 1234<u>0</u>5. The digit to the right of the tens place is 5. Since this is in the category '5 or more', add one to the digit in the tens place (0 becomes 1). The rounded number is 123 410.

c   Underline the digit in the hundred thousands place: <u>1</u>23 405. The digit to the right of the hundred thousands place is 2. Since this is in the category 'less than 5', it remains unchanged. The rounded number is 100 000.

PRACTICE 22

Round the number 9 026 913 to the following places.

a   thousands   b   tens   c   millions

When rounding a number, if the underlined digit is a 9 and the critical digit is 5 or more, then the underlined digit is rounded up to 10. What this means in practice is that the underlined digit is changed to a zero (0) and the next digit to the left is rounded up by one.

EXAMPLE 23

a   Round the number 3 089 695 to the thousands place.

b   Round the number 3 997 421 to the ten thousands place.

c   Round the number 3 997 421 to the hundred thousands place.

Solutions

a   Underline the digit in the thousands place: 3 08$\underline{9}$ 695. The digit to the right of the thousands place is a 6. Since this is in the '5 or more' category, the underlined 9 is rounded up to a 10. This means that the underlined 9 is changed to a zero (0) and the next digit to the left (8) is rounded up to a 9. The rounded number is 3 090 000.

b   Underline the digit in the ten thousands place: 3 9$\underline{9}$7 421. The digit to the right of the ten thousands place is 7 and since this is in the category '5 or more', the underlined 9 is rounded up to a 10. This means that the underlined 9 is changed to a 0 and the next digit to the left (the 9 in the hundred thousands place) is rounded up to a 10. This means in turn that the 9 in the hundred thousands place is replaced by a zero (0) and the next digit to the left (the 3 in the millions place) is rounded up to a 4. The rounded number is 4 000 000.

c   Underline the digit in the hundred thousands place: 3 $\underline{9}$97 421. The digit to the right of the hundred thousands place is 9 and since this is in the category '5 or more', the underlined 9 is replaced by a zero (0) and the next digit to the left (the 3 in the millions place) is rounded up to a 4. The rounded number is 4 000 000. Note that for this particular number (but not for every number), rounding to the nearest ten thousand gives the same result as rounding to the nearest hundred thousand.

PRACTICE 23

a   Round the number 249 526 to the thousands place.

b   Round the number 3 999 951 to the hundreds place.

c   Round the number 3 909 991 to the tens place.

2.8   SIGNIFICANT FIGURES

The accuracy required when using numbers depends upon the context in which the number is used. For example, if we were describing someone's wealth in terms of millions of dollars, a couple of hundred dollars more, or less, would not make any substantial difference. While it would be possible to specify that we want to round someone's wealth to the nearest hundred thousand dollars, this is only effective if we know just how wealthy someone is. If we decided to round to the nearest hundred thousand dollars but the wealth being measured is in the billions of dollars, rounding to the nearest hundred thousand dollars would be almost irrelevant. In some circumstances, particularly where we are unsure of the magnitude of the number being approximated, it is useful to have a rounding technique that works in a different way. We need a technique that uses for approximation only the figures (or digits) that provide the important information about the size of a number. These digits are known as **significant figures**.

> **Significant figure** An approximation to a specified number of digits, starting from the highest place value digits in the number

Suppose we want to write the number that is closest to 46 892 using, at most, two non-zero digits. This number would be 47 000. (The third digit, 8, is greater than 4, so the 6 is rounded up.) Therefore, we can say that 46 892 to two significant figures (s. f.) is 47 000. The number 1 349 029 to two significant figures would be 1 300 000. (The third digit, 4, is less than 5, so the 3 remains as is and the number is rounded down.)

If we need to write a number to two significant figures, we look at the first three digits, with the third digit being the critical digit that determines whether the second digit is rounded up or remains the same. Similarly, to write a number to three significant figures we look at the first four digits with the fourth digit being the critical digit, and so on. In general, we always look at one more digit than the number of significant figures required.

EXAMPLE 24

Write 654 319 to:

a   2 s. f.   b   4 s. f.   c   1 s. f.

Solutions

a   To write the number to 2 s. f. we need to look at the first 3 digits. The 3rd digit is a 4, so the 2nd digit, the 5, remains a 5, and the number is rounded to 650 000.

b   To write the number to 4 s. f. we need to look at the first 5 digits. The 5th digit is a 1, so the 4th digit, the 3, remains a 3, and the number is rounded to 654 300.

c   To write the number to 1 s. f. we need to look at the first 2 digits. The 2nd digit is a 5, so the 1st digit, the 6, is rounded up to 7, and the number is rounded to 700 000.

EXAMPLE 25

Write 4 990 013 to:

a   2 s. f.   b   1 s. f.   c   4 s. f.

Solutions

a   To write the number to 2 s. f. we need to look at the first 3 digits. The 3rd digit is a 9, so the 2nd digit, the 9, is rounded up to 10, which means the 2nd digit becomes zero and the 1st digit is rounded up to a 5. The number is rounded to 5 000 000.

b   To write the number to 1 s. f. we need to look at the first 2 digits. The 2nd digit is a 9, so the 1st digit, the 4, is rounded up to 5, and the number is rounded to 5 000 000. Note that because the 2nd and 3rd digits are both 9, rounding to 1 s. f. produces the same result as rounding to 2 s. f.

c   To write the number to 4 s. f. we need to look at the first 5 digits. The 5th digit is a zero, so the 4th digit, also a zero, remains a zero, and the number is rounded to 4 990 000.

2.9   CHAPTER SUMMARY

SKILL OR CONCEPT	DEFINITION OR DESCRIPTION	EXAMPLES
Whole number	A number without a fractional or decimal part, made up of one or more digits.	$24, 0, 987645$
Place value	The rightmost digit in a whole number is in the ones place, the next digit to the left is in the tens place, the next digit to the left is in the hundreds place, and so on, with the place value of each digit being ten times the place value of the digit next to it on the right.	$14352 = 1 \times 10000 + 4 \times 1000 + 3 \times 100 + 5 \times 10 + 2 \times 1$
Adding or subtracting whole numbers	• Write numbers vertically lining up according to place value. • Add or subtract the digits working from right to left; subtract bottom digit from top digit. • Carry if the sum of two digits is greater than 9; borrow if the difference when subtracting the bottom digit is negative.	$8641 + 237 =$ $2307 - 259 =$ $\begin{array}{r} 8641 \\ +587 \\ \hline 9228 \end{array}$ $\begin{array}{r} 2307 \\ -259 \\ \hline 2046 \end{array}$
Multiplying whole numbers	• Write numbers vertically lining up according to place value. • Working from right to left, multiply each digit in the top number by the rightmost digit in the bottom number. • Repeat the above step for each digit in the bottom number, working from right to left. • Add up the results of all the multiplication steps to get the final answer.	$1681 \times 356 =$ $\begin{array}{r} 1681 \\ \times 356 \\ \hline 10086 \\ 84050 \\ 504300 \\ \hline 598436 \end{array}$
Dividing whole numbers	• Divide the dividend by the divisor. • Carry out the division digit by digit working from left to right. • If you use short division, just note the carry as you go. • If you use long division, write out the multiplication and subtraction needed to calculate the carry before moving on to the next digit.	Short division: $1520 \div 46 = 33$ remainder 2 $\begin{array}{r} 3 \ \ 3 \ \ R\,2 \\ 46\,\overline{)\,1\ 5\ 2_{14}0} \end{array}$ Long division: $1520 \div 46 = 33$ remainder 2 $\begin{array}{r} 3 \ \ 3 \ \ R\,2 \\ 46\,\overline{)\,1\ 5\ 2\ 0} \\ -1\ 3\ 8 \ \downarrow \\ \hline 1\ 4\ 0 \\ -1\ 3\ 8 \\ \hline 2 \end{array}$

SKILL OR CONCEPT	DEFINITION OR DESCRIPTION	EXAMPLES
Order of operations	BEDMAS (Brackets, Exponents, Division and Multiplication, Addition and Subtraction) specifies the order in which operations should be carried out.	$(4 + 5) \times 9 - 21 \div 3$ (brackets first) $= 9 \times 9 - 21 \div 3$ (then multiplication) $= 81 - 21 \div 3$ (then division) $= 81 - 7$ (then subtraction) $= 74$
Rounding whole numbers	• Underline the digit in the place value to which the number is being rounded. • The digit to the right of the underlined digit is the *critical digit*. • If the critical digit is 5 or more, add one to the underlined digit. • If the critical digit is less than 5, the underlined digit remains unchanged. • All other digits to the right of the underlined digit become zero.	Round 852 567 to the nearest thousand. 852 567 = 853 000 (nearest thousand) Round 34 969 to the nearest hundred. 34 969 = 35 000 (nearest hundred; 9 rounds up to 10, so 4 becomes 5 and 9 becomes 0)
Significant figures (s. f.)	• Starting from the leftmost digit, underline the *n*th digit where you are rounding to *n* significant figures. • The digit to the right of the underlined digit is the critical digit. • If the critical digit is 5 or more, add one to the underlined digit. • If the critical digit is less than 5, the underlined digit remains unchanged. • All other digits to the right of the underlined digit become zero.	Write 9 853 468 to 4 s. f. 9 853 468 = 9 853 000 (4 s. f.) Write 239 969 to 3 s. f. 239 969 = 240 000 (9 rounds up to 10, so 3 becomes 4 and 9 becomes 0)

2.10 REVIEW QUESTIONS

A SKILLS

1 Write in words the place value of the underlined digit in the following numbers.

  a 1 3<u>6</u>5 419   b 2576<u>34</u>1

  c <u>2</u>394   d 41 608 18<u>9</u>

2 Underline the digit with the specified place value in the given number.

  a thousands in 135 419   b tens in 3 411 728

  c millions in 32 964 176

3 Carry out the following additions or subtractions.

  a 2254 − 1032   b 813 + 164   c 4305 + 15296

  d 2731 − 889   e 30048 − 12567   f 49991 + 202436

4 Carry out the following multiplications or divisions.

  a 81 × 63   b 486 ÷ 3   c 7 × 891

  d 9076 ÷ 12   e 486 × 316   f 1302 ÷ 21

  g 48 × 79   h 1351 ÷ 7   i 42 × 4015

  j 6238 ÷ 9   k 5179 × 742   l 2104 ÷ 24

5 Carry out the following calculations.

  a 5 + (22 − 6) × 2   b 7 + 84 ÷ 2 × (14 − 8)

  c (45 − 3 × 9) ÷ 3 × 5 + 60 ÷ 4   d 45 − 3 × 9 ÷ 3 × 5 + 60 ÷ 4

6 Round the following numbers to the required place value.

  a 19863 (tens)   b 5823789 (hundreds)

  c 26767456 (hundred thousands)   d 98996 (tens)

7 Write the following numbers to the required number of significant figures.

  a 36072 (1 s. f.)   b 2041005 (2 s. f.)

  c 124881 (3 s. f.)   d 9009965 (4 s. f.)

B   APPLICATIONS

1   In a cricket match, the number of runs made by each of the eleven players was as follows: 45, 13, 68, 143, 14, 28, 35, 44, 32, 3 and 5. What was the total number of runs made by the team?

2   Approximate data for the total rainfall for Adelaide during 2014 is shown in the table (Source: www.bom.gov.au).

   a   What was the total rainfall for each season (spring, summer, autumn, winter)?

   b   Round the rainfall amounts in part (a) to the nearest ten.

   c   What was the total rainfall for the year?

   d   The average of a set of numbers is the sum of the numbers divided by how many numbers there are in the set. Use your calculations from part (a) to find the average monthly rainfall.

   e   Express the average rainfall for spring and winter to one significant figure.

MONTH	RAINFALL (MM)
Jan	10
Feb	98
Mar	19
Apr	51
May	64
Jun	104
Jul	100
Aug	21
Sep	31
Oct	5
Nov	24
Dec	6

3   The price of a stock opened at $70 and dropped by $3 per hour for the next four hours, before rising by $7 during the next hour, then dropping by $4 per hour for the next two hours, and finally rising by $12 before the day's trading closed. What was the closing price of the stock?

4   Tom had $248 in his bank account. After buying eight DVDs, each at the same price, he had $120 left. How much did each DVD cost?

5   The population of Springfield was 24 950. It decreased by 326 each year for five years. What was its population after five years?

6   The formula for the area of a rectangle is *length × width*. If a soccer field has a length of 105 metres and a width of 68 metres, what is its area?

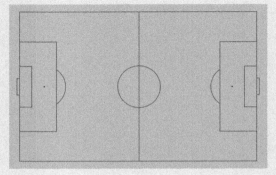

7   On a recent trip, James drove 2496 kilometres. The fuel consumption for the trip was 192 litres. How many kilometres per litre did the car average?

8   Macy's interest-free student loan is $1350. If she pays off $75 per month, how long will it take her to pay it off completely?

9   An irrigation subcontractor is installing water pipes. If she is installing 744 metres of pipe and each piece of pipe is 3 metres long, how many pieces of pipe will she need?

10 To supply food for a party, Jack needs to buy four pizzas at $26 each, six bottles of drink at $2 each, a packet of paper plates for $6 and a packet of napkins at $4. What is the total cost of the party? Express your answer to two significant figures to obtain a rough budget for the party.

11 Dr Dianati accidentally misread his travel itinerary and as a result his plane fare will cost $709 instead of the original $488. How much extra does Dr Dianati have to pay?

12 How many seconds are there in one day?

13 A customer has asked for 850 pencils. The store has 63 boxes of pencils in stock. There are 14 pencils in each box. Does the store have enough pencils to supply to the customer?

14 A store chain is placing an order for chocolate bars. They estimate the chain will sell 5500 of this chocolate bar in a month.
The chocolate bars come in boxes of 24.
How many boxes does the store need to buy to meet expected sales?

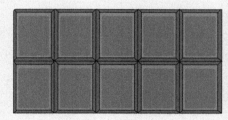

15 A group of West Coast Eagles supporters is going to Melbourne by bus to see the Eagles in a football final. There are 531 supporters making the trip. Each bus seats 48 people. How many buses are needed to transport all the supporters?

16 There are 324 students studying Biology, 499 students studying Chemistry and 250 students studying Physics. Estimate how many students are studying these sciences in total by rounding your answer to the nearest hundred.

17 There are 849 students studying Law, 4099 students studying Education, 1256 students studying Business and 467 students studying Engineering. Estimate how many students in total are studying in these areas by expressing each number to two significant figures and then calculating a total based on your estimated numbers.

18 Five houses in the Sunnyvale area sold for the following prices: $524 000, $306 500, $290 000, $610 100 and $738 000. Work out a rough average for these house prices by rounding each price to the nearest ten thousand, adding the rounded numbers, and dividing by the number of houses sold.

3

WHOLE NUMBER PROPERTIES

CHAPTER CONTENT

- ›› Introduction to whole number properties
- ›› Factors and multiples
- ›› Primes and composites
- ›› Divisibility tests
- ›› Highest Common Factor (HCF)
- ›› Lowest Common Multiple (LCM)
- ›› Prime factorisation
- ›› Using prime factorisation to find HCF and LCM
- ›› Square numbers
- ›› Square roots

CHAPTER OBJECTIVES

- ›› Determine the factors and multiples of a number
- ›› Classify prime and composite numbers
- ›› Use common tests of divisibility
- ›› Calculate the highest common factor (HCF) of two or more numbers
- ›› Calculate the lowest common multiple (LCM) of two or more numbers
- ›› Generate the prime factorisation of a whole number
- ›› Use prime factorisation to solve HCF, LCM and square root problems
- ›› Recognise and calculate square numbers
- ›› Calculate the square root of a perfect square

SLIPPERY PROBLEM

Two oil tankers carry 850 and 680 litres of oil, respectively. The oil needs to be packed into containers (drums) for transport. In order to reduce costs, a shipping company wants to know the largest size of oil drum that is able to hold the oil from each tanker without leaving any residual oil or any drums only partly full (that is, the total amount of oil from each tanker is able to be stored exactly in the drums). The largest size of drum able to be used can be worked out by calculating the highest common factor of 850 and 680. The highest common factor of the two numbers is 170, so the largest container that can be used is 170 litres.

3.1 INTRODUCTION TO WHOLE NUMBER PROPERTIES

The properties of whole numbers are used extensively in practical situations. For example:

- Trevor baked 20 chocolate chip cookies to be shared between four people. How many cookies should each person get?
- A shop sells dining room chairs in sets of six. In order to transport complete sets of chairs, which of the following quantities of chairs—100, 24, 40, 30, 25, 12, 48—should be loaded onto the truck?
- A backyard pool is roughly 4 metres wide, 6 metres long and 2 metres deep, and stabiliser needs to be added to the swimming pool in readiness for summer. The amount of stabiliser needed depends on the capacity of the pool. How would the pool owner calculate the capacity of the pool?
- On a nine-hole golf course, a golfer made the following scores: par, birdie, bogey, par, par, par, double birdie and bogey. A birdie is one under par, a double birdie two under par, and a bogey one over par. What would the golfer's final score be?

All of these questions use properties of whole numbers. The first situation involving the cookies is answered by dividing 20 by 4 to get 5 cookies per person. Both 4 and 5 are *factors* of 20. The second situation can be solved by identifying *multiples* of 6. All numbers that are multiples of 6 will allow chairs to be transported in complete sets. The numbers in the list that are multiples of 6 are 24, 30, 12 and 48.

The capacity of a swimming pool is its *volume* in litres, and the volume of a rectangular pool is *length × width × depth*. In this case, an estimate of the pool capacity would be $4 \times 6 \times 2 = 48$ cubic metres written as 48 m^3.

Finally, the golfer's score is usually written as a positive or negative integer—a directed number. The golfer's final score would be calculated as follows: $0 - 1 + 1 + 0 + 0 + 0 - 2 + 1 = -1$.

All of these properties of whole numbers—factors, multiples, the indices used in calculations of volume and directed numbers, as well as more properties—are the focus of this chapter.

3.2 FACTORS AND MULTIPLES

Factors

In addition to sharing chocolate chip cookies fairly or packing up chairs in sets of six, **factors** and multiples have numerous practical applications.

A factor is a whole number value that divides into another number without a remainder. For example, here are all of the factors of 20: 1, 2, 4, 5, 10 and 20. These numbers are the only numbers that divide into 20 leaving no remainder.

Tiffany Winn

Making connections
The calculation of capacity and volume are covered in Chapter 12.

Factor
A whole number that multiplies with another whole number with the result being the product. A factor can also be defined as a whole number that divides into another whole number without a remainder.

Factors come in pairs. When each pair of factors is multiplied, the result is the number of which they are factors. In the example above, 1 and 20 are one set of factor pairs because $1 \times 20 = 20$, 2 and 10 are another set of factor pairs as $2 \times 10 = 20$, and 4 and 5 is a third set of factor pairs as $4 \times 5 = 20$. In other words:

- 1 and 20 are factors of 20: 1 is a factor of 20 because it divides into 20 without a remainder ($20 \div 1 = 20$) and the result of the division, 20, is the factor that pairs with 1
- 2 and 10 are factors of 20: 2 is a factor of 20 because it divides into 20 without a remainder ($20 \div 2 = 10$) and the result of the division, 10, is the factor that pairs with 2
- 4 and 5 are factors of 20: 4 is a factor of 20 because it divides into 20 without a remainder ($20 \div 4 = 5$) and the result of the division, 5, is the factor that pairs with 4

There are no whole numbers between 4 and 5, so there are no more factors of 20.

When finding factors of a number, one strategy is to start with 1 and the number itself. Since a number is always divisible by 1 and itself, 1 and the number are a pair of factors. Then try 2 and if 2 is a factor, find its pair. Next try 3 and if 3 is a factor, find its pair. Continue until you have eliminated the possibility of more factors, noting that if you are finding pairs of factors, the gap between paired factors will slowly reduce, which limits how far you need to search for more factors.

EXAMPLE 1

List all of the factors of:

a   24   b   90   c   15

Solutions

a   Using the method outlined above:

- The first pair of factors is 1 and 24.
- Check if 2 is a factor. Since 2 divides into 24 without a remainder and $24 \div 2 = 12$, the next pair of factors is 2 and 12.
- Check if 3 is a factor. Since 3 divides into 24 without a remainder and $24 \div 3 = 8$, the next pair of factors is 3 and 8.
- Check if 4 is a factor. Since 4 divides into 24 without a remainder and $24 \div 4 = 6$, the next pair of factors is 4 and 6.
- Check if 5 is a factor. It does not divide evenly into 24, so it is not a factor.
- As we have already found that 6 is a factor, we have now found all the factors of 24, which are 1, 2, 3, 4, 6, 8, 12 and 24.

b Using the method outlined above:
- The first pair of factors is 1 and 90.
- Check if 2 is a factor. Since 2 divides into 90 without a remainder and $90 \div 2 = 45$, the next pair of factors is 2 and 45.
- Check if 3 is a factor. Since 3 divides into 90 without a remainder and $90 \div 3 = 30$, the next pair of factors is 3 and 30.
- Check if 4 is a factor. Since 4 does not divide evenly into 90, it is not a factor.
- Check if 5 is a factor. Since 5 divides into 90 without a remainder and $90 \div 5 = 18$, the next pair of factors is 5 and 18.
- Check if 6 is a factor. Since 6 divides into 90 without a remainder and $90 \div 6 = 15$, the next pair of factors is 6 and 15.
- Check if 7 is a factor. Since 7 does not divide evenly into 90, it is not a factor.
- Check if 8 is a factor. Since 8 does not divide evenly into 90, it is not a factor.
- Check if 9 is a factor. Since 9 divides into 90 without a remainder and $90 \div 9 = 10$, the next pair of factors is 9 and 10.
- As we have already found that 10 is a factor, we have now found all factors of 90, which are 1, 2, 3, 5, 6, 9, 10, 15, 18, 30, 45 and 90.

c Using the method outlined above:
- The first pair of factors is 1 and 15.
- Check if 2 is a factor. Since 2 does not divide evenly into 15, it is not a factor.
- Check if 3 is a factor. Since 3 divides into 15 without a remainder and $15 \div 3 = 5$, the next pair of factors is 3 and 5.
- Check if 4 is a factor. Since 4 does not divide evenly into 15, it is not a factor.
- As we have already found that 5 is a factor, we have now found all factors of 15, which are 1, 3, 5 and 15.

Note that an even number can never be a factor of an odd number, so we could save ourselves some time by recognising that we do not have to test 2 and 4 as possible factors of 15; we can simply jump over them.

PRACTICE 1

List all the factors of the following numbers.

a 21   **b** 144   **c** 84

EXAMPLE 2: APPLICATION

A sports coach wants to split 24 players into at least three teams of equal size to play short games at practice. What size teams could be made?

Solution

Since the factors of 24 are 1 and 24, 2 and 12, 3 and 8, and 4 and 6, and the coach needs at least 3 teams, the coach could have 3 teams of 8 or 4 teams of 6.

PRACTICE 2: APPLICATION

A teacher has a class of 30 students that she wants to split into even groups, without leaving anyone out. What size groups could she potentially have?

Multiples

Multiple
A whole number that can be divided by another number without leaving a remainder

A **multiple** is a whole number that can be divided by another number without leaving a remainder. For example, 18 is a multiple of 6 because 18 can be divided by 6 without leaving a remainder: $18 \div 6 = 3$.

Note that the multiples of a number are the answers in the times table for that number. For example, the list of numbers that make up the six times table up to 6 times 10 is as follows: 6, 12, 18, 24, 30, 36, 42, 48, 54 and 60. Any number that is a multiple of 6 will be in the six times table.

EXAMPLE 3

Write out the first twelve multiples of:

a 5   b 7

Solutions

a The first twelve multiples of 5 are 5, 10, 15, 20, 25, 30, 35, 40, 45, 50, 55 and 60. After starting with 5, each successive multiple is calculated by adding 5 to the previous multiple. The list of multiples represents the 5 times table up to 12 × 5.

b The first twelve multiples of 7 are 7, 14, 21, 28, 35, 42, 49, 56, 63, 70, 77 and 84. After starting with 7, each successive multiple is calculated by adding 7 to the previous multiple. The list of multiples represents the 7 times table up to 12 × 7.

PRACTICE 3

Write out the first twelve multiples of:

a 3   b 8

EXAMPLE 4

a Is 42 a multiple of 7?   b Is 401 a multiple of 3?   c Is 216 a multiple of 9?

Solutions

a Since 42 can be divided exactly by 7 (42 ÷ 7 = 6), 42 is a multiple of 7.

b Since 401 cannot be divided exactly by 3 (401 ÷ 3 = 133, remainder 2), 401 is not a multiple of 3.

c Since 216 can be divided exactly by 9 (216 ÷ 9 = 24), 216 is a multiple of 9.

PRACTICE 4

a Is 72 a multiple of 6?   b Is 634 a multiple of 8?   c Is 426 a multiple of 3?

EXAMPLE 5: APPLICATION

Eggs are being packaged in containers of 6. If I have 328 eggs, will they fit into egg cartons without any left over?

Solution

Since there is a remainder when 328 is divided by 6 (328 ÷ 6 = 54, remainder 4), there will be 4 leftover eggs if they are put into half-dozen containers.

PRACTICE 5: APPLICATION

Circle which numbers in the following list make an exact number of weeks, without any days left over: 7, 18, 24, 56, 22, 107, 198 and 28.

3.3   PRIMES AND COMPOSITES

What if we were to design a game of chance based on Australian Football League (AFL) results. Our suggestion is that if the winning team's score is a prime number and the losing team's score is not a prime number, then you win; and if the winning team's score is not a prime number and the losing team's score is a prime number, then we win. Would you take up this game? In this section, we will investigate prime and composite numbers and revisit this proposition at the end of the discussion.

A **prime number** is a whole number that is greater than 1 with only two whole number factors, which are 1 and itself. Between 2 and 10, the following numbers are prime: 2, 3, 5 and 7. Each of these numbers is only divisible by 1 and itself. Those whole numbers greater than 1 that are not prime numbers are called composite numbers.

Prime number
A whole number
greater than 1
with only two
whole number
factors: 1 and
itself

Tiffany Winn

A **composite number** is a whole number that is greater than 1 that can be divided exactly by whole numbers other than 1 and itself. Between 2 and 10, the following numbers are composite: 4, 6, 8 and 9. As well as being divisible by 1 and itself, each of these numbers is divisible by at least one other number: 4 is divisible by 2, 6 is divisible by 2 and 3, 8 is divisible by 2 and 4, and 9 is divisible by 3.

The Sieve of Eratosthenes (named after the Greek mathematician) is an ancient but simple way to identify prime and composite numbers from 1 to 100. To use the Sieve of Eratosthenes we use the table of numbers below, and the following steps:

1 Draw a cross through number 1, because it is neither prime nor composite.

2 Colour in the squares of all the multiples of 2, except 2.

3 Colour in the squares of all the multiples of 3, except 3 (if a square is already coloured in, just leave it).

4 Colour in the squares of all the multiples of 5, except 5 (if a square is already coloured in, just leave it).

5 Colour in the squares of all the multiples of 7, except 7 (if a square is already coloured in, just leave it).

When you have finished all five steps, if you have done the colouring correctly, the squares left uncoloured will be all the prime numbers between 2 and 100 (inclusive). The coloured squares will represent all the composite numbers between 2 and 100 (inclusive).

1	2	3	4	5	6	7	8	9	10
11	12	13	14	15	16	17	18	19	20
21	22	23	24	25	26	27	28	29	30
31	32	33	34	35	36	37	38	39	40
41	42	43	44	45	46	47	48	49	50
51	52	53	54	55	56	57	58	59	60
61	62	63	64	65	66	67	68	69	70
71	72	73	74	75	76	77	78	79	80
81	82	83	84	85	86	87	88	89	90
91	92	93	94	95	96	97	98	99	100

Below we have shown the table after using the five steps detailed above. All the white squares (except 1 which is crossed out) are prime numbers. All the coloured squares are composite numbers. Squares have different shades of green according to which step they were coloured in, as per the following key:

Step 2	Step 3	Step 4	Step 5

	2	3	4	5	6	7	8	9	10
1	2	3	4	5	6	7	8	9	10
11	12	13	14	15	16	17	18	19	20
21	22	23	24	25	26	27	28	29	30
31	32	33	34	35	36	37	38	39	40
41	42	43	44	45	46	47	48	49	50
51	52	53	54	55	56	57	58	59	60
61	62	63	64	65	66	67	68	69	70
71	72	73	74	75	76	77	78	79	80
81	82	83	84	85	86	87	88	89	90
91	92	93	94	95	96	97	98	99	100

EXAMPLE 6

Label the following numbers as prime, composite or neither.

a 42   b 1   c 19   d 159   e 127

Solutions

a   42 is a composite number. This is easy to see because it is an even number so will be divisible by 2. As well as 1 and itself, 42 is divisible by 2, 3, 6, 7, 14 and 21.

b   1 is neither prime nor composite; prime and composite numbers must be greater than 1.

c   19 is a prime number. It is only divisible by 1 and itself.

d   159 is a composite number. It is divisible by 1, 3, 53 and 159.

e   127 is a prime number. It is only divisible by 1 and itself.

PRACTICE 6

Label the following numbers as prime, composite or neither.

a 138   b 91   c 73   d 1   e 109

Tiffany Winn

Returning to our game proposition at the start of the section: we would have a greater chance of winning than you would. Looking at the table of numbers from 1 to 100 above, you can see that prime numbers occur more frequently in the lower numbers than in the higher numbers, and a winning team is more likely to have a higher score.

3.4 DIVISIBILITY TESTS

Using divisibility tests is an efficient way of determining whether one number is divisible by another. For example, if we are interested in whether the number 1407 is divisible by 3, we do not actually have to do the division to find out. The divisibility test for 3 involves adding up the digits in the number. In this case, when we add up the digits in 1407, we get $1 + 4 + 0 + 7 = 12$. Since 12 (the result of adding up the digits) is divisible by 3, we know that 1407 will be divisible by 3. The table below highlights those numbers between 2 and 10 (inclusive) for which there are divisibility tests. Notice that 7 is the only number for which there is no test.

RULES FOR DIVISIBILITY		
A number is divisible by ...	Divisible	Not Divisible
2 if it is an even number (last digit is 0, 2, 4, 6 or 8)	1804	908 453
3 if the digits in the number add up to a multiple of 3	53 694 ($5 + 3 + 6 + 9 + 4 = 27$ and 27 divisible by 3)	9076 ($9 + 0 + 7 + 6 = 22$ and 22 not divisible by 3)
4 if the last two digits form a number that is divisible by 4	41 672 (72 divisible by 4)	11 086 (86 not divisible by 4)
5 if the last digit in the number is 0 or 5	25 365	20 398
6 if the number is divisible by both 2 and 3	475 392 (even number and $4 + 7 + 5 + 3 + 9 + 2 = 30$, which is divisible by 3)	2091 (not even) 8534 (even number but not divisible by 3)
8 if the last three digits form a number that is divisible by 8	5864 (864 is divisible by 8)	90 532 (532 is not divisible by 8)
9 if the digits in the number add up to a multiple of 9	3177 ($3 + 1 + 7 + 7 = 18$ and 18 divisible by 9)	3903 ($3 + 9 + 0 + 3 = 15$ and 15 not divisible by 9)
10 if the last digit in the number is 0	8796 450	8 456 342

Note that a number cannot be divisible by any even number if it is not divisible by 2. For example, a number cannot be divisible by 4 or 8 if it is not divisible by 2. Another way of understanding this is to realise that all multiples of 4 and 8—and in fact all multiples of even numbers—are even. Similarly, a number cannot be divisible by 9 if it is not divisible by 3, because 3 is a factor of 9. However, care must be taken with the logic used to decide divisibility of one factor based on another factor. For example, a number may be divisible by 2 and 4 but not 8; it is only when a number is *not* divisible by 2 that conclusions can be drawn about its divisibility with respect to 4 and 8.

EXAMPLE 7

Determine the divisibility of the following numbers, as specified.

a   Is 2688 divisible by any of 2, 3, 6 or 7?

b   Is 52 915 divisible by any of 2, 3, 5 or 10?

c   Is 66 240 divisible by any of 4, 8 or 9?

Solutions

a   2688 is:
   * Divisible by 2 because it is even (ends in 8)
   * Divisible by 3 because $2 + 6 + 8 + 8 = 24$ and $24 \div 3 = 8$ (no remainder)
   * Divisible by 6 because it is divisible by 2 and by 3
   * Divisible by 7 because $2688 \div 7 = 384$ (no remainder)

b   52 915 is:
   * Not divisible by 2 because it is not even (ends in 5)
   * Not divisible by 3 because $5 + 2 + 9 + 1 + 5 = 22$ and 22 is not divisible by 3 ($22 \div 3 = 7$ remainder 1)
   * Divisible by 5 because it ends in 5
   * Not divisible by 10 because it does not end in 0 (ends in 5)

c   66 240 is:
   * Divisible by 4 because the number formed by the last two digits (40) is divisible by 4 ($40 \div 4 = 10$)
   * Divisible by 8 because the number formed by the last three digits (240) is divisible by 8 ($240 \div 8 = 30$) (no remainder)
   * Divisible by 9 because $6 + 6 + 2 + 4 + 0 = 18$ and 18 is divisible by 9 ($18 \div 2 = 9$)

PRACTICE 7

Determine the divisibility of the following numbers, as specified.

a   Is 52 664 divisible by any of 2, 3, 6 or 8?

b   Is 82 950 divisible by any of 4, 5, 7 or 10?

c   Is 22 368 divisible by any of 3, 4, 8 or 9?

3.5   HIGHEST COMMON FACTOR (HCF)

Remember that the factors of a number are numbers that when multiplied together result in that number; therefore, the factors of a number are also the whole numbers that divide the number without a remainder. One particular factor of a number that we need to consider is what is known as the **Highest Common Factor (HCF)**.

Tiffany Winn

Highest Common Factor (HCF)
The largest number that is a factor of two or more numbers

EXAMPLE 8

Find the HCF of each pair of numbers.

a   8 and 36   b   24 and 42

Solutions

a   The factors of 8 are 1, 2, 4 and 8.

The factors of 36 are 1, 2, 3, 4, 6, 9, 12, 18 and 36.

Looking through both lists, the largest number that is a factor of both numbers is 4.

So, the HCF of 8 and 36 is 4.

b   The factors of 24 are 1, 2, 3, 4, 6, 8, 12 and 24.

The factors of 42 are 1, 2, 3, 6, 7, 14, 21 and 42.

Looking through both lists, the largest number that is a factor of both numbers is 6.

So, the HCF of 24 and 42 is 6.

PRACTICE 8

a   Find the HCF of 12 and 15.   b   Find the HCF of 18 and 39.

EXAMPLE 9: APPLICATION

A large rectangular room has a floor 8 m by 14 m. What is the largest size of square tile that can be evenly laid on the floor without any overhanging tiles?

Solution

Finding the HCF of the length and width of the room will tell us the side length of the largest square tile. Remember that a factor of a number is a number that divides into that number without any remainder. In this case, what that means is that a square tile with side length that is a factor of both numbers will fit evenly on the floor without any overhanging pieces.

The factors of 8 are 1, 2, 4 and 8.

The factors of 14 are 1, 2, 7 and 14.

Looking through both lists, the largest number that is a factor of both numbers is 2.

So, the HCF of 8 and 14 is 2, and the largest square tile that can be laid on the floor is 2 m by 2 m.

PRACTICE 9: APPLICATION

Two drums contain 30 and 48 litres of bleach and methylated spirits, respectively. A business wants to move the liquids into smaller containers for storage. What is the biggest container that can be used to hold the contents of each of the drums exactly, without any leftover liquid?

EXAMPLE 10: APPLICATION

Tim has a bag of 36 grape-flavoured lollies, Anthony has a bag of 54 orange-flavoured lollies and Leyton has a bag of 42 lime-flavoured lollies. They need to divide their lollies into bags with each bag containing the same number of each flavour lolly. So, for example, if one bag contains 4 grape, 5 orange and 3 lime lollies, then every bag must contain 4 grape, 5 orange and 3 lime lollies, and every lolly must be used. What is the greatest number of lolly bags they will be able to make up? How many of each kind of lolly will be in each bag?

Solution

Finding the HCF will tell us how many of each kind of lolly will go in each bag, because the HCF of all three numbers represents the largest factor that they all have in common. Remember that a factor of a number divides into that number without any remainder. In this case, what that means is that to make up lolly bags using all the lollies without any left over, the total number of lolly bags must be a factor of all three types of lolly. Further, the greatest number of lolly bags that it is possible to make without any lollies left over will be equal to the HCF of all three numbers.

The factors of 36 are 1, 2, 3, 4, 6, 9, 12, 18 and 36.

The factors of 54 are 1, 2, 3, 6, 9, 18, 27 and 54.

The factors of 42 are 1, 2, 3, 6, 7, 14, 21 and 42.

The highest number that is common to all of the lists is 6, so the HCF of 36, 54 and 42 is 6. This means the greatest number of lolly bags that can be made up is 6.

There are 36 grape-flavoured lollies, so there are $36 \div 6 = 6$ grape-flavoured lollies in each bag.

There are 54 orange-flavoured lollies, so there are $54 \div 6 = 9$ orange-flavoured lollies in each bag.

There are 42 lime-flavoured lollies, so there are $42 \div 6 = 7$ lime-flavoured lollies in each bag.

PRACTICE 10: APPLICATION

Beethoven Uni, Strauss Uni and De Sousa Uni bands are marching together, one band after the other in a parade. The Beethoven band has 56 members, Strauss has 40 and De Sousa has 96. The band director wants to arrange the band members into rows where each row, regardless of the band, has the same number of members. This will allow the bands to march one behind the other in columns where each column is of equal length. What is the largest number of band members that can be in each row if each row is to have an equal number of members? How many rows will each university have of band members?

Tiffany Winn

3.6   LOWEST COMMON MULTIPLE (LCM)

Remember that a multiple of a number is a whole number that can be divided exactly by the original number without leaving a remainder. Another way of thinking about this is that a multiple is the result of multiplying the original number by a whole number. One particular type of multiple that we should consider is the **Lowest Common Multiple (LCM)**.

EXAMPLE 11

Find the LCM of each pair of numbers.

a   8 and 36   b   24 and 42

Solutions

a   The multiples of 8 are 8, 16, 24, 32, 40, 48, 56, 64, 72, 80, 88, 96, 104, …
    The multiples of 36 are 36, 72, 108, …
    Looking through both lists, the smallest number that is a multiple of both numbers is 72.
    So, the LCM of 8 and 36 is 72.

b   The multiples of 24 are 24, 48, 72, 96, 120, 144, 168, 192, 216, 240, …
    The multiples of 42 are 42, 84, 126, 168, …
    Looking through both lists, the smallest number that is a multiple of both numbers is 168.
    So, the LCM of 24 and 42 is 168.

PRACTICE 11

a   Find the LCM of 12 and 15.   b   Find the LCM of 18 and 39.

EXAMPLE 12: APPLICATION

Two traffic lights change from green to red at intervals of 20 and 35 seconds, respectively. After how many minutes will they both change at the same time?

Solution

Finding the LCM of both numbers will tell us when the traffic lights change at the same time, because the LCM of both numbers represents the smallest multiple that they both have in common—the smallest interval on which they are both able to change. Remember that a multiple is the result of multiplying the original number by a whole number. In this case, both traffic lights will change at the same time at any point where they have a common multiple; we are looking for the lowest common multiple.

The multiples of 20 are 20, 40, 60, 80, 100, 120, 140, 160, 180, 200, ...

The multiples of 35 are 35, 70, 105, 140, ...

Looking through both lists, the smallest number that is a multiple of both numbers is 140.

So, the LCM of 20 and 35 is 140. This means that the traffic lights will change at the same time after 140 seconds, or 2 minutes and 20 seconds.

PRACTICE 12: APPLICATION

Two friends, Rosie and Anna, are working together at a Henley Beach cafe today. Rosie works every 5 days, and Anna works every 6 days. How many days do they have to wait until they next get to work together?

EXAMPLE 13: APPLICATION

Gemma is buying hot dogs and hot dog rolls for a picnic. For some unknown reason, hot dogs come in packages of 8 and hot dog rolls in packages of 10. Gemma wants to have the same number of hot dogs and rolls. What is the least number of hot dogs and rolls she can buy? How many packages of hot dogs and rolls will she need?

Solution

Finding the LCM of both numbers will tell us the smallest number of hot dogs and rolls Gemma can buy, because the LCM of both numbers represents the smallest multiple that they both have in common—the smallest quantity which is common to both packages of 8 and packages of 10. Remember a multiple is the result of multiplying the original number by a whole number. In this case, the number of hot dogs and rolls bought will be the same for any number, which is a common multiple; we are looking for the lowest common multiple.

The multiples of 8 are 8, 16, 24, 32, 40, 48, 56, 64, 72, 80, ...

The multiples of 10 are 10, 20, 30, 40, ...

Looking through both lists, the smallest number that is a multiple of both numbers is 40.

So, the LCM of 8 and 10 is 40. This means that the smallest number of hot dogs and rolls Gemma can buy if she wants the same number of each will be 40. This will mean buying $40 \div 8 = 5$ packages of hot dogs and $40 \div 10 = 4$ packages of hot dog rolls.

PRACTICE 13: APPLICATION

Boxes that are 12 cm tall are being packaged together with boxes that are 16 cm tall. What is the smallest container that will fit both stacks of the 12 cm tall boxes and stacks of the 16 cm tall boxes snugly with no gap at the top?

3.7 PRIME FACTORISATION

Prime factorisation occurs when a number is written as the product of its prime factors. For example, the prime factorisation of 24 is $2 \times 2 \times 2 \times 3$, because $24 = 2 \times 2 \times 2 \times 3$ and 2 and 3 are both prime numbers. As another example, the prime factorisation of 90 is $2 \times 3 \times 3 \times 5$ because $90 = 2 \times 3 \times 3 \times 5$ and 2, 3 and 5 are prime numbers.

Prime factorisation trees provide a useful way of working out the prime factorisation of a number. A prime factorisation tree can be built up step by step by breaking a number into factors, as shown in the figures below of three examples of prime factorisation trees for the number 24. It is only necessary to do one tree to find the prime factorisation tree for a particular number. The diagram below includes three trees to highlight that it does not matter which pairs of factors you start with, or the order in which any factorisation is carried out, the resulting prime factorisation will be the same. In the case of 24, the prime factorisation is $24 = 2 \times 2 \times 2 \times 3$, as shown by the numbers at the end leaves of each tree. The key to doing a prime factorisation tree is to successively choose pairs of factors that multiply together to give the number at the root of their branch, and to stop whenever a prime number is reached. So, the prime numbers form the leaves of the tree and any other numbers must be further factorised.

When factorising any number, the divisibility test rules highlighted in Section 3.4 provide a useful reference point. For example, any number with digits that add up to a multiple of 3 will factorise by 3. In the case of 24, the starting factor can be 2, 3 or 4.

Example 1: Prime factorisation tree for 24, starting with $24 = 6 \times 4$:

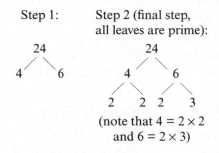

Example 2: Prime factorisation tree for 24, starting with $24 = 3 \times 8$:

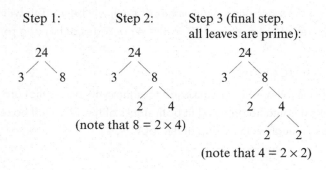

Example 3: Prime factorisation tree for 24, starting with 24 = 2 × 12:

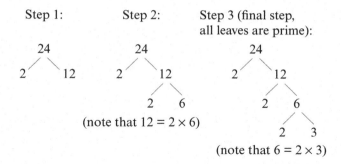

Step 1:   Step 2:   Step 3 (final step,
                                  all leaves are prime):

(note that 12 = 2 × 6)   (note that 6 = 2 × 3)

All of the above trees show that the prime factorisation of 24 = 2 × 2 × 2 × 3.

EXAMPLE 14

a   Draw a prime factorisation tree for 90. Use this tree to write the prime factorisation of 90.

b   Draw a prime factorisation tree for 120. Use this tree to write the prime factorisation for 120.

Solutions

a   The divisibility test rules tell us that since 90 ends in 0, it is divisible by 10. This can be used as a starting point for the tree.

A prime factorisation tree for 90, starting with 90 = 9 × 10, is:

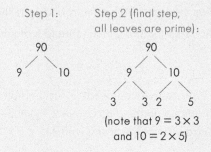

Step 1:   Step 2 (final step,
                 all leaves are prime):

(note that 9 = 3 × 3
and 10 = 2 × 5)

So, the prime factorisation of 90 is 90 = 2 × 3 × 3 × 5.

Tiffany Winn

b The divisibility test rules tell us that since 120 ends in 0, it is divisible by 10. This can be used as a starting point for the tree.

A prime factorisation tree for 120, starting with $120 = 12 \times 10$, is:

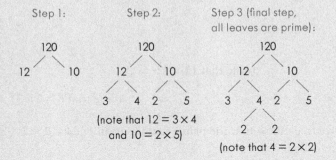

So, the prime factorisation of 120 is $120 = 2 \times 2 \times 2 \times 3 \times 5$.

PRACTICE 14

a Draw a prime factorisation tree for 36. Use this tree to write the prime factorisation of 36.

b Draw a prime factorisation tree for 208. Use this tree to write the prime factorisation of 208. (Hint: any even number has a factor of 2)

EXAMPLE 15

a Draw a prime factorisation tree for 243. Use this tree to write the prime factorisation of 243.

b Draw a prime factorisation tree for 651. Use this tree to write the prime factorisation of 651.

Solutions

a Use the divisibility test rules to work out what 243 is divisible by. It is not even, so it is not divisible by 2, but adding up the digits gives $2 + 4 + 3 = 9$, so 243 is divisible by 9. A prime factorisation tree for 243, starting with $243 = 9 \times 27$, is:

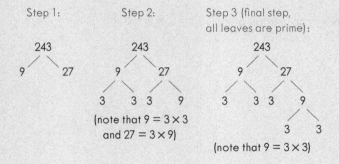

So, the prime factorisation of 243 is $243 = 3 \times 3 \times 3 \times 3 \times 3$.

b The digits in 651 add up to 12 (6 + 5 + 1 = 12), and 12 is divisible by 3, so 651 is divisible by 3. Dividing 3 into 651 gives 217. Going through the divisibility test rules, there is no rule that works for 217, but dividing 217 by 7 gives 31 with no remainder, so 217 is divisible by 7. Both 7 and 31 are prime numbers, which can be verified by attempting to divide them by any possible factors.

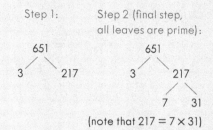

Step 1:

Step 2 (final step, all leaves are prime):

(note that 217 = 7 × 31)

So, the prime factorisation of 651 is 651 = 3 × 7 × 31.

PRACTICE 15

a Draw a prime factorisation tree for 825. Use this tree to write the prime factorisation of 825.

b Draw a prime factorisation tree for 392. Use this tree to write the prime factorisation of 392.

3.8 USING PRIME FACTORISATION TO FIND HCF AND LCM

Prime factorisation is a useful tool for solving problems that require finding either a Lowest Common Multiple (LCM) or a Highest Common Factor (HCF). These types of problems can be solved by writing out a list of factors or multiples, but they can get particularly difficult and time-consuming for larger numbers. Even with finding, say, the LCM of numbers such as 24 and 42, as in Example 11, writing a list of multiples is time-consuming. Another strategy is needed and prime factorisation is one such strategy.

The key to using prime factorisation to find HCF and LCM are the following rules:

• HCF = product of lowest number of shared factors
• LCM = product of highest number of each factor

Note that the *highest* common factor requires multiplying the *lowest* number of *shared* factors; the *lowest* common multiple requires multiplying the *highest* number of *each* factor.

Tiffany Winn

EXAMPLE 16

Use prime factor trees to work out the prime factorisations of 8 and 36. Then use these prime factorisations to work out the HCF and LCM of 8 and 36.

Solution

Prime factorisation trees for 8 and 36 are:

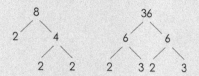

So, the prime factorisations of 8 and 36 are:

$8 = 2 \times 2 \times 2$ and $36 = 2 \times 2 \times 3 \times 3$

Working out the HCF of 8 and 36 requires finding the product of the lowest number of shared (prime) factors. The prime factors of 8 and 36 are 2 and 3, as shown above. The only shared factor is 2; 3 is a factor of 36 but not of 8. There are three 2's as prime factors of 8 and two 2's as prime factors of 36; the lowest number of 2's is two. So, the HCF of 8 and 36 is $2 \times 2 = 4$.

Working out the LCM of 8 and 36 requires finding the product of the highest number of each (prime) factor. The prime factors of 8 and 36 are 2 and 3. There are three 2's as prime factors of 8 and two 2's as prime factors of 36; the highest number of 2's is three. There are no 3's as prime factors of 8 and two 3's as prime factors of 36, so the highest number of 3's is two. So, the LCM of 8 and 36 is $2 \times 2 \times 2 \times 3 \times 3 = 72$.

Note that in Example 8 we worked out the HCF of 8 and 36 and in Example 11 we worked out the LCM of 8 and 36 using lists of factors. The answers using the prime factorisation method are the same as those obtained using a list of factors.

PRACTICE 16

Draw the prime factorisation trees for 12 and 15. Use these to find the prime factorisations of 12 and 15. Then use these prime factorisations to find the HCF and LCM of 12 and 15. Compare your answers with those obtained in Practice 8 and Practice 11.

EXAMPLE 17

Use prime factor trees to write the prime factorisation of 24 and 42. Then use these prime factorisations to work out the HCF and LCM of 24 and 42.

Solution

Prime factorisation trees for 24 and 42 are:

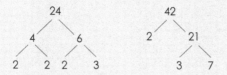

So, the prime factorisations of 24 and 42 are:

$24 = 2 \times 2 \times 2 \times 3$ and $42 = 2 \times 3 \times 7$

Working out the HCF of 24 and 42 requires finding the product of the lowest number of shared (prime) factors. The prime factors of 24 and 42 are 2, 3 and 7. The shared factors are 2 and 3; 7 is a factor of 42 but not of 24. There are three 2's as prime factors of 24 and there is one 2 as a prime factor of 42; the lowest number of 2's is one. There is one 3 as a prime factors of 24 and there is one 3 as a prime factor of 42; the lowest number of 3's is one. So, the HCF of 24 and 42 is $2 \times 3 = 6$.

Working out the LCM of 24 and 42 requires finding the product of the highest number of each (prime) factor. The prime factors of 24 and 42 are 2, 3 and 7. There are three 2's as prime factors of 24 and one 2 as a prime factor of 42; the highest number of 2's is three. There is one 3 as a prime factor of 24 and one 3 as a prime factor of 24; the highest number of 3's is one. There are no 7's as prime factors of 24 and one 7 as a prime factor of 42, so the highest number of 7's is one. So, the LCM of 24 and 42 is $2 \times 2 \times 2 \times 3 \times 7 = 168$.

Note that in Example 8 we worked out the HCF of 24 and 42 and in Example 11 we worked out the LCM of 24 and 42 using lists of factors. The answers using the prime factorisation method are the same as those obtained using a list of factors.

PRACTICE 17

Draw the prime factorisation trees for 18 and 39 and use these to find the prime factorisations of 18 and 39. Then use these prime factorisations to find the HCF and LCM of 18 and 39. Compare your answers with those obtained in Practice 8 and Practice 11.

EXAMPLE 18

Draw the prime factorisation trees for 210 and 216. Use these to find the prime factorisations of 210 and 216. Then use these prime factorisations to find the HCF and LCM of 210 and 216.

Solution

Prime factorisation trees for 210 and 216 are:

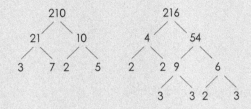

So, the prime factorisations of 210 and 216 are:

$210 = 2 \times 3 \times 5 \times 7$ and $216 = 2 \times 2 \times 2 \times 3 \times 3 \times 3$

Working out the HCF of 210 and 216 requires finding the product of the lowest number of shared (prime) factors. The prime factors of 210 and 216 are 2, 3, 5 and 7. The shared factors are 2 and 3; 5 and 7 are factors of 210 but not of 216. There is one 2 as a prime factor of 210 and three 2's as prime factors of 216; the lowest number of 2's is one. There is one 3 as a prime factor of 210 and three 3's as prime factors of 216; the lowest number of 3's is one. So, the HCF of 210 and 216 is $2 \times 3 = 6$.

Working out the LCM of 210 and 216 requires finding the product of the highest number of each (prime) factor. The prime factors of 210 and 216 are 2, 3, 5 and 7. There is one 2 as a prime factor of 210 and three 2's as prime factors of 216; the highest number of 2's is three. There is one 3 as a prime factor of 210 and three 3's as prime factors of 216; the highest number of 3's is three. There is one 5 as a prime factor of 210 and no 5's as prime factors of 216, so the highest number of 5's is one. There is one 7 as a prime factor of 210 and no 7's as prime factors of 216, so the highest number of 7's is one. So, the LCM of 210 and 216 is $2 \times 2 \times 2 \times 3 \times 3 \times 3 \times 5 \times 7 = 7560$.

PRACTICE 18

Draw the prime factorisation trees for 246 and 297. Use these to find the prime factorisations of 246 and 297. Then use these prime factorisations to find the HCF and LCM of 246 and 297.

EXAMPLE 19

Draw the prime factorisation trees for 264, 132 and 308. Use these to find the prime factorisations of 264, 132 and 308. Then use these prime factorisations to find the HCF and LCM of 72, 132 and 308.

Solution

Prime factorisation trees for 264, 132 and 308 are:

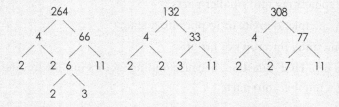

So, the prime factorisations of 264, 132 and 308 are:

$264 = 2 \times 2 \times 2 \times 3 \times 11$

$132 = 2 \times 2 \times 3 \times 11$

$308 = 2 \times 2 \times 7 \times 11$

Working out the HCF of 264, 132 and 308 requires finding the product of the lowest number of shared (prime) factors. The prime factors of 264, 132 and 308 are 2, 3, 7 and 11. The shared factors are 2 and 11; 3 is a factor of 264 and 132 but not 308, and 7 is a factor of 308 but not 132 and 264. There are three 2's as prime factors of 264, two 2's as prime factors of 132 and two 2's as prime factors of 308; the lowest number of 2's is two. There is one 11 as a prime factor of each of 264, 132 and 308; the lowest number of 11's is one. So, the HCF of 264, 132 and 308 is $2 \times 2 \times 11 = 44$.

Working out the LCM of 264, 132 and 308 requires finding the product of the highest number of each (prime) factor. The prime factors of 264, 132 and 308 are 2, 3, 7 and 11. There are three 2's as prime factors of 264, two 2's as prime factors of 132 and two 2's as prime factors of 308; the highest number of 2's is three. There is one 3 as a prime factor of 264, one 3 as a prime factors of 132 and no 3's prime factors of 308; the highest number of 3's is one. There are no 7's as prime factors of 264 or 132 but there is one 7 as a prime factor of 308; the highest number of 7's is one. There is one 11 as a prime factor of each of 264, 132 and 308; the highest number of 11's is one. So, the LCM of 264, 132 and 308 is $2 \times 2 \times 2 \times 3 \times 7 \times 11 = 1848$.

PRACTICE 19

Draw the prime factorisation trees for 315, 405 and 450. Use these to find the prime factorisations of 315, 405 and 450. Then use these prime factorisations to find the HCF and LCM of 315, 405 and 450.

When to use HCF or LCM

Initially, working out whether an application question requires you to find HCF or LCM can be tricky. However, there are some key words that will help you differentiate.

If a question includes any of these words—break, greatest, largest, most or biggest—then it is likely to be a HCF question, because the *Highest* Common Factor (also known as the *Greatest* Common Factor) is the *largest/greatest/biggest* of common factors.

In addition, if a problem asks you to:

- split or break something into smaller parts
- find how many people to invite or to put into a team
- or arrange something into rows or groups,

then it is likely to be a HCF question, because finding factors (unlike finding multiples) involves splitting a number into parts.

On the other hand, if a question includes any of these words—least or smallest—then it is likely to be a LCM question, because the *Lowest* Common Multiple is the *smallest* multiple of two or more numbers.

In addition, if a problem asks you to:

- find when two or more events happen at the same time
- buy or get multiple things,

then it is likely to be a LCM question, because both examples involve finding common multiples of existing numbers.

Each of the following examples and practice question pairs includes part (a) and part (b), one of which requires HCF and the other LCM. It is your job to work out which to use.

EXAMPLE 20: APPLICATION

a A school teacher needs to divide 153 Year 7 students and 162 Year 8 students into teams, each with the same combination of Year 7 and Year 8 students and nobody left out. What is the greatest number of teams that can be formed? How many Year 7 and Year 8 students would be on each team if the greatest number of possible teams were formed?

b Three friends work together at a café. The way their shifts are organised means that Benita works every three days, Zoe every six days and Cherie every five days. The friends want to plan a catch up when they are all working together. How often will this be?

Solutions

a This question is about dividing groups into teams, so it is a HCF question. The first step for both HCF and LCM questions is to draw a prime factorisation tree for both numbers. Prime factorisation trees for 153 and 162 are:

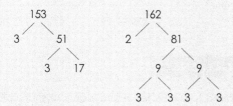

So, the prime factorisations of 153 and 162 are:

$153 = 3 \times 3 \times 17$

$162 = 2 \times 3 \times 3 \times 3 \times 3$

Since we are looking for the HCF of the two numbers, we need the product of the lowest number of shared factors. The only shared factor is 3. The lowest number of 3's is two. So, the HCF of 153 and 162 is 9, and this means that the greatest number of teams that can be made (according to the specifications in the question) is 9. There are 153 Year 7 students, so there will be $153 \div 9 = 17$ Year 7 students on each team, and $162 \div 9 = 18$ Year 8 students on each team.

b This question involves three events—all three friends being at work—at the same time, so it is a LCM question. The first step for both HCF and LCM questions is to draw a prime factorisation tree for both numbers. Prime factorisation trees for 3, 5 and 6 are:

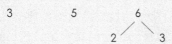

(Note that 3 and 5 are already prime, so their tree only consists of a single number.)

So, the prime factorisations of 3, 5 and 6 are:

$3 = 3$

$5 = 5$

$6 = 2 \times 3$

Since we are looking for the LCM of the three numbers, we need the product of the highest number of each factor. The factors are 2, 3 and 5. The highest number of 2's is one, of 3's is one and of 5's is one. So, the LCM of 3, 5 and 6 is $2 \times 3 \times 5 = 30$. This means that every 30 days the friends will work on the same day and be able to catch up together.

PRACTICE 20: APPLICATION

a Anna is making a game board that is 32 × 88 cm. She wants to cover the game board in equal-sized square tiles. What is the largest tile she can use?

b A series of traffic lights changes every 20, 30 and 40 seconds. How often will the lights change at the same time?

EXAMPLE 21: APPLICATIONS

a Marianne has just moved to a new town and wants to share plates of baked goods with her neighbours. She has 84 cookies and 48 brownies to share, and wants to split them equally among the plates with no food left over. What is the greatest number of plates she can make to share? How many cookies and brownies will be on each plate?

b The Dudley Street Bakery sells mini chicken and mushroom pies cheaper in boxes of 18. Across town, the Semaphore Pines Bakery also sells mini chicken and mushroom pies, but their cheaper boxes have 48 pies in each. If Leyton wants to buy the same number of pies from each bakery for a blind taste test, what is the smallest number of pies he will have to buy from each bakery?

Solutions

a This question is about breaking something into smaller parts (food into individual plates), so it is a HCF question. The first step for both HCF and LCM questions is to draw a prime factorisation tree for both numbers. Prime factorisation trees for 84 and 48 are:

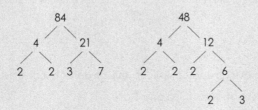

So, the prime factorisations of 84 and 48 are:

$84 = 2 \times 2 \times 3 \times 7$

$48 = 2 \times 2 \times 2 \times 2 \times 3$

Since we are looking for the HCF of the two numbers, we need the product of the lowest number of shared factors. The shared factors are 2 and 3. The lowest number of 2's is two and the lowest number of 3's is one. So, the HCF of 84 and 48 is $2 \times 2 \times 3 = 12$, and this means that the greatest number of plates of food that can be made (according to the specifications in the question) is 12. There are 84 cookies, so there will be $84 \div 12 = 7$ cookies on each plate, and $48 \div 12 = 4$ brownies on each plate.

b This question is about buying multiple items, so it is a LCM question. The first step for both HCF and LCM questions is to draw a prime factorisation tree for both numbers. Prime factorisation trees for 18 and 48 are:

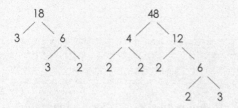

So, the prime factorisations of 18 and 48 are:

$18 = 2 \times 3 \times 3$

$48 = 2 \times 2 \times 2 \times 2 \times 3$

Since we are looking for the LCM of the two numbers, we need the product of the highest number of each factor. The prime factors are 2 and 3. The highest number of 2's is four. The highest number of 3's is two. So, the LCM of 18 and 48 is $2 \times 2 \times 2 \times 2 \times 3 \times 3 = 144$, and this means that Leyton will need to buy 144 pies to have the same amount from each bakery. Pies from Dudley Street Bakery come in boxes of 18, so Leyton will need to buy $144 \div 18 = 8$ boxes of pies from Dudley Street Bakery. Pies from Semaphore Pines Bakery come in boxes of 48, so Leyton will need to buy $144 \div 48 = 3$ boxes of pies from Semaphore Pines Bakery.

PRACTICE 21: APPLICATIONS

a Scarlett has toy train track pieces that are 15 centimetres long, Amelia's pieces are 6 centimetres long and Tom's pieces are 9 centimetres long. If they each build a long, straight track, what is the shortest length of track they can build so that they all build tracks of the same length? How many pieces of track will each child use?

b Three acrobatic teams from different countries are combining to give a performance at the opening ceremony of the next Olympic Games. The Hong Kong team has 96 members, the team from China 330 members, and the team from Korea 144 members. For their entrance into the stadium, the teams want to organise themselves into rows where each row has the same number of members. This will allow the teams to enter the stadium one behind the other in columns where each column is of equal length. What is the largest number of team members that can be in each row if each row is to have an equal number of members? How many rows will each country's team take up?

3.9   SQUARE NUMBERS

The concept of square numbers is perhaps easiest to understand in the context of laying tiles. Think about laying tiles on a floor. The figure below shows all the square tiles that could be cut from the sheet shown:

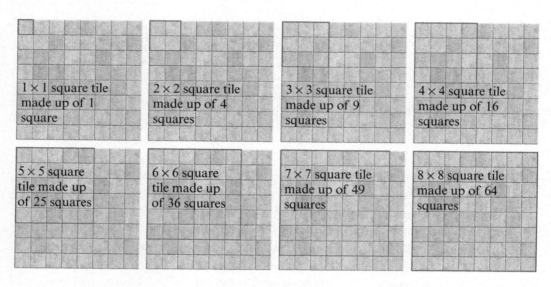

1 × 1 square tile made up of 1 square

2 × 2 square tile made up of 4 squares

3 × 3 square tile made up of 9 squares

4 × 4 square tile made up of 16 squares

5 × 5 square tile made up of 25 squares

6 × 6 square tile made up of 36 squares

7 × 7 square tile made up of 49 squares

8 × 8 square tile made up of 64 squares

Perfect square
The product of a whole number multiplied by itself

As the figure above shows, there are some numbers that are **perfect squares**, and others that are not. For example, 1, 4, 9, 25, 36, 49 and 64 are all perfect squares. They all represent the number of smaller, equal-sized square tiles that make up a larger, square tile and they can all be calculated by multiplying a number by itself:

- $1 = 1 \times 1$
- $4 = 2 \times 2$
- $9 = 3 \times 3$
- $16 = 4 \times 4$
- $25 = 5 \times 5$
- $36 = 6 \times 6$
- $49 = 7 \times 7$
- $64 = 8 \times 8$

On the other hand, for example, 5 is not a perfect square. There is no way to make a larger square tile out of five smaller square tiles, and there is no whole number that, when multiplied by itself, will give five.

EXAMPLE 22

Calculate the squares of the following numbers.

a 5   b 9   c 22   d 13   e 10

Solutions

a $5 \times 5 = 25$, so the square of 5 is 25

b $9 \times 9 = 81$, so the square of 9 is 81

c $22 \times 22 = 484$, so the square of 22 is 484

d $13 \times 13 = 169$, so the square of 13 is 169

e $10 \times 10 = 100$, so the square of 10 is 100

PRACTICE 22

Calculate the squares of the following numbers.

a 4   b 11   c 3   d 15   e 22

Making connections See Chapter 2 for help with long multiplication if you need more information about how to do the multiplication in these examples.

3.10 SQUARE ROOTS

Taking the **square root** of a number is the opposite of squaring a number. When we take the square root of a number, we are finding another number that when multiplied by itself, will give the original number.

The symbol for square root is $\sqrt{}$. So, for example, in order to state that the square root of 25 is 5, you would write $\sqrt{25} = 5$. To state that the square root of 64 is 8, you would write $\sqrt{64} = 8$.

A square root can often be found by trial and error. For example, if searching for the square root of 64, you might first try 5, but since $5 \times 5 = 25$ and 25 is less than 64, you would then look at a larger number. If you next tried 9, you would work out that $9 \times 9 = 81$ and 81 is larger than 64, so you would know the square root of 64 had to be 6, 7 or 8. Since 81 is much closer to 64 than 25, trying 8 next would be a good strategy.

Another strategy that can be used to find a square root of a perfect square is prime factorisation. A number that is a perfect square will have all of its prime factors in pairs. The square root of the number is found by finding the product of one of each of the pairs of prime factors, as shown in the following example.

Square root A value that, when multiplied by itself, gives the number

Tiffany Winn

EXAMPLE 23

Use prime factorisation to determine whether the following numbers are perfect squares. If the number is a perfect square, find its square root.

a   324   b   250   c   208   d   441

Solutions

a   A prime factorisation tree for 324 is:

So, the prime factorisation for 324 is $2 \times 2 \times 3 \times 3 \times 3 \times 3$.

This means that 324 is a perfect square, because all its prime factors come in pairs:

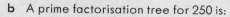

- First pair: 2×2
- Second pair: 3×3
- Third pair: 3×3

The square root of 324 is calculated by finding the product of one of each of the pairs of factors. This means finding the product of 2 (from first pair), 3 (from second pair) and 3 (from third pair): $2 \times 3 \times 3 = 18$, so 18 is the square root of 324.

b   A prime factorisation tree for 250 is:

So, the prime factorisation for 250 is $2 \times 5 \times 5 \times 5$.

This means that 250 is not a perfect square, because its prime factors do not all come in pairs: there is only one 2, and there are three 5's. The square root of 250 will therefore not be a whole number and cannot be calculated using this method.

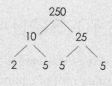

c   A prime factorisation tree for 208 is:

So, the prime factorisation for 441 is $2 \times 2 \times 2 \times 2 \times 13$.

This means that 208 is not a perfect square, because even though its prime factor 2 comes in pairs, 13 does not come in a pair. The square root of 208 will therefore not be a whole number and cannot be calculated using this method.

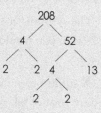

d   A prime factorisation tree for 441 is:

So, the prime factorisation for 441 is $3 \times 3 \times 7 \times 7$.

This means that 441 is a perfect square, because all its prime factors come in pairs:

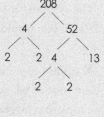

- First pair: 3×3
- Second pair: 7×7

The square root of 441 can be calculated by finding the product of one of each of the pairs of factors. This means finding the product of 3 (from first pair) and 7 (from second pair): $3 \times 7 = 21$, so 21 is the square root of 441.

CRITICAL THINKING

Which two whole numbers will the square root of 250 lie between?

PRACTICE 23

Use prime factorisation to determine whether the following numbers are perfect squares. If the number is a perfect square, find its square root.

a 180   b 196   c 169   d 224

EXAMPLE 24

Calculate the square roots of the following perfect squares, either by using prime factorisation or trial and error.

a 121   b 1   c 49   d 196   e 256

Solutions

a Divisibility tests tell us that 121 is not divisible by 2 or any even number (not even), nor 3 or 9 (digits add up to 4 which is not divisible by 3 and not divisible by 9). Trial and error tells us it is not divisible by 7 (121 ÷ 7 = 17 remainder 2), but that it is divisible by 11 (121 ÷ 11 = 11). Since 11 × 11 = 121, 11 is the square root of 121:

$$\sqrt{121} = 11$$

b 1 × 1 = 1, so the square root of 1 is 1:

$$\sqrt{1} = 1$$

c Divisibility tests tell us that 49 is not divisible by 2 or any even number (not even), nor 3 or 9 (digits add up to 11 which is not divisible by 3 and not divisible by 9). Trial and error tells us it is divisible by 7 (49 ÷ 7 = 7). Since 7 × 7 = 49, the square root of 49 is 7:

$$\sqrt{49} = 7$$

d A prime factorisation tree for 196 is:

So, the prime factorisation for 196 is 2 × 2 × 7 × 7. This means:

$$\sqrt{196} = 2 \times 7 = 14$$

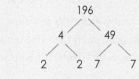

e A prime factorisation for 256 is:

So, the prime factorisation for 256 is 2 × 2 × 2 × 2 × 2 × 2 × 2 × 2. This means:

$$\sqrt{265} = 2 \times 2 \times 2 \times 2 = 16$$

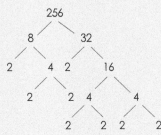

PRACTICE 24

Calculate the square roots of the following perfect squares, either by using prime factorisation or trial and error.

a 4   b 676   c 0   d 1089   e 225

EXAMPLE 25: APPLICATION

Jade wants to put wire around her square-shaped garden to stop her dogs digging up the garden. If the wire is only sold in whole metres and the area of the garden is 196 m^2, how much wire will she need to buy? (Hint: area of a square = side length × side length)

Solution

The amount of wire needed will be 4 × side length of the garden.

The side length of the garden will be the square root of the area (196) because area of a square = side length × side length.

A prime factorisation tree will help us find the square root of 196 (assuming 196 is a perfect square). A prime factorisation tree for 196 is:

So, the prime factorisation of 196 is 2 × 2 × 7 × 7. This means the square root of 196 is 2 × 7 = 14, and the side length of the garden will be 14 m. The amount of wire needed to fence the garden is 4 × 14 = 56 m.

PRACTICE 25: APPLICATION

Jackson wants to put wire around his square-shaped garden to stop his dogs digging up the garden. If the wire is only sold in whole metres and the area of the garden is 484 m^2, how much wire will he need to buy? (Hint: area of a square = side length × side length)

Making connections Calculating the area of a square and other geometric figures will be covered in Chapter 12.

3.11   CHAPTER SUMMARY

SKILL OR CONCEPT	DEFINITION OR DESCRIPTION	EXAMPLES
Factor	A whole number that divides into another whole number without a remainder.	3 is a factor of 24 because $24 \div 3 = 8$ (exactly, no remainder)
Finding all factors of a whole number	Finding factors in pairs by starting from 1 (which pairs with the number itself) and checking 2, 3, 4 and so on up to a point where it is clear there can be no more factors because all possibilities have been covered.	Find all factors of 24: Start at 1, this pairs with 24. 2 is a factor (24 is even), pairs with $24 \div 2 = 12$ 3 is a factor ($2 + 4 = 6$, which is divisible by 3), pairs with $24 \div 3 = 8$ 4 is a factor, pairs with $24 \div 4 = 6$ 5 is not a factor There are no more factors as 4 and 6 have already been tested. So, all the factors of 24 are: 1, 2, 3, 4, 6, 8, 12 and 24
Multiple	A whole number that can be divided exactly by another number without leaving a remainder. A multiple is the result of multiplying the original number by a whole number.	36 is a multiple of 12 because $36 \div 12 = 3$ (exactly, no remainder) 36 is also a multiple of 4 because $36 \div 4 = 9$ (exactly, no remainder) 36 is not a multiple of 7 because $36 \div 7 = 5$ remainder 1
Identifying multiples of a whole number	• To test if a value is a multiple, divide it by the whole number in question. • If the value can be divided by the whole number without leaving a remainder, then it is a multiple. • If the value, when divided by the whole number leaves a remainder, then it is not a multiple.	Is 56 a multiple of 7? Yes, because $56 \div 7 = 8$ (exactly, no remainder). Is 44 a multiple of 8? No, because $44 \div 8 = 5$ remainder 4.
Prime number	A whole number greater than 1 that is only divisible by 1 and itself.	Examples: 2, 3, 5, 7 and 11.
Composite number	A whole number greater than 1 that can be divided exactly by whole numbers other than 1 and itself.	Examples: 4, 6, 8, 9 and 10.
Divisibility tests	A number is: • divisible by 2 if even • divisible by 3 if the number formed by the sum of the digits is divisible by 3 • divisible by 4 if the last two digits are divisible by 4 • divisible by 5 if it ends in 0 or 5 • divisible by 6 if it is divisible by 2 and 3 • divisible by 8 if the last three digits are divisible by 8 • divisible by 9 if the number formed by the sum of the digits is divisible by 9 • divisible by 10 if ends in 0.	• 494 divisible by 2; 301 is not • 750 divisible by 3; 863 is not • 9524 divisible by 4; 934 is not • 9855 divisible by 5; 9076 is not • 9872 divisible by 8; 865 is not • 8568 divisible by 9; 7967 is not • 670 divisible by 10; 4058 is not

Tiffany Winn

SKILL OR CONCEPT	DEFINITION OR DESCRIPTION	EXAMPLES
Highest Common Factor (HCF)	The HCF of two or more numbers is the largest number that is a factor of all of the numbers.	The factors of 36 are 1, 2, 3, 4, 6, 9, 12, 18 and 36. The factors of 56 are 1, 2, 4, 7, 8, 14, 28 and 56. The HCF of 36 and 56 is 4.
Lowest Common Multiple (LCM)	The LCM of two or more numbers is the smallest number that is a multiple of all of the numbers.	The multiples of 36 are 36, 72, 108, 144, 180, 216, 254, 290, 326, 360, 396, 432, 468, 504, 540 and so on. The multiples of 56 are 56, 112, 168, 224, 280, 336, 392, 448, 504, 560 and so on. The LCM of 36 and 56 is 504.
Prime factorisation	A whole number written as a product of its prime factors and can be calculated using prime factorisation trees. There is only one prime factorisation for any whole number.	Two different prime factorisation trees for 36 (note they result in the same prime factorisation): Prime factorisation of $36 = 2 \times 2 \times 3 \times 3$
Finding HCF of two or more numbers with prime factorisation	Work out the prime factorisation of each number; the HCF of all numbers will be the product of the lowest number of each shared prime factor.	Prime factorisation trees to determine prime factorisation of 36 and 56: Prime factorisation of: $$36 = 2 \times 2 \times 3 \times 3$$ $$56 = 2 \times 2 \times 2 \times 7$$ The only shared factor is 2. HCF is the lowest number of the shared factors $2 \times 2 = 4$
Finding LCM of two or more numbers using prime factorisation	Work out the prime factorisation of each number; the LCM of all numbers will be the product of the highest number of each prime factor.	Prime factorisation trees to determine prime factorisation of 36 and 56: Prime factorisation of: $$36 = 2 \times 2 \times 3 \times 3$$ $$56 = 2 \times 2 \times 2 \times 7$$ Prime factors are 2, 3 and 7. Highest number of 2's is three and of 3's is two and of 7's is one. LCM: $2 \times 2 \times 2 \times 3 \times 3 \times 7 = 504$

SKILL OR CONCEPT	DEFINITION OR DESCRIPTION	EXAMPLES
Perfect square	A perfect square results when a whole number is multiplied by itself.	Examples: • 9 because $9 = 3 \times 3$ • 25 because $25 = 5 \times 5$ • 121 because $121 = 11 \times 11$
Square root	The square root of a number is a value that, when multiplied by itself, gives the number. The symbol for square root is √.	Examples: • $3 = \sqrt{9}$, since $3 \times 3 = 9$ • $3 = \sqrt{25}$, since $5 \times 5 = 25$ • $11 = \sqrt{121}$, since $11 \times 11 = 121$
Using prime factorisation to find the square root of a perfect square	All of the prime factors of a perfect square come in pairs. The square root of the number will be the product of one of each of the pairs of the prime factors.	Prime factorisation tree for 36 is: 36 4 9 2 2 3 3 Prime factorisation of 36 is $2 \times 2 \times 3 \times 3$ Taking the product of one of each of the pairs of factors gives $\sqrt{36} = 2 \times 3 = 6$

Tiffany Winn

3.12 REVIEW QUESTIONS

A SKILLS

1   Find all factors of the following numbers.

    a 28   **b** 92   **c** 225   **d** 400   **e** 153

2   Work out which of the numbers from Question 1 are perfect squares, if any.

3   List the first ten multiples of:

    a 11   **b** 7   **c** 9   **d** 22   **e** 13

4   For each of the following pairs of numbers, indicate whether the first number is a multiple of the second.

    a 637; 13   **b** 343; 7   **c** 300; 9   **d** 508; 4   **e** 262; 8

5   Categorise the following numbers as prime, composite or neither.

    a 19   **b** 1   **c** 4023   **d** 692   **e** 0

6   Work out whether the first number is divisible by the second number without doing the actual division, and explain how you got your answer.

    a 8208; 9   **b** 4020; 10   **c** 6442; 8   **d** 2095; 2   **e** 9876; 6

7   Use a prime factorisation tree to find the prime factorisation of the following numbers.

    a 190   **b** 17   **c** 846   **d** 245   **e** 15

8   Find the Highest Common Factor (HCF) of each of the following collections of numbers by using prime factorisation.

    a 56 and 120   **b** 64 and 147   **c** 40 and 132

    d 5, 60 and 75   **e** 12, 36 and 90

9   Check your answer to 8(a) and 8(e) by making a list of all the factors of the numbers in each question.

10 Find the Lowest Common Multiple (LCM) of each of the following collections of numbers by using prime factorisation.

    a 56 and 120   **b** 64 and 147   **c** 40 and 132

    d 5, 60 and 75   **e** 12, 36 and 90

11 Check your answer to 10(a) and 10(e) by listing the multiples of the numbers in each question until a common multiple is found.

12 Find the square of the following numbers.

    a 19   **b** 17   **c** 23   **d** 20   **e** 15

13 Using prime factorisation, find the square root of the following numbers or show that the square root cannot be found.

a 24   b 900   c 289   d 441   e 425

B APPLICATIONS

1 A train arrives at equal intervals over a twelve-hour period and there are four trains in the day. How often does the train arrive?

2 Flyte Street florist is making up a bunch of bouquets from 18 roses and 30 tulips. The store wants all the bouquets to include both roses and tulips and be identical. What is the greatest number of bouquets the store can make?

3 A sports coach has a group of 40 players that she wants to split into at least even groups, without leaving anyone out. What size groups could she potentially have?

4 A farmer is planting trees in an orchard. She wants to plant 50 almond trees and 30 apple trees in rows. She wants each row to be a mix of apple and almond trees and wants each row to be the same, for example, if one row has 7 almond and 4 apple trees, then each row should have 7 almond and 4 apple trees. What is the maximum number of trees she can plant per row?

5 Coffee bags are being packaged in containers of 8. If I have 348 coffee bags, will they fit into containers without any coffee bags left over?

6 A radio station gave away a $50 prize to every 50th caller and free concert tickets to every 15th caller. How many callers will there be before someone receives both $50 and a concert ticket?

7 Isobel is training for a triathlon. She runs every third day, swims every fifth day and bikes every second day. If Isobel runs, swims and bikes today, in how many days will she next do all three different types of exercise?

8 A basketball league based in Midden has 32 forwards and 80 guards registered to play. Tom, who is running the league, needs to include each player on a team. As well as each team having the same number of players, Tom wants each team to have the same number of forwards as other teams, and similarly for guards. If Tom creates the maximum number of teams possible, how many guards will be on each team?

9 Ivan's dad is buying hot dogs and hot dog buns for an extended family get together. Hot dogs come in packs of 12 and hot dog buns come in packs of 8. Ivan's dad wants at least 30 hot dogs and he wants the same number of hot dogs and hot dog buns. What is the smallest total number of hot dogs that Ivan's dad can buy?

10 Kevin is going to plant 68 gum trees and 51 wattle trees. He wants to plant the trees in rows so that each row has the same number of trees and is made up of only one type of tree. What is the maximum number of trees Kevin can have in each row?

11 Soraya is buying paper cups and plates. Cups come in packages of 12, and plates come in packages of 20. She wants to buy the same number of cups and plates, but also wants to buy the least number of packages possible. How much will Soraya have to pay if each package of cups is $3 and each package of plates is $5? Explain how you got your answer.

12 If I am making up a square-shaped board that has a side length of 20 cm, what will the area of the board be? (Hint: area of a square = side length × side length)

13 My square game board has an area of 576 cm². How long will each side be? If my game board is subdivided into 64 squares of equal size, what will be the side length and area of each of these 64 squares?

14 A square coffee table has a side length of 45 cm. What will be the area of the table?

15 You want to decorate the walls of your square-shaped bedroom with a border along the top. If the area of the room is 250 000 cm², how long will the border need to be?

4

DIRECTED NUMBERS

CHAPTER CONTENT

» Introduction to directed numbers

» Adding and subtracting directed numbers

» Multiplying and dividing directed numbers

» Mixed operations and directed numbers

CHAPTER OBJECTIVES

» Understand the concept of number direction as well as magnitude

» Position directed numbers on a number line

» Add and subtract directed numbers

» Multiply and divide directed numbers

» Apply the correct order to directed number operations

IS IT PAR, A BIRDIE OR A BOGEY?

When a golfer finishes a round of golf, their score for the round is either a plus or a minus score. This is because golf scores are calculated according to whether the golfer is above or below the expected number of shots for each hole. Therefore, calculating golf scores requires understanding the use of both positive and negative numbers.

The table on the next page shows a sample set of scores for a round of nine holes of golf. The expected number of strokes required for each hole is called 'par', and shown in the 'Par' column. The golfer's actual score, how many strokes the golfer took to complete the hole, is shown in the 'My Score' column. The 'My Golf Score' shows the difference between the Par and My Score columns. For example, for hole number 1, the Par and My Score columns are the same, so the difference is zero. However, for hole number 3, the golfer took one more stroke than expected, that is one more than par, to complete the hole, so the My Golf Score column shows +1 (this is called a bogey). For hole 6, the golfer took two less strokes than expected to complete the hole, so the My Golf Score column shows −2. Overall, the golfer's score for the round is obtained by adding the golf scores for individual holes together:

$$0 + 0 + 1 + 0 - 1 - 2 + 1 - 1 + 0 = -2$$

The golfer's overall score for the round is −2, or 'two under par'. Interestingly, unlike many other sports, in golf a lower score is better than a higher score; for example, a score of −2 is better than a score of +2.

HOLE NUMBER	PAR	MY SCORE	MY GOLF SCORE
1	3	3	0
2	3	3	0
3	4	5	+1
4	5	5	0
5	3	2	−1
6	4	2	−2
7	2	3	+1
8	4	3	−1
9	3	3	0

4.1   INTRODUCTION TO DIRECTED NUMBERS

Integers
The set of numbers that include the counting numbers (1, 2, 3, 4, …), zero (0), and the negative of the counting numbers (−1, −2, −3, −4, …)

Directed numbers are numbers that have both a magnitude (size) and a direction. So, for example, the numbers +2 and −2 have the same magnitude but different directions: +2 has a positive direction and −2 has a negative direction. Often a number with a positive direction is written without the plus sign, so +2 is usually written as just 2. In contrast, a negative number must have its sign to be recognised as negative. The set of numbers made up of the counting numbers, zero, and the negative of the counting numbers is called **integers**.

Integers (directed numbers) can be represented on a number line, where the convention is that a negative direction represents distance to the left of zero and a positive direction represents distance to the right of zero, as shown below:

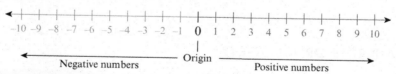

A key feature of integers is that the equivalent positive and negative numbers are opposites. So, for example, the opposite of 3 is −3 and the opposite of 8 is −8. According to the same principle, the opposite of −3 is − (−3), which is just 3. So, the negative of a negative number is its positive equivalent.

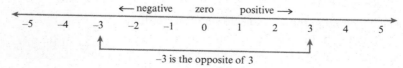

Whether we realise it or not, many of the numbers we use in everyday life are directed numbers. Think about, for example, the following situations:

1   Two cars are driving on a highway. One car is driving at 10 km/h over the speed limit. Another is driving at 15 km/h under the speed limit. Which directed numbers would correspond to each of these speeds? What is the difference between the speeds of the two cars?

2   My bank account has $540 in it when I leave to go food shopping and then I spend $128. I stop and pay a bill for $58 and buy lunch for $17. The next day I am paid $320 for some work I did and I then go to buy a new quilt for $189. Which directed numbers correspond to each of these transactions? How much money do I have left in my bank account?

3   Yesterday the temperature was five degrees above zero but today it is three degrees below zero. Which directed numbers would correspond to each of these temperatures? What is the difference between the two temperatures?

4   On a nine-hole golf course, a golfer made the following scores: par, birdie, bogey, par, par, par, double birdie, bogey and birdie. Note that a birdie is one under par, a double birdie two under par, and a bogey one over par. What would the golfer's final score be? How would that score usually be written?

All of these questions use directed numbers. The first question about the speed of cars can be solved by recognising that a speed of 10 km/h over the speed limit can be written as +10 and a speed of 15 km/h under the speed limit as −15. The difference between the two speeds is obtained by subtracting one number from the other: 10 − (−15). As will be discussed in section 4.2, subtracting a negative number is the equivalent of adding a positive number, so the difference between the two speeds is 25 km/h.

The second question can be solved by treating withdrawals as negative numbers and deposits as positive numbers, with the initial balance being a credit, so a positive number: +540. The money left in the account is 540 − 128 − 58 − 17 + 320 − 189 = $468.

The third question is similar to the first question except that it concerns weather rather than car speeds. A temperature of five degrees above zero can be written as +5, four below zero as −4, and the difference between the two is + 5 − (−4) = 9 degrees.

The last question involves a golf situation similar to that described in the previous section. The golfer's scores can be written as: 0, −1, +1, 0, 0, 0, −2, +1, and −1. The golfer's final score is 0 − 1 + 1 + 0 + 0 + 0 − 2 + 1 − 1 = −2, or 2 under par.

4.2   ADDING AND SUBTRACTING DIRECTED NUMBERS

The first key rule to remember when adding and subtracting directed numbers is that minus (−) means move to the left on a number line, whereas add (+) means move to the right on a number line.

For example, using a number line to solve −5 + 4 involves starting at −5 and moving 4 units to the right, because we are adding, as shown below:

−5 + 4 = −1

Tiffany Winn

Similarly, solving $3 - 5$ involves starting at 3 but this time moving 5 to the left, because we are subtracting, as shown below:

$3 - 5 = -2$

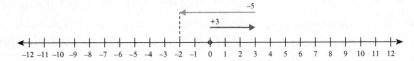

Solving a problem involving the addition and/or subtraction of more than two numbers simply requires doing each addition and/or subtraction in turn, from left to right, as shown below for the example of $4 - 7 + 6 - 8 - 2$:

$4 - 7 + 6 - 8 - 2 = -7$

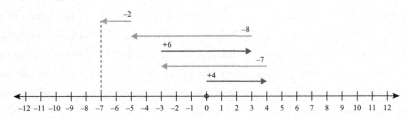

EXAMPLE 1

Draw number lines that show the following calculations.

a $4 - 9$ b $-2 - 3$ c $-4 + 8$ d $4 + 6$ e $12 - 5 + 2 - 8$

Solutions

a $4 - 9 = -5$, as shown in the number line below:

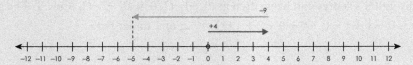

b $-2 - 3 = -5$, as shown in the number line below:

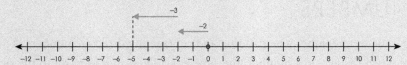

c $-4 + 8 = 4$, as shown in the number line below:

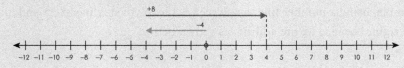

d 4 + 6 = 10, as shown in the number line below:

e 12 − 5 + 2 − 8 = 1, as shown in the number line below:

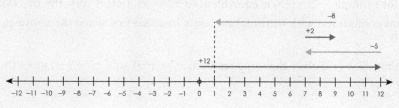

PRACTICE 1

Draw number lines that show the following calculations.

a −8 − 4   b −1 + 6   c 10 − 13

d 3 + 3   e −12 + 5 − 2 + 8

Now let's consider a question that involves subtracting a negative number, such as 4 − (−9). When thinking about this type of question it is important to understand that the first − sign indicates 'subtract', and the second − sign (that is attached to the 9) means 'negative'. Solving this question requires understanding the remaining key rules for adding and subtracting directed numbers. When positive and/or negative signs are written side-by-side *without a number in-between* (as in the example above), then the first sign indicates the operation, either addition or subtraction, and the second sign indicates whether the number being added or subtracted is positive or negative. We can combine the first add/subtract sign and the second positive/negative sign into a single resulting sign as shown in the table.

FIRST SIGN	SECOND SIGN	RESULTING SIGN
+	+	+
+	−	−
−	+	−
−	−	+

Let's use this table to consider the following examples:

- Using the first row in the table, −2 + (+3) simplifies to −2 + 3, which is then calculated to be 1 by starting at zero and moving 2 to the left and then 3 to the right on a number line.

- Using the second row in the table, −2 + (−3) simplifies to −2 − 3, which is then calculated to be −5 by starting at zero and moving 2 to the left and then 3 more to the left on a number line.

Tiffany Winn

- Using the third row in the table, −2 − (+3) simplifies to −2 − 3, which is −5.
- Using the last row in the table, −2 − (−3) simplifies to −2 + 3, which is 1.

Notice in these examples the second value that is being either added or subtracted from the first value is written in a set of brackets. This is a useful way of discerning between the operation (addition or subtraction) and the sign of the second value (positive or negative). However, because there is only one number inside the brackets, the brackets are not necessary, for example, −2 − (−3) is equivalent to −2 − −3. In this way, the BEDMAS order of operations is observed with the brackets being handled first when the above equations are solved.

Notice that −2 + (+3) gives the same result as −2 − (−3), and −2 + (−3) gives the same result as −2 − (+3).

Making connections

To review BEDMAS revisit Chapter 2.

EXAMPLE 2

Simplify these calculations and then draw a number line that shows how they can be worked out.

a   −8 − (−4)   b   −1 + (−6)   c   −3 + (9 − 1)

d   5 − (+7)   e   −11 + (+3) − (−4) + 8 + (−2)

Solutions

a   As a first step, −8 − (−4) is simplified to −8 + 4, by removing the brackets and replacing the two negatives that are together (*without a number in-between*) with a positive. Then, −8 + 4 can be solved with the use of a number line:

$$-8 - (-4) = -8 + 4$$
$$= -4$$

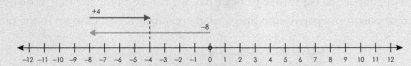

b   Because the + and − that are together (*without a number in-between*), the brackets are removed and they are replaced by a single minus sign, −1 + (−6) = −1 − 6. This calculation can then be worked out using a number line:

$$-1 + (-6) = -1 - 6$$
$$= -7$$

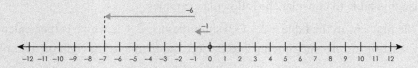

c According the order of operations BEDMAS, the calculation inside the brackets needs to be done first: $-3 + (9 - 1) = -3 + (+8)$.

Next, for the + and + that are together (*without a number in-between*) the brackets are removed and they are replaced by a single plus sign, $-3 + (+8) = -3 + 8$. This calculation can then be worked out using a number line:

$$-3 + (9 - 1) = -3 + (+8)$$
$$= -3 + 8$$
$$= 5$$

d As a first step, $5 - (+7)$ can be simplified to $5 - 7$, by removing the brackets and replacing the − and + signs that are together (*without a number in-between*) with a negative. Then, $5 - 7$ can be solved with the use of a number line:

$$5 - (+7) = 5 - 7$$
$$= -2$$

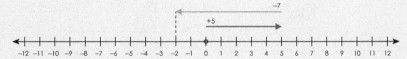

e Firstly, wherever there are two signs together (*without a number in-between*), the brackets are removed and they are replaced by the appropriate single sign. Using the rules from the table, $-11 + (+3) - (-4) + 8 + (-2) = -11 + 3 + 4 + 8 - 2$. This calculation can then be worked out using a number line:

$$-11 + (+3) - (-4) + 8 + (-2) = -11 + 3 + 4 + 8 - 2$$
$$= 2$$

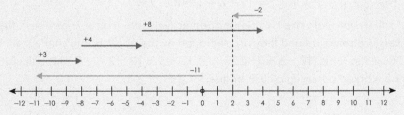

PRACTICE 2

Simplify these calculations and then draw a number line that shows how they can be worked out.

a $5 - (-7)$ **b** $-2 - (-8)$ **c** $4 + (-9 + 2)$

d $-5 + (-5)$ **e** $-10 + (-1) + 7 - (+8) - (-15)$

EXAMPLE 3

Simplify these calculations.

a $-3 - (-6) + 6 - 1 - (-2 - (-7))$ b $12 - (7 - 2 + 1) + (-10) - 2$

c $-11 + (+21) - (+4) + 3$ d $-10 - (-1) + 7 + (+8) - 15$

Solutions

a The calculation inside the final set of brackets needs to be done first according to BEDMAS. This calculation requires two steps: first, removing the double minus sign ($- -$) and replacing it with a positive (+) sign, and second, doing the actual calculation. Both steps are shown here: $(-2 - (-7)) = (-2 + 7) = 5$

So, in the question as a whole, $(-2 - (-7))$ is now replaced with 5:

$-3 - (-6) + 6 - 1 - (-2 - (-7)) = -3 - (-6) + 6 - 1 - 5$

Next, wherever there are two signs together (*without a number in-between*), the brackets are removed and they are replaced by the appropriate single sign. Using the rules from the table, $-3 - (-6) + 6 - 1 - 5 = -3 + 6 + 6 - 1 - 5 = 3$. This calculation can then be worked out using a number line:

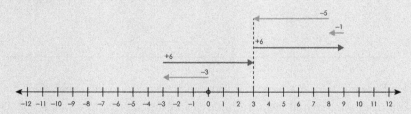

b The calculation inside the middle set of brackets needs to be done first according to BEDMAS. So, in $(7 - 2 + 1)$, first we do $7 - 2$, which gives +5, and then we calculate $+5 + 1$ which gives 6.

In the calculation as a whole, $(7 - 2 + 1)$ is now replaced by 6:

$12 - (7 - 2 + 1) + (-10) - 2 = 12 - 6 + (-10) - 2$

Next, wherever there are two signs together (*without a number in-between*), the brackets are removed and they are replaced by the appropriate single sign. Using the rules from the table, $12 - 6 + (-10) - 2 = 12 - 6 - 10 - 2 = -6$. This calculation can then be worked out using a number line:

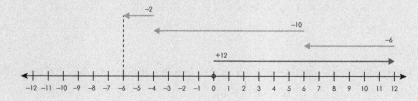

c Wherever there are two signs together (*without a number in-between*), the brackets are removed and they are replaced by the appropriate single sign. Using the rules from the table, $-11 + (+21) - (+4) + 3 = -11 + 21 - 4 + 3 = 9$. This calculation can then be worked out using a number line:

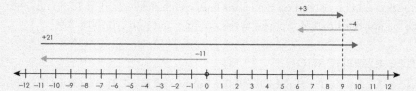

d Wherever there are two signs together (*without a number in-between*), the brackets are removed and they are replaced by the appropriate single sign. Using the rules from the table, $-10 - (-1) + 7 + (+8) - 15 = -10 + 1 + 7 + 8 - 15 = -9$. This calculation can then be worked out using a number line:

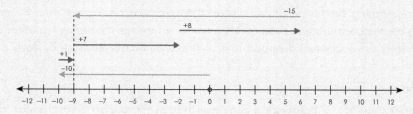

PRACTICE 3

Simplify these calculations.

a $4 + (-5) - 2 + (-(8 - (-2) + (-5) + (+3)))$

b $-(-2) + 9 - 13 - 4 - (-14)$

c $4 + (-7) - (-11 + 4 - (-7)) + 0 - 9$

d $3 + (-4) - 13 + 6$

EXAMPLE 4: APPLICATION

Tania's credit card bill was $426. She paid off $230, bought a new shirt on the credit card for $43, paid off $350, and bought a birthday gift for a friend for $35. How much does she now owe on the credit card?

Solution

When Tania owes $426 on the credit card, it means she is $426 in debt, and this can be written as a negative number: -426. Each time she pays something on the credit card, this negative amount gets smaller; therefore, a payment toward what is owed is added to the negative number to decrease the amount owing. In contrast, a further purchase on the credit card increases the amount owing and is subtracted from the existing balance. This results in the following equation to solve this problem:

$$-426 + 230 - 43 + 350 - 35$$

This equation can be solved on a number line, which shows us that Tania now has a credit balance on her card of $76. So, she owes nothing and is in credit by $76.

PRACTICE 4: APPLICATION

The base of Mount Kea in Hawaii is 6000 m below sea level. The peak of the mountain is 10204 m above its base. What is the elevation of Mount Kea above sea level?

EXAMPLE 5: APPLICATION

Early one morning, the temperature on Mount Hotham was 9°C below zero. By early morning the temperature had risen by 4°C. By mid-afternoon it had risen another 5°C, but then by early evening it had dropped 10°C. What was the temperature by early evening?

Solution

The temperature of 9°C below zero can be written as −9, and provides the starting temperature. A rise in temperature means that the temperature is added to the previous temperature, whereas a drop in temperature means that the temperature is subtracted, resulting in the following equation to solve this problem:

$$-9 + 4 + 5 - 10$$

This can be solved on a number line, giving a result of −10°C:

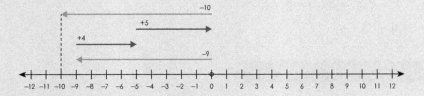

PRACTICE 5: APPLICATION

Over the last five years, an electricity company posted the following profits or losses of $13140 profit, $4098 loss, $22918 loss, $13954 profit and $11 123 profit. What was the total profit or loss for the company over the five years?

4.3 MULTIPLYING AND DIVIDING DIRECTED NUMBERS

The rules for multiplying and dividing directed numbers are reasonably easy to follow. The key rule with multiplying and dividing directed numbers is to do the calculation as if the signs were not there; that is, treat the calculation as if it were simply a calculation involving whole numbers, and then apply the appropriate sign once the calculation is complete. For example, the calculation

FIRST SIGN	SECOND SIGN	RESULTING SIGN
+	+	+
+	−	−
−	+	−
−	−	+

-7×8 is done first by multiplying 7×8 to get 56. The second step is then to note that the multiplication involves one positive and one negative number (one + sign and one − sign). Using the same key rule for combining signs as for addition and subtraction, one positive (+) and one negative (−) combine to make a negative (−), as shown in this table, so $-7 \times 8 = -56$. Similarly, if we were to calculate $400 \div -5$, we would start with $400 \div 5 = 80$ and because a positive number (400) is divided by a negative number (−5), the result is a negative number:

$$400 \div -5 = -80$$

Note that the application of these rules for combining signs is different for multiplication and division as compared to addition and subtraction. With addition and subtraction, the rules for combining signs are only applied if the two signs are next to each other *without a number in-between*. However, for multiplication and division, the calculation is done without regard for the signs, and then the two signs, which do have numbers in-between, are combined according to the given rules in the table to obtain a single sign.

For multiplication and division, as for addition and subtraction, the normal BEDMAS rules still apply, and the same rules for replacing two signs that are next to each other without a number in-between still apply.

EXAMPLE 6

Perform the following calculations.

a 4×-5 b $-24 \div 3$ c $-121 \div -11$ d -9×-7 e -10×4

Solutions

a To calculate 4×-5:

- First calculate $4 \times 5 = 20$.
- Then note that there is one positive number (4) and one negative number (−5), so the result will be negative.
- The answer is $4 \times -5 = -20$.

Tiffany Winn

b To calculate $-24 \div 3$:

- First calculate $24 \div 3 = 8$.
- Then note that there is one negative number (-24) and one positive number (3), so the result will be negative.
- The answer is $-24 \div 3 = -8$.

c To calculate $-121 \div -11$:

- First calculate $121 \div 11 = 11$.
- Then note that there are two negative numbers ($-121, -11$), so the result will be positive.
- The answer is $-121 \div -11 = 11$.

d To calculate -9×-7:

- First calculate $9 \times 7 = 63$.
- Then note that there are two negative numbers ($-9, -7$), so the result will be positive.
- The answer is $-9 \times -7 = 63$.

e To calculate -10×4:

- First calculate $10 \times 4 = 40$.
- Then note that there is one negative number (-10) and one positive number (4), so the result will be negative.
- The answer is $-10 \times 4 = -40$.

PRACTICE 6

Perform the following calculations.

a -3×-12 **b** $48 \div -6$ **c** -7×-7

d 12×-5 **e** $-18 \div -3$

EXAMPLE 7

Simplify the following calculations.

a $-5 \times 11 \times (-24 \div 6 \div -2)$ **b** $-108 \div -12 \times 4$

c $-105 \div -(-5 \times 3) \div -1$ **d** $-4 \times (8 \times (96 \div -12))$

e $(-0 \div 12) \times (-33 \div 3)$

Solutions

a $-5 \times 11 \times (-24 \div 6 \div -2)$

$= -5 \times 11 \times 2$

$= -55 \times 2$

$= -110$

Brackets first, within brackets go left to right, $-24 \div 6 = -4$ and then $-4 \div -2 = 2$

Both multiplications have equal precedence, so go left to right first $-5 \times 11 = -55$

And then $-55 \times 2 = -110$

b $-108 \div -12 \times 4$

$= 9 \times 4$

$= 36$

Multiplication and division have equal precedence, so go left to right $-108 \div -12 = 9$

And then $9 \times 4 = 36$

c $-105 \div -(-5 \times 3) \div -1$

$= -105 \div -(-15) \div -1$

$= -105 \div 15 \div -1$

$= -7 \div -1$

$= 7$

Brackets first, $-5 \times 3 = -15$

Two minus signs together without a number in-between replaced by a plus sign

Both division operations equal precedence, so go left to right, $-105 \div 15 = -7$

And then $-7 \div -1 = 7$

d $-4 \times (8 \times (96 \div -12))$

$= -4 \times (8 \times -8)$

$= -4 \times -64$

$= 256$

Innermost brackets first, $96 \div -12 = -8$

Brackets next, $8 \times -8 = -64$

And then $-4 \times -64 = 256$

e $(-0 \div 12) \times (-33 \div 3)$

$= 0 \times -11$

$= 0$

Brackets first, $-0 \div 12 = -0 = 0$ and $-33 \div 3 = -11$

And then $0 \times -11 = -0 = 0$

PRACTICE 7

Simplify the following calculations.

a $-100 \div (-24 \div 6 \times -5) \times -3$

b $-10 \times (-12 \div 4) \times 4$

c $(-12 \times 3) \div -9 \div 2 \times 5$

d $-4 \times (88 \div (-48 \div -12))$

e $-3 \times (-49 \div -7)$

4.4 MIXED OPERATIONS AND DIRECTED NUMBERS

When calculations with directed numbers involve a mix of addition, subtraction, multiplication or division, the key rule to remember is BEDMAS, which guides the order in which operations are carried out. BEDMAS stands for:

1 **B**rackets
2 **E**xponents
3 **D**ivision and **M**ultiplication (equal precedence, from left to right)
4 **A**ddition and **S**ubtraction (equal precedence, from left to right).

EXAMPLE 8

Calculate the value of the following equations.

a $6 + 3 \times 11 - (-5 - (-8))$

b $-4 - 22 \div (-9 + 6 + (-8)) \times 3$

c $-(12 + 4) \div (-13 - (-7 + (-2)))$

d $-4 - (4 - (-1) + (-9)) \times 3 \div -2 + 5$

e $(-33 - (14 - (-4))) \div -(-12 + 7)$

Solutions

a $6 + 3 \times 11 - (-5 - (-8))$ Brackets first, $-5 - (-8) = -5 + 8 = 3$

$= 6 + 3 \times 11 - 3$ Multiplication before addition or subtraction, $3 \times 11 = 33$

$= 6 + 33 - 3$ Add and subtract have equal precedence, so go left to right, $6 + 33 = 39$

$= 39 - 3$ And then $39 - 3 = 36$

$= 36$

b $-4 - 22 \div (-9 + 6 + (-8)) \times 3$ Brackets first, $-9 + 6 + (-8) = -9 + 6 - 8 = -11$

$= -4 - 22 \div -11 \times 3$ Division and multiplication before subtraction, division and multiplication take equal precedence, so go left to right, $22 \div -11 = -2$

$= -4 - (-2) \times 3$ Multiplication before subtraction, $-2 \times 3 = -6$

$= -4 - -6$ Two minus signs together without a number in-between are replaced by a plus sign

$= -4 + 6$ And then $-4 + 6 = 2$

$= 2$

c In this question, brackets are completed first but which set of brackets is evaluated first does not matter except that inner brackets must be evaluated before their enclosing brackets.

$-(12 + 4) \div (-13 - (-7 + (-2)))$ Brackets first, inner brackets before their enclosing brackets, so $-7 + (-2) = -7 - 2 = -9$

$= -(12 + 4) \div (-13 - -9)$ Two minus signs without a number in-between are replaced by a plus sign

$= -(12 + 4) \div (-13 + 9)$ Outer brackets are next, $12 + 4 = 16$ and $-13 + 9 = -4$

$= -16 \div -4$ And then $-16 \div -4 = 4$

$= 4$

d $-4 - (4 - (-1) + (-9)) \times 3 \div -2 + 5$ Brackets first, $+ 4 - (-1) + (-9) = 4 + 1 - 9 = -4$

$= -4 - -4 \times 3 \div -2 + 5$ Two minus signs without a number in-between replaced by plus sign

$= -4 + 4 \times 3 \div -2 + 5$ Multiplication and division before addition and subtraction, multiplication and division have equal precedence, so proceed left to right

$= -4 + 12 \div -2 + 5$ Division before addition, $12 \div -2 = -6$

$= -4 + -6 + 5$ Plus and minus sign together without a number in-between replaced by a single minus sign

$= -4 - 6 + 5$ Addition and subtraction have equal precedence, so proceed left to right $-4 - 6 = -10$

$= -10 + 5$ And then $-10 + 5 = -5$

$= -5$

e In this question, brackets are completed first but which set of brackets is evaluated first does not matter except that inner brackets must be evaluated before their enclosing brackets.

$(-33 + (14 - (-4))) \div -(-12 + 7)$ Brackets first, inner brackets before their enclosing brackets, $14 - (-4) = 14 + 4 = 18$

$= (-33 + 18) \div -(-12 + 7)$ Outermost brackets next, $-33 + 18 = -15$ and $-12 + 7 = -5$

$= -15 \div - -5$ Two minus signs together replaced by a single plus sign

$= -15 \div +5$ And then $-15 \div +5 = -15 \div 5 = -3$

$= -3$

PRACTICE 8

Calculate the value of the following equations.

a $-12 + 48 \div -6 - (-5 + (-8))$

b $-4 - (-4 + 28) \div -6 \times 3 \times 2$

c $-140 \div (-16 - (-2 + (-7)) \times 4)$

d $(-4 - 8) \div -2 \times (4 \times -1 + (-5)) \times + 5$

e $(-13 - (-7)) \div 2 \times -(-12 + 7)$

4.5   CHAPTER SUMMARY

SKILL OR CONCEPT	DEFINITION OR DESCRIPTION	EXAMPLES
Directed number	A directed number has direction indicated by a negative or positive sign as well as magnitude.	• −4 is four units to the left of 0 on number line • +4 is four units to the right of 0 on the number line
Adding and subtracting directed numbers	Calculations are carried out by moving either left (subtraction) or right (addition) on a number line. The usual BEDMAS rules apply. When two signs are **together without a number in-between**, they combine into a single sign according to the following rules: • + and + combines to be + • + and − combines to be − • − and + combines to be − • − and − combines to be +	$3 - 5 = -2$ $5 - (-3 + 1)$ $= 5 - (-2)$ $= 5 + 2$ $= 7$ $-4 + (-2)$ $= -4 - 2$ $= -6$ $-11 + (-2) - (-23) + 2$ $= -11 - 2 + 23 + 2$ $= -13 + 23 + 2$ $= 10 + 2$ $= 12$
Multiplying and dividing directed numbers	The usual BEDMAS rules apply. Multiplication or division of two directed numbers is carried out by performing the respective operation on the numbers while ignoring their signs. Then, the sign is incorporated into the result according to the following rules: • + and + combines to be + • + and − combines to be − • − and + combines to be − • − and − combines to be +	To calculate $-5 \times +4$ • Calculate $5 \times 4 = 20$ • One negative number (−5) and one positive number (4) means result is negative • Result is −20 To calculate $4 \times (-20 \div -2)$ • Brackets first to calculate $20 \div 2 = 10$ • Two negative numbers (−20 and −2) combine to make a positive • Result of $-20 \div -2$ is 10 • Now calculate 4×10 • $4 \times 10 = 40$ • Result is $4 \times (-20 \div -2) = 40$
Combined operations with directed number	• BEDMAS rules apply • Do multiplication and division, and addition and subtraction according to their appropriate rules	To calculate $15 - 4 \times (-3 + 8 \div -2)$ • Brackets first, so calculate $-3 + 8 \div -2$ • Division before addition, so: $-3 + 8 \div -2$ $= -3 + -4$ $= -3 - 4$ $= -7$ • Now calculate $15 - 4 \times -7$ • Now multiplication before subtraction, so: $15 - 4 \times -7$ $= 15 - -28$ $= 15 + 28$ $= 43$ • Result $15 - 4 \times (-3 + 8 \div -2) = 43$

4.6 REVIEW QUESTIONS

A SKILLS

1 Carry out the following calculations.

a $-3 + (-4)$ b $-2 + 6$ c $22 + (-11)$

d $40 - (-36)$ e $92 - (+12)$ f $-15 + (-12)$

g $-12 + 12$ h $13 - (+4)$ i $-13 + (-4)$

j $45 - (-45)$

2 Carry out the following calculations.

a $-6 + 3 - 11 - (-5 - (-9))$ b $16 + -7 - (-5) + 4 - 2$

c $-10 + 31 - (+6) - 4$ d $21 - (-7 + 2 - 1) + (-12) - 6$

e $-1 + 7 + (8 - (-2) + (-3)) - 15$ f $-4 - (2 - (3 + (-4)))$

g $-12 + 6 - (8 + 4 - (-2))$ h $1 - (-7 - (-2) + 5) - 3$

i $(2 - (-3) - (4 + (-7)))$ j $13 + (-4) - (6 - (12 - (-4)))$

3 Carry out the following calculations.

a $-5 \times +8$ b $42 \div -7$ c $-300 \div +10$

d $7 \times +8$ e -6×-7 f $-99 \div -3$

g $54 \div -9$ h 4×-4 i $-24 \div -2$

j -32×-10

4 Carry out the following calculations.

a $-8 \times 2 \div 4$ b $36 \div -3 \times 2$ c $7 \times -7(-8 \div -2)$

d $72 \div 12 \times 3$ e $8 \div 4 \times -2$ f $-13 \times -3 \times +2$

g $48 \times -3 \times (8 \times 4 \div 2)$ h $35 \div -7 \times (-2 \times 4)$ i $5 \times 6 \div 3$

j $-20 \times (-12 \div 2)$ k $6 \times (-28 \div 7)$ l $88 \div +11 \times -2$

5 Carry out the following calculations.

a $-5 + (-22 - 6) \times 2$ b $7 + 84 \div -2 \times (14 - (-8))$

c $(-42 - 3 \times -9) \div 3 \times 5 + (-60) \div 4$ d $45 - 3 \times 9 \div 3 \times 5 + 60 \div 4$

e $16 + (-12 + (-6) \div 3) \times 2$ f $-18 + 32 - (14 - 8) \div 2 \times 4$

g $(-20 + (-4) \times 4) \div 6 \times 2 - 44 \div -11$ h $73 - 8 \times 3 \div 4 \times 2 - 84 \div 7$

i $5 + (-4) \times +2 \times (18 - (+6))$ j $28 + (-16 - (-24)) \div 2 \times (6 \div 3)$

k $25 \div -(32 - 3 \times 9) \times 5 + 60 \div -4 \times -3$ l $45 - 4 \times (-9 \div 3) \times -4 - (+5)$

Tiffany Winn

B APPLICATIONS

1   Each play in American football consists of four
    'downs'. A team needs to advance 10 yards
    over the four downs. In a recent play, the
    New England Patriots gained 4 yards, lost 9
    yards, gained 3 yards and gained 11 yards. Did
    they progress the required 10 yards?

2   Before going on a cruise, Sam lost 5 kg but
    then he gained 3 kg on the cruise. When
    he returned Sam lost 1 kg but then gained
    another 6 kg. How much below or above his original weight was he then?

3   An elevator starting on the first floor of a building went up 15 floors, then down 8 floors,
    down 2 floors, up 22 floors, down 25 floors and up 12 floors. Which floor did it stop at?

4   A confectionary store lost $18 300 in one year of business but then made a profit for five
    years in succession of $3000, $4100, $2500, $3500 and $5000. Did they manage to recover
    their loss over the five years?

5   Rose had $139 in her bank account. She bought three DVDs at $13 each and six gifts at $18
    each. What was the balance in her account after this shopping?

6   After diving to 105 m below sea level, a diver rises by 8 m per minute for 12 minutes. Where is
    the diver relative to the surface of the water?

7   The population of a country town was 15 800. It went into decline and decreased by 340
    each year over four years and then by 520 in the next year. Then a mining boom occurred
    and the town population increased by 1614 in one year. What was the town population after
    the mining boom compared to the original population of 15 800?

8   The highest recorded temperature in Alice Springs is 45.2°C and the lowest recorded
    temperature −7.5°C. What is the difference between the highest and lowest recorded
    temperatures?

9   A cooking business had the following profit/loss records since it began operation: lost
    $4300 for the first year, made a profit of $3500 each year for the next three years, lost
    $2096, made a profit of $5620, lost $1225 each year for four years, made a profit of $1400
    each year for two years. Overall, has the cooking business been profitable and by how
    much?

10 The temperature of a chemical reaction commenced at −12°C and then dropped by 6°C every minute for 8 minutes, then a catalyst was added and the temperature rose by 9°C every minute for 15 minutes. What was the temperature of the chemical reaction at that time?

11 During the course of a day, a stock dropped 35 points, rose 28 points, dropped 40 points, rose 23 points, rose 18 points and dropped 8 points. By how many points did the stock rise or fall overall during the day?

12 Tim noticed his account was in debit by $228 so he deposited $300. Then he spent $14 each on three boxes of chocolates and $29 on fuel. How much money was left in his account?

5

FRACTIONS

CHAPTER CONTENT

» Introduction to fractions
» Types of fractions
» Equivalent fractions
» Comparing fractions

» Simplifying fractions
» Adding and subtracting fractions
» Multiplying and dividing fractions

CHAPTER OBJECTIVES

» Read and write common fractions and mixed numbers
» Convert between common fractions and mixed numbers
» Write fractions in their simplest form

» Identify equivalent fractions
» Compare fractions
» Add and subtract fractions
» Multiply and divide fractions
» Solve applied problems using fractions

FRACTIONS AND MUSIC

Music is written using symbols to represent information about the composition. Each bar of music represents a measure that equals one whole. All of the notes that are recorded in each bar must, therefore, add up to one. Notes are written in different forms to represent their fraction of the whole in each bar.

The notes can then be combined in various ways, for example:

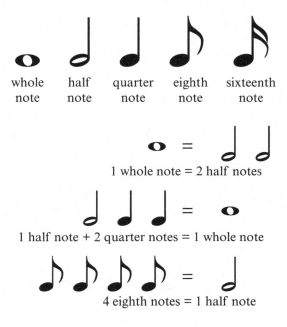

whole note   half note   quarter note   eighth note   sixteenth note

1 whole note = 2 half notes

1 half note + 2 quarter notes = 1 whole note

4 eighth notes = 1 half note

But whatever the combination, the sum of the notes in each bar must always be equal to one.

5.1   INTRODUCTION TO FRACTIONS

What is a fraction?

The term **fraction** can mean a number of things. Sometimes fractions are used to describe parts of a whole; for example, $\frac{4}{5}$ of the population means 4 out of every 5 people. A fraction can also be used to represent the division of two whole numbers; for example, $\frac{3}{2}$ means 3 divided by 2.

A fraction is any number that can be written in the form $\frac{a}{b}$, where a and b are whole numbers and b is not zero. So, according to this definition, $\frac{2}{3}$, $\frac{5}{2}$, $\frac{4}{1}$ and $\frac{0}{2}$ can all be called fractions.

The word fraction comes from the Latin word *fractus*, which means 'broken into parts' and this gives a sense of what fractions are all about.

> **Fraction**
> Any number that can be written in the form $\frac{a}{b}$, where a and b are whole numbers and b is not zero

CRITICAL THINKING

Use the numbers 0, 1, 2, 3, 5 and 6 just once to write three fractions: one that is less than 1, one that is more than 1 and one that is equal to zero.

Kathy Brady

Each part of a fraction has a special name:

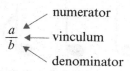

- The *denominator* represents how many equal parts are contained in the whole.
- The *numerator* is how many equal parts of the whole the fraction represents.
- This *vinculum*, or *fraction line*, stands for 'out of' or 'divided by'.

How can fractions be represented?

Fractions can be represented as parts of a whole; for example, $\frac{2}{5}$ of this circle is shaded green.

Fractions can be represented as parts of a set; for example, $\frac{2}{5}$ of the circles are green.

Fractions can be represented as the result when two whole numbers are divided; for example, when 2 chocolate bars are divided equally between 5 people. This means that each person receives $\frac{2}{5}$ of a chocolate bar.

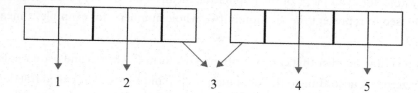

Fractions can be represented on a number line; for example, the figure below shows the fractions with a denominator of 5. Here we can see that the whole number 2 can also be written as $\frac{10}{5}$, and so 3 could be written as $\frac{15}{5}$.

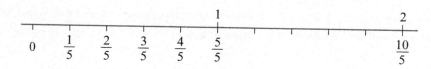

Making connections *A fraction can also be represented as a decimal number or as a percentage. These will be discussed in Chapter 6.*

EXAMPLE 1

What fraction does the shaded part of this diagram represent?

Solution

The whole in this diagram is divided into five equal parts, so the denominator will be 5. Because three of those parts are shaded, the numerator will be 3. Therefore, the fraction represented by the shaded parts is $\frac{3}{5}$.

PRACTICE 1

What fraction represents the shaded triangles in this set?

EXAMPLE 2

Use a diagram to represent $\frac{7}{9}$ of a set.

Solution

We could use any type of object to represent the set, as long as they are all the same. In this example, circles will be used. Because the denominator is 9, there are nine circles in the set of which 7 will be shaded.

PRACTICE 2

Use a diagram to illustrate $\frac{3}{7}$ of an object.

EXAMPLE 3: APPLICATION

There are 83 female and 97 male students studying Biology this semester. What fraction of the students studying Biology are female?

Solution

We need to find what fraction of the students studying Biology are female, so firstly we need to find the total number of students studying Biology, which is $83 + 97 = 180$. As there are 83 female students studying Biology, the fraction of female students studying Biology is $\frac{83}{180}$.

PRACTICE 3: APPLICATION

Professor Williams has 92 students in his class. Of these, 48 are first years and the other students are all in second year. What fraction of his class are second-year students?

Kathy Brady

5.2   TYPES OF FRACTIONS

There are two different types of fractions: common fractions and mixed numbers.

Common fractions

A common fraction is a fraction written in the form $\frac{a}{b}$. When the numerator a is less than the denominator b, the value of the fraction will be less than 1. For example, $\frac{2}{7}$, $\frac{10}{11}$ and $\frac{1}{6}$ all have a value of less than 1 because the numerators are less than the respective denominators.

On the other hand, when the numerator a is greater than the denominator b, the value of the fraction will be greater than 1, for example, $\frac{5}{4}$, $\frac{12}{7}$ and $\frac{9}{1}$. You may have heard or read of fractions in this form being referred to as 'improper' fractions. However, there is nothing improper at all about fractions in this form just because they are 'top heavy', rather they are simply fractions that have a value that is greater than 1.

Mixed numbers

Sometimes it is more appropriate to write a fraction that has a value greater than 1 as a combination of a whole number and a common fraction, for example, $3\frac{2}{5}$ or $5\frac{1}{6}$.

This means that a number that is greater than 1 can, therefore, be written either as a 'top heavy' common fraction or as a mixed number. Diagrams can be used to illustrate these two different forms of the same number as in Example 4.

EXAMPLE 4

Use diagrams to illustrate that $3\frac{2}{5} = \frac{17}{5}$.

Solution

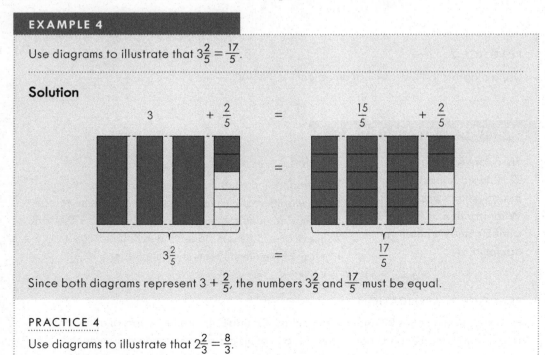

Since both diagrams represent $3 + \frac{2}{5}$, the numbers $3\frac{2}{5}$ and $\frac{17}{5}$ must be equal.

PRACTICE 4

Use diagrams to illustrate that $2\frac{2}{3} = \frac{8}{3}$.

Converting mixed numbers to common fractions

We can see from Example 4 that each rectangle represents 1 whole and also 5-fifths. From this, the total number of *fifths* in $3\frac{2}{5}$ is $(3 \times 5) + 2$ or 17. We can use this relationship to form a rule for converting mixed numbers to common fractions.

1   Multiply the denominator of the fraction part by the whole number part in the mixed number.

2   Add the numerator of the fraction part to this product.

3   Write this value over the denominator of the fraction part.

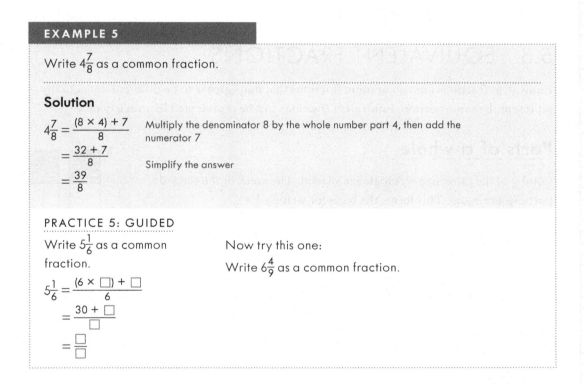

EXAMPLE 5

Write $4\frac{7}{8}$ as a common fraction.

Solution

$4\frac{7}{8} = \dfrac{(8 \times 4) + 7}{8}$ Multiply the denominator 8 by the whole number part 4, then add the numerator 7

$= \dfrac{32 + 7}{8}$ Simplify the answer

$= \dfrac{39}{8}$

PRACTICE 5: GUIDED

Write $5\frac{1}{6}$ as a common fraction.

$5\frac{1}{6} = \dfrac{(6 \times \square) + \square}{6}$

$= \dfrac{30 + \square}{\square}$

$= \dfrac{\square}{\square}$

Now try this one:

Write $6\frac{4}{9}$ as a common fraction.

Converting common fractions to mixed numbers

When a number is written as a mixed number, it is possible to understand the size of the value more easily than when it is written as a common fraction. For example, we know that $12\frac{3}{8}$ lies between 12 and 13. But it is not easy to form this conclusion from its equivalent common fraction form $\frac{99}{8}$. Therefore, it is also useful to be able to convert common fractions to mixed numbers using the following steps.

1   Divide the numerator by the denominator.

2   Write the resultant quotient.

3   Then form the fraction part by writing any remainder from the division over the denominator.

Kathy Brady

EXAMPLE 6

Write $\frac{67}{7}$ as a mixed number.

Solution

$\frac{67}{7} = 67 \div 7$ Divide the numerator by the denominator, 9 Rem 4
                                      Write the whole number quotient and then the
$\quad = 9\frac{4}{7}$ remainder over the denominator

PRACTICE 6

Write $\frac{33}{8}$ as a mixed number.

5.3   EQUIVALENT FRACTIONS

Equivalent fractions are two or more fractions that may appear to be different but actually represent the same portion. Equivalent fractions can be represented in a variety of ways.

Parts of a whole

$\frac{1}{3}$ and $\frac{2}{6}$ of the same size rectangle are shaded. The areas of the shaded portions are equal. This forms the basis for writing $\frac{1}{3} = \frac{2}{6}$.

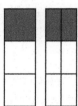

Parts of a set

On the left, $\frac{1}{3}$ of the circles are green. On the right, $\frac{2}{6}$ of the circles are green. The number of green circles in each case is 2, so $\frac{1}{3}$ and $\frac{2}{6}$ of the same size set of circles are green. Therefore, it is possible to write $\frac{1}{3} = \frac{2}{6}$.

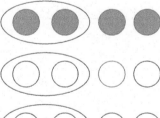

Number line

Here the number line has been divided in two ways – into thirds and sixths and we can see that $\frac{1}{3} = \frac{2}{6}$. Using the examples above, we can see that $\frac{1}{3} = \frac{2}{6}$ and notice that $\frac{1 \times 2}{3 \times 2} = \frac{2}{6}$.

In general, to form equivalent fractions we multiply both the numerator and the denominator of the given fraction by the same whole number.

EXAMPLE 7

Write two fractions that are equivalent to $\frac{2}{5}$.

Solution

To form equivalent fractions, we need to multiply the denominator and the numerator by the same number. But which number should we use? Let's try 3 and then 7.

$$\frac{2}{5} = \frac{2}{5} \times \frac{3}{3} = \frac{6}{15} \qquad \text{and} \qquad \frac{2}{5} = \frac{2}{5} \times \frac{7}{7} = \frac{14}{35}$$

So this means that $\frac{2}{5} = \frac{6}{15} = \frac{14}{35}$.

In fact, we could have selected any two numbers to multiply $\frac{2}{5}$ by. This means that for $\frac{2}{5}$, or any other fraction for that matter, there are an infinite number of other fractions that are equivalent.

PRACTICE 7

Write three fractions that are equivalent to $\frac{4}{5}$.

EXAMPLE 8

Write a fraction that is equivalent to $\frac{5}{6}$ with a denominator of 48.

Solution

We know that $\frac{5}{6} \times \frac{\square}{\square} = \frac{\square}{48}$. So looking at the denominators we know that $6 \times 8 = 48$, and because we need to multiply the numerator and the denominator by the same number we must multiply 5×8 in the numerators to get:

$$\frac{5}{6} \times \frac{8}{8} = \frac{40}{48}$$

PRACTICE 8: GUIDED

Write a fraction that is equivalent to $\frac{7}{8}$ with a denominator of 24.

$$\frac{7}{8} = \frac{\square}{24}$$

$$\frac{7}{8} = \frac{7}{8} \times \frac{\square}{\square}$$

$$= \frac{7}{8} \times \frac{3}{3}$$

$$= \frac{\square}{24}$$

Now try these:

a $\frac{4}{9} = \frac{\square}{63}$

b $\frac{3}{5} = \frac{\square}{60}$

CRITICAL THINKING

A very useful property of equivalent fractions is that their cross-products are always equal. Look at this example:

$$\frac{1}{3} \cross \frac{2}{6}$$

In this case, $1 \times 6 = 6$ and $2 \times 3 = 6$.

a   Can you think of a way to show that this property will always work?

b   Take a look at the three equivalent fractions shown here. Notice that the numerators and the denominators of these fractions are comprised of the digits 1, 2, 3, 4, 5, 6, 7, 8 and 9 — each appearing just once.

$$\frac{3}{6} = \frac{9}{18} = \frac{27}{54}$$

c   Check that these fractions are equivalent by making sure their cross-products are equal.

d   Write the following trios of equivalent fractions that use the same nine digits just once.

$$\frac{3}{21} = \frac{\square}{\square} = \frac{\square}{\square} \qquad\qquad \frac{2}{4} = \frac{\square}{\square} = \frac{\square}{\square}$$

5.4   COMPARING FRACTIONS

In some situations, it is necessary to compare the size of fractions. As an example, if $\frac{5}{9}$ of the trains in a city's public transport system run on time, and $\frac{4}{7}$ of the buses are on time, how would we decide which is the most efficient mode of transport? To do this we would need to compare the fractions.

When fractions have a common denominator this is easy to achieve, for example, see the diagram to the right:

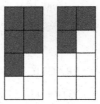

Clearly the greater shaded area is on the left where $\frac{5}{8}$ is shaded and the smaller shaded area is on the right where $\frac{3}{8}$ is shaded. Because $\frac{5}{8}$ and $\frac{3}{8}$ have the same denominator we can compare their size by looking at the numerator. Since 5 is greater than 3, we can say that $\frac{5}{8}$ is greater than $\frac{3}{8}$ and write this as $\frac{5}{8} > \frac{3}{8}$. The symbol > means 'greater than'.

To compare fractions

1   If the fractions have a common denominator, compare their numerators.

2   If the fractions have different denominators, write them as equivalent fractions with the same denominator, and then compare the numerators.

EXAMPLE 9

Compare the fractions given in the introduction to this section: $\frac{5}{9}$ and $\frac{4}{7}$.

Solution

These fractions have different denominators. Therefore, we need to first write them as equivalent fractions with a common denominator.

One way to decide on what the common denominator should be is to use the product of the denominators: $9 \times 7 = 63$.

$$\frac{5}{9} = \frac{5}{9} \times \frac{7}{7} = \frac{35}{63} \qquad \text{and} \qquad \frac{4}{7} = \frac{4}{7} \times \frac{9}{9} = \frac{36}{63}$$

Now we compare the numerators. As $36 > 35$, we know that $\frac{4}{7} > \frac{5}{9}$. We can also now answer the question posed above that buses are the most efficient means of transport.

PRACTICE 9
Compare $\frac{5}{8}$ and $\frac{7}{11}$.

EXAMPLE 10

Compare $\frac{7}{16}$ and $\frac{5}{12}$.

Solution

In this example, we could start by finding the product of the denominators, which would be $16 \times 12 = 192$. Now that does not look like a very friendly number to have as a denominator! In this example, there is a better denominator that we can use, and that is the lowest common multiple (LCM) of the two denominators 16 and 12.

The LCM of 16 and 12 is 48. We now need to write the equivalent fractions that have denominators of 48.

$$\frac{7}{16} = \frac{7}{16} \times \frac{3}{3} = \frac{21}{48} \qquad \text{and} \qquad \frac{5}{12} = \frac{5}{12} \times \frac{4}{4} = \frac{20}{48}$$

Therefore, $\frac{7}{16} > \frac{5}{12}$.

PRACTICE 10: GUIDED
Compare $\frac{5}{6}$ and $\frac{7}{8}$.

To find LCM

$$\frac{5}{6} = \frac{5}{6} \times \frac{4}{\Box} = \frac{\Box}{24} \qquad \frac{7}{8} = \frac{7}{8} \times \frac{\Box}{\Box} = \frac{21}{\Box}$$

$$\frac{\Box}{\Box} > \frac{\Box}{\Box}$$

Now try this one:

Compare $\frac{5}{16}$ and $\frac{7}{20}$.

Making connections
Check Chapter 2 review for how to find the LCM of two or more numbers.

In Example 10, we used the LCM to determine the common denominator that would be used to compare the fractions.

In general, for two or more fractions their lowest common denominator (LCD) is the LCM of their denominators.

In Example 11, notice how we use the LCD to order fractions from smallest to largest.

EXAMPLE 11

Write these fractions in order from smallest to largest: $\frac{3}{5}, \frac{7}{10}, \frac{5}{8}$.

Solution

Because these fractions have different denominators we need to find equivalent fractions with a common denominator. We will use the LCD as that denominator.

To find the LCM of 5, 8, 10 using multiples:

$\quad$ 5 = 5, 10, 15, 20, 25, 30, 35, 40

$\quad$ 8 = 8, 16, 24, 32, 40

$\quad$ 10 = 10, 20, 30, 40

So, LCM = 40 and this will be the LCD for the three fractions.

To write equivalent fractions with a denominator of 40:

$$\frac{3}{5} = \frac{3}{5} \times \frac{8}{8} = \frac{24}{40} \qquad \frac{5}{8} = \frac{5}{8} \times \frac{5}{5} = \frac{25}{40} \qquad \frac{7}{10} = \frac{7}{10} \times \frac{4}{4} = \frac{28}{40}$$

So, in order from smallest to largest the fractions are $\frac{3}{5}, \frac{5}{8}, \frac{7}{10}$.

PRACTICE 11

Arrange $\frac{2}{3}, \frac{3}{8}, \frac{5}{12}$ in order from largest to smallest.

EXAMPLE 12: APPLICATION

A serving of regular strawberry yoghurt contains approximately $\frac{4}{25}$ of the daily recommended dietary intake (RDI) of calcium and a serving of regular natural yoghurt contains $\frac{1}{5}$ of the calcium RDI. Which of the two kinds of yoghurt provides the best source of calcium?

Solution

In order to compare the fractional quantities of calcium, we need to write the two fractions as equivalent fractions with a common denominator.

Since 5 is a factor of 25 that will be the common denominator.

For natural yoghurt, $\frac{1}{5} = \frac{1}{5} \times \frac{5}{5} = \frac{5}{25}$ of the RDI of calcium is provided. Comparing this to strawberry yoghurt that provides $\frac{4}{25}$ of the RDI of calcium, natural yoghurt is the best source of calcium.

(Source: www.foodstandards.gov.au/science/monitoringnutrients/nutrientables/nuttab/pages/nuttab2010.aspx)

Making connections
In Chapter 16 there is an example of how these probabilities are calculated.

Making connections
Check Chapter 3 to review factors, multiples and finding the HCF of two or more numbers.

PRACTICE 12: APPLICATION

In a lecture on probability, a student learns that when rolling a pair of dice, the probability of getting a sum of 8 is $\frac{5}{36}$ and the probability of getting a sum of 9 is $\frac{1}{9}$. Which has the greater probability, 8 or 9?

5.5   SIMPLIFYING FRACTIONS

Fractions are in *simplest form* or *lowest terms* if the only common factor of the denominator and numerator is 1. For example, $\frac{4}{7}$ is in simplest form because 1 is the only factor of both 4 and 7, but $\frac{8}{18}$ is not because 2 is also a factor of both 8 and 18. We use the relationship between factors and multiples to reduce fractions to simplest form in a process called *cancelling*.

To simplify a fraction using cancellation you will need to find a common factor of both the numerator and the denominator. This could be the highest common factor (HCF) or just any common factor.

Cancelling using the HCF

This is the most efficient way to simplify a fraction. First you must identify the HCF of both the numerator and the denominator, and then rewrite the fraction as multiples of the HCF.

EXAMPLE 13

Simplify $\frac{36}{45}$ to the lowest terms by cancelling using the HCF.

Solution

The HCF of 36 and 45 is 9. Therefore, the fraction can be rewritten as:

$$\frac{36}{45} = \frac{4 \times 9}{5 \times 9}$$

Now the HCF of 9 can be cancelled in both the numerator and the denominator:

$$\frac{4 \times \cancel{9}}{5 \times \cancel{9}} = \frac{4 \times 1}{5 \times 1} = \frac{4}{5}$$

Notice that the only common factor of 4 and 5 is 1, so this fraction is in lowest terms.

PRACTICE 13

Simplify $\frac{30}{18}$ to the lowest terms by cancelling using the HCF.

Cancelling using any common factors

Sometimes the HCF is not obvious. In this case, you can use any number that is a common factor of the numerator and the denominator to cancel. This method does not always give simplest terms immediately but it has the advantage of reducing the size of the numerator and the denominator so that it becomes easier to identify further factors. For instance, if the numerator and denominator are both even then you could start with the common factor of 2.

EXAMPLE 14

Simplify $\frac{28}{42}$ to the lowest terms.

Solution

Notice that both 28 and 42 are even; therefore, 2 will be a common factor. So cancelling by 2 would be a useful step.

$$\frac{28}{42} = \frac{14 \times 2}{21 \times 2} = \frac{14 \times 1}{21 \times 1} = \frac{14}{21}$$

Here we can see that 7 is a common factor of 14 and 21, which means that we can cancel again by 7.

$$\frac{14}{21} = \frac{2 \times 7}{3 \times 7} = \frac{2 \times 1}{3 \times 1} = \frac{2}{3}$$

Note that in this case we could have achieved the same cancellation in one step if we had recognised that the HCF was $2 \times 7 = 14$, but this was not quite so obvious.

PRACTICE 14: GUIDED

Simplify $\frac{210}{390}$ to the lowest terms.

Cancellation by the first common factor

$$\frac{210}{390} = \frac{21 \times \square}{\square \times 10} = \frac{\square}{\square}$$

Cancellation by the second common factor

$$\frac{\square}{39} = \frac{7 \times \square}{\square \times 3} = \frac{\square}{\square}$$

Now try these:

Simplify each fraction to the lowest terms.

a $\frac{40}{96}$

b $\frac{45}{30}$

5.6   ADDING AND SUBTRACTING FRACTIONS

When adding or subtracting fractions, one of two possibilities will occur: the fractions will all have a common denominator or they will have different denominators. Each of these cases requires slightly different approaches to finding a solution.

Adding or subtracting fractions with a common denominator

It is only possible to add or subtract things that are in the same terms. Fractions can be considered to be in the same terms when they have a common denominator. In this case, it is possible to add (or subtract) fractions using the following steps:

1   Add or subtract the numerators.

2   Write as a fraction over the common denominator.

3   Simplify the answer to the lowest terms.

Beware! Do not add the denominators.

EXAMPLE 15

Calculate $\frac{7}{15} + \frac{3}{15}$.

Solution

The two fractions have a common denominator of 15, so they can be added because they are in the same terms.

$$\frac{7}{15} + \frac{3}{15} = \frac{7+3}{15} \quad \longleftarrow \text{ The numerators are added}$$
$$\quad \longleftarrow \text{ The denominator stays the same}$$

$$= \frac{10}{15}$$

We now need to write this fraction in simplest terms.

$$\frac{10}{15} = \frac{2 \times 5}{3 \times 5} = \frac{2 \times 1}{2 \times 1} = \frac{2}{3}$$

PRACTICE 15

Calculate $\frac{5}{8} - \frac{1}{8}$.

Adding or subtracting fractions with different denominators

Adding or subtracting fractions with different denominators is not as straightforward because before the fractions can be added (or subtracted), they must first be rewritten as equivalent fractions with the same denominator. The following steps are used:

1   Decide what the common denominator will be.

2   Rewrite the fractions as equivalent fractions using the common denominator.

3   Add or subtract the numerators.

4   Write as a fraction over the common denominator.

5   Simplify the answer to the lowest terms.

Beware! Do not add the denominators.

EXAMPLE 16

Calculate $\frac{7}{15} + \frac{4}{3}$.

Solution

These fractions do not have the same denominator, so they cannot be added until they are rewritten as equivalent fractions with the same denominator. What will this common denominator be? As we saw in Example 10, the best common denominator to use will be the LCM of the two denominators. In this case, the denominators are 15 and 3, so the LCM of these numbers is 15. Since $\frac{7}{15}$ already has this denominator, we will only need to rewrite $\frac{4}{3}$ as an equivalent fraction with a denominator of 15.

$$\frac{4}{3} = \frac{4}{3} \times \frac{5}{5} = \frac{20}{15}$$

So, now the fractions both have a denominator of 15 and the question has become:

$$\frac{7}{15} + \frac{20}{15} = \frac{7+20}{15}$$ 　　　The numerators are added

$$= \frac{27}{15}$$ 　　　The denominator stays the same

Now we need to write this fraction in simplest terms.

$$\frac{27}{15} = \frac{9 \times 3}{5 \times 3} = \frac{9 \times 1}{5 \times 1} = \frac{9}{5}$$

PRACTICE 16
Calculate $\frac{5}{12} - \frac{1}{3}$.

EXAMPLE 17

Calculate $\frac{5}{8} + \frac{3}{10}$.

Solution

These fractions do not have the same denominator, so they cannot be added until they are rewritten as equivalent fractions with the same denominator. The best common denominator will be the LCM of 8 and 10, which is 40. So before adding, each fraction needs to be rewritten with a denominator of 40.

$$\frac{5 \times 5}{8 \times 5} + \frac{3 \times 4}{10 \times 4} = \frac{25}{40} + \frac{12}{40}$$

$$= \frac{25 + 12}{40}$$ 　　　Add the numerators, and the denominator stays the same

$$= \frac{37}{40}$$

This fraction is in the lowest possible terms, so no further cancellation is required.

PRACTICE 17
Calculate $\frac{7}{4} + \frac{3}{10}$.

EXAMPLE 18

Calculate $\frac{11}{12} - \frac{3}{8}$.

Solution

These fractions need to be rewritten with a common denominator before the subtraction can take place. The common denominator will be the LCM of 12 and 8, which is 24.

$$\frac{11 \times 2}{12 \times 2} - \frac{3 \times 3}{8 \times 3} = \frac{22}{24} - \frac{9}{24}$$
$$= \frac{22 - 9}{24}$$
$$= \frac{13}{24}$$

This fraction is in the lowest possible terms, so no further cancellation is required.

PRACTICE 18

Calculate $\frac{7}{8} - \frac{4}{5}$.

Adding mixed numbers

When adding fractions that are in mixed number form, an extra step is required to handle the whole number parts of the mixed numbers, as follows:

1 Add the whole numbers first.
2 Write the fractional parts with a common denominator.
3 Add the fractional parts and reduce them to the lowest terms. Then rewrite as a mixed number if necessary.
4 Combine the sum of the whole numbers and the sum of the fractional part.

EXAMPLE 19

Calculate $1\frac{2}{3} + 3\frac{3}{4}$.

Solution

$1\frac{2}{3} + 3\frac{3}{4}$ can be considered as $1 + \frac{2}{3} + 3 + \frac{3}{4}$. This is why we can add the whole number parts and the fractional parts separately.

$$1\frac{2}{3} + 3\frac{3}{4} = 4 + \frac{2}{3} + \frac{3}{4} \qquad \text{Add the whole number parts}$$
$$= 4 + \frac{2 \times 4}{3 \times 4} + \frac{3 \times 3}{4 \times 3} \qquad \text{Common denominator for fractional parts is LCM of 3 and 4, which is 12}$$
$$= 4 + \frac{8}{12} + \frac{9}{12}$$
$$= 4 + \frac{17}{12} \qquad \text{Add the numerators, and the denominators stay the same}$$
$$= 4 + 1\frac{5}{7} \qquad \text{Rewrite the fractional parts as a mixed number}$$
$$= 5\frac{5}{7} \qquad \text{Combine whole number parts and fractional parts}$$

PRACTICE 19: GUIDED

Calculate $4\frac{3}{5} + 2\frac{5}{7}$.

$$4\frac{3}{5} + 2\frac{5}{7} = \Box + \Box + \frac{3}{5} + \frac{5}{7}$$

$$= \Box + \frac{3 \times \Box}{5 \times \Box} + \frac{5 \times \Box}{7 \times \Box}$$

$$= \Box + \frac{21}{\Box} + \frac{\Box}{35}$$

$$= \Box + \frac{\Box}{35}$$

$$= \Box + \Box\frac{\Box}{35}$$

$$= \Box\frac{\Box}{\Box}$$

Now try this one.

Calculate $3\frac{1}{3} + 2\frac{7}{8}$.

EXAMPLE 20: APPLICATION

A laboratory technician has been recording the growth of a plant under experimental conditions.

In week one the plant grew $1\frac{7}{8}$ cm, in week two the plant grew another $1\frac{5}{6}$, and in week three it grew a further $\frac{5}{16}$. What was the total growth recorded?

Solution

The total growth will be the sum of the weekly totals; that is, $1\frac{7}{8} + 1\frac{5}{6} + \frac{5}{16}$.

$$1\frac{7}{8} + 1\frac{5}{6} + \frac{5}{16} = 2 + \frac{7}{8} + \frac{5}{6} + \frac{5}{16} \qquad \text{Add the whole number parts}$$

$$= 2 + \frac{7 \times 6}{8 \times 6} + \frac{5 \times 8}{6 \times 8} + \frac{5 \times 3}{16 \times 3} \qquad \text{Common denominator for fractional parts is LCM of 8, 6 and 16, which is 48}$$

$$= 2 + \frac{42}{48} + \frac{40}{48} + \frac{15}{48}$$

$$= 2 + \frac{57}{48} \qquad \text{Add the numerators and the denominator stays the same}$$

$$= 2 + 1\frac{9}{48} \qquad \text{Rewrite fractional part as a mixed number}$$

$$= 2 + 1\frac{3}{16} \qquad \text{Reduce fractional part to lowest terms by cancelling by 3}$$

$$= 3\frac{3}{16} \text{ cm} \qquad \text{Combine whole number parts and fractional parts. Include the units}$$

The total growth recorded was $3\frac{3}{16}$ cm.

PRACTICE 20: APPLICATION

Henry does a lot of walking getting around campus. On Tuesday, his walk from the carpark to the science building is $\frac{7}{10}$ km, then from the science building to the refectory is $1\frac{2}{5}$ km and back to the carpark from the refectory is $1\frac{1}{6}$ km. How far does Henry walk around campus on Tuesday?

Subtracting mixed numbers—method 1

There are two techniques that can be used to subtract mixed numbers depending on the values that are to be calculated. The first technique is the same approach as that used to add mixed numbers, that is:

1   Subtract the whole numbers first.

2   Write the fractional parts with a common denominator.

3   Subtract the fractional parts and reduce to the lowest terms. Then rewrite as a mixed number, if necessary.

EXAMPLE 21

Calculate $3\frac{3}{4} - 1\frac{2}{3}$.

Solution

$3\frac{3}{4} - 1\frac{2}{3}$ can be considered as $3 + \frac{3}{4} - 1 - \frac{2}{3}$; that is, all parts of the second value are subtracted from the first value (as you would expect in any subtraction calculation).

$$3\frac{3}{4} - 1\frac{2}{3} = 2 + \frac{3}{4} - \frac{2}{3} \qquad \text{Subtract the whole number parts}$$

$$= 2 + \frac{3 \times 3}{4 \times 3} - \frac{2 \times 4}{3 \times 4} \qquad \text{Common denominator for fractional parts is LCM of 3 and 4, which is 12}$$

$$= 2 + \frac{9}{12} - \frac{8}{12}$$

$$= 2 + \frac{1}{12} \qquad \text{Subtract the numerators, and the denominator stays the same}$$

$$= 2\frac{1}{12} \qquad \text{Combine whole number parts and fractional parts}$$

PRACTICE 21

Calculate $3\frac{11}{15} - 1\frac{3}{10}$.

Subtracting mixed numbers—method 2

But ... Method 1 only works when the fractional part of the first mixed number is larger than the fractional part of the second mixed number. If this is not the case, or if you are uncertain about which fractional part is larger, then Method 2 will need to be used. This involves:

1   First converting the mixed numbers to common fractions.

2   Then writing the fractions over a common denominator.

3   Carry out the subtraction.

4   Convert back to a mixed number.

EXAMPLE 22

Calculate $3\frac{1}{3} - 1\frac{3}{4}$.

Solution

Since the fractional part of the second mixed number, $\frac{3}{4}$, is greater than $\frac{1}{3}$ which is the fractional part of the first mixed number, we need to use Method 2.

$$3\frac{1}{3} - 1\frac{3}{4} = \frac{(3 \times 3) + 1}{3} - \frac{(4 \times 1) + 3}{4}$$ Convert each mixed number to a common fraction

$$= \frac{10}{3} - \frac{7}{4}$$

$$= \frac{10 \times 4}{3 \times 4} - \frac{7 \times 3}{4 \times 3}$$ Common denominator for fractional parts is LCM of 3 and 4, which is 12

$$= \frac{40}{12} - \frac{21}{12}$$

$$= \frac{19}{12}$$ Subtract the numerators, and the denominator stays the same

Convert back to a mixed number

$$= 1\frac{7}{12}$$

PRACTICE 22: GUIDED

Calculate $5\frac{1}{4} - 4\frac{7}{12}$.

$$5\frac{1}{4} - 4\frac{7}{12} = \frac{\square}{4} - \frac{\square}{12}$$

$$= \frac{21 \times \square}{\square \times \square} - \frac{\square}{12}$$

$$= \frac{\square}{12} - \frac{\square}{12}$$

$$= \frac{\square}{12}$$

$$= \frac{\square}{\square} \text{ (in lowest terms)}$$

Now try this one.

Calculate $4\frac{3}{5} - 1\frac{2}{3}$.

5.7   MULTIPLYING AND DIVIDING FRACTIONS

When multiplying or dividing fractions, unlike adding or subtracting, we do not need to worry about having a common denominator. This means that in most cases multiplication or division of fractions is reasonably straightforward.

Making connections In Chapter 12 there are further explanations of calculating area.

Multiplying common fractions

How might we carry out the multiplication $\frac{2}{3} \times \frac{4}{5}$?

We could use a diagram to represent the multiplication $\frac{2}{3} \times \frac{4}{5}$ and our understanding of how to calculate the area of a rectangle using $L \times W$ to generate a method.

In this diagram, each small rectangle is $\frac{1}{15}$ of the large rectangle. The area of the shaded region is therefore 8 lots of $\frac{1}{15}$ (the total number of shaded small rectangles), which is $\frac{8}{15}$.

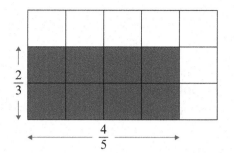

Since area is calculated using $L \times W$, which in this case is $\frac{2}{3} \times \frac{4}{5}$, we can see that this is equal to $\frac{8}{15}$.

So, if $\frac{2}{3} \times \frac{4}{5} = \frac{8}{15}$, we can see that the product of the two numerators on the LHS is equal to the numerator on the RHS, and the product of the two denominators on the LHS is equal to the denominator on the RHS.

In general, to multiply two (or more) fractions we multiply the numerators to form the resultant numerator and multiply the denominators to form the resultant denominator, and then reduce to the lowest terms.

EXAMPLE 23

Calculate $\frac{3}{4} \times \frac{10}{7}$.

Solution

$$\frac{3}{4} \times \frac{10}{7} = \frac{3 \times 10}{4 \times 7}$$ Multiply the numerators and multiply the denominators

$$= \frac{30}{28}$$ Reduce to lowest terms by cancelling by 2

$$= \frac{15}{14}$$

In this example, the answer has been left as a common fraction rather than being converted to a mixed number. This is perfectly acceptable.

PRACTICE 23

Calculate $\frac{8}{7} \times \frac{5}{12}$.

In Example 23, we multiplied the numerators and denominators first and then cancelled the resultant fraction to the lowest terms. Sometimes, however, it is more efficient to carry out the cancelling first. Doing this means that we can work with smaller numbers but still get the same answer.

When cancelling before multiplying, we can divide *any* numerator and *any* denominator by a common factor.

Beware! Always make sure that when cancelling you are using one numerator and one denominator.

EXAMPLE 24

Calculate $\frac{8}{21} \times \frac{3}{40}$.

Solution

In this example, we can see that if the denominators were to be multiplied first, then the resultant denominator would be quite a large number. By cancelling first, and then multiplying, we would be able to work with smaller numbers in the following way.

In $\frac{8}{21} \times \frac{3}{40}$, because one of the numerators is 3 and one of the denominators is 21, each of these can be cancelled down by dividing each by 3. This would give:

$$\frac{8}{21} \times \frac{\cancel{3}}{40} = \frac{8}{7} \times \frac{1}{40} \qquad \text{Divide the numerator 3 by 3 and the denominator 21 by 3}$$

Some further cancellation can occur because the numerator 8 and the denominator 40 also have a common factor of 8. This would give:

$$\frac{\cancel{8}}{7} \times \frac{1}{\cancel{40}} = \frac{1}{7} \times \frac{1}{5} \qquad \text{Divide the numerator 8 and the denominator 40 by 8}$$

$$= \frac{1}{35} \qquad \text{Multiply the fractions, with result in lowest terms}$$

PRACTICE 24

Calculate $\frac{9}{10} \times \frac{35}{12}$ by cancelling first and then multiplying.

CRITICAL THINKING

An unknown fraction that has *not* been simplified (reduced to the lowest terms) is multiplied by $\frac{3}{5}$. The resultant answer in the lowest terms is $\frac{6}{35}$.

What are three possible alternatives that the unknown fraction could be?

Dividing common fractions

When it comes to dividing fractions, the rules 'invert and multiply' or 'multiply by the reciprocal' are possibly among the best remembered from school days. But what do these rules mean and how do they work?

We can start to understand how the division of fractions works by looking at the relationship between multiplication and division. Let's consider firstly what happens when $12 \div 4$ and the relationship that occurs between the division and multiplication.

Making connections
In Chapter 2 review the meanings of the terms dividend, divisor and quotient.

12	÷	4	=	3	4	×	3	=	12
Dividend		Divisor		Quotient	Divisor		Quotient		Dividend

Now in the same way, let's consider $\frac{1}{2} \div \frac{1}{3}$.

$\frac{1}{2}$	$\div$	$\frac{1}{3}$	$=$	?	$\frac{1}{3}$	$\times$	?	$=$	$\frac{1}{2}$
Dividend		Divisor		Quotient	Divisor		Quotient		Dividend

This relationship implies that there is a number such that $\frac{1}{3} \times \square = \frac{1}{2}$. So how do we find '$\square$'. Let's call the unknown, m. So:

$\frac{1}{3} \times m = \frac{1}{2}$

If the LHS is multiplied by $\frac{3}{1}$, then this will cancel with the $\frac{1}{3}$ to leave just the m. Of course, in any equation, what we do to the LHS we must also do to the RHS.

$\frac{\cancel{3}}{\cancel{1}} \times \frac{\cancel{1}}{\cancel{3}} \times m = \frac{3}{1} \times \frac{1}{2}$

Therefore, by cancelling and multiplying the fractions we can get $m = \frac{3}{2}$.

So what? Notice that when we eliminated the $\frac{1}{3}$ on the LHS we multiplied by its inverse, or by what is known as its *reciprocal*, which in this case is $\frac{3}{1}$. Every number that can be written in the form $\frac{a}{b}$ has a reciprocal that is $\frac{b}{a}$. So this means that the quotient for the division $\frac{1}{2} \div \frac{1}{3}$ was found when the dividend, $\frac{1}{2}$, was multiplied by the reciprocal of the divisor, $\frac{3}{1}$.

In general, when dividing fractions, multiply the first named fraction (the dividend) by the reciprocal of the second named fraction (the divisor) and reduce to the lowest terms, if needed.

EXAMPLE 25

Calculate $\frac{7}{16} \div \frac{3}{4}$.

Solution

$\frac{7}{16} \div \frac{3}{4} = \frac{7}{16} \times \frac{4}{3}$ Write first fraction, replace division with a multiplication sign and write reciprocal of second fraction

$= \frac{7}{16} \times \frac{\cancel{4}}{3}$ Cancel by dividing the numerator of 4 and the denominator of 16 by 4

$= \frac{7}{4} \times \frac{1}{3}$ Multiply the fractions

$= \frac{7}{12}$

PRACTICE 25

Calculate $\frac{3}{5} \div \frac{6}{15}$.

Multiplying and dividing mixed numbers

When the question involves mixed numbers, there is one extra step before multiplying or dividing. That is, to firstly convert the mixed numbers to common fractions, and then proceed as in Examples 24 and 25.

Kathy Brady

EXAMPLE 26

Calculate $4\frac{4}{5} \times 3\frac{3}{4}$.

Solution

Before carrying out the multiplication, the mixed numbers must be converted to common fractions.

$$4\frac{4}{5} \times 3\frac{3}{4} = \frac{(5 \times 4) + 4}{5} \times \frac{(4 \times 3) + 3}{4}$$
$$= \frac{24}{5} \times \frac{15}{4}$$
$$= \frac{6}{5} \times \frac{15}{1} \qquad \text{Divide the numerator 24 and the denominator 4 by 4}$$
$$= \frac{6}{1} \times \frac{3}{1} \qquad \text{Divide the numerator 15 and the denominator 5 by 5}$$
$$= \frac{18}{1} \qquad \text{Any fraction with a denominator of 1 is a whole number}$$
$$= 18$$

PRACTICE 26

Calculate $5\frac{1}{4} \times 1\frac{1}{9}$.

EXAMPLE 27

Calculate $2\frac{1}{4} \div 1\frac{7}{8}$.

Solution

Before carrying out the division the mixed numbers must be converted to common fractions.

$$2\frac{1}{4} \div 1\frac{7}{8} = \frac{(4 \times 2) + 1}{4} \div \frac{(8 \times 1) + 7}{8} \qquad \text{Convert each mixed number to a common fraction}$$
$$= \frac{9}{4} \div \frac{15}{8} \qquad \text{Write the first fraction, and then multiply by the reciprocal of the second fraction}$$
$$= \frac{9}{4} \times \frac{8}{15}$$
$$= \frac{3}{4} \times \frac{8}{5} \qquad \text{Divide the numerator 9 and the denominator 15 by 3}$$
$$= \frac{3}{1} \times \frac{2}{5} \qquad \text{Divide the numerator 8 and the denominator 4 by 4}$$
$$= \frac{6}{5} \qquad \text{Multiply the fractions}$$
$$= 1\frac{1}{5} \qquad \text{Convert back to a mixed number}$$

PRACTICE 27: GUIDED

Calculate $1\frac{1}{8} \div 7\frac{1}{2}$.

$$1\frac{1}{8} \div 7\frac{1}{2} = \frac{(8 \times \square) + \square}{8} \div \frac{(\square \times 7) + 1}{\square}$$
$$= \frac{9}{\square} \div \frac{15}{2}$$
$$= \frac{9}{\square} \times \frac{\square}{15}$$

$$= \frac{3}{8} \times \frac{2}{\square}$$
$$= \frac{\square}{\square} \times \frac{1}{5}$$
$$= \frac{\square}{\square}$$

Now try this one.

Calculate $5\frac{1}{4} \div 1\frac{1}{8}$.

5.8 CHAPTER SUMMARY

SKILL OR CONCEPT	DEFINITION OR DESCRIPTION	EXAMPLES
Fraction	Any number written in the form $\frac{a}{b}$, where a and b are whole numbers and b is not zero.	$\frac{5}{7}, \frac{11}{3}$
Common fraction	A fraction written in the form $\frac{a}{b}$. When the numerator a is less than the denominator b, the value of the fraction will be less than 1. When the numerator a is greater than the denominator b, the value of the fraction will be greater than 1.	$\frac{2}{7}, \frac{10}{11}$ $\frac{5}{4}, \frac{12}{7}$
Mixed number	A fraction that has a value greater than 1 that is written as a combination of a whole number and a common fraction.	$3\frac{2}{5}, 5\frac{1}{6}$
Converting a mixed number to a common fraction	• Multiply the denominator of the fraction part by the whole number part in the mixed number. • Add the numerator of the fraction part to this product. • Write this value over the denominator of the fraction part.	$3\frac{5}{8} = \frac{(8 \times 3) + 5}{8}$ $= \frac{24 + 5}{8}$ $= \frac{29}{8}$
Converting a common fraction to a mixed number	• Divide the numerator by the denominator. • Write the resultant quotient. • Then form the fraction part by writing any remainder from the division over the denominator.	$\frac{17}{4} = 4\frac{1}{4}$
Equivalent fractions	Two or more fractions that represent the same value.	$\frac{2}{3}, \frac{6}{9}$
To form equivalent fractions	Multiply both the numerator and the denominator of the given fraction by the same whole number.	$\frac{3}{5} = \frac{3 \times 4}{5 \times 4} = \frac{12}{20}$
To compare fractions	If the fractions have a common denominator, compare their numerators. If the fractions have different denominators, write them as equivalent fractions with the same denominator, and then compare the numerators.	$\frac{3}{7}, \frac{5}{7}$ $3 < 5$, therefore $\frac{3}{7} < \frac{5}{7}$ $\frac{1}{5}, \frac{2}{15} \rightarrow \frac{3}{15}, \frac{2}{15}$ $3 > 2$, therefore $\frac{1}{5} > \frac{2}{15}$
Lowest common denominator (LCD)	The LCD of two or more fractions is the lowest common multiple of their denominators.	The LCD of $\frac{9}{10}$ and $\frac{3}{25}$ is 50.
Lowest terms	A fraction is in the lowest terms if the only common factor of the denominator and numerator is 1.	$\frac{4}{11}$
Simplifying or cancelling fractions	• Identify the highest common factor (HCF) of the numerator and the denominator. • Divide each by the HCF.	$\frac{30}{18}$ HCF is 6, so $\frac{30}{18} = \frac{5}{3}$

SKILL OR CONCEPT	DEFINITION OR DESCRIPTION	EXAMPLES
Adding or subtracting fractions with a common denominator	• Add or subtract the numerators. • Write as a fraction over the common denominator. • Simplify the answer to the lowest terms.	$\frac{3}{8} + \frac{1}{8} = \frac{4}{8} = \frac{1}{2}$ $\frac{5}{8} - \frac{3}{8} = \frac{2}{8} = \frac{1}{4}$
Adding or subtracting fractions with different denominators	• Decide on the common denominator. • Rewrite the fractions as equivalent fractions using the common denominator. • Add or subtract the numerators and write as a fraction over the common denominator. • Give the answer in the lowest terms.	$\frac{3}{8} + \frac{5}{6} = \frac{9}{24} + \frac{20}{24} = \frac{29}{24}$ $\frac{9}{10} - \frac{1}{15} = \frac{27}{30} - \frac{2}{30} = \frac{25}{30} = \frac{5}{6}$
Adding mixed numbers	• Add the whole numbers first. • Write the fractional parts with a common denominator. • Add the fractional parts, reduce to lowest terms and then rewrite as a mixed number, if necessary.	$1\frac{2}{7} + 4\frac{1}{3} = 5 + \frac{2}{7} + \frac{1}{3}$ $= 5 + \frac{6}{21} + \frac{7}{21}$ $= 5\frac{13}{21}$
Subtracting mixed numbers	• Convert the mixed numbers to common fractions. • Write the fractions over a common denominator. • Carry out the subtraction. • Convert back to a mixed number.	$4\frac{1}{8} - 2\frac{1}{2} = \frac{33}{8} - \frac{5}{2}$ $= \frac{33}{8} - \frac{20}{8}$ $= \frac{13}{8}$ $= 1\frac{5}{8}$
Multiplying common fractions	Multiply the numerators and multiply the denominators, and then reduce to lowest terms.	$\frac{5}{12} \times \frac{3}{5} = \frac{15}{60} = \frac{1}{4}$
Reciprocal	Every number written in the form $\frac{a}{b}$ has a reciprocal; that is, $\frac{b}{a}$.	The reciprocal of $\frac{3}{5}$ is $\frac{5}{3}$.
Dividing common fractions	Multiply the dividend by the reciprocal of the divisor and reduce to the lowest terms.	$\frac{15}{16} \div \frac{5}{6} = \frac{15}{16} \times \frac{6}{5} = \frac{3}{8} \times \frac{3}{1} = \frac{9}{8}$
Multiplying or dividing mixed numbers	Before carrying out the multiplication or division, convert mixed numbers to common fractions.	$2\frac{1}{3} \times 3\frac{3}{5} = \frac{7}{3} \times \frac{18}{5}$ $= \frac{7}{1} \times \frac{6}{5}$ $= \frac{42}{5}$ $= 8\frac{2}{5}$ $1\frac{1}{15} \div 3\frac{1}{3} = \frac{16}{15} \div \frac{10}{3}$ $= \frac{16}{15} \times \frac{3}{10}$ $= \frac{8}{5} \times \frac{1}{5}$ $= \frac{8}{25}$

5.9   REVIEW QUESTIONS

A SKILLS

1   Convert the following mixed numbers to common fractions.

   a $2\frac{3}{4}$ b $10\frac{2}{3}$

2   Convert the following common fractions to mixed numbers.

   a $\frac{27}{8}$ b $\frac{74}{9}$

3   For each fraction, write an equivalent fraction using the given numerator or denominator.

   a $\frac{5}{8} = \frac{\square}{32}$ b $\frac{3}{12} = \frac{12}{\square}$

4   Insert > or < between the following pairs of fractions.

   a $\frac{2}{5} \square \frac{3}{7}$ b $\frac{7}{10} \square \frac{5}{9}$

5   Arrange the following sets of fractions in order from smallest to largest.

   a $\frac{17}{20}, \frac{7}{10}, \frac{4}{5}$ b $\frac{11}{6}, \frac{5}{6}, \frac{7}{8}, \frac{7}{12}$

6   Reduce the following fractions to the lowest terms.

   a $\frac{220}{60}$ b $\frac{60}{84}$

7   Add the following fractions.

   a $\frac{8}{15} + \frac{3}{10}$ b $\frac{1}{2} + \frac{3}{7} + \frac{5}{8}$ c $2\frac{3}{4} + 1\frac{5}{8}$ d $3\frac{2}{5} + 2\frac{3}{7}$

8   Subtract the following fractions.

   a $\frac{7}{8} - \frac{3}{5}$ b $\frac{15}{12} - \frac{7}{30}$ c $4\frac{1}{3} - 2\frac{2}{5}$ d $3\frac{3}{10} - 1\frac{11}{15}$

9   Multiply the following fractions.

   a $\frac{3}{7} \times \frac{14}{9}$ b $\frac{5}{36} \times \frac{21}{40}$ c $3\frac{1}{11} \times 2\frac{1}{2}$ d $1\frac{3}{5} \times 1\frac{1}{3} \times 1\frac{7}{8}$

10 Divide the following fractions.

   a $\frac{2}{3} \div \frac{4}{27}$ b $\frac{15}{44} \div \frac{5}{11}$ c $\frac{1}{3} \div 5\frac{5}{9}$ d $6\frac{3}{7} \div 1\frac{4}{21}$

B   APPLICATIONS

1   Between 1901 and 2014, a Nobel Prize has been awarded to 860 individuals and 22 organisations. What fraction of the Nobel Prize winners were organisations? (Source: nobelprize.org)

2   In a piece of text written in English the letter G will occur $\frac{1}{50}$ of the time and the letter B will occur $\frac{3}{20}$ of the time. Which of these letters occurs most frequently?

Kathy Brady

3   In 2014, the amount of electricity generated
    from renewable sources in Australia amounted
    to approximately 32 000 gigawatt-hours. Of this
    amount, $\frac{3}{10}$ was generated by wind, $\frac{3}{20}$ by solar, and $\frac{2}{25}$
    by bioenergy. What fraction of the total electricity
    generated from renewable sources comes from these
    three sources combined? (Source: cleanenergycouncil.
    org.au)

4   The perimeter (the total distance around) of the triangle is $6\frac{5}{6}$ cm. If two sides of the triangle
    measure $3\frac{1}{2}$ cm in total, what is the length of the third side?

5   In one Australian electorate $\frac{5}{7}$ of the population is of voting age and on the electoral roll.
    If in an election, $\frac{9}{10}$ of those on the electoral roll cast a vote, what fraction of the whole
    population of the electorate voted?

6   The length of a pool table approved for tournament play is $2\frac{2}{5}$ metres. The width of the table
    is required to be half of the length. (Source: wpa-pool.com)

    a   What is the width of the approved pool table?

    b   What is the area of the surface of this pool table?

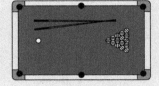

    (Hint: *area = length × width*)

7   Degrees in Celsius (C) can be obtained from degrees in Fahrenheit (F) by subtracting
    32° from the Fahrenheit measurement and then dividing by $\frac{9}{5}$. What Celsius temperature
    corresponds to a Fahrenheit temperature of 68°?

8   The nutrition information on a box of ready-to-eat breakfast cereal indicates that 100
    grams of the cereal provides 10 grams of dietary fibre, which is $\frac{2}{5}$ of the recommended daily
    allowance (RDA) for dietary fibre. What is the RDA for dietary fibre? (Source: nutritiondata.
    self.com)

6

DECIMALS AND PERCENTAGES

CHAPTER CONTENT

- » Introduction to decimals
- » Decimal place value
- » Adding and subtracting decimals
- » Multiplying decimals
- » Dividing decimals
- » Mixed decimal operations

- » Rounding decimals
- » Significant figures
- » Introduction to percentages
- » Applications of percentages
- » Fractions, decimals and percentages

CHAPTER OBJECTIVES

- » Determine the place value of digits in decimal numbers
- » Add and subtract decimals
- » Multiply and divide decimals
- » Apply order of operations to decimal calculations
- » Record values to a specified number of decimal places or signficant figures

- » Perform calculations using the percentage relationship
- » Solve applied problems using decimals and percentages
- » Convert between fractions, decimals and percentages

BLOOD ALCOHOL CONCENTRATION

Anyone that is a driver, who sometimes likes to have a drink or two at a social function, is well aware of remaining under the legal limit for Blood Alcohol Concentration (BAC), which is 0.05. But what does this well used decimal number, 0.05, mean?

BAC is a measure of the alcohol in your body, expresed as grams of alcohol per 100 millilitres of blood. So a BAC of 0.05 indicates that every 100 millilitres of your blood contains 0.05 grams of alcohol. But what does this mean with regard to how many alcoholic drinks can be consumed to drive legally? The rule of thumb is that one standard drink will raise your BAC by 0.02, and an average healthy person will take about one hour to process one standard drink. This means, for example, if you were to consume three standard drinks within one hour your BAC would be raised by 0.6 and you would need to wait at least one hour to be able to legally and safely drive. (Source: betterhealth.vic.gov.au)

6.1   INTRODUCTION TO DECIMALS

Decimals are commonly used as an alternative way of writing a fraction. We encounter decimals frequently in everyday life, for example, $28.35 or 2.4 metres. A decimal comprises a whole number part on the left-hand side of a decimal point and a fractional part on the right-hand side of the decimal point.

Whole number part $\longrightarrow$ 6.42 $\longleftarrow$ Fractional part

$\uparrow$

Decimal point

Making connections
Powers, or index notation, are discussed further in Chapter 8.

The word decimal emphasises the importance of the number 10 in this notation. The fractional part of a decimal number will always have a power of 10 as its denominator, for example, 10, 100 or 1000.

6.2   DECIMAL PLACE VALUE

Just as with whole numbers, each digit to the right of the decimal point in a decimal number has a specific place value. Let's look at this using monetary amounts, for example, $43.25. If we break this amount into whole dollars and cents we get:

$$\$43.25 = \$43 + 25 \text{ cents}$$

The 25 cents in this amount can also be considered as 2×10 cents $+ 5 \times 1$ cents. Since 10 cents is $\frac{1}{10}$ of a dollar, and 1 cent is $\frac{1}{100}$ of a dollar, the amount $43.25 can be written using place value notation as:

$$43.25 = 4 \times 10 + 3 \times 1 + 2 \times \frac{1}{10} + 5 \times \frac{1}{100}$$

Decimal place values can be extended even further with each digit in a decimal number having a specific place value, as per the examples in this place value chart.

PLACE VALUE AND DECIMALS									
ten thousands	thousands	hunderds	tens	ones	and	tenths	hunderdths	thousandths	ten thousandths
				3	.	2	5		
		1	4	5	.	1	0	6	8
			2	4	.	0	7	9	

EXAMPLE 1

Write 24.097 in place value form.

Solution

The decimal fraction component of this number extends to one-thousands as per the place value chart above.

$$24.097 = 2 \times 10 + 4 \times 1 + 0 \times \frac{1}{10} + 9 \times \frac{1}{100} + 7 \times \frac{1}{1000}$$

PRACTICE 1

Write 145.1068 in place value form.

CRITICAL THINKING

For each question provide your answer with a reason to support your response.

Is 3.091 larger or smaller than 3.109?

Is 0.12 twelve tenths or twelve hundredths?

Is 6.3 the same as 6.03?

Looking at decimal place value also provides a way to compare and order two or more decimal numbers.

EXAMPLE 2

Compare 8.5029 and 8.51.

Solution

To compare decimals we need to write numbers vertically lining up the decimal points and then comparing each digit at each place value working from left to right. On the first occasion that the digits with the same place value differ, the decimal number with the largest digit at that place value is the largest number.

The 8's are equal ⟶ 8.5029

8.51 ⟵ But 1 > 0

↑
The 5's are equal

Therefore, 8.51 > 8.5029. Notice that the decimal with more digits is not necessarily larger.

PRACTICE 2

Write 6.1, 6.16 and 6.02 in order from smallest to largest.

Making connections
Check Chapter 2 to review borrowing and carrying when adding or subtracting numbers.

6.3   ADDING AND SUBTRACTING DECIMALS

Decimals are added and subtracted in much the same way as whole numbers by adding or subtracting the digits in each place value working from right to left, borrowing and carrying as required.

To add or subtract decimal numbers:

1   Write the numbers vertically lining up the decimal points.

2   Fill extra zeros in the place values as required.

3   Add or subtract the digits working from right to left.

4   Insert the decimal point in the answer in line with the other decimal points.

EXAMPLE 3

Calculate 21.312 + 7.25.

Solution

Decimal points are
lined up vertically

$$
\begin{array}{r}
21.312 \\
+\ \ 7.250 \\
\hline
28.562
\end{array}
$$

An extra zero is added to fill
the thousandths place value in 7.25

Note that the decimal point in the answer lines up with
the decimal points in the numbers being added.

PRACTICE 3

Calculate 475.226 + 25.1.

EXAMPLE 4

Calculate 4.626 + 3 + 2.254.

Solution

In this example, 3, which is a whole number, is one of the
numbers being added. Whole numbers have no fractional
part but can still be written in decimal number form with
zeros in the place values of the fractional part.

$$
\begin{array}{r}
4.626 \\
3.000 \\
+\ 2.254 \\
\hline
9.880 \\
=\ 9.88
\end{array}
$$

Extra 0 at the end of
the answer can be dropped

PRACTICE 4

Calulate 5.428 + 7 + 11.372.

EXAMPLE 5

Calculate 349.01 − 26.8.

Solution

Decimal points are
lined up vertically

↓

$$
\begin{array}{r}
349.01 \\
-\ \ 26.80 \\
\hline
322.21
\end{array}
$$

An extra zero is added to fill
the hundredths place value in 26.8

PRACTICE 5

Calculate 394.02 − 46.191.

EXAMPLE 6: APPLICATION

A hamburger, with beef pattie, cheese, lettuce and sauce, purchased from a traditional takeaway shop contains 1.64 milligrams of niacin (Vitamin B3) for every 100 grams. The same hamburger purchased from a fast food chain contains 1.8 milligrams of niacin for every 100 grams. How much more niacin is contained in the fast food hamburger? (Source: foodstandards.gov.au)

Solution

We need the difference between the niacin content of each hamburger, which means a subtraction is required.

$$
\begin{array}{r}
1.80 \\
-\ 1.64 \\
\hline
0.16
\end{array}
$$

An extra zero is added to fill
the hundredths place value in 1.8

When the answer is less than one
a zero is usually added to the ones
place to the left of the decimal point

So, the fast food hamburger has 0.16 milligrams more niacin than the traditional takeaway hamburger.

PRACTICE 6: APPLICATION

In 1912, the first official world record holder of the men's 100 metre sprint was Donald Lippincott with a time of 10.6 seconds. The current world record holder is Usain Bolt with a time of 9.58 seconds. How much faster is Usain Bolt? (Source: iaaf.org).

6.4 MULTIPLYING DECIMALS

We will commence thinking about multiplying decimals by recalling the language of multiplication.

$$
\begin{array}{r}
43 \quad \longleftarrow \text{Factor} \\
\times\ 6 \quad \longleftarrow \text{Factor} \\
\hline
258 \quad \longleftarrow \text{Product}
\end{array}
$$

When multiplying decimals, we proceed just as we would for whole numbers using the following steps.

1 Carry out the multiplication ignoring the decimal points.

2 Count the total number of decimal places in the factors.

3 Count in from the right end of the product the number of decimal places in the factors and insert the decimal point.

EXAMPLE 7

Calculate 2.03 × 0.18

Solution

The problem needs to be written out so that the digits holding the same place value are lined up vertically. This will mean that the decimal points will also be lined up vertically.

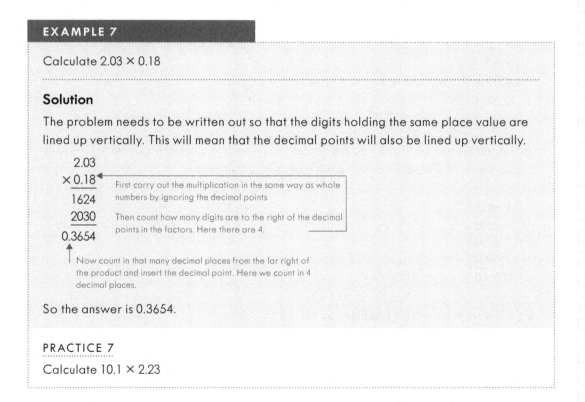

```
   2.03
 × 0.18
 ──────
   1624
   2030
 ──────
 0.3654
```

First carry out the multiplication in the same way as whole numbers by ignoring the decimal points

Then count how many digits are to the right of the decimal points in the factors. Here there are 4.

Now count in that many decimal places from the far right of the product and insert the decimal point. Here we count in 4 decimal places.

So the answer is 0.3654.

PRACTICE 7

Calculate 10.1 × 2.23

CRITICAL THINKING

Find two decimals with a product that is greater than their sum.

Now, find two decimals with a product that is less than their sum.

In general, how can we describe the relationship between two decimals when comparing their sums and products?

EXAMPLE 8: APPLICATION

A woman weighing 70 kilograms will burn 45.5 kilojoules per minute when jogging. How many kilojoules will she burn in 1.2 hours of jogging?

Solution

If the woman burns 45.5 kilojoules per minute, first we need to calculate how many minutes there are in the total time of jogging.

$$1.2 \text{ hours} = 1.2 \times 60 \text{ minutes}$$

$$
\begin{array}{r}
60.0 \\
\times\, 1.2 \\
\hline
1200 \\
6000 \\
\hline
72.00
\end{array}
$$

We need to add a 0 following the decimal point in 60, so that place values and decimal points line up vertically

There are 2 digits to the right of the decimal points in the factors, so the decimal point is inserted 2 decimal places from the right in the product

Therefore, the woman jogs for 72 minutes. Now we need to calculate how many kilojoules are burned in 72 minutes by carrying out a second multiplication.

$$
\begin{array}{r}
72.0 \\
\times\, 45.5 \\
\hline
3600 \\
36000 \\
288000 \\
\hline
3276.00
\end{array}
$$

There are 2 digits to the right of the decimal points in the factors, so the decimal point is inserted 2 decimal places from the right in the product

So, the woman burns 3276 kilojoules on her jog.

PRACTICE 8: APPLICATION

A nurse is required to administer the drug Solumedrol to a child weighing 34.1 kilograms. The dose required is 1.5 milligrams per kilogram of the child's weight. How many milligrams of Solumedrol will be administered in each dose?

6.5   DIVIDING DECIMALS

Before thinking about dividing decimals we need to recall the language of division.

$$\begin{array}{r} 43 \leftarrow \text{Quotient} \\ \text{Divisor} \longrightarrow 6\overline{)258} \leftarrow \text{Dividend} \end{array}$$

Whole number divisor

If the divisor is a whole number, then we will carry out the division just as we would for non-decimal numbers. The only thing that needs to be remembered is to insert the decimal point in the quotient directly above the decimal point in the dividend.

EXAMPLE 9

Calculate 405.27 ÷ 3

Solution

$$\begin{array}{r} 135.09 \\ 3\overline{)405.27} \end{array}$$

Once the decimal point is reached in the divisor, the decimal point is inserted in the quotient directly above and then the division continues

PRACTICE 9

Calculate 869.61 ÷ 7

Decimal divisor

If the divisor is a decimal number, then the division is carried out using the following steps.

1   Create a divisor that is a whole number by moving the decimal point to the right end.

2   In the dividend, move the decimal point an equal number of places to the right.

3   Divide the new dividend by the new quotient in the same way as the previous example.

EXAMPLE 10

Calculate 22.75 ÷ 0.7

Solution

We could begin by writing this division in fractional form $\frac{22.75}{0.7}$.

We now need to create a denominator that is a whole number. This will occur if we multiply the number by 10. Using the principle of equivalent fractions, the numerator must also be multiplied by 10, as follows:

$$\frac{22.75}{0.7} \times \frac{10}{10} = \frac{227.5}{7}$$

Kathy Brady

The decimal point in the divisor has moved one decimal place to the right to change the number to a whole number, and the decimal point in the dividend has also moved one decimal place to the right.

So, in standard division form, the question has then become $227.5 \div 7$, which can be solved in the same way as Example 9.

$$
\begin{array}{r}
32.5 \\
7 \overline{)\ 227.5}
\end{array}
$$

PRACTICE 10: GUIDED

Calculate $8.634 \div 0.03$

In fraction form, this division can be written as:

$$\frac{8.634}{0.03} \times \frac{\square}{100} = \frac{863.4}{\square}$$

So, in standard division form, the question becomes $\square \div 3$.

Once the division is carried out, the quotient is $\square 87.\square$

Now try this one.

Calculate $0.00225 \div 0.0009$

6.6   MIXED DECIMAL OPERATIONS

The order of operations that is required for whole number calculations are also applied to mixed decimal operations.

The rules for mixed decimal operations are as follows.

1   Carry out all calculations in brackets.

2   Evaluate any exponents.

3   Carry out all divisions and multiplications in order from left to right.

4   Carry out all additions and subtractions in order from left to right.

Making Connections Check Chapter 2 to the order of operations: BEDMAS.

EXAMPLE 11

Evaluate $38.27 - \dfrac{8.49}{0.3}$.

Solution

In this example, the division $\dfrac{8.49}{0.3}$ must be carried out first. In this division, a whole number denominator needs to be created by multiplying both the denominator and the numerator by 10:

$$\frac{8.49}{0.3} \times \frac{10}{10} = \frac{84.9}{3}$$

We can now carry out the division:

$$\begin{array}{r} 28.3 \\ 3\overline{)84.9} \end{array}$$

The final step in the order of operations is to carry out the subtraction:

$$\begin{array}{r} 38.27 \\ - \;\; 28.30 \\ \hline 9.97 \end{array}$$

PRACTICE 11

Evaluate $5.84 \div 0.4 + 3.1$.

EXAMPLE 12

Evaluate $0.8(1.3 + 2.9) - 0.5^2$.

Solution

In this example, according to the order of operations, the brackets must be calculated first, so $1.3 + 2.9 = 4.2$. Next the exponent 0.5^2 needs to be evaluated. Remember that $0.5^2 = 0.5 \times 0.5$.

$$\begin{array}{r} 0.5 \\ \times \;\; 0.5 \\ \hline 0.25 \end{array}$$

There are 2 digits to the right of the decimal points in the factors, so the decimal point is inserted 2 decimal places from the right in the product

Now the question has become $0.8(4.2) - 0.25$.

Here $0.8(4.2)$ means that multiplication is required:

$$\begin{array}{r} 0.8 \\ \times \;\; 4.2 \\ \hline 16 \\ 320 \\ \hline 3.36 \end{array}$$

There are 2 digits to the right of the decimal points in the factors, so the decimal point is inserted 2 decimal places from the right in the product

And finally, the subtraction is carried out: 3.36 − 0.25

$$\begin{array}{r} 3.36 \\ -\ 0.25 \\ \hline 3.11 \end{array}$$

PRACTICE 12

Evaluate $1.2^2 + (1 + 0.2)(1 − 0.2)$.

6.7   ROUNDING DECIMALS

Sometimes when carrying out a calculation, only an estimate or an approximation is required. In these cases, rounding a decimal is a technique that is used to limit the number of decimal places. Rounding can occur to any given number of decimal places.

To round a decimal, follow these steps.

1   Underline the digit in the place value to which the number is being rounded.

2   Look at the digit to the right of the underlined digit, which is called the *critical digit*.

3   If the critical digit is 5 or more, add one to the underlined digit and then drop all other digits to the right of the underlined digit.

4   If the critical digit is less than 5, the underlined digit remains unchanged, and then drop all other digits to the right of the underlined digit.

EXAMPLE 13

Round 549.1683 to:

a   one decimal place

b   two decimal places

c   three decimal places

Solutions

a   In the number, we underline the digit in the tenths decimal place 549.1683. The critical digit is the digit to the right: 6. Since 6 is more than 5, we need to add one to the underlined digit and then drop all other digits to the right. So we get 549.2.

b   Underline the digit in the hundredths decimal place 549.1683. The critical digit is 8, which is more than 5, so we need to add one to the underlined digit to get 549.17.

c   549.1683 becomes 549.168 to three decimal places because 3, the critical digit, is less than 5 and therefore the underlined 8 remains unchanged.

Round 1.2356 to:

a one decimal place   **b** two decimal places   **c** three decimal places

6.8 SIGNIFICANT FIGURES

The accuracy required when using a number depends upon the context in which the number is being used. For example, if we were describing someone's wealth in terms of millions of dollars, a couple of hundred dollars more or less will not make any substantial difference. So in some circumstances only the figures (or digits) that provide the important information about the size of a number might be used, and these digits are known as significant figures.

Suppose we want to write the number that is closest to 468 using, at most, two non-zero digits. This number would be 470. Therefore, we can say that 468 to two significant figures is 470. The term significant figure is often abbreviated to s. f. So to write a number to 2 s. f. we need to look at the first three digits, and to write a number to 3 s. f. we need to look at the first four digits and so on. In general, we always look at one more digit than the number of significant figures required.

EXAMPLE 14

Write 6543.19 to 2 s. f.

Solution

To write the number to 2 s. f. we need to look at the first three digits. So we are considering the number to be 6540.00. The 3, 1 and 9 are effectively ignored. The next step is to round the number up or down by looking at the final non-zero digit. In this case this digit is 4, which is less than 5, so the number is rounded down. So to 2 s. f. 6543.19 becomes 6500.

PRACTICE 14

Write 36.482 to 3 s. f.

When there are zeros at the beginning of a decimal number, they are ignored.

EXAMPLE 15

Write 0.004693 to 2 s. f.

Solution

Ignore the zeros at the beginning of the number, then look at the first three non-zero digits; that is, 0.00469. Now round up or down by looking at the final digit. Because this digit is 9, which is greater than 5, the number is rounded up. So to 2 s. f. 0.004693 becomes 0.0047.

PRACTICE 15

Write 0.047316 to 1 s. f.

6.9   INTRODUCTION TO PERCENTAGES

We are all familiar with the use of percentage notation to present or summarise information in the media or online such as:

* 37.3% of Australians aged 14 years and over consume alcohol on a weekly basis. (Source: druginfo.adf.org.au)
* Of technicians and trade workers in the Australian workforce, 14.3% are female and 85.7% are male. (Source: wgea.gov.au)
* 1.5% of all births in Australia are twins. (Source: twins.org.au)
* Queensland comprises 22.5% of the area of Australia. (Source: ga.gov.au)
* The Australian $1 coin comprises 92% copper, 6% aluminium and 2% nickel. (Source: ramint.gov.au)

What are percentages?

The term 'per cent' means per 100 and the symbol % is used when we are describing a value as a part of 100. Therefore, 50% means 50 per 100 or 50 parts of 100. For example, 50% of this box is green.

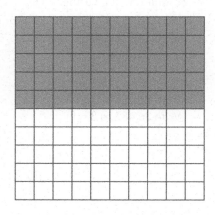

The percentage relationship

The percentage relationship involves three numbers: the per cent, the base and the part. For example:

$$50\% \text{ of } 12 \text{ is } 6$$

Per cent   Base   Part

This relationship shows that the per cent of a base is the part. In this example, 50% is the per cent. The base is 12, which is the number we are taking the percentage of. The part of the base being found is 6.

Percentage questions or problems involve being given two of three numbers in the percentage relationship and then needing to find the other one: the part, the base or the per cent.

Finding the part

When finding the part, the per cent and the base are provided and the relationship 'the per cent of the base is the part' is used. This relationship can be described using a formula in the following way:

$$\frac{\%}{100} \times \frac{base}{1} = part$$

EXAMPLE 16

What is 75% of 16?

Solution

The given per cent is 75 and the base is 16. So the question can be set up using the percentage relationship formula as:

$$\frac{75}{100} \times \frac{16}{1} = part$$

We can now use the cancellation and multiplication of fractions to find the part.

$$\frac{75}{100} \times \frac{16}{1} = \frac{3}{4} \times \frac{16}{1} \qquad \text{Divide the numerator 75 and the denominator 100 by 25}$$
$$= \frac{3}{1} \times \frac{4}{1} \qquad \text{Divide the numerator 16 and the denominator 4 by 4}$$
$$= 12 \qquad \text{And stating this result as the percentage relationship, we get that 75% of 16 is 12.}$$

PRACTICE 16

What is 24% of 300?

Finding the per cent

When finding the per cent, the part and the base are given. We start by writing what fraction the part is of the base. Then to write this as a percentage we need to multiply this fraction by 100, by turning it into a part of 100. This can be written as a formula in the following way:

$$\frac{part}{base} \times \frac{100}{1} = per\ cent$$

EXAMPLE 17

What per cent of 50 is 30?

Solution

The base is 50 and the part is 30, so using the formula above:

$$\frac{30}{50} \times \frac{100}{1} = per\ cent$$

Again we use cancellation and multiplication of fractions to find the answer.

$$\frac{30}{50} \times \frac{100}{1} = \frac{3}{5} \times \frac{100}{1} \qquad \text{Divide the numerator 30 and the denominator 30 by 10}$$

$$= \frac{3}{1} \times \frac{20}{1} \qquad \text{Divide the numerator 100 and denominator 5 by 5}$$

$$= 60\%$$

Because this answer is a per cent we need to add the percentage sign. And stating this result using the percentage relationship we can say that 30 is 60% of 50.

PRACTICE 17

What per cent of 36 is 24?

Finding the base

When finding the base, the per cent and the part are provided, and we need to use a rearrangement of the percentage relationship as a starting point. In this rearrangement, the part of the base is written as a fraction and then it is turned into an equivalent fraction using 100 as the denominator.

$$\frac{part}{base} = \frac{given\%}{100}$$

EXAMPLE 18

15% of what number is 24?

Solution

The part is 24 and the given per cent is 15%. We commence by placing these values in the formula above:

$$\frac{24}{base} = \frac{15}{100}$$

We now use the property that the cross-products of equivalent fractions are always equal, in order to manipulate this expression.

$$24 \times 100 = 15 \times base$$

Now we can see that if each side is divided by 15, we are able to the isolate the value that is the base.

$$\frac{24 \times 100}{15} = base$$

In this situation, we use cancellation to simplify this expression before calculating the base.

$$\frac{24 \times 100}{15} = \frac{8 \times 100}{5} \quad \text{Divide the numerator 24 and the denominator 15 by 3}$$

$$= \frac{8 \times 20}{1} \quad \text{Divide the numerator 100 and the denominator 5 by 5}$$

Using the percentage relationship, we can say that 24 is 15% of 160.

PRACTICE 18

20% of what distance is 45 kilometres?

Making connections
Check the Critical Thinking activity in Chapter 5 to review this property of equivalent fractions.

EXAMPLE 19: APPLICATION

In a laboratory class of 40 students, 16 of the students are female. What percentage is male?

Solution

In this question we are finding a per cent. The base is the number of students in the class, 40, and the part looks like it is 16 female students. But wait a minute! We are not being asked to find the percentage of female students, rather the percentage of male students. So the first thing to do is to calculate the number of male students. This is $40 - 16 = 24$. This means our part is actually 24. We need to use the relationship:

$$\frac{part}{base} \times \frac{100}{1} = per\ cent$$

Using this formula we have:

$$\frac{24}{40} \times \frac{100}{1} = per\ cent$$

Kathy Brady

By cancelling and multiplying we get:

$$\frac{24}{40} \times \frac{100}{1} = \frac{3}{5} \times \frac{100}{1}$$ Divide the numerator 24 and the denominator 40 by 8

$$= \frac{3}{1} \times \frac{20}{1}$$ Divide the numerator 100 and the denominator 5 by 5

$$= 60\%$$

Therefore, 60% of the class is male.

PRACTICE 19: APPLICATION

If 20% of a student's wage is paid on rent, and the rent is $90 per week. What is the student's weekly wage?

(Hint: You will need to start by asking yourself what am I trying to find: base, part or per cent?)

6.10   APPLICATIONS OF PERCENTAGES

In everyday life, percentage is one of the most commonly used numerical tools. Just a few examples include mark-up or discounts, commissions, simple and compound interest, tax rates, percentage increase or decrease, depreciation and appreciation, and interpreting pie charts. In this section, we will look at how some of the more common applications of percentages work.

Mark-ups and discounts

Mark-ups and discounts are used in retail to determine the selling price of an item. The price of an item increases when a mark-up is applied and the price decreases when the item is discounted. Mark-ups and discounts are expressed as a percentage of the original price.

EXAMPLE 20: APPLICATION

What is the selling price if 15% discount is taken off the marked price of $25?

Solution

The selling price is calculated by firstly determining what is 15% of $25, as follows:

$$\frac{15}{100} \times \frac{25}{1} = \frac{3}{20} \times \frac{25}{1}$$ Divide the numerator 15 and the denominator 100 by 5

$$= \frac{3}{4} \times \frac{5}{1}$$ Divide the numerator 25 and the denominator 20 by 5

$$= \frac{15}{4}$$

This fraction answer represents a dollar value so it must now be rewritten using a division:

$$4 \overline{)15.00} \quad 3.75$$

The amount discounted is \$3.75. Therefore, the selling price is \$25.00 − \$3.75 = \$21.25.

PRACTICE 20: APPLICATION

A television normally selling for \$600 is advertised with a 35% discount. What is the sale price?

CRITICAL THINKING

Which would be the best deal?

* An item with a 25% discount
* The same item with a 20% discount and then a further 5% off the already marked down price.

Can you explain the difference between the two selling prices when it appears that each have a 25% discount?

EXAMPLE 21: APPLICATION

What is the selling price if a \$75 item is marked up by 40%?

Solution

As in previous example, the selling price is calculated by firstly determining 40% of \$75.

$$\frac{40}{100} \times \frac{75}{1} = \frac{40}{4} \times \frac{3}{1} \qquad \text{Divide the numerator 75 and the denominator 100 by 25}$$
$$= \frac{10}{1} \times \frac{3}{1} \qquad \text{Divide the numerator 40 and the denominator 4 by 4}$$
$$= 30$$

Therefore, the item is marked up by \$30 and the selling price will be \$75 + \$30 = \$105.

PRACTICE 21: APPLICATION

A coffee shop owner buys coffee at \$16 a packet and adds on 30% profit. What is the marked-up sale price of the coffee?

Percentage increase or decrease

This application of percentages involves using a per cent to describe an increase or a decrease in a value. To calculate percentage increase or decrease follow these steps.

1 Calculate the difference between the new and the original values.

2 Calculate what per cent this difference is of the original value.

3 Express this per cent in terms of whether it is an increase or decrease.

EXAMPLE 22: APPLICATION

A microwave oven originally priced at $270 is sold at a discount for $180. What is the percentage decrease from the original price?

Solution

We first need to calculate the difference between the original price and the discounted price.

$270 − $180 = $90

Now we need to determine what per cent this difference is of the original price.

$$\frac{90}{270} \times \frac{100}{1} = \frac{\cancel{9}}{27} \times \frac{100}{1} \qquad \text{Divide the numerator 90 and the denominator 270 by 10}$$
$$= \frac{1}{3} \times \frac{100}{1} \qquad \text{Divide the numerator 9 and the denominator 27 by 9}$$
$$= \frac{100}{3}$$

Finally, we need to convert this fraction to a mixed number in order to be able to write it as a per cent.

$$\frac{100}{3} = 33\frac{1}{3}$$

Therefore, the microwave oven had a $33\frac{1}{3}$% decrease from the original price.

PRACTICE 22: APPLICATION

A tennis racquet that normally sells for $180 is on sale for $153. What percentage of the original price is the discount?

EXAMPLE 23: APPLICATION

Last year a family paid $2000 for health insurance. This year their health insurance is $2400. By what percentage has it increased?

Solution

As in the last example, the first step is to calculate the difference between the original price and the new price.

$2400 − $2000 = $400

Now calculate what per cent the difference is of the original price.

$$\frac{400}{2000} \times \frac{100}{1} = \frac{4}{20} \times \frac{100}{1} \quad \text{Divide the numerator 400 and the denominator 2000 by 100}$$
$$= \frac{4}{1} \times \frac{5}{1} \quad \text{Divide the numerator 100 and the denominator 20 by 20}$$
$$= 20$$

This means that there has been a 20% increase in the family's health insurance.

PRACTICE 23: APPLICATION

A weekly magazine's circulation increased from 40 000 to 55 000 in one year. By what percentage did the circulation increase?

Simple interest

When money is invested, interest is earned on the deposit; on the other hand, when money is borrowed, interest must be paid on the loan. The amount of interest earned or charged depends on three things: the amount of money invested or borrowed, which is known as the principal; the annual rate of interest; and the length of time applied to the loan or the investment. The simple interest is calculated in the same way regardless of whether interest is charged or interest is earned using the following formula:

simple interest = principal × rate of interest (as a %) × number of years

EXAMPLE 24: APPLICATION

A customer deposited $1600 into an account paying 5% simple interest each year. What was the account balance at the end of two years?

Solution

The three values we need for the formula are the principal (amount deposited) of $1600, the interest rate of 5% and the number of years, which is 2. Inserting these values into the formula we get:

$$\text{Simple interest} = \frac{1600}{1} \times \frac{5}{100} \times \frac{2}{1} \quad \text{Divide the numerator 1600 and the denominator 100 by 100}$$
$$= \frac{16}{1} \times \frac{5}{1} \times \frac{2}{1}$$
$$= 160$$

This means that the deposit has earned $160 interest at the end of two years and the account balance will be the principal plus the interest earned $1600 + $160 = $1760.

PRACTICE 24: APPLICATION

A student borrowed $2000 from a friend, agreeing to pay 4% simple interest each year. He promised to repay the entire amount at the end of 3 years. How much money must be repaid to the friend?

Compound interest

Most financial institutions pay their customers compound interest on their balances. When compound interest is calculated, it includes both the principal and the previous interest that has been earned.

EXAMPLE 25: APPLICATION

A bank customer has $10000 deposited in a savings account that pays 10% interest, compounded annually. There are no withdrawals and no deposits. What is the balance at the end of 3 years?

Solution

When determining compound interest, the calculations are divided into year-by-year applications of the simple interest formula.

Interest earned at the end of Year 1

The principal is $10000 and the interest rate is 10%.

$$\text{Interest} = \frac{10\,000}{1} \times \frac{10}{100} \times \frac{1}{1} \quad \text{Divide the numerator 10\,000 and the denominator 100 by 100}$$
$$= \frac{100}{1} \times \frac{10}{1} \times \frac{1}{1}$$
$$= 1000$$

This means that $1000 interest was earned in the first year and the new balance at the end of the first year will be the principal plus interest earned $10000 + $1000 = $11000.

Interest earned at the end of Year 2

The balance at the end of Year 1, $11000, becomes the new principal for Year 2.

$$\text{Interest} = \frac{11\,000}{1} \times \frac{10}{100} \times \frac{1}{1} \quad \text{Divide the numerator 11\,000 and the denominator 100 by 100}$$
$$= \frac{110}{1} \times \frac{10}{1} \times \frac{1}{1}$$
$$= 1100$$

We can see that more interest has been earned in the second year. This is because the interest is being calculated both on the original principal and the interest earned in the first year. The balance at the end of the second year will be $11 000 + $1100 = $12 100.

Interest earned at the end of Year 3

The balance at the end of Year 2, $12 100, is the new principal for Year 3.

$$\text{Interest} = \frac{12\,100}{1} \times \frac{10}{100} \times \frac{1}{1} \quad \text{Divide the numerator 12\,100 and the denominator 100 by 100}$$

$$= \frac{121}{1} \times \frac{10}{1} \times \frac{1}{1}$$

$$= 1210$$

The power of compound interest can be seen here, in that more interest is earned in the third year. Therefore, the balance at the end of the three-year investment will be:
$12 100 + $1210 = $13 310

PRACTICE 25: APPLICATION

An investor deposits $20 000 in an account that pays 5% interest, compounded annually. What is the amount in the account after 3 years?

6.11 FRACTIONS, DECIMALS AND PERCENTAGES

An important numeric skill is being able to convert fractions, decimals or percentages into one of the other forms that is an equivalent value. There are six different conversions that we will consider.

Converting a percentage to a decimal

1 Drop the % sign from a given per cent and then divide that number by 100.
2 Remember to divide by 100 we move the decimal point two places to the left.

EXAMPLE 26

Convert 38.6% to a decimal.

Solution

$38.6\% = \frac{38.6}{100} = 0.386$ To divide by 100 the decimal point is moved 2 places to the left

PRACTICE 26

Convert 2.2% to a decimal.

Converting a decimal to a percentage

1   Multiply the decimal by 100 and insert % sign.

2   Remember to multiply by 100 we move the decimal point two places to the right.

EXAMPLE 27

Convert 0.765 to a percentage.

Solution

We multiply by 100 and insert % sign.

$0.765 = 0.765 \times 100\% = 76.5\%$ To multiply by 100 the decimal point is moved 2 places to the right

PRACTICE 27

Convert 1.083 to a percentage.

Converting a fraction to a percentage

1   Multiply the fraction by 100 and insert a % sign.

2   Cancel the fraction multiplication to simplest terms if possible.

EXAMPLE 28

Convert $\frac{7}{50}$ to a percentage.

Solution

$\frac{7}{50} = \frac{7}{50} \times \frac{100}{1}\%$ Divide the numerator 100 and the denominator 50 by 50

$= \frac{7}{1} \times \frac{2}{1}\%$

$= 14\%$

PRACTICE 28

Convert $\frac{11}{20}$ to a percentage.

Converting a percentage to a fraction

1   Drop the percentage sign and write the given number as a fraction over 100.

2   Simplify the fraction to lowest terms if possible.

EXAMPLE 29

Convert 16% to a fraction.

....

Solution

$16\% = \frac{16}{100} = \frac{4}{25}$ Divide the numerator 16 and the denominator 100 by 4

PRACTICE 29

Convert 44% to a fraction.

Converting a decimal to a fraction

1 Write the fractional part of the decimal as the numerator. Any leading or trailing zeros can be ignored.
2 Write the denominator as a 1 followed by as many zeros as there are decimal places in the decimal number.
3 Simplify the fraction, if possible.

EXAMPLE 30

Convert 0.44 to a fraction.

....

Solution

$0.44 = \frac{44}{100}$ There are 2 decimal places so the denominator will be a 1 followed by 2 zeros

$\quad\;\; = \frac{22}{25}$ Divide the numerator 44 and the denominator 100 by 4

PRACTICE 30

Convert 2.85 to a fraction.

EXAMPLE 31

Convert 0.0120 to a fraction.

Solution

$0.0120 = \frac{12}{1000}$ There are 3 decimal places so the denominator will be a 1 followed by 3 zeros.
Leading and trailing zeros in the decimal are ignored when writing the numerator

$\quad\quad\;\; = \frac{3}{250}$ Divide the numerator 12 and the denominator 1000 by 4

PRACTICE 31

Convert 6.060 to a fraction.

Converting a fraction to a decimal

1   If the denominator is a power of 10, the decimal can be written directly.

2   If not, divide the numerator of the fraction by the denominator.

3   In the division, insert a decimal point to the right of the dividend (which is the numerator).

4   Carry out the division, placing the decimal point in the quotient directly above the decimal point in the dividend.

EXAMPLE 32

Convert $\frac{52}{1000}$ to a decimal.

Solution

This fraction has a denominator that is a power of 10 so the decimal can be written directly. There are three zeros in the denominator, which means that the decimal number will have three decimal places.

$$\frac{52}{1000} = 0.052$$

PRACTICE 32

Convert $\frac{37}{10}$ to a decimal.

EXAMPLE 33

Convert $1\frac{15}{8}$ to a decimal.

Solution

Because $1\frac{5}{8} = 1 + \frac{5}{8}$ we only need to covert $\frac{5}{8}$ to a decimal and then add 1 to that value. To convert $\frac{5}{8}$ to a decimal we need to divide the denominator into the numerator making sure to include a decimal point to the far right of the dividend (which is the numerator).

$$8\overline{)5.}$$

Now we carry out the division being careful to write the decimal point in the quotient directly above the decimal point in the dividend.

$$\begin{array}{r} 0.625 \\ 8\overline{)5.000} \end{array}$$

Therefore, $1\frac{5}{8} = 1.625$

PRACTICE 33

Convert $2\frac{7}{25}$ to a decimal.

Terminating and recurring decimals

So far all of the examples of converting fractions into decimals we have seen have resulted in a *terminating decimal*; that is, eventually the division will conclude with no remainder. That is, it has a finite number of digits. However, not all fractions convert to terminating decimals. In fact, most fractions do not. If a fraction cannot convert to a terminating decimal, it will result in an infinite cycle of repeated digits in the decimal part that is called a *recurring decimal*.

EXAMPLE 34

Write $\frac{4}{7}$ as a decimal.

Solution

As in Example 33, we need to divide the numerator by the denominator to convert this fraction to a decimal. Remember to insert the decimal point to the far right of the dividend.

$$7\overline{)4.}$$

Now we carry out the division:

$$\begin{array}{r} 0.57142857... \\ 7\overline{)4.00000000...} \end{array}$$

This division does not conclude, rather it keeps going infinitely. But there is a pattern in the decimal part of the quotient with repeating cycles of 571428. This is a recurring decimal and we use the notation $0.\overline{571428}$ to indicate that it is recurring.

PRACTICE 34
Write $\frac{3}{11}$ as a decimal.

6.12 CHAPTER SUMMARY

SKILL OR CONCEPT	DEFINITION OR DESCRIPTION	EXAMPLES
Decimal number	Comprises a whole number part on the LHS of a decimal point and a fractional part on the RHS of the decimal point.	4.56, 0.137
Decimal place value	The first digit to the right of the decimal point is tenths, the second is hundredths, the third is thousandths and so on.	$1.435 = 1 + 4 \times \frac{1}{10} + 3 \times \frac{1}{100} + 5 \times \frac{1}{1000}$
Comparing decimals	Write numbers vertically lining up decimal points, then compare the digits at each place value working from left to right. On the first occasion that the digits in the same place value differ, the decimal with the largest digit at that place value is the largest number.	6.573 6.57219 Digits in ones, tenths and hundredths place value are the same. Digits in thousandths place value differ: $3 > 2$, therefore $6.573 > 6.57219$
Adding or subtracting decimals	• Write numbers vertically lining up decimal points. • Fill in extra zeros in place values as required. • Add or subtract the digits working from right to left. • Insert a decimal point in the answer in line with the other decimal points.	$8.641 + 2.37 =$ $\quad\ \ 8.641$ $+\ \ 2.370$ $\overline{\quad 11.011}$ $8.54 - 2.307 =$ $\quad\ \ 8.540$ $-\ \ 2.307$ $\overline{\quad\ \ 6.233}$
Multiplying decimals	• Carry out the multiplication ignoring the decimal points. • Count the total number of decimal places in the factors. • Count in from the far right end of the product the number of decimal places in the factors and insert the decimal point.	$16.81 \times 3.56 =$ $\quad\ \ 16.81$ $\times\ \ \ 3.56$ $\overline{\quad 10086}$ $\quad 84050$ $\ 504300$ $\overline{\ 59.8436}$
Dividing decimals with whole number divisor	Carry out division as for non-decimal numbers and insert a decimal point in the quotient directly above the decimal point in the dividend.	$533.61 \div 7 =$ $\qquad 76.23$ $7\overline{)\ 533.61}$
Dividing decimals with decimal number divisor	Create a whole number divisor by moving the decimal point to the right end. In the dividend, move the decimal point an equal number of places to the right. Divide the new dividend by the new quotient as above.	$37.47 \div 0.3 =$ $\qquad 12.49$ $3\overline{)\ 37.47}$

SKILL OR CONCEPT	DEFINITION OR DESCRIPTION	EXAMPLES
Mixed decimal operations	• Carry out all calculations in brackets. • Evaluate any exponents. • Carry out all divisions and multiplications in order from left to right. • Carry out all additions and subtractions in order from left to right.	$2.5^2 - \dfrac{0.42}{0.6} + (0.74 + 1.081)$ $= 2.5^2 - \dfrac{0.42}{0.6} + 1.821$ $= 6.25 - \dfrac{0.42}{0.6} + 1.821$ $= 6.25 - 0.7 + 1.821$ $= 7.377$
Rounding decimals	• Underline the digit in the place value to which the number is being rounded. • The digit to the right of the underlined digit is the **critical digit.** • If the critical digit is 5 or more, add one to the underlined digit . • If the critical digit is less than 5, the underlined digit remains unchanged. • Drop all other digits to the right of the underlined digit.	Round 85.2567 to two decimal places $85.2\underline{5}67 = 85.26$ (2 d.p.) Round 0.34149 to three decimal places $0.34\underline{1}49 = 0.341$ (3 d.p.)
Significant figures (s. f.)	• Write a number with one more digit than the number of s. f. required, ignoring all other trailing digits. • Round number up or down to the required number of s. f. by looking at the final non-zero digit. • Leading zeros at the beginning of a decimal number are ignored.	Write 85.2567 to 3 s. f. $85.2567 \approx 85.25$ $= 85.3$ (3 s. f.) Write 0.00046219 to 3 s. f. $0.00046219 \approx 0.0004621$ $= 0.000462$
Percentage	A fraction with a denominator of 100. The symbol % is used when we are describing a value as a part of 100.	$\dfrac{35}{100} = 35\%$
Percentage relationship	The base is the value that we are calculating the percentage of. The percentage relationship is that the per cent of a base is a part.	25% of 40 (base) is 10 (part)
Finding the part	$\dfrac{\%}{100} \times \dfrac{base}{1} = part$	What is 20% of 80? $\dfrac{20}{100} \times \dfrac{80}{1} = \dfrac{1}{5} \times \dfrac{80}{1} = 16$
Finding the per cent	$\dfrac{part}{base} \times \dfrac{100}{1} = per\ cent$	What per cent of 65 is 13? $\dfrac{13}{65} \times \dfrac{100}{1} = \dfrac{1}{5} \times \dfrac{100}{1} = 20\%$
Finding the base	$\dfrac{part}{base} = \dfrac{given\ \%}{100}$	15% of what value is 12? $\dfrac{12}{base} = \dfrac{15}{100}$ $12 \times 100 = 15 \times base$ $\dfrac{12 \times 100}{15} = \dfrac{4 \times 100}{5}$ $\dfrac{4 \times 20}{1} = 80$
Mark-ups and discounts	Mark-ups and discounts are expressed as a percentage of the original price.	What is the selling price if a $75 item has a 12% discount? $\dfrac{12}{100} \times \dfrac{75}{1} = \dfrac{3}{25} \times \dfrac{75}{1} = \dfrac{3}{1} \times \dfrac{3}{1} = 9$ Discount = $9, Selling price = $66

Kathy Brady

SKILL OR CONCEPT	DEFINITION OR DESCRIPTION	EXAMPLES
Percentage increase or decrease	• Calculate the difference between the new and the original values. • Calculate what percentage this difference is of the original value. • Express this percentage in terms of whether it is an increase or a decrease.	By what percentage does $25 increase to $45? Difference = $45 − $25 = $20 $\frac{20}{25} \times \frac{100}{1} = \frac{4}{5} \times \frac{100}{1} = \frac{4}{1} \times \frac{20}{1} = 80$ 80% increase
Simple interest	simple interest = principal $\times$ interest (%) $\times$ years	What is the simple interest on $4000 at 3% for 5 years? simple interest $= \frac{4000}{1} \times \frac{3}{100} \times \frac{5}{1}$ $= \frac{40}{1} \times \frac{3}{1} \times \frac{5}{1}$ $= \$600$
Converting a percentage to a decimal	• Drop the % sign from given per cent and then divide that number by 100. • Remember to divide by 100 we move the decimal point two places to the left.	$18.5\% = \frac{18.5}{100} = 0.185$
Converting a decimal to a percentage	• Multiply the decimal by 100 and insert % sign. • To multiply by 100 move decimal point two places to the right.	$0.372 = 0.372 \times 100\% = 37.2\%$
Converting a fraction to a percentage	• Multiply fraction by 100 and insert a % sign. • Cancel the fraction multiplication to simplest terms if possible.	$\frac{7}{25} = \frac{7}{25} \times \frac{100}{1} = \frac{7}{1} \times \frac{4}{1} = 28\%$
Converting a percentage to a fraction	• Drop the % sign and write the given number as a fraction over 100. • Simplify to lowest terms if required.	$18\% = \frac{18}{100} = \frac{9}{50}$
Converting a decimal to a fraction	• Write fractional part of the decimal as the numerator, ignoring leading or trailing zeros. • Write the denominator as a 1 followed by as many zeros as there are decimal places in the number. • Simplify the fraction, if possible.	$0.036 = \frac{36}{1000} = \frac{9}{250}$
Converting a fraction to a decimal	• If the denominator is a power of 10, write the decimal directly. • If not, divide the numerator by the denominator. • In the division, insert a decimal point to the right of the dividend (which is the numerator). • Carry out the division, placing the decimal point in the quotient directly above the decimal point in the dividend.	$\frac{17}{1000} = 0.017$ $\frac{9}{8} = 8\overline{)9.} = 8\overline{)9.000} = 1.125$ (quotient 1.125)
Terminating decimal	A decimal with a finite number of digits.	0.345, 1.782
Recurring decimal	A decimal with digits that repeat infinitely.	$0.33333... = 0.\overline{3}$ $0.411111... = 0.4\overline{1}$ $1.2478324783... = 1.\overline{24783}$

6.13 REVIEW QUESTIONS

A SKILLS

1 Arrange the following numbers in order from highest to lowest.

36.83, 38.36, 33.86, 38.63, 36.38, 33.68

2 Carry out the following additions or subtractions.

a 13.65 + 4.19

b 2.576 + 34.1 + 17.28

c 23.94 − 17.66

d 41.608 − 18.9

3 Carry out the following multiplications or divisions.

a 5.4 × 6.3

b 8.19 × 1.67

c 43.05 ÷ 0.3

d 2.768 ÷ 0.008

4 Carry out the following calculations.

a $(8.1 \times 6.3) + 5.9^2$

b $(4.95 \div 0.3)(16.5 - 7.8)$

c $7.4 + (0.92 \div 0.4)^2$

5 Round the following numbers to the required number of decimal places.

a 19.863 (1 d.p.)   b 581.3749 (2 d.p.)   c 67.67456 (3 d.p.)

6 Write the following numbers to the required number of significant figures.

a 3507.1 (1 s.f.)   b 0.02515 (2 s.f.)   c 115.381 (3 s.f.)

7 Solve to find the part.

a What is 5% of 80?

b What is 60% of 145?

8 Solve to find the per cent.

a What per cent of 800 is 96?

b What per cent of 300 is 225?

9 Solve to find the base.

a 3% of what is 51?

b 7% of what is 98?

10 Convert to a percentage.

a 0.628   b 1.41   c $\frac{11}{25}$ d $\frac{5}{40}$

11 Convert to a fraction.

a 45%   b 27.5%   c 0.016   d 0.456

12 Convert to a decimal.

a 20.8%   b 146%   c $\frac{13}{5}$ d $\frac{9}{11}$

Kathy Brady

B   APPLICATIONS

1   There have been only 18 Australian Test batsmen who have scored
    over 5000 runs. The batting averages of four of them are Ricky
    Ponting 51.85, Allan Border 50.56, Steve Waugh 51.06 and Matthew
    Hayden 50.73.

    a   Arrange these players in ascending order of batting averages.

    b   When he retired in the 2015 season, Michael Clarke's batting
        average was 50.79. Where would he be placed in this list?
        (Source: stats.espncricinfo.com)

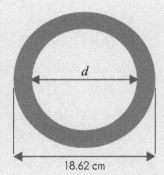

2   The rainfall totals (in millilitres) for Tasmania in the period February
    to May 2015 are provided in the table.

February	34.5 mm
March	115.8 mm
April	40.9 mm
May	192.5 mm

(Source: bom.gov.au)

    a   What was the total rainfall in Tasmania in that period?

    b   What would the total be when expressed to two significant
        figures?

3   The cross-section of a concrete pipe is shown in the diagram.
    If the total diameter of the pipe is 18.62 cm and the thickness of
    each concrete wall is 1.4 cm, what is the diameter, d, across the
    centre of the pipe?

4   A miniature house has been
    built at 0.125 of a full size house.
    This means that in the miniature
    house everything is exactly 0.125
    times smaller than its real-world
    equivalent. The miniature house
    has a table in the dining area. The
    miniature house builder also has
    the same style of table in her own
    dining room. If her table is 90 cm
    high, how high is the table in the
    miniature house?

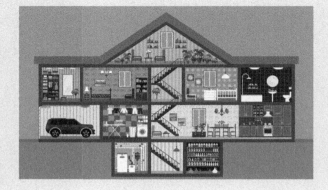

5 The average of a set of numbers is the sum of the numbers divided by how many numbers there are in the set. Using the rainfall data for Tasmania given in Question 2, find the average rainfall for the months February to May 2105. Write your answer to one decimal place.

6 A sporting goods store purchases treadmills from their supplier for $800. They then mark-up the cost price of the treadmill by 30%.

a What is the selling price of the treadmill? During the New Years' clearance sales the store offers a 20% discount on the regular selling price.

b What is the selling price of the treadmill during the sales period?

c How much profit or loss does the store make selling at the discounted price?

7 In 2009, the number of road fatalities in Queensland was 7.5 per 100 000 persons. In 2010, this number had reduced to 5.5 per 100 000 persons. What was the percentage decrease in road fatalities in Queensland between 2009 and 2010? Round your answer two decimal places. (Source: abs.gov.au/ausstats)

8 In 1990, carbon dioxide (CO_2) emissions in China were recorded at 2.2 metric tons per capita. Twenty years later, in 2010, CO_2 emissions in China had risen to 6.2 metric tons per capita. What was the percentage increase in CO_2 emissions in China in that period of time? Round your answer to two decimal places. (Source: data.worldbank.org)

9 A student arranges finance to buy a small car. The finance company charges 12% simple interest each year. If the student borrows $3000 over a three-year loan period, how much will the student have to repay?

10 Another student has received a small inheritance of $6000, when a close relative passed away. They invested it in a term deposit for 2 years earning 5% compound interest per year.

a What will be the balance of that account at the end of the term?

b What would the difference have been if the account had paid simple interest?

11 In a 24-hour period, Ali spent 0.4 of his time sleeping, 15% of his time watching TV, and $\frac{1}{8}$ of his time eating. In the remaining hours he was studying. How many hours of study did he do?

Kathy Brady

7

RATIO AND PROPORTION

CHAPTER OBJECTIVES

» Understand the concept of ratio

» Express ratios in simplest terms

» Convert between different representations of a ratio

» Understand the concepts of rate and unit rate

» Express rates in simplest terms

» Convert a rate to a unit rate

» Understand the concept of proportion

» Determine unknown quantities in a proportion

DELICIOUS TIMES THREE

Anyone who has ever done any cooking will have used ratios and proportions at some time. For example, in the figure, the left-hand side shows a recipe for chocolate chip cookies and the right-hand side shows the recipe needed for a triple batch of cookies, to make roughly three times as many cookies. When tripling the recipe, we must triple the amount of each ingredient in the recipe if we still want the recipe to work. For example:

- Three times as much flour: $1\frac{1}{2} \times 3 = 4\frac{1}{2}$ cups
- Three times as many eggs: $1 \times 3 = 3$ eggs
- Three times as much brown sugar: $\frac{3}{8} \times 3 = \frac{9}{8} = 1\frac{1}{8}$ cups
- Three times as much butter: $\frac{2}{3} \times 3 = \frac{6}{3} = 2$ cups

Chocolate Chip Cookies	Tripled version of recipe
$\frac{3}{8}$ cup sugar	$1\frac{1}{8}$ cups sugar
$\frac{3}{8}$ cup brown sugar	$1\frac{1}{8}$ cups brown sugar
$\frac{2}{3}$ cup butter	2 cups butter
1 egg	3 eggs
1 tsp vanilla	3 tsp vanilla
$1\frac{1}{2}$ cups flour	$4\frac{1}{2}$ cups flour
$\frac{1}{2}$ tsp baking soda	$1\frac{1}{2}$ tsp baking soda
$\frac{1}{2}$ tsp salt	$1\frac{1}{2}$ tsp salt
$\frac{1}{2}$ cup chopped nuts	$1\frac{1}{2}$ cup chopped nuts
150 g chocolate chips	450 g chocolate chips

× 3 for every ingredient

- Three times as many nuts: $\frac{1}{2} \times 3 = \frac{3}{2} = 1\frac{1}{2}$ cups

and similarly for the remaining ingredients.

The reason each ingredient must be tripled is to *maintain the quantity of each ingredient relative to other ingredients*. For example, for the original recipe I use $\frac{3}{8}$ cup of sugar and $\frac{3}{8}$ cup of brown sugar. This means the amount of sugar and brown sugar in the recipe is the same. When I double or triple the recipe, the amount of sugar as compared to brown sugar must always be the same. If you look at the calculations above, you will see that in the tripled recipe, there are $1\frac{1}{8}$ cups of brown sugar, so there will also be $1\frac{1}{8}$ cups of sugar in the tripled recipe.

In mathematical terms, what we are doing when we double or triple a recipe is maintaining the *ratio* of one ingredient to another and keeping all ingredients in *proportion*.

7.1   RATIOS

A **ratio** describes the relationship between two quantities or values. For example, if there are 9 students doing chemistry and 11 students doing English, then we can say that the ratio of chemistry students to English students is 9 : 11. As in this example, often ratios are written using a colon ':', and when reading the colon in a ratio, we say 'is to' or just 'to'.

The ratio 10 is to 5 can be written as 10 : 5. However, ratios are best expressed in simplest terms. In this example, since both numbers in the ratio share a common factor, the ratio can be expressed in simplest terms by dividing each number in the ratio by the Highest Common Factor (HCF). The HCF of 10 and 5 is 5, so we get:

$$10 \div 5 = 2$$
$$5 \div 5 = 1$$

The simplest form of the ratio 10 : 5 is 2 : 1.

So, if a cake uses 100 grams of butter and 500 grams of flour, then the ratio of butter to flour is 100 : 500. However, this can be written more simply as 1 : 5 by dividing both sides of the ratio by the HCF, which in this case is 100.

Ratio
A comparison between two or more quantities that have the same units

Making connections
To review HCF revisit Chapter 3.

EXAMPLE 1

Express the ratio 30 to 4 in simplest terms.

Solution

The ratio can first be written as 30 : 4.

Both numbers in the ratio can be divided by their HCF, which is 2.

Since $30 \div 2 = 15$ and $4 \div 2 = 2$, the ratio can be expressed in simplest form as 15 : 2.

Tiffany Winn

PRACTICE 1

Express the ratio 15 to 6 in simplest terms.

EXAMPLE 2

Express the following ratios in simplest terms.

Solutions

a   6 to 21

a   The ratio 6 to 21 can be expressed as 6 : 21.

However, 6 and 21 have a common factor (HCF = 3), so $6 \div 3 = 2$ and $21 \div 3 = 7$.

The ratio $6 : 21 = 2 : 7$ in simplest terms.

b   11 to 3

b   The ratio 11 to 3 can be expressed as 11 : 3.

This ratio is already in simplest terms as the numbers in the ratio have no common factors.

c   48 to 8

c   The ratio 48 to 8 can be expressed as 48 : 8.

However, 48 and 8 have common factors (HCF = 8), so $48 \div 8 = 6$ and $8 \div 8 = 1$.

The ratio $48 : 8 = 6 : 1$ in simplest terms.

PRACTICE 2

Express the following ratios in simplest terms.

a   60 to 15   b   7 to 56   c   9 to 19

EXAMPLE 3: APPLICATION

Pure gold is 24-karat gold. Gold that is less than 24 karats is composed of some gold and other metals, which are added for hardness. What is the ratio of gold to the other metal in 14-karat gold?

Solution

If there are 14 parts of gold, then there will be $24 - 14 = 10$ parts of another metal.

The ratio of gold to the other metal is 14 : 10. This ratio can be simplified by dividing both numbers by their HCF (HCF = 2) to obtain $14 : 10 = 7 : 5$.

PRACTICE 3: APPLICATION

If a distance of 2 cm on a particular map corresponds to a distance of 5000 cm in real-life, what is the simplest form of the ratio that expresses the scale of the map?

There are many different ways of writing a ratio. For example, the ratio 1 : 2 can be written as:

- 1 : 2
- $\frac{1}{2}$
- 1 is to 2
- 0.5
- 50%

In other words: $1 : 2 = \frac{1}{2} = 1$ is to $2 = 0.5 = 50\%$.

Consider a class that has 10 female and 20 male students. The ratio of females to males is 10 : 20. If we divide both sides of the ratio by the highest common factor, which is 10, then we obtain the simplest version of the ratio, which is 1 : 2 or 1 is to 2. This means that for every female student there are two male students. The ratio can also be written as a fraction in the form $\frac{a}{b}$. Since the number of females is half the number of males, the ratio can also be written as $\frac{10}{20}$, which can be simplified to $\frac{1}{2}$ (since the number of females is half the number of males). Another way of writing $\frac{1}{2}$ is 0.5 and yet another way of writing $\frac{1}{2}$ is 50%, so these forms are also acceptable ways of writing the ratio 1 : 2.

Making connections
To review writing fractions in simplest form revisit Chapter 5, and to review converting fractions to decimals and percentages revisit Chapter 6.

EXAMPLE 4

Express the ratio 35 to 140 as a fraction in simplest form, as a percentage and as a decimal.

Solution

Simplest form

The ratio 35 to 140 can be written as 35 : 140 and as the fraction $\frac{35}{140}$.

The HCF of 35 and 140 is 35, and $35 \div 35 = 1$ and $140 \div 35 = 4$, so we have 35 : 140 = 1 : 4.

The simplest form of 35 : 140, written as a fraction, is $\frac{1}{4}$.

As a percentage

To convert the ratio 1 : 4 to a percentage, one method is to express it as a fraction, multiply by 100, and then do the division indicated by the resulting fraction:

$$\frac{1}{4} \times \frac{100}{1} = \frac{1 \times 100}{4 \times 1} = \frac{100}{4}$$

Since $\frac{100}{4} = 100 \div 4 = 25$, the ratio 35 : 140 or 1 : 4 is equivalent to 25%.

As a decimal

To convert the ratio 1 : 4 to a decimal, the simplest method is to use the percentage and move the decimal point two places to the left: 1 : 4 = 25% = 0.25.

PRACTICE 4

Express the ratio 200 to 8 as a fraction in simplest form, as a percentage and as a decimal.

Note that when converting a ratio to a fraction, if the HCF of the two numbers is difficult to spot, the ratio can be simplified by using any common factor, and then simplified again using other common factors until the ratio is in simplest form.

For example, one common factor of both 35 and 140 is 5 and $35 \div 5 = 7$ and $140 \div 5 = 28$, so we have $35 : 140 = 7 : 28$. The numbers 7 and 28 both share a common factor (7) and $7 \div 7 = 1$ and $28 \div 7 = 4$, so we have $35 : 140 = 7 : 28 = 1 : 4$.

Making connections
Refer to section 6.7 for more explanation on how to round decimals.

EXAMPLE 5

Express the following ratios in the form specified, approximating to three decimal places if needed.

a   60 is to 18 as a fraction in simplest form

b   3 is to 11 as a percentage

Solutions

a   The ratio $60 : 18$ can be written as a fraction as $\frac{60}{18}$. Since 60 and 18 have common factors, the fraction needs to be reduced to be in simplest form. One common factor of 60 and 18 is 6, $60 \div 6 = 10$ and $18 \div 6 = 3$, so $\frac{60}{18} = \frac{10}{3}$. The numbers 10 and 3 have no common factors, so the ratio $60 : 18$ can be written as a fraction in simplest form as $\frac{10}{3}$.

b   The ratio $3 : 11$ can be written as a fraction as $\frac{3}{11}$. To convert $\frac{3}{11}$ to a percentage, we multiply by 100 and do the resultant division:

$$\frac{3}{11} \times \frac{100}{1} = \frac{3 \times 100}{11 \times 1} = \frac{300}{11} \text{ (recurring)}$$

$$11\overline{)3 \quad 0 \,_80.\,_30 \,_80 \,_30 \,_80}$$

Since $\frac{300}{11} = 27.\overline{27}$, the ratio $3 : 11$ is equivalent to 27.27% (2 d.p.).

PRACTICE 5

Express the following ratios in the form specified, approximating to three decimal places if needed.

a   18 is to 144 as a fraction in simplest form

b   5 is to 12 as a decimal

EXAMPLE 6

Express the following ratios as a fraction in simplest form, as a decimal and as a percentage, rounding to two decimal places if needed.

a   2 is to 15   b   250 is to 40   c   36 is to 99

Solutions

a The ratio 2 is to 15 can be written as the fraction $\frac{2}{15}$. It is already in simplest form because 2 and 15 do not have any common factors.

To convert $\frac{2}{15}$ to a decimal, we do the division represented by the fraction:

$$\begin{array}{r} 0.\;1\;\;3\;\;3 \text{ (recurring)} \\ 15\overline{)\,2\,._5 0\;_5 0\;_5 0} \end{array}$$

Since $2 \div 15 = 0.1\overline{3}$, the ratio $2 : 15 = 0.13$ (2 d.p.)

Converting from a decimal to a percentage requires moving the decimal point two places to the right. So, the ratio $2 : 15 = 13.33\%$ (2 d.p.)

b The ratio $250 : 40$ can be written as the fraction $\frac{250}{40}$. This fraction can be simplified by dividing by the HCF of 10, $250 \div 10 = 25$ and $40 \div 10 = 4$, so $\frac{250}{40} = \frac{25}{4}$. This fraction is in simplest form because 25 and 4 have no common factors.

To convert $\frac{25}{4}$ to a percentage, multiply by 100 and do the resultant division:

$$\frac{25}{4} \times \frac{100}{1} = \frac{25 \times 100}{4 \times 1} = \frac{2500}{4}$$

$$\begin{array}{r} 6\;\;2\;\;5 \\ 4\overline{)\,2\;\;5\;_1 0\;_2 0} \end{array}$$

So, the ratio $250 : 40 = 625\%$.

Converting from a percentage to a decimal requires moving the decimal point two places to the left, so the ratio $250 : 40 = 6.25$.

c The ratio 36 is to 99 can be written as the fraction $\frac{36}{99}$. The numbers 36 and 99 have HCF 3, $36 \div 3 = 12$ and $99 \div 3 = 33$, so we have $\frac{36}{99} = \frac{12}{33}$ in simplest form.

To convert $\frac{12}{33}$ to a decimal, we do the division represented by the fraction:

$$\begin{array}{r} 0.\;3\;\;6\;\;3\;\;6 \text{ (recurring)} \\ 33\overline{)\,1\;\;2.\;0\;_{21} 0\;_{12} 0\;_{21} 0} \end{array}$$

So, the ratio $36 : 99 = 0.\overline{36}$.

Converting from a decimal to a percentage requires moving the decimal point two places to the right. So, the ratio $36 : 99 = 36.36\%$ (2 d.p.).

PRACTICE 6

Express the following ratios as a fraction in simplest form, as a decimal and as a percentage.

a 63 is to 14   **b** 25 is to 305   **c** 498 is to 12

Making
connections
*Revisit Chapter 5
to review how to
compare the size
of two fractions
using common
denominators.*

EXAMPLE 7: APPLICATION

In a class of 24 students, 15 are males. In another class of 35 students, 22 are males. In which class is the ratio of boys to girls the greatest?

Solution

In the first class, the ratio of males to females is 15 : 9, since $24 - 15 = 9$. As a fraction in simplest form, $\frac{15}{9} = \frac{5}{3}$. In the second class, the ratio of males to females is 21 : 14, since $35 - 21 = 14$. As a fraction in simplest form, $\frac{21}{14} = \frac{3}{2}$.

Determining which class has the greater ratio of males to females requires deciding which fraction $\left(\frac{5}{3} \text{ or } \frac{3}{2}\right)$ is greater. This can either be done by writing each fraction over a common denominator or converting each fraction to a decimal or a percentage for easy comparison. Using the first method, $\frac{5}{3} = \frac{10}{6}$ and $\frac{3}{2} = \frac{9}{6}$, so the class with a greater male : female ratio is the first class.

PRACTICE 7: APPLICATION

In one box of 12 doughnuts, there are 4 raspberry doughnuts. In another box of 30 doughnuts, there are 12 raspberry doughnuts. Which box of doughnuts has the greatest percentage of raspberry doughnuts?

EXAMPLE 8: APPLICATION

A 50-kilogram bag of lawn fertiliser contains 10 kilograms of nitrogen and 15 kilograms of potash. Another 30-kilogram bag of lawn fertiliser contains 6 kilograms of nitrogen and 8 kilograms of potash. Which bag has the greater ratio of nitrogen to potash?

Solution

The first fertiliser has a 10 : 15 ratio of nitrogen to potash, which is $\frac{10}{15} = \frac{2}{3}$. The second fertiliser has a ratio of nitrogen to potash of 6 : 8, which is $\frac{6}{8} = \frac{3}{4}$.

Expressing both ratios as fractions with a common denominator gives $\frac{2}{3} = \frac{8}{12}$ and $\frac{3}{4} = \frac{9}{12}$, which indicates that the second fertiliser has a greater ratio of nitrogen to potash.

PRACTICE 8: APPLICATION

One poster has a width of 21 centimetres and a height of 45 centimetres. Another poster has a width of 36 centimetres and a height of 75 centimetres. For which poster is the width relatively longest compared to the height?

Remember that a ratio expresses a comparison between quantities that have the same units. Be careful with units that are different multiples of the same base unit. For example, what is the ratio of 1 litre to 2 millilitres?

The ratio of 1 litre to 2 millilitres (1 L to 2 mL) is NOT 1 : 2. The units of both quantities must be the same in order to determine the ratio. Since 1 L = 1000 mL, the ratio is:

$$1000 \text{ mL} : 2 \text{ mL}$$
$$= 1000 : 2$$
$$= 500 : 1$$

EXAMPLE 9

Express the following ratios in simplest form.

a 400 mg to 5 kg   b 50 km to 20 cm   c 80 mm to 0.05 km

d 3.85 kg to 1200 g   e 50 µL to 0.0004 L

Solutions

a First, the units in the ratio must be the same. Since 5 kg = 5000 g = 5000000 mg, the ratio can be written as 400 : 5000000. Both numbers have a common factor of 100. Since 400 ÷ 100 = 4 and 5000000 ÷ 100 = 50000, the ratio can be written as 4 : 50000.

   However, 4 is a factor of 50000, so the ratio is still not in simplest form. Since 4 ÷ 4 = 1 and 50000 ÷ 4 = 12500, the simplest form of the ratio 400 mg is to 5 kg is 1 : 12500.

b First, both numbers in the ratio need to be expressed in common units, so first convert kilometres to centimetres by multiplying by 1000 and then by 100 :

   50 km = 50000 m = 5000000 cm

   So, the ratio of 50 km to 20 cm is 5000000 : 20. The HCF of both numbers is 20, and 5000000 ÷ 20 = 250000 and 20 ÷ 20 = 1, so the ratio of 50 km to 20 cm in simplest form is 250000 : 1.

c First, both numbers in the ratio need to be expressed in common units, so first convert kilometres to metres and then to millimetres by multiplying by 1000 and then by 1000 again:

   0.05 km = 0.05 × 1000 m = 50 m

   50 m = 50 × 1000 mm = 50000 mm

   So, the ratio of 80 mm to 0.05 km is 80 : 50000. Since 40 is a common factor, and 80 ÷ 40 = 2 and 50000 ÷ 40 = 1250, the ratio of 80 : 50000 = 2 : 1250. The HCF of both 2 and 1250 is 2, and since 2 ÷ 2 = 1 and 1250 ÷ 2 = 625, the ratio of 80 mm to 0.05 km in simplest form is 1 : 625.

d First, both numbers in the ratio need to be expressed in common units, so start with converting kilograms to grams by multiplying by 1000 : 3.85 kg = 3850 g

So, the ratio of 3.85 kg to 1200 g is 3850 : 1200. One common factor of both numbers is 50. Since 3850 ÷ 50 = 77 and 1200 ÷ 50 = 24, the ratio of 3850 : 1200 = 77 : 24. The numbers 77 and 24 have no common factors, so the ratio 3.85 kg to 1200 g in simplest form is 77 : 24.

e First, both numbers in the ratio need to be expressed in common units, so convert litres (L) to microlitres (μL) by multiplying by 1 000 000:

$$0.0004 \text{ L} = 0.0004 \times 1\,000\,000\ \mu L$$
$$= 400\ \mu L$$

So, the ratio of 50 μL to 0.0004 L is 50 : 400. The HCF of both numbers is 50, and 50 ÷ 50 = 1 and 400 ÷ 50 = 8, so the ratio of 50 μL to 0.0004 L in simplest form is 1 : 8.

Making connections
Visit Chapter 12 to review further metric unit conversions.

PRACTICE 9

Express the following ratios in simplest form.

a 2 km to 400 cm   **b** 50 L to 45 kL   **c** 0.03189 ML to 201 mL

d 1.5 g to 0.12 μg   **e** 2.84 dm to 2.96 cm

EXAMPLE 10: APPLICATION

A given map displays a scale of 4 cm to 1 km. What is the ratio of the map to the actual distance? Express the ratio in simplest form.

Solution

Since both units need to be the same, one unit must be converted to the other or both to a common unit.

1 km = 1000 m = 100 000 cm

So, the ratio of the map to the actual distance is 4 : 100 000. Since 4 and 100 000 share a common factor of 4, and 4 ÷ 4 = 1 and 100 000 ÷ 4 = 25 000, the ratio of the map to the actual distance in simplest form is 1 : 25 000.

PRACTICE 10: APPLICATION

A given map displays a scale of 2 cm to 300 m. What is the ratio of the map to the actual distance? Express the ratio in simplest form.

EXAMPLE 11: APPLICATION

Apples were listed as costing 55c each and bananas as $1.40. How much more expensive were bananas relative to apples? Round your answer to one decimal place if needed.

Solution

The cost of each item needs to be represented using the same unit, so one strategy is to convert $1.40 to cents: $1.40 = 140c.

Then, the cost of bananas relative to apples is represented by the ratio 140 : 55.

Since 5 is a common factor, the ratio can be simplified to 28 : 11.

There are no further common factors, so 28 : 11 is the simplest form of the ratio. To obtain the cost of bananas relative to apples, do the division represented by the ratio:

$$11 \overline{)2\,8.^{6}0\,_{5}0\,^{6}0\,_{6}0\,_{5}0} = 2.5\,4\,5\,4 \text{ (recurring)}$$

Rounding to one decimal place, the cost of bananas is approximately 2.5 times the cost of apples.

PRACTICE 11: APPLICATION

Oranges cost 90c each and mangoes cost $1.60 each. How much cheaper are oranges compared to mangoes? Round your answer to one decimal place if needed.

7.2   RATES

A **rate** is a ratio that expresses a comparison between two quantities of different units. There are many situations in day-to-day living that require the use of rates. If a car drives an average of 12 kilometres on 1 litre of fuel, then the fuel efficiency of the car is written as the rate 12 kilometres per litre. An AFL football team that has an average of 200 kicks in a game is described as having a rate of 200 kicks per game. If a person's heart is beating 65 beats in a minute, their pulse rate is 65 beats per minute. We are able to recognise the use of a rate through the inclusion of the term 'per'.

Rate
A special ratio that expresses a comparison between two quantities of different units

EXAMPLE 12

Write in simplest form the rates described in the following situations, rounding to one decimal place if needed.

Solutions

a On average, I drive 55 km each hour.

a This rate can be expressed as 55 km *per* hour. It is already in simplest form because one of the quantities (hours) has a value of one.

b One kilogram of bananas costs $2.99.

b This rate can be expressed as $2.99 *per* kg. The rate is in simplest form.

c Travelling 120 km I used 11 L of fuel.

c This rate can be expressed as 120 km *per* 11 L and it is in simplest form because the two numbers 120 and 11 do not have any common factors.

d 150 mg of medication needs to be infused over a period of 45 minutes.

d This rate can be expressed as 150 mg *per* 45 minutes. This can be simplified using the highest common factor 15 to be 10 mg *per* 3 minutes.

e I earned $4580 in 6 weeks work.

e This rate can be expressed as $4580 *per* 6 weeks. Both numbers have a highest common factor of 2, so the ratio simplifies to $2295 *per* 3 weeks.

PRACTICE 12

Write in simplest form the rates described in the following situations.

a On average, I drive 90 km in two hours.

b Five kilograms of nectarines costs $45.

c I used 22 L of fuel on a trip of 265 km.

d 330 mg of medication needs to be infused over 90 minutes.

e I earned $14 584 in 8 weeks work.

Unit rate
A rate in which the second named value is one

A ratio is often simplified so that it is expressed using the lowest possible whole numbers. Rather than just simplifying to the lowest possible whole numbers, rates are often simplified to **unit rates**, as this allows easier comparison between rates.

A unit rate can be obtained by dividing the numbers given in the ratio. As one example, a heartbeat of 180 beats in 3 minutes simplifies to a unit rate of 60 beats per minute (meaning 60 beats in 1 minute) because $180 \div 3 = 60$. As another example, a distance of 370 km travelled in 4 hours simplifies to 92.5 km per hour (meaning 92.5 km in one hour) because $370 \div 4 = 92.5$.

EXAMPLE 13

A patient has 150 mL of medication infused every 60 minutes. Express this as a unit rate.

Solution

This rate can be expressed as 150 mL per 60 minutes. This rate can be simplified by dividing both numbers by their HCF of 10: 150 mL per 60 minutes = 15 mL per 6 minutes.

Now, to simplify to a unit rate, we do the division indicated by the ratio:

$$15 \text{ mL per 6 minutes} = \frac{15 \text{ mL}}{6 \text{ minutes}}$$

$$\begin{array}{r} 2.5 \\ 6\overline{)15.30} \end{array}$$

Since 15 ÷ 6 = 2.5, 150 mL per 60 minutes expressed as a unit rate is 2.5 mL per minute.

PRACTICE 13

On a 900 km round trip, our car went through 75 L. What is the fuel efficiency of our car, expressed as a distance per litre?

EXAMPLE 14

Express the following as unit rates, simplifying to two decimal places if necessary.

a   A patient has 140 mL of medication infused every 90 minutes.

b   On a recent trip, my car used 16 L of fuel to drive 200 km.

c   Over 12 weeks of healthy eating, Anna's weight dropped by 18 kg.

d   A gym membership costs $350 for 12 months.

e   It costs $4.25 for a 725 g box of cereal.

Solutions

a   140 mL of medication infused every 90 minutes is expressed as the rate: 140 mL per 90 minutes

Dividing both numbers by a common factor of 10 gives:

140 mL per 90 minutes = 14 mL per 9 minutes

Now, to simplify to a unit rate, we do the division indicated by the ratio:

$$14 \text{ mL per 9 minutes} = \frac{14 \text{ mL}}{9 \text{ minutes}}$$

$$\begin{array}{r} 1.555 \text{ (recurring)} \\ 9\overline{)14.{}_50{}_50{}_50} \end{array}$$

140 mL per 90 minutes expressed as a unit rate is 1.56 mL per minute (2 d.p.).

b　16 L of fuel to drive 200 km is expressed as the rate: 16 L per 200 km

Dividing both numbers by a common factor of 4 gives: 16 L per 200 km = 4 L per 50 km

And dividing again by a common factor of 2 gives: 4 L per 50 km = 2 L per 25 km

Now, to simplify to a unit rate, we do the division indicated by the ratio:

$$2 \text{L per } 25 \text{ km} = \frac{2 \text{ L}}{25 \text{ km}}$$

$$25 \overline{)\,2.\,0\,0}\quad^{0.\,0\,8}$$

Since $2 \div 25 = 0.08$, 16 L of fuel to drive 200 km is expressed as the unit rate 0.08 L per km. Incidentally, fuel consumption is typically expressed as L per 100 km, rather than L per km, because it is more readable; therefore, 0.08 L per km = 8 L per 100 km.

c　Losing 18 kg in 12 weeks can be expressed as the rate: 18 kg per 12 weeks

Dividing both numbers by HCF 6 gives: 18 kg per 12 weeks = 3 kg per 2 weeks

Now, to simplify to a unit rate, we do the division indicated by the ratio:

$$3 \text{ kg per } 2 \text{ weeks} = \frac{3 \text{ kg}}{2 \text{ weeks}}$$

$$2 \overline{)\,3.\,0}\quad^{1.\,5}$$

So, 18 kg in 12 weeks expressed as a unit rate is 1.5 kg per week.

d　A cost of $350 for 12 months can be expressed as the rate: $350 per 12 months

Dividing both numbers by a common factor of 2 gives: $350 per 12 months = $175 per 6 months

Now, to simplify to a unit rate, we do the division indicated by the ratio:

$$\$175 \text{ per } 6 \text{ months} = \frac{\$175}{6 \text{ months}}$$

$$6 \overline{)\,1\,\,7\,{}_5 5.\,{}_1 0\,\,{}_4 0\,\,{}_4 0}\quad^{2\,\,9.\,1\,\,6\,\,6 \text{ (recurring)}}$$

$350 for 12 months expressed as a unit rate is $29.17 per month (2 d.p.).

e　$4.25 for 725 g can be expressed as the rate: $4.25 per 725 g

In order to facilitate simplification using division by common factors, we will express the price in cents: $4.25 per 725 g = 425c per 725 g

Dividing by HCF of 25 gives: 425c per 725 g = 17c per 29 g

Now, to simplify to a unit rate, we do the division indicated by the ratio:

$$17c \text{ per } 29 \text{ g} = \frac{17c}{29 \text{ g}}$$

$$29 \overline{)\,1\,\,7.\,0\,{}_{25} 0\,{}_{18} 0}\quad^{0.\,5\,\,8\,\,6 \text{ (3 d.p.)}}$$

Rounded to two decimal places $4.25 for a 725-gram box of cereal expressed as a unit rate is 0.59c per gram (in other words, a bit more than half a cent per gram). Incidentally, supermarket prices are often expressed as a price per 100 g since this is more readable. In this case, 0.59c per gram = 59c per 100 g (2 d.p.).

PRACTICE 14

Express the following as unit rates, simplifying to two decimal places if necessary.

a A parachutist falls at a rate of 14 kilometres per 30 minutes.

b A baseball player made 18 hits in 50 appearances at the plate.

c Parking costs $25 for 8 hours.

d A hotel charges $1526 for 7 nights accommodation.

e It costs $4.80 for a 500 g jar of peanut butter.

EXAMPLE 15: APPLICATION

a Boxes of a particular breakfast cereal are sold in two sizes, a 220 g box costs $2.65 and a 380 g box costs $4.84. Which size box is better value?

b A 750 g block of cheese costs $9.80 and a 1 kg block of the same cheese costs $13.50. What is the unit price of each block? Which block is better value?

Solutions

a In each case, we simplify using common factors and then do the division represented by the ratio to obtain the unit rate:

$$\frac{\$2.65}{220\,g} = \frac{265c}{220\,g} = \frac{53c}{44\,g}$$

(common factor of 5)

$$\frac{\$4.84}{380\,g} = \frac{484c}{380\,g} = \frac{121c}{95\,g}$$

(common factor of 4)

$$1.\,2\;\;0\;\;4 \text{ Remainder } 34$$
$$44\overline{)5\;\;3._90\;_{20}0\;_{20}0}$$

$$1.\,2\;\;7\;\;3 \text{ Remainder } 65$$
$$95\overline{)1\;\;2\;\;1._{26}0\;_{70}0\;_{35}0}$$

The unit price of the 220 g box is 1.20c per gram (2 d.p.) and the unit price of the 380 g box is 1.27c per gram (2 d.p.). Since 1.20 < 1.27, the smaller box has a lower unit price and is better value.

b In each case, we simplify using common factors, and then do the division represented by the ratio to obtain the unit rate:

$$\frac{\$9.80}{750\,g} = \frac{980c}{750\,g} = \frac{98c}{75\,g}$$

(common factor of 10)

$$\frac{\$13.50}{1\,kg} = \frac{1350c}{1000\,g} = \frac{27c}{20\,g}$$

(common factor of 50)

$$1.\,3\;\;0\;\;6 \text{ Remainder } 50$$
$$75\overline{)9\;\;8._{23}0\;_50\;_{50}0}$$

$$1.\,3\;\;5$$
$$20\overline{)2\;\;7._70\;_{10}0}$$

Note that the two rates must both use the same units in order to be compared, so for the larger block of cheese, 1 kg is converted to 1000 g.

The unit price of the 750 g block is 1.31c per gram (2 d.p.) and the unit price of the 1 kg block is 1.35c per gram. Since 1.31 < 1.35, the smaller block has a lower unit price and is better value.

Tiffany Winn

7.3   PROPORTION

What do we mean when we say that a model ship is 'in proportion' to the original?

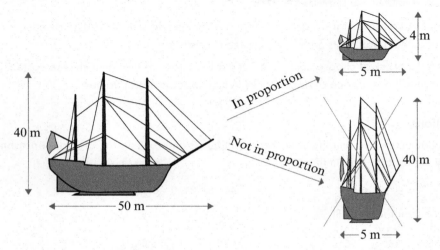

Proportion
Two or more relationships can be expressed as ratios that are equal

Making connections
For a refresher on equivalent fractions, see Chapter 5.

To be 'in **proportion**' means that the dimensions can be written as two ratios that are equal. For example, dividing both 40 and 50 by a common factor of 10 shows us that $\frac{40}{50}$ is the same as $\frac{4}{5}$, as shown in the diagram above, because $\frac{40}{50}$ and $\frac{4}{5}$ are equivalent fractions. In fact, any two equivalent fractions represent a relationship that is in proportion.

Perhaps the easiest way of testing whether two ratios are equal or 'in proportion' is to write the ratios as fractions and cross-multiply to obtain the cross-products. If the two cross-products are equal, then the ratios are equal, as shown below.

The cross-products are obtained as follows:

1   Multiply the numerator of the first fraction by the denominator of the second: $30 \times 4 = 120$

2   Multiply the denominator of the first fraction by the numerator of the second fraction: $40 \times 3 = 120$

$$\frac{30}{40} \overset{?}{\times} \frac{3}{4}$$

EXAMPLE 16

Determine if the following pairs of ratios are equal.

a $\frac{11}{44}$ and $\frac{3}{12}$ b $3:8$ and $51:136$ c $18:45$ and $2:5$

d $1:3$ and $30:100$ e $2:9$ and $14:61$

Solutions

a The two cross-products are obtained as follows:

$$\frac{11}{44} \overset{?}{\times} \frac{3}{12}$$

1 Multiply the numerator of the first fraction by the denominator of the second: $11 \times 12 = 132$

2 Multiply the denominator on the first fraction by the numerator of the second fraction: $44 \times 3 = 132$.

The cross-products are equal, so the ratios are equal or 'in proportion'.

Note that another way of testing if the ratios are equal would be to convert them both to equivalent fractions: $\frac{11}{44} \times \frac{3}{3} = \frac{33}{132}$ and $\frac{3}{12} \times \frac{11}{11} = \frac{33}{132}$, so the ratios are equal.

b Calculate the cross-products as follows:

$$\frac{3}{8} \overset{?}{\times} \frac{51}{136}$$

The cross-products are:

$8 \times 51 = 408$

$3 \times 136 = 408$

The cross-products are equal, so the ratios are equal.

c Calculate the cross-products as follows:

$$\frac{2}{5} \overset{?}{\times} \frac{18}{45}$$

The cross-products are:

$45 \times 2 = 90$

$18 \times 5 = 90$

The cross-products are equal, so the ratios are equal.

d Calculate the cross-products as follows:

$$\frac{1}{3} \overset{?}{\times} \frac{30}{100}$$

The cross-products are:

$1 \times 100 = 100$

$3 \times 30 = 90$

The cross-products are not equal, so the ratios are not equal.

e   Calculate the cross-products as follows:

$$\frac{2}{9} \overset{?}{\times} \frac{14}{61}$$

The cross-products are:
   $2 \times 61 = 122$
   $9 \times 14 = 126$
The cross-products are not equal, so the ratios are not equal.

PRACTICE 16

Determine if the following pairs of ratios are equal.

a $\frac{4}{7}$ and $\frac{64}{112}$ b $30:56$ and $5:8$ c $13:78$ and $1:6$

d $9:11$ and $110:132$ e $9:4$ and $72:32$

EXAMPLE 17: APPLICATION

A model ship is 24 cm long and 9 cm tall. The original is 16 m long and 6 m tall. Is the model in proportion to the original?

Solution

We can write the ratio of the length of each ship to its height as $\frac{24}{9}$ for the model and $\frac{16}{6}$ for the original. Note that we do not need to convert the units for either model to be the same as the other, rather what is important is that the units are the same within each individual ratio.

Now we can cross-multiply:

$$\frac{24}{9} \overset{?}{\times} \frac{16}{6}$$

The cross-products are:
   $24 \times 6 = 144$
   $9 \times 16 = 144$
The cross-products are equal, so the ratios are in proportion.

PRACTICE 17: APPLICATION

A man is 178 cm tall and 40 cm wide. His son is much shorter and thinner at 128 cm tall and 32 cm wide. Is the son in proportion to his father?

Sometimes instead of just determining whether two given ratios are equal, it is useful to be able to find a ratio that is equal to another one. For example, when reading a map with a scale of 1 : 8000, we know that each unit on the map represents 8000 units in real-life. If the distance between two points on the map is 2.5 cm, we can still use the cross-multiplication method to work out how far apart the points are in real-life. In the expression below, the d represents the real-life distance and the equation reminds us that the ratio representing the scale of the map (1 : 8000) is equal to the ratio representing the map distance between two points as compared to their real-life distance (2.5 : d). The real-life distance is denoted by d because we have not yet worked it out:

$$\frac{1}{8000} = \frac{2.5}{d}$$

To solve this problem, we cross-multiply and set each of the cross-products equal to each other:

$$1 \times d = 8000 \times 2.5$$
$$d = 20\,000 \text{ cm}$$

We can then convert that distance to metres for readability, so $20\,000$ cm $= 200$ m.

EXAMPLE 18

Find the value of x in the following.

Solutions

a   4 is to 20 as x is to 90

a   Since the two ratios are in proportion, the fractions represented by the two ratios must be equal to each other:

$$\frac{4}{20} = \frac{x}{90}$$

For the fractions to be equal, their cross-products must be equal, so we have:

$$20 \times x = 4 \times 90$$
$$20 \times x = 360$$
$$x = 18$$

b   18 is to x as 15 is to 4

b   Since the two ratios are in proportion, the fractions represented by the two ratios must be equal:

$$\frac{18}{x} = \frac{15}{4}$$

For the fractions to be equal, their cross-products must be equal, so we have:

$$18 \times 4 = x \times 15$$
$$72 = 15 \times x$$
$$x = \frac{72}{15}$$
$$x = 4.8$$

c x is to 84 as 12 is to 70

c   Since the two ratios are in proportion, the fractions represented by the two ratios must be equal:

$$\frac{x}{84} = \frac{12}{70}$$

For the fractions to be equal, their cross-products must be equal, so we have:

$$x \times 70 = 84 \times 12$$
$$x \times 70 = 1008$$
$$x = \frac{1008}{70}$$
$$x = 14.4$$

Making connections
To review solving algebraic equations visit Chapter 13

d 72 is to 112 as 27 is to x

d Since the two ratios are in proportion, the fractions represented by the two ratios must be equal:

$$\frac{72}{112} = \frac{27}{x}$$

For the fractions to be equal, their cross-products must be equal, so we have:

$$72 \times x = 112 \times 27$$
$$72 \times x = 3024$$
$$x = \frac{3024}{72}$$
$$x = 42$$

PRACTICE 18

Find the value of x in the following.

a 9 is to 30 as 21 is to x

b x is to 24 as 21 is to 9

c 28 is to 35 as x is to 45

d 28 is to x as 15 is to 6

EXAMPLE 19: APPLICATION

My computer takes 9 seconds to download a 15 megabyte piece of software. At this rate, how long will my computer take to download a 32 megabyte piece of software?

Solution

The ratio of seconds taken to size of download for the first piece of software is $\frac{9}{15}$. For the second piece of software, we know the size of the download, but not the time taken, so we could write this ratio as $\frac{\square}{32}$. If the time for the downloads are in proportion, then we can say that:

$$\frac{9}{15} = \frac{\square}{32}$$

Now if we cross-multiply we get:

$$9 \times 32 = 288$$
$$15 \times \square$$

If the download times are in proportion, then we know that $15 \times \square = 288$.

Therefore, the unknown download time will be $\frac{288}{15} = 19.2$ seconds.

PRACTICE 19: APPLICATION

I need a metal alloy made up of one-third lead and two-thirds tin for an electronics device that I am building. If the alloy I have has 87 parts of lead and 164 parts of lead, is it in the correct proportion for what I need?

EXAMPLE 20: APPLICATION

You need to make up a jerry can of two-stroke fuel for a dirt bike. This requires a 25 : 1 mix of petrol and oil. You have 5 L of petrol. How much oil is required?

Solution

Since the fuel has a petrol : oil ratio of 25 : 1, this will be equal to the ratio of the actual amounts of petrol and oil used, as shown below, with x denoting the unknown quantity of oil required:

$$\frac{25}{1} = \frac{5}{x}$$

For the fractions to be equal, their cross-products must be equal, so we have:

$$25 \times x = 1 \times 5$$
$$25 \times x = 5$$
$$x = \frac{1}{5}$$
$$x = 0.2$$

The amount of oil needed is 0.2 L or 200 mL.

PRACTICE 20: APPLICATION

A 4 cm by 6 cm photo (width 4 cm, height 6 cm) needs to be enlarged so that its width is 18 cm. What will the height of the enlarged photo be so that the dimensions of the photo remain in proportion?

EXAMPLE 21: APPLICATION

A medication tablet contains paracetamol and codeine phosphate in the ratio 5 : 3. If the tablet contains a total of 800 mg of medication, how much of each substance (paracetamol and codeine phosphate) are in one tablet?

Solution

We know that the ratio of paracetamol and codeine phosphate in the medication is 5 : 3, but we have been given the *total* amount of medication (paracetamol and codeine phosphate combined). So we use the paracetamol to codeine phosphate ratio to calculate the ratio of paracetamol to total medication, and that of codeine phosphate to total medication:

Paracetamol : total medication $= 5 : 8$ (because in the original ratio $5 + 3 = 8$)

Codeine phosphate : total medication $= 3 : 8$ (because in the original ratio $5 + 3 = 8$)

Now we can say that 5 : 8 is equal to the amount of paracetamol in a tablet compared to the total amount of medication in a tablet:

$$\frac{5}{8} = \frac{p}{800}$$

For the fractions to be equal, their cross-products must be equal, so we have:

$$8 \times p = 5 \times 800$$
$$8 \times p = 4000$$
$$p = \frac{4000}{8}$$
$$p = 500$$

The amount of paracetamol in one tablet is 500 mg. Since the total amount of medication (paracetamol + codeine phosphate) in one tablet is 800 mg, the amount of codeine in one tablet is then $800 - 500 = 300$ mg.

PRACTICE 21: APPLICATION

If 120 L of oil takes 8 hours to flow through a pipe, how long will 270 L of oil take to flow through the same pipe?

EXAMPLE 22: APPLICATION

If the distance between two towns on a map is 7 cm and the scale of the map is 2 : 130 000, how far apart are the towns in real-life?

Solution

Since the map has a scale of 2 : 130 000 and the ratio of the distance on the map compared to the real-life distance for any given town will be equal to ratio of the scale of the map:

$$\frac{2}{130\,000} = \frac{7}{d}$$

For the fractions to be equal, their cross-products must be equal, so we have:

$$2 \times d = 130\,000 \times 7$$
$$2 \times d = 910\,000$$
$$d = \frac{910\,000}{2}$$
$$d = 455\,000$$

The real-life distance between the towns is 455 000 cm, which equates to 4550 m or 4.55 km.

PRACTICE 22: APPLICATION

On a given day, the stock exchange listed 1 Australian dollar as being worth 0.7 US dollars. If an Australian travelling in the US bought a shirt that cost $18 US dollars, how much did the shirt cost him in Australian dollars?

EXAMPLE 23: APPLICATION

Many model trains are built using the HO scale, where trains and tracks are built at $\frac{1}{87}$ of their actual size. If a train carriage has an actual length of 14 m, what will be its model size? Approximate to two decimal places.

Solution

Since the model trains are built at a ratio of 1 : 87 of their actual size, this ratio will be equal to the ratio of the model size compared to the actual size of any train engine or carriage:

$$\frac{1}{87} = \frac{t}{14}$$

For the fractions to be equal, their cross-products must be equal, so we have:

$$87 \times t = 1 \times 14$$
$$87 \times t = 14$$
$$t = \frac{14}{87}$$
$$t = 0.16 \text{ (2 d.p.)}$$

The length of the model train carriage is 0.16 m or 16 cm (2 d.p.).

PRACTICE 23: APPLICATION

At a particular time on a particular day, a person who is 1.65 m tall casts a shadow of 2.4 m. They are standing next to a tree that casts a shadow of 8.4 m. Using the fact that the height and shadow lengths of the person and the tree will be in proportion, work out the height of the tree.

EXAMPLE 24: APPLICATION

A jug of cordial is made up in a ratio of 1 : 8. If the jug contains 1 L of cordial, how much cordial syrup was used to make it? How much water?

Solution

This is an example that involves dilutions. When dealing with dilutions, a ratio of 1 : 8 means that the amount of cordial syrup relative to the whole amount of liquid in the diluted solution is in the ratio 1 : 8, as shown in the picture below. This means that a solution of cordial made up in the ratio 1 : 8 has 1 part cordial to 7 parts water.

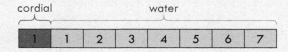

So, to work out the amount of cordial syrup used to make the jug of cordial using
1 L = 1000 mL, we use the following proportional equation:

$$\frac{1}{8} = \frac{c}{1000}$$

For the fractions to be equal, their cross-products must be equal, so we have:

$$8 \times c = 1 \times 1000$$
$$8 \times c = 1000$$
$$c = \frac{1000}{8}$$
$$c = 125 \text{ mL}$$

So, 125 mL of cordial was used to make the jug of 1000 mL of cordial.

The amount of water used was 1000 − 125 = 875 mL.

PRACTICE 24: APPLICATION

A jug of cordial is made up in a ratio of 2 : 15. If the jug contains 1 L of cordial, how much cordial syrup was used to make it? How much water?

EXAMPLE 25: APPLICATION

A jug of cordial is made up in a ratio of 3 : 20. If the jug contains 1.5 L of cordial, how much cordial syrup was used to make it? How much water?

Solution

Remembering that when working with dilutions, a ratio of 3 : 20 means that the amount of cordial syrup relative to the whole amount of liquid in the diluted solution is in the ratio 3 : 20, and converting 1.5 L to 1500 mL, we use the following proportional equation:

$$\frac{3}{20} = \frac{c}{1500}$$

For the fractions to be equal, their cross-products must be equal, so we have:

$$20 \times c = 3 \times 1500$$
$$20 \times c = 4500$$
$$c = \frac{4500}{20}$$
$$c = 225 \text{ mL}$$

So, 225 mL of cordial was used to make the jug of 1500 mL of cordial.

The amount of water used was 1500 − 225 = 1275 mL.

PRACTICE 25: APPLICATION

A jug of cordial is made up in a ratio of 3 : 22. If the jug contains 2 L of cordial, how much cordial syrup was used to make it? How much water?

7.4   CHAPTER SUMMARY

SKILL OR CONCEPT	DEFINITION OR DESCRIPTION	EXAMPLES
Ratio	A ratio expresses a comparison between two or more quantities which have the same units. It is often written using a colon ':'	• 1 : 2 is a ratio meaning '1 is to 2 …' • If a bowl contains 5 apples and 4 oranges, the ratio of apples to oranges is 5 : 4.
Writing ratios	A ratio can be written as: • two numbers separated by a colon • two numbers separated by the words 'is to' • a fraction • a decimal • a percentage.	The ratio 3 : 4 can be written as: • 3 : 4 • 3 is to 4 • $\frac{3}{4}$ • 0.75 • 75%
Converting between different ways of writing a ratio	The two numbers in a ratio can be written as a fraction by: • making the first number the numerator and the second number the denominator • then simplifying the fraction by using common factors • then converting to either a decimal or a percentage.	The ratio 14 : 16 can be written as the fraction $\frac{14}{16}$. This fraction simplifies to $\frac{7}{8}$ when both numbers are divided by 2. Doing the division indicated by the fraction gives $7 \div 8 = 0.875$ (decimal form) and converting 0.875 to a percentage gives 87.5%.
Rate	A rate is a special ratio that expresses a comparison between two quantities of different units and is often written using the term 'per'.	A car that travels 120 kilometres in 2 hours has a rate of 120 kilometres per 2 hours or, in simplest form, 60 kilometres per hour.
Unit Rate	A unit rate is a rate where the second number (the denominator, when the rate is written as a fraction) is one.	The rate 120 kilometres per 2 hours is equivalent to the unit rate 60 kilometres per hour.
Converting a rate to a unit rate	A unit rate can be obtained from a given, non-unit rate by either: • converting the given rate to a fraction with a denominator of 1 or • dividing the numbers in the given rate.	The rate 120 kilometres per 2 hours converts to the unit rate 60 kilometres per hour because $\frac{120 \text{ km}}{2 \text{ h}} = \frac{60 \text{ km}}{1 \text{ h}}$. The rate $9 for 8 hours converts to the unit rate $\frac{\$9}{8 \text{ h}} = \frac{\$1.125}{1 \text{ h}} = \$1.125$ per hour because $9 \div 8 = 1.125$.
Proportion	Being in proportion means two or more relationships can be expressed as ratios that are equal.	The ratio 4 : 8 is equal to the ratio 1 : 2 because $\frac{4}{8} = \frac{1}{2}$. The two ratios are said to be 'in proportion'.
Determining if ratios are in proportion	Two ratios are equal, and in proportion, if their cross-products are equal.	$\frac{30}{40} \overset{?}{\diagup\!\!\!\!\diagdown} \frac{3}{4}$ Cross-products are: $40 \times 3 = 120$ $30 \times 4 = 120$ The cross-products are equal, so ratios 30 : 40 and 3 : 4 are equal and in proportion.
Determining the unknown quantity in a proportion	The unknown quantity in a proportion can be found by using a symbol to represent the missing (unknown) quantity, setting the cross-products of the ratios to be equal, and solving the resulting expression for the unknown quantity.	$\frac{30}{40} \overset{?}{\diagup\!\!\!\!\diagdown} \frac{y}{4}$ Cross products are: $30 \times 4 = 120$ $40 \times y = 40y$ Equate the cross-products and solve the equation: $120 \div 40$ $y = 3$

Tiffany Winn

7.5   REVIEW QUESTIONS

A SKILLS

1   Express the following ratios in simplest terms.

   a 80 to 15   **b** 21 to 105   **c** 492 to 303   **d** 90 to 3   **e** 100 to 450

   f 7 to 63   **g** 12 to 144   **h** 777 to 111   **i** 6 to 39   **j** 8 to 108

   k 19 to 51   **l** 49 to 7   **m** 201 to 342   **n** 435 to 2   **o** 4 to 628

2   Express the following ratios as a fraction in simplest form, as a percentage (to one decimal place) and as a decimal (up to three decimal places as required).

   a 21 is to 24   **b** 285 is to 33   **c** 400 is to 80

   d 13 is to 39   **e** 14 is to 35   **f** 44 is to 121

3   Express the following rates in simplest form.

   a 200 emails in 30 days

   b 15 square metres of carpet for $1375

   c 2 litres of pumpkin soup makes 8 serves

   d 3 nurses for every 15 patients

   e 4 million hits on a website in 12 months

4   Express the following rates as unit rates, rounding to two decimal places if needed.

   a 268 heart beats in 5 minutes

   b 204 shots on basket but only 36 successful baskets

   c $9.69 for 3 kilograms of apples

   d 1208 kilojoules in 4 chocolate chip cookies

   e $96 for a 30 day gym membership

5   Determine whether the following ratios are in proportion.

   a 3 is to 6 and 2 is to 5   **b** 8 is to 1 and 4 is to $\frac{1}{2}$

   c 23 is to 69 and 26 is to 80   **d** 14 is to 36 and 21 is to 54

   e 36 is to 144 and 102 is to 406   **f** 10 is to 55 and 18 is to 99

6   Find the missing value x in the following proportions, rounding to two decimal places if needed.

   a 18 is to 15 as 6 is to x **b** 7 is to 2 as x is to 15   **c** 70 is to x as 21 is to 6

   d 9 is to 40 as 7.2 is to x **e** x is to 19 as 30 is to 15   **f** x is to 1.8 as 56 is to 48

B APPLICATIONS

1 Express the following ratios in simplest form.

a There are 45 males and 20 females in a university class.

b When shopping, I bought twice as many oranges as apples.

c Out of 500 handmade chocolates, 4 were substandard and put into the seconds basket.

d The price to earnings ratio of a particular stock was 84 : 6.

e Lilian typically takes 45 minutes and Jade 42 minutes to complete the same task.

2 Express the following rates in simplest form.

a A parachutist falls at a rate of 8 kilometres per 15 minutes.

b I walked an average of 22 kilometres per day on a four-day hike.

c A baseball player made 22 hits in 60 appearances at the plate.

d Parking costs $36 for 9 hours.

e A hotel charges $2015 for 5 nights accommodation.

3 Express the following rates as unit rates, rounding to two decimal places if needed.

a A plane covers the 655 kilometres from Adelaide to Melbourne in 50 minutes.

b It costs $53 to buy 50 litres of fuel.

c I scored 432 points towards my GPA in 32 units of study.

d 800 milligrams of medication needs to be infused every 120 minutes.

e I watched 480 minutes of television over 5 days.

4 What is the ratio of gold to the other metal in 18-karat gold?

Hint: Pure gold is 24-karat gold. In 18-karat gold, 18 out of 24 parts are gold and the other parts are another metal, which is added for hardness.

5 Elsa bought a block of chocolate that cost $4.80 for 80 grams. What is the unit price of the chocolate?

6 If a distance of 4 centimetres on a particular map corresponds to a distance of 22000 centimetres in real-life, what is the simplest form of the ratio that expresses the scale of the map?

7 If I bought a packet of 5 notebooks that cost $7.99, what is the unit price of the notebooks? Round your answer to two decimal places.

8 You need to make up a jerry can of two-stroke fuel for a lawnmower. This requires a 30 : 1 mix of petrol and oil. You have 5 litres of petrol. How much oil is required?

Tiffany Winn

9 In a class of 32 students, 15 are females. In another class of 38 students, 22 are males. In which class is the ratio of females to males the greatest?

10 A 4 cm by 6 cm photo (width 4 cm, height 6 cm) needs to be enlarged so that its height is 27 cm. What will the width of the enlarged photo need to be so that the dimensions of the photo remain in proportion?

11 In a local council election, 2480 people voted for the incumbent, 2014 for the opponent, and 216 cast a protest vote. Express the ratio of the incumbent's vote to the total number of votes, and the ratio of the opponent's vote to the total number of votes, in simplest form.

12 If approximately 1500 people can ride a roller coaster in an hour, and each ride takes 1.5 minutes, how many people are able to go on each ride?

13 If a bathtub containing 60 litres of water empties in 7 minutes, at what rate per minute does the water empty out of the bath?

14 A particular brand of cereal is sold in 250 g, 425 g and 700 g packets. The 250 g packet costs $2.60, the 425 g packet costs $4.60 and the 700 g packet costs $6.90. What is the unit price of cereal for each packet? Which packet is the best value?

15 A ship is 14 m long and 5 m tall. If I am making a model ship that is to the scale of the original ship and I want my model to be 24 cm tall, how long should my model be?

16 A medication tablet contains ibuprofen and codeine phosphate in the ratio 23 : 2. If the tablet contains a total of 200 milligrams of medication, how much of each substance (ibuprofen and codeine phosphate) are in one tablet?

17 AA batteries cost $2.68 for a packet of 4, $4.20 for a packet of 8 and $5.90 for a packet of 10. What is the unit price of the batteries in each case? Which packet is the cheapest per battery?

18 At a particular time on a particular day, a person who is 1.8 m tall casts a shadow of 2.4 m. They are standing next to a tree that casts a shadow of 10.2 m. Using the fact that the height and shadow lengths of the person and the tree will be in proportion, work out the height of the tree.

8

EXPONENTS AND LOGARITHMS

CHAPTER CONTENT

» Introduction to exponents

» Laws of exponents

» Negative and fractional exponents

» Scientific notation

» Introduction to logarithms

» Laws of logarithms

CHAPTER OBJECTIVES

» Write and evaluate expressions in exponential form

» Apply the laws of exponents

» Evaluate expressions involving negative and fractional exponents

» Write values using scientific notation

» Write and evaluate expressions in logarithmic form

» Apply the laws of logarithms

» Use exponential and logarithmic expressions in applied contexts

---

PAPER FOLDING A MOON SHOT

It is a commonly held assumption that it is not possible to fold a piece of paper more than seven times. This is probably true for an ordinary sheet of photocopy paper being folding by an average person. There have been, however, a number of recorded attempts that have resulted in up to 12 folds using very long, very thin paper and many helpers.

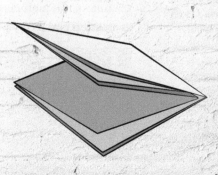

If, hypothetically, we could fold an ordinary piece of paper an unlimited number of times, how many times would we need to fold it to reach the Moon? If the distance from Earth to the Moon is approximately 384 000 kilometres, you would think that an enormous number of folds would be required.

So, let's start by thinking about a regular ream of paper that contains 500 sheets. The ream package is about 5 centimetres high, so we can calculate that each sheet will be $\frac{5}{100} = \frac{1}{100} = 0.01$ centimetres thick. When we start with an unfolded piece of paper, it is one page thick. So when it is folded once, it becomes 2 pages thick. If we fold the paper a second time, it will then be 4 pages thick. A third fold will make it 8 pages thick. Can you see what is happening here? Each fold doubles the thickness. A fourth fold will be 16 pages thick and a fifth fold will be 32 pages thick. At this point, the height of the folded paper will be $32 \times 0.01 = 0.32$ centimetres. After 9 folds, the height of the folded paper is more than the height of the original ream. After 20 folds, it is more than 10 kilometres high.

After 41 folds, we are over halfway to the Moon, and one more fold will double that again and see us on the Moon. So the number of times we need to fold a Moon shot is 42. Not probably as many as we might have expected! This is an example of the power of exponents. Very small values can become very large quickly because, as in this case, we compounded the doubling over and over.

8.1   INTRODUCTION TO EXPONENTS

Exponents provide a convenient abbreviated way to write repeated multiplication. Consider, for example, the following:

$$4 \times 4 \times 4 \times 4 \times 4 \times 4 \times 4 \times 4$$

Here we can see that 4 is multiplied by itself 8 times. Therefore, in this multiplication there are 8 factors, all of which are 4.

Exponents, also known as indices, provide us with a way of writing repeated multiplications using a shorter, simpler format, known as index notation. Using exponents, the above multiplication can be written in index notation as:

$4^8 \longleftarrow$ Exponent

Base

Here the number 4, which is being multiplied, is referred to as the **base**, and the number 8, the exponent, tells us how many times 4 is multiplied, that is the number of factors that will occur in the multiplication. The way that we would read this expression is '4 to the power of 8' or '4 raised to the power of 8'.

EXAMPLE 1

Write the following using index notation.

a $3 \times 3 \times 3 \times 3 \times 3$

Solutions

a   Here, 3 is the value being multiplied, which is the base. The number of times the multiplication is repeated is 5, which is the exponent.

$$3 \times 3 \times 3 \times 3 \times 3 = 3^5$$

b $\frac{1}{2} \times \frac{1}{2} \times \frac{1}{2} \times \frac{1}{2}$

b The base is not necessarily a whole number. In this example, the number being multiplied is a fraction. So, the base is $\frac{1}{2}$ and the number of factors is 4, which is the exponent.

$$\frac{1}{2} \times \frac{1}{2} \times \frac{1}{2} \times \frac{1}{2} = \left(\frac{1}{2}\right)^4$$

c $0.6 \times 0.6 \times 0.6$

c Again in this example, the number being multiplied is not a whole number, rather it is the decimal fraction 0.6, which will be the base. There are 3 factors, so the exponent will be 3.

$$0.6 \times 0.6 \times 0.6 = (0.6)^3$$

d $(-4) \times (-4) \times (-4) \times (-4)$

d The base is not necessarily a positive number. In this example, the number being multiplied is negative. So, the base is (-4) and the number of factors is 4, which will be the exponent.

$$(-4) \times (-4) \times (-4) \times (-4) = (-4)^4$$

PRACTICE 1

Write the following using index notation.

a $7 \times 7 \times 7 \times 7 \times 7 \times 7 \times 7$

b $\frac{2}{3} \times \frac{2}{3} \times \frac{2}{3}$

c $0.15 \times 0.15 \times 0.15 \times 0.15$

d $(-3) \times (-3) \times (-3) \times (-3) \times (-3)$

EXAMPLE 2

Determine the value of the following.

Solutions

a 6^4

a In this example, the base is 6, which is the number to be multiplied. The exponent is 4, which is the number of factors.

$$6^4 = 6 \times 6 \times 6 \times 6 = 1296$$

b $\left(\frac{1}{2}\right)^6$

b Here the base is $\frac{1}{2}$ and the exponent is 6.

$$\left(\frac{1}{2}\right)^6 = \frac{1}{2} \times \frac{1}{2} \times \frac{1}{2} \times \frac{1}{2} \times \frac{1}{2} \times \frac{1}{2} = \frac{1}{64}$$

c $(0.3)^3$

c In this example, the base is 0.3 and the exponent is 3.

$$(0.3)^3 = 0.3 \times 0.3 \times 0.3 = 0.027$$

d $(-2)^5$

d Here the base is (-2) and the exponent is 5.

$$(-2)^5 = (-2) \times (-2) \times (-2) \times (-2) \times (-2) = -32$$

Making connections
Revisit Chapter 4 to review the rules for multiplying negative numbers.

Kathy Brady

PRACTICE 2

Determine the value of the following.

a 5^3 b $\left(\dfrac{1}{3}\right)^4$ c $(1.2)^3$ d $(-3)^5$

Bases and exponents—special cases

Some special bases and exponents provide us with useful properties when handling index notation.

1 Let's consider 1 as a base. For example:

$$1^3 = 1 \times 1 \times 1 = 1 \qquad \text{or} \qquad 1^7 = 1 \times 1 \times 1 \times 1 \times 1 \times 1 \times 1 = 1$$

On the basis of these examples, we can see that whenever 1 is a base, no matter what the exponent, the resultant multiplication will always be equal to 1.

2 The other important special base is 10. The properties of this base are particularly important as they underpin our number system. For example:

$$10^2 = 10 \times 10 = 100 \qquad \text{or} \qquad 10^4 = 10 \times 10 \times 10 \times 10 = 10\,000$$

Using these examples, we can see that whenever the base is 10, the number of zeros in the resultant product is equal to the exponent on the base, 10.

3 There are also two special exponents. The first is an exponent of 1. Consider, for example, 5^1. If the exponent tells us how many factors there are in the resultant product, then $5^1 = 5$. Likewise:

$$9^1 = 9 \qquad \text{or} \qquad 14^1 = 14$$

In each case, we can see that when the number has an exponent of 1 it is equal to itself.

Any value raised to the power of 1 equals itself: $a^1 = a$

4 Finally, we will look at what happens when we have an exponent of 0. To consider the properties of this particular exponent let's look at a pattern:

$$10^4 = 10 \times 10 \times 10 \times 10 = 10\,000$$
$$10^3 = 10 \times 10 \times 10 = 1000$$
$$10^2 = 10 \times 10 = 100$$
$$10^1 = 10$$
$$10^0 = ?$$

In this pattern, we can see that when we move from 10^4 to 10^3 we divide by 10, then again when we move from 10^3 to 10^2 we divide by 10 again, and that this pattern continues at each step. So, if we were to divide by 10 at the last step moving from 10^1 to 10^0, we find that $10^0 = 1$. From this pattern, we can see that when we raise any non-zero value to the power of 0, the result will always be equal to 1.

Any non-zero value, raised to the power of 0 will always be equal to 1: $a^0 = 1$

CRITICAL THINKING

Use the powers of 2 to replicate this pattern to determine that $2^0 = 1$.

EXAMPLE 3

Determine the value of the following.

Solutions

a 1^{19}

a   In this example, the base is 1. By definition, when the base is 1, regardless of the exponent, the result will always be 1.

$$1^{19} = 1$$

b 10^{12}

b   When the base is 10, the resultant multiplication will have as many zeros as is the value of the exponent.

$$10^{12} = 1\,000\,000\,000$$

c $\left(\frac{1}{4}\right)^1$

c   The value of the base is not always a whole number, such as in this example where the base is a fraction. The same property will apply, in that any value raised to the power of 1 will be itself.

$$\left(\frac{1}{4}\right)^1 = \frac{1}{4}$$

d $(0.5)^0$

d   We have a decimal value as the base here, but again the same properties of exponents will apply. In this case, any non-zero value raised to the power of zero is equal to 1.

$$(0.5)^0 = 1$$

PRACTICE 3

Determine the value of the following.

a 1^{101} b 10^{10} c $(-3.765)^1$ d $\left(\frac{9}{13}\right)^0$

8.2   LAWS OF EXPONENTS

There are three laws that we are able to use to help us simplify expressions and examples that involve exponents.

Product rule for exponents

We will commence with what is known as the **product rule**.

Consider the product:

$$2^3 \times 2^5$$

This can be expanded to:

$$(2 \times 2 \times 2) \times (2 \times 2 \times 2 \times 2 \times 2)$$

In this expanded form, we have a repeated multiplication of 2 with 8 factors, which we have established can be written as 2^8. Therefore:

$$2^3 \times 2^5 = 2^8$$

Notice here that the sum of the two exponents on the LHS (3 + 5) is equal to the exponent on the RHS (8).

Let's consider another product that involves exponents: $5^2 \times 5^8$. In expanded form:

$$(5 \times 5) \times (5 \times 5 \times 5 \times 5 \times 5 \times 5 \times 5 \times 5)$$

Here we have a base of 5 with 10 factors in the repeated multiplication. Therefore:

$$5^2 \times 5^8 = 5^{10}$$

Notice again here that the sum of the two exponents on the LHS (2 + 8) is equal to the exponent on the RHS (10). These two examples illustrate the product rule for exponents. In general terms:

$$a^x \times a^y = a^{x+y}$$

EXAMPLE 4

Simplify the following using the product rule, if possible.

Solutions

a $3^4 \times 3^5$

a   We can use the product rule here because each of the values being multiplied has the same base.

$$3^4 \times 3^5 = 3^{4+5} = 3^9$$

b 4×4^6

b   The base in each of the values being multiplied is 4. Note that 4 can be written with an exponent as 4^1.

$$4^1 \times 4^6 = 4^{1+6} = 4^7$$

c $2^3 \times 2^6 \times 2^2$

c   The product rule can also be used in a sequence of steps when more than two terms are being multiplied.

$$2^3 \times 2^6 \times 2^2 = 2^3 \times 2^{6+2} = 2^3 \times 2^8 = 2^{3+8} = 2^{11}$$

d $5^2 \times 7^2$

d   In this example, the bases are not the same value, therefore we cannot use the product rule to simplify the expression.

PRACTICE 4

Simplify the following using the product rule, if possible.

a $5^3 \times 5^9$

b $(-3)^2 \times (-3)^7$

c $6^4 \times 6 \times 6^5$

d $4^3 \times (-4)^3$

EXAMPLE 5: APPLICATION

A computer's memory storage can be measured in terms of **bits**, **bytes** and **megabytes**. A bit is the smallest unit of data that a computer uses. A byte is equal to 2^3 bits. A megabyte is equal to 2^{20} bytes. How many bits are there in a megabyte? Write the answer as a power of 2.

Solution

If one byte is equal to 2^3 bits and there are 2^{20} bytes in a megabyte, then the number of bits in a megabyte will be $2^3 \times 2^{20} = 2^{23}$.

PRACTICE 5: APPLICATION

The hard drive storage in most computers is usually expressed in terms of gigabytes. A gigabyte is equal to 2^{10} megabytes. How many bytes are there in one gigabyte? Write the answer as a power of 2.

Kathy Brady

Quotient rule for exponents

The second law of exponents is the **quotient rule**. The quotient rule relates to dividing values in exponential form. Let's look at a couple of examples. Consider the division $\frac{2^5}{2^3}$. This can be expanded to:

$$\frac{2 \times 2 \times 2 \times 2 \times 2}{2 \times 2 \times 2}$$ Once the expression has been expanded, we can cancel 2s in both the numerator and the denominator.

This will leave $2 \times 2 = 2^2$ in the numerator. So, $\frac{2^5}{2^3} = 2^2$.

Notice here that the difference between the exponents on the LHS $(5 - 3)$ is equal to the exponent on the RHS (2).

Let's look at another example: $\frac{5^8}{5^2}$. We start by writing it in expanded form:

$$\frac{5 \times 5 \times 5 \times 5 \times 5 \times 5 \times 5 \times 5}{5 \times 5}$$ Again, once the expression has been expanded, we can cancel 5s in both the numerator and the denominator.

This will leave $5 \times 5 \times 5 \times 5 \times 5 \times 5 = 5^6$ in the numerator. So, $\frac{5^8}{5^2} = 5^6$.

Also notice here that the difference between the exponents on the LHS $(8 - 2)$ is equal to the exponent on the RHS (6). These two examples illustrate the quotient rule for exponents. In general terms:

$$\frac{a^x}{a^y} = a^{x-y}$$

EXAMPLE 6

Simplify the following using the quotient rule, if possible.

a $\frac{7^{10}}{7^6}$

b $\frac{4^8}{4}$

c $\frac{3^5}{6^5}$

Solutions

a   We can use the quotient rule here because each of values being divided has the same base.
$$\frac{7^{10}}{7^6} = 7^{10-6} = 7^4$$

b   The base in each of the values being multiplied is 4. Note here that 4 can be written with an exponent as 4^1.
$$\frac{4^8}{4} = 4^{8-1} = 4^7$$

c   In this example, the bases are not the same value, therefore we cannot use the quotient rule to simplify the expression.

PRACTICE 6

Simplify the following using the quotient rule, if possible.

a $\frac{8^9}{8^3}$ b $\frac{6}{6^0}$ c $\frac{2^2}{(-2)^2}$

Exponential expressions can also be simplified using a combination of both the product rule and the quotient rule.

EXAMPLE 7

Simplify the following using the product rule and the quotient rule.

Solutions

a $\dfrac{3^5 \times 3^7}{3^2 \times 3^4}$

a In this example, we need to firstly use the product rule to simplify both the numerator and the denominator:
$$\frac{3^5 \times 3^7}{3^2 \times 3^4} = \frac{3^{5+7}}{3^{2+4}} = \frac{3^{12}}{3^6}$$
Now we can complete the division using the quotient rule:
$$\frac{3^5 \times 3^7}{3^2 \times 3^4} = \frac{3^{5+7}}{3^{2+4}} = \frac{3^{12}}{3^6} = 3^{12-6} = 3^6$$

b $\dfrac{7^6 \times 7^5}{7^7 \times 7^3}$

b Firstly start with using the product rule to simplify the numerator and the denominator:
$$\frac{7^6 \times 7^5}{7^7 \times 7^3} = \frac{7^{6+5}}{7^{7+3}} = \frac{7^{11}}{7^{10}}$$
Now use the quotient rule:
$$\frac{7^6 \times 7^5}{7^7 \times 7^3} = \frac{7^{6+5}}{7^{7+3}} = \frac{7^{11}}{7^{10}} = 7^{11-10} = 7^1 = 7$$
Note that $7^1 = 7$, because as we noted in the previous section any value raised to the power of 1 is itself.

c $\dfrac{4^3 \times 4^4}{4^5 \times 4^2}$

c $\dfrac{4^3 \times 4^4}{4^5 \times 4^2} = \dfrac{4^{3+4}}{4^{5+2}} = \dfrac{4^7}{4^7} = 4^0 = 1$

PRACTICE 7: GUIDED

Simplify the following using the product rule and quotient rule: $\dfrac{4^6 \times 4^8}{4^3 \times 4^9}$

Now try these:

a $\dfrac{5^9 \times 5^2}{5^8 \times 5^6}$ **b** $\dfrac{2^{11} \times 2^4}{2^6 \times 2^8}$

$$\frac{4^6 \times 4^8}{4^3 \times 4^9} = \frac{4^{6+\square}}{\square^{3+9}} = \frac{\square^{14}}{4^{\square}} = \square^{14-12} = 4^{\square} = 16$$

Kathy Brady

Power rule for
exponents
When a value
in exponential
format is raised
to another
power the
exponent in
the resultant
expression
is found by
finding the
product of the
two original
exponents. In
general terms:
$(a^x)^y = a^{xy}$

Power rule for exponents

The third law of exponents is known as the **power rule**. Let's look at a couple of examples to see how the power rule works. Consider $(2^4)^3$; here the value 2^4 is being raised to a power of 3. This means that 2^4 is the base in this expression and there are 3 factors in the repeated multiplication:

$$2^4 \times 2^4 \times 2^4$$

This can now be expanded to:

$$(2 \times 2 \times 2) \times (2 \times 2 \times 2) \times (2 \times 2 \times 2)$$

In expanded form, we have a repeated multiplication of 2 with 12 factors, which can be written in exponential form as 2^{12}. Therefore:

$$(2^4)^3 = 2^{12}$$

Notice that the product of the exponents on the LHS (4×3) is equal to the exponent on the RHS (12).

Let's looks at another example: $(5^2)^4$. Here the value 5^2 is being raised to a power of 4. This means that 5^2 is the base in this expression and there are 4 factors in the repeated multiplication:

$$5^2 \times 5^2 \times 5^2 \times 5^2$$

This can now be expanded to:

$$(5 \times 5) \times (5 \times 5) \times (5 \times 5) \times (5 \times 5)$$

In expanded form, we have a repeated multiplication of 5 with 8 factors, which can be written in exponential form as 5^8. Therefore:

$$(5^2)^4 = 5^8$$

Notice that the product of the exponents on the LHS (2×4) is equal to the exponent on the RHS (8). These examples demonstrate the power rule for exponents. In general terms:

$$(a^x)^y = a^{xy}$$

EXAMPLE 8

Use the power rule to determine the unknown exponents in the following.

Solutions

a $(4^3)^5 = 4^\square$

a In this example, on the LHS 4^3 is raised to a power of 5. This means on the RHS the exponent will be the product $(3 \times 5) = 15$.

b $(7^4)^0 = 7^\square = \square$

b In this example, on the LHS 7^4 is raised to a power of 0. Therefore, on the RHS the exponent will be the product $(4 \times 0) = 0$. This means that the expression on the LHS will be 7^0, which is equal to 1 because any value raised to a power of 0 is always equal to 1.

c $(9^5)^\square = 9^{20}$

c In this example, we know that the resultant power on the RHS is 20. On the LHS we know one of the required exponents, is 5. This means that $(5 \times \square) = 20$. Therefore, the unknown exponent is 4.

d $(2^\square)^7 = 2^{21}$

d In this example, we can see that $(\square \times 7) = 21$. Therefore, the unknown exponent is 3.

PRACTICE 8

Use the power rule to determine the unknown exponents in the following.

a $(3^6)^3 = 3^\square$ **b** $(5^4)^\square = 5^{28}$ **c** $(2^\square)^7 = 1$

8.3 NEGATIVE AND FRACTIONAL EXPONENTS

So far we have only considered exponents that are whole positive numbers or zero. However, exponents can also be negative or a fraction. Each of these has a specific meaning and applications.

Negative exponents

To understand what a **negative exponent** means we need to revisit the quotient rule. Consider the division $\frac{2^3}{2^5}$. This can be expanded to:

$$\frac{2 \times 2 \times 2}{2 \times 2 \times 2 \times 2 \times 2}$$

Once the expression has been expanded, we can cancel 2s in both the numerator and the denominator. This leaves $2 \times 2 = 2^2$ in the denominator.

$$\frac{2^3}{2^5} = \frac{1}{2^2} \qquad \text{This is the reciprocal of } 2^2.$$

Now if we were to apply the quotient rule in this case, we would get $\frac{2^3}{2^5} = 2^{3-5} = 2^{-2}$.

So, since $\frac{2^3}{2^5} = \frac{1}{2^2}$ and $\frac{2^3}{2^5} = 2^{-2}$, we can say that $2^{-2} = \frac{1}{2^2}$.

So a negative exponent is equal to the reciprocal of the same exponential expression with a positive exponent. In general terms:

$$a^{-x} = \frac{1}{a^x}$$

Exponential expressions are considered to be fully simplified when they contain only positive exponents.

Negative exponent
A negative exponent in an exponential expression is equal to the reciprocal of the same expression with a positive exponent. In general terms:
$a^{-x} = \frac{1}{a^x}$

Making Connections
To review the meaning of a reciprocal revisit Chapter 5.

EXAMPLE 9

Fully simplify the following.

a 3^{-4}

b $(-7)^{-2}$

c 5^{-1}

d $2^5 \times 2^{-7}$

e $\dfrac{4^3}{4^6}$

f $(6^2)^{-3}$

Solutions

a This example has a negative exponent so to be fully simplified we need to write it with a positive exponent. There are two steps that are completed simultaneously. Write the expression as a reciprocal with a numerator of 1 and change the negative exponent to positive.

$$3^{-4} = \frac{1}{3^4}$$

b In this example, we have a negative base and a negative exponent. It is important to note that the sign on the base does not change. It will remain negative. We only need to ensure that the exponent is written as a positive.

$$(-7)^{-2} = \frac{1}{(-7)^2}$$

c With a positive exponent $5^{-1} = \frac{1}{5^1}$. As we have already noted, any value written to the power of 1 is itself.

$$5^{-1} = \frac{1}{5^1} = \frac{1}{5}$$

d In this example, we need to use the product rule first, so $2^5 \times 2^{-7} = 2^{5+(-7)} = 2^{-2}$. Now we need to fully simplify by writing the expression with a positive exponent.

$$2^5 \times 2^{-7} = 2^{5+(-7)} = 2^{-2} = \frac{1}{2^2}$$

e Here we need to apply the quotient rule first: $\frac{4^3}{4^6} = 4^{3-6} = 4^{-3}$. Now we need to fully simplify by writing the expression with a positive exponent.

$$\frac{4^3}{4^6} = 4^{3-6} = 4^{-3} = \frac{1}{4^3}$$

f In this example, we need to use the power rule first: $(6^2)^{-3} = 6^{2 \times (-3)} = 6^{-6}$. Now we need to fully simplify by writing the expressions with a positive exponent.

$$(6^2)^{-3} = 6^{2 \times (-3)} = 6^{-6} = \frac{1}{6^6}$$

PRACTICE 9

Fully simplify the following.

a 5^{-3}

b $(-4)^{-4}$

c 9^{-1}

d $6^4 \times 6^{-9}$

e $\dfrac{7^5}{7^8}$

f $(2^{-5})^3$

Fractional exponents

Here we will explore **fractional exponents**. To gain some understanding of what fractional exponents mean we will return to using the product rule.

We will start by considering a fractional exponent of $\frac{1}{2}$ with a base of 3. This is written as $3^{\frac{1}{2}}$. If we multiply $3^{\frac{1}{2}}$ by itself, we have $3^{\frac{1}{2}} \times 3^{\frac{1}{2}}$ and using the product rule to simplify this multiplication we get:

$$3^{\frac{1}{2}} \times 3^{\frac{1}{2}} = 3^{\frac{1}{2}+\frac{1}{2}} = 3^1 = 3$$

Now we also know that $\sqrt{3} \times \sqrt{3} = 3$.

Since $3^{\frac{1}{2}} \times 3^{\frac{1}{2}} = 3$ and $\sqrt{3} \times \sqrt{3} = 3$, we can form the conclusion that the fractional exponent $\frac{1}{2}$ has the same meaning as taking a square root of the base. In general terms:

$$a^{\frac{1}{2}} = \sqrt{a}$$

Square roots are not the only type of roots that can be taken when considering a particular number. We can also have what is known as a cube root. Let's use an example to explain further the idea of a cube root. We know that $2 \times 2 \times 2 = 8$ so we say that the cube root of 8 is 2. That is, the number that needs to be multiplied by itself *three* times to get a result of 8 is 2. We would write this as $\sqrt[3]{8} = 2$. In general terms: $\sqrt[3]{a}$

Now let's consider the fractional exponent $\frac{1}{3}$. This time we will use this exponent with a base of 5, which is written as $5^{\frac{1}{3}}$. If we multiply $5^{\frac{1}{3}}$ by itself 3 times, we get: $5^{\frac{1}{3}} \times 5^{\frac{1}{3}} \times 5^{\frac{1}{3}}$. Using the product rule to simplify this multiplication:

$$5^{\frac{1}{3}} \times 5^{\frac{1}{3}} \times 5^{\frac{1}{3}} = 5^{\frac{1}{3}+\frac{1}{3}+\frac{1}{3}} = 5^1 = 5$$

This provides an interesting outcome. What has happened is that when we multiplied $5^{\frac{1}{3}}$ by itself 3 times, the result returned us to the base number 5. This observation allows us to form the conclusion that the exponent $\frac{1}{3}$ acts as a cube root. In general terms:

$$a^{\frac{1}{3}} = \sqrt[3]{a}$$

EXAMPLE 10

Without using a calculator evaluate the following.

Solutions

a $36^{\frac{1}{2}}$

a The exponent of $\frac{1}{2}$ indicates that we need to find to square root of 36.

$$\sqrt{36} = 6$$

b $27^{\frac{1}{3}}$

b The exponent of $\frac{1}{3}$ indicates that we need to find the cube root of 27; that is, find a number that when multiplied by itself three times results in 27. This number is 3.

$$27^{\frac{1}{3}} = \sqrt[3]{27} = 3$$

Fractional exponent $\frac{1}{2}$
The fractional exponent $\frac{1}{2}$ represents the square root of the base. In general terms: $a^{\frac{1}{2}} = \sqrt{a}$

Making Connections
To review the concept of square roots revisit Chapter 3.

Fractional exponent $\frac{1}{3}$
The fractional exponent $\frac{1}{3}$ represents the cube root of the base. In general terms: $a^{\frac{1}{3}} = \sqrt[3]{a}$

c $64^{-\frac{1}{2}}$

c In this example, we have an exponent that is both negative and fractional. The best approach is to take two steps. The first step is to write the expression with a positive exponent:

$$64^{-\frac{1}{2}} = \frac{1}{64^{\frac{1}{2}}}$$

Now because the positive exponent is $\frac{1}{2}$ we know that we need to find the $\sqrt{64}$ in the denominator, which is 8.

$$64^{-\frac{1}{2}} = \frac{1}{64^{\frac{1}{2}}} = \frac{1}{\sqrt{64}} = \frac{1}{8}$$

d $125^{-\frac{1}{3}}$

d We take the same two sets steps in this example, starting with writing the expression with a positive exponent:

$$125^{-\frac{1}{3}} = \frac{1}{125^{\frac{1}{3}}}$$

Because the positive exponent is $\frac{1}{3}$ we need to find $\sqrt[3]{125}$. This is the value that when multiplied by itself three times results in 125. This number is 5.

$$125^{-\frac{1}{3}} = \frac{1}{125^{\frac{1}{3}}} = \frac{1}{\sqrt[3]{125}} = \frac{1}{5}$$

PRACTICE 10

Without using a calculator evaluate the following.

a $81^{\frac{1}{2}}$ **b** $64^{\frac{1}{3}}$ **c** $121^{-\frac{1}{2}}$ **d** $1000^{-\frac{1}{3}}$

EXAMPLE 11

Fully simplify the following expressions.

a $(4^6)^{\frac{1}{3}}$

b $5^{\frac{1}{2}} \times 5^{-1}$

Solutions

a In this example, we need to use the power rule.

$$(4^6)^{\frac{1}{3}} = 4^{6 \times \frac{1}{3}} = 4^2$$

b Here we need to use the product rule first.

$$5^{\frac{1}{2}} \times 5^{-1} = 5^{\frac{1}{2} + (-1)} = 5^{\frac{1}{2} - 1} = 5^{-\frac{1}{2}}$$

Now we need to write this expression with a positive exponent.

$$5^{\frac{1}{2}} \times 5^{-1} = 5^{\frac{1}{2} + (-1)} = 5^{\frac{1}{2} - 1} = 5^{-\frac{1}{2}} = \frac{1}{5^{\frac{1}{2}}}$$

Finally, to fully simplify this expression we need to rewrite the fractional exponent as a square root.

$$5^{\frac{1}{2}} \times 5^{-1} = 5^{\frac{1}{2} + (-1)} = 5^{\frac{1}{2} - 1} = 5^{-\frac{1}{2}} = \frac{1}{5^{\frac{1}{2}}} = \frac{1}{\sqrt{5}}$$

c $\dfrac{3^{-\frac{1}{2}}}{3^{\frac{1}{2}}}$

c In this example, we need to use the quotient rule first.

$$\frac{3^{-\frac{1}{2}}}{3^{\frac{1}{2}}} = 3^{-\frac{1}{2}-\frac{1}{2}} = 3^{-1}$$

Next we need to write the expression with a positive exponent.

$$\frac{3^{-\frac{1}{2}}}{3^{\frac{1}{2}}} = 3^{-\frac{1}{2}-\frac{1}{2}} = 3^{-1} = \frac{1}{3^{1}} = \frac{1}{3}$$

d $\left(2^{-\frac{1}{3}}\right)^{9}$

d Here we start by using the power rule.

$$\left(2^{-\frac{1}{3}}\right)^{9} = 2^{-\frac{1}{3} \times 9} = 2^{-3}$$

Now we finish by writing the expression with a positive exponent.

$$\left(2^{-\frac{1}{3}}\right)^{9} = 2^{-\frac{1}{3} \times 9} = 2^{-3} = \frac{1}{2^{3}}$$

PRACTICE 11

Fully simplify the following expressions.

a $\left(5^{\frac{1}{2}}\right)^{8}$ **b** $4^{-\frac{1}{2}} \times 4$ **c** $\dfrac{2^{\frac{1}{2}}}{2}$ **d** $\left(7^{12}\right)^{-\frac{1}{3}}$

EXAMPLE 12: APPLICATION

To calculate the current value of a second-hand car the following formula is used:

Current value = New value $(1.5)^{-\text{years old}}$

A particular car costs \$45 000 purchased new. What will the car be worth after one year?

Solution

First, we should substitute the given values into the formula:

Current value = $45000(1.5)^{-1}$

The value 1.5 is raised to the power of −1, so we need to write this part of the expression as a reciprocal with a numerator of 1 and change the negative exponent to positive:

Current value = $45000 \times \dfrac{1}{(1.5)^{1}} = \dfrac{45000}{1.5}$

Finally, we can calculate the current value:

Current value = $45000 \div 1.5 = \$30000$

PRACTICE 12: APPLICATION

The current value of another second-hand car can be calculated using this formula:

Current value = New value$(1.6)^{-\text{years old}}$

If the new value of a car is \$64 000, what is the value of the car after 2 years?

Kathy Brady

8.4   SCIENTIFIC NOTATION

Very large numbers, such as 467 000 000 000, or very small numbers, such as 0.000000000871, are very hard to say verbally. Additionally, when writing numbers such as these it would be easy to make a mistake by either leaving out or adding in zeros. Very large or very small numbers can, however, be written in an alternative shorter, simpler form using exponents and index notation. This way of writing numbers is known as scientific notation.

Numbers written using scientific notation have three components: the co-efficient, the base and the exponent. Here is an example of how these three components appear:

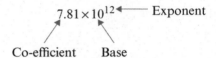

$$7.81 \times 10^{12} \longleftarrow \text{Exponent}$$

Co-efficient   Base

There are some limitations or rules that apply to the components of a number written in scientific notation:

- The co-efficient must be greater than or equal to 1 but less than 10.
- The base is always 10.
- The exponent is a positive or negative whole number.

The importance of the exponent in a number written in scientific notation is that it indicates how many places the decimal point in the co-efficient must be jumped when converting between standard number and scientific notation.

Converting from standard numbers to scientific notation

There are two steps involved in converting a standard number to scientific notation. In this first example, we will use a very large number.

1   The first step is to move the decimal point so that we have a number that is greater than or equal to 1 but less than 10. For example:

$$\underset{}{93\,000\,000_{|}}$$

In whole numbers, the decimal point is, by convention, not written, but if it were it would appear to the right of the last digit in the 'ones' place. So, in this example the decimal point must be moved until it sits between the 9 and the 3 to give a resultant number of 9.3, which will be the co-efficient.

2 Now the co-efficient must be multiplied by 10 raised to the power of the number of places the decimal point was moved. Notice that because the standard number is greater than 1, the decimal point has been moved *to the left* and the exponent will be *positive*. In this example, we moved 7 decimal places. So written in scientific notation:

$$93\,000\,000 = 9.3 \times 10^7$$

Note that if the original standard number is greater than or equal to 1, we will always move the decimal point to the left and the exponent will always be positive.

Now let's look at writing a very small number, that is a number less than 1, in scientific notation.

1 The first step is to move the decimal point so that we have a number that is greater than or equal to 1 but less than 10. In this example, the decimal point will need to sit between the 5 and the 1 to give a resultant number of 5.144, which will be the co-efficient.

$$0.00005144$$

2 Now the co-efficient must be multiplied by 10 raised to the power of the number of places the decimal point was moved. Notice that because the standard number is less than 1, the decimal point has been moved *to the right* and the exponent will be *negative*. In this example, we moved 5 decimal places. So written in scientific notation:

$$0.00005144 = 5.144 \times 10^{-5}$$

EXAMPLE 13

Write the following numbers using scientific notation.

a 6 289 000 000

Solutions

a Step 1: Create the co-efficient by moving the decimal point to the left to a position that creates a number greater than or equal to 1 but less than 10.

$$6\,289\,000\,000$$

Step 2: Multiply the co-efficient of 6.289 by 10 raised to the power of the number of places the decimal point has moved (9 places). Because we moved the decimal point to the left, the exponent will be positive.

$$6\,289\,000\,000 = 6.289 \times 10^9$$

b 100 800

b Step 1: Create the co-efficient by moving the decimal point to the left to a position that creates a number greater than or equal to 1 but less than 10.

$$1\underset{\blacktriangle}{00\,800}$$

Step 2: Multiply the co-efficient of 1.008 by 10 raised to the power of 5, which is the number of places the decimal point has moved to the left.

$$100\,800 = 1.008 \times 10^5$$

c 0.000000981

c Step 1: Create the co-efficient by moving the decimal point to the right to a position that creates a number that is greater than or equal to 1 but less than 10.

$$0.000000981$$

Step 2: Multiply the co-efficient of 9.81 by 10 raised to the power of the number of places the decimal point has moved (7 places). Because we moved the decimal point to the right, the exponent will be negative.

$$0.000000981 = 9.81 \times 10^{-7}$$

d 0.0001034

d Step 1: Create the co-efficient by moving the decimal point to the right to a position that creates a number that is greater than or equal to 1 but less than 10.

$$0.0001034$$

Step 2: Multiply the co-efficient of 1.034 by 10 raised to the power of -4, which is the number of places the decimal point has moved to the right.

$$0.0001034 = 1.034 \times 10^{-4}$$

PRACTICE 13

Write the following numbers using scientific notation.

a 17 518 000   **b** 0.0000039

c 1 000 200   **d** 0.006531

Converting from scientific notation to standard numbers

When converting from scientific notation to standard numbers we reverse the previous steps. If the exponent is positive, such as converting 3.45×10^4 to a standard number, the steps would be:

1 Examine the exponent to determine the direction to move the decimal point. In this case, the exponent is positive, so we know that the standard number will be greater than 1.

2 As the exponent is 4, we move the decimal point 4 places *to the right* (back to where it came from when the number in scientific notation was formed).

$$3.4500$$

$$3.45 \times 10^4 = 34\,500$$

Note again that by convention in this whole number we do not actually include the decimal point to the right of the digit in the 'ones' place.

Now let's look at another example where the exponent is negative, such as converting 6.73×10^{-7} to a standard number. The steps would be:

1 Examine the exponent to determine the direction to move the decimal point. In this case, the exponent is negative, so we know that the standard number will be less than 1.

2 As the exponent is -7, we move the decimal point 7 places *to the left* (back to where it came from when the number in scientific notation was formed).

$$0.0000006.73$$

$$6.73 \times 10^{-7} = 0.000000673$$

EXAMPLE 14

Write the following as a standard number.

a 4.091×10^7

Solutions

a Step 1: As the exponent is positive, we know that the number is greater than 1 and that we must move the decimal point *to the right*.

Step 2: As the exponent is 7, we move the decimal point 7 places to the right.

$$4.0910000$$

$$4.091 \times 10^7 = 40\,910\,000$$

---

b 1.0004×10^5

b   Step 1: As the exponent is positive, we know that the number is greater than 1 and that we must move the decimal point to the right.

Step 2: As the exponent is 5, we move the decimal point 5 places to the right.

$$1.00040\_$$

$$1.0004 \times 10^5 = 100\,040$$

c 7.6×10^{-6}

c   Step 1: As the exponent is negative, we know that the number is less than 1 and that we must move the decimal point to the left.

Step 2: As the exponent is -6, we move the decimal point 6 places to the left.

$$0.000007.6$$

$$7.6 \times 10^{-6} = 0.0000076$$

d 2.1058×10^{-4}

d   Step 1: As the exponent is negative, we know that the number is less than 1 and that we must move the decimal point to the left.

Step 2: As the exponent is -4, we move the decimal point 4 places to the left.

$$0.0002.1058$$

$$2.1058 \times 10^{-4} = 0.00021058$$

PRACTICE 14

Write the following as a standard number.

a 5.37×10^5

b 1.09×10^{-4}

c 3.7639×10^6

d 4.684×10^{-3}

EXAMPLE 15: APPLICATION

The mass of an electron is approximately 9.1094×10^{-28} grams. Write this as a standard number.

Solution

Because the exponent is -28, we move the decimal point 28 places to the left.

$$0.00000000000000000000000000009.1094$$

So, as a standard number the mass of an electron is:

0.00000000000000000000000000091094 grams

This example is an excellent illustration of the benefits of using scientific notation to accurately write very small numbers.

PRACTICE 15: APPLICATION

The mass of the Sun is 1.989×10^{30} kilograms. Write this as a standard number.

Calculations using scientific notation

We have seen how using scientific notation provides shorter and simpler ways of writing very large or very small numbers. The other advantage of using scientific notation is that through the use of the laws of indices carrying out calculations that involve multiplication or division are also simplified.

EXAMPLE 16

Evaluate the following, writing the answer in scientific notation.

a $4 \times 10^3 \times 6 \times 10^4$

Solutions

a In this example, one number written using scientific notation (4×10^3) is being multiplied by another number also written using scientific notation (6×10^4).

The first thing to do is rearrange the expression in the following way:

$$4 \times 10^3 \times 6 \times 10^4 = 4 \times 6 \times 10^3 \times 10^4$$

In doing this, we have grouped the numbers and the values written using index notation. This allows us to work with each group separately to do the multiplication:

$$4 \times 6 \times 10^3 \times 10^4 = 24 \times 10^{3+4} = 24 \times 10^7$$

The product rule has been used to multiply $10^3 \times 10^4$, where the exponents have been added. However, our answer 24×10^7 does not use correct scientific notation as the co-efficient (24) is greater than 10.

One more step is required to fix this.

$$24 \times 10^7 = 2.4 \times 10 \times 10^7$$

In this step, the value 24 has been rewritten as the product of 2.4 and 10, which creates a co-efficient that is greater then or equal to 1 but less than 10. As a result of this step, we now need to carry out one more application of the product rule to multiply 10×10^7 (note that the exponent on the 10 is 1, but by convention we do not usually write this).

So, $2.4 \times 10 \times 10^7 = 2.4 \times 10^8$ and, therefore:

$$4 \times 10^3 \times 6 \times 10^4 = 2.4 \times 10^8$$

b $3.1 \times 10^{-4} \times 7.2 \times 10^6$

b   This example is approached in the same way, by firstly rearranging the expression:

$$3.1 \times 10^{-4} \times 7 \times 10^6 = 3.1 \times 7.2 \times 10^{-4} \times 10^6$$

Now we multiply the decimal numbers and the values in index notation by applying the product rule:

$$3.1 \times 7.2 \times 10^{-4} \times 10^6 = 22.32 \times 10^{-4+6} = 22.32 \times 10^2$$

Since the co-efficient 22.32 is not a number that is greater than or equal to 1 but less than 10, we need to rewrite it as follows:

$$22.32 \times 10^2 = 2.232 \times 10 \times 10^2$$

And then carry out one more application of the product rule:

$$2.232 \times 10 \times 10^2 = 2.232 \times 10^{1+2} = 2.232 \times 10^3$$

So, $3.1 \times 10^{-4} \times 7.2 \times 10^6 = 2.232 \times 10^3$

PRACTICE 16: GUIDED

Evaluate $6.5 \times 10^{-6} \times 9.1 \times 10^6$ writing the answer in scientific notation.

$$6.5 \times 10^{-6} \times 9.1 \times 10^3 = 6.5 \times \square \times 10^{-6} \times 10^{\square}$$
$$= 59.15 \times 10^{-6+\square}$$
$$= 59.15 \times 10^{-3}$$
$$= 5.915 \times \square \times 10^{-3}$$
$$= \square \times 10^{1+(\square)}$$

Therefore, $6.5 \times 10^{-6} \times 9.1 \times 10^3 = 5.915 \times 10^{\square}$

Now try these:

a $4.7 \times 10^5 \times 3.8 \times 10^2$

b $3.5 \times 10^4 \times 6.6 \times 10^{-6}$

EXAMPLE 17: APPLICATION

A dam that is contaminated by *E. Coli* contains 2.4×10^6 litres of water. The recorded *E. Coli* levels in the dam are 8.1×10^3 per litre. How many *E. Coli* bacteria are in the dam?

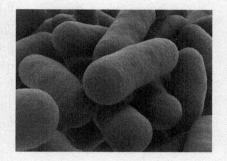

Solution

If there are 8.1×10^3 *E. Coli* bacteria per litre of water and 2.4×10^6 litres of water in the dam, then the total number of *E. Coli* will be the product $8.1 \times 10^3 \times 2.4 \times 10^6$.

$$8.1 \times 10^3 \times 2.4 \times 10^6 = 19.44 \times 10^{3+6}$$
$$= 19.44 \times 10^9$$
$$= 1.944 \times 10 \times 10^9$$
$$= 1.944 \times 10^{1+9}$$
$$= 1.944 \times 10^{10}$$

Therefore, there are 1.944×10^{10} *E. Coli* bacteria in the dam.

PRACTICE 17: APPLICATION

The mass of a proton is approximately 1.7×10^{-24} grams. One atom of iron has 26 protons. What is the mass of protons in 10^6 atoms of iron?

EXAMPLE 18

Evaluate the following writing the answer in scientific notation.

a $\dfrac{6 \times 10^3}{4 \times 10^4}$

Solutions

a   We approach these division examples in a similar way to multiplication, in that the first step is to separate the numbers from the values written in index notation in the following way:

$$\frac{6 \times 10^3}{4 \times 10^4} = \frac{6}{4} \times \frac{10^3}{10^4}$$

This allows us to work with each group separately to do the division:

$$\frac{6}{4} \times \frac{10^3}{10^4} = 1.5 \times 10^{3-4} = 1.5 \times 10^{-1}$$

The quotient rule has been used to divide $\frac{10^3}{10^4}$, where the exponents have been subtracted.

Since the co-efficient in 1.5×10^{-1} is a number greater than or equal to 1 and less than 10, this answer is already written in scientific notation and no further steps are required.

b $\dfrac{7.2 \times 10^{-4}}{3 \times 10^6}$

b   This example is approached in the same way, by firstly re-arranging the expression:

$$\frac{7.2 \times 10^{-4}}{3 \times 10^6} = \frac{7.2}{3} \times \frac{10^{-4}}{10^6}$$

Now we divide the numbers and the values in index notation by applying the quotient rule:

$$\frac{7.2}{3} \times \frac{10^{-4}}{10^6} = 2.4 \times 10^{-4-6} = 2.4 \times 10^{-10}$$

Since the co-efficient in 2.4×10^{-10} is a number greater than or equal to 1 and less than 10, this answer is already written in scientific notation and no further steps are required.

PRACTICE 18

Evaluate the following writing the answer in scientific notation.

a $\dfrac{8 \times 10^5}{6 \times 10^{-2}}$

b $\dfrac{9.3 \times 10^{-2}}{3.6 \times 10^4}$

EXAMPLE 19: APPLICATION

Mars is approximately 2.3×10^8 kilometres from the Sun. If the speed of light is approximately 3×10^5 kilometres per second, how long does it take for light to travel from the Sun to Mars?

Solution

In this example, we need to find how many lots of 3×10^5 kilometres there are in the distance 2.3×10^8 kilometres to determine the number of seconds it takes for the light to travel. Therefore, a division is required.

$$\frac{2.3 \times 10^8}{3 \times 10^5} = \frac{2.3}{3} \times \frac{10^8}{10^5} = 0.77 \times 10^3$$

This answer is not in scientific notation as the co-efficient 0.77 is less than 1. To achieve this, we need to move the decimal point one place to the left and then multiply by 10^{-1} as per the conversion rules above:

$$0.77 \times 10^3 = 7.7 \times 10^{-1} \times 10^3 = 7.7 \times 10^2$$

Therefore, it will take 7.7×10^2 or about 770 seconds for light to travel from the Sun to Mars.

A Petri dish contains 6.4×10^{10} bacterial organisms. If the recorded count of the organism is 8×10^8 per square centimetre, what is the area of the Petri dish?

8.5   INTRODUCTION TO LOGARITHMS

We are familiar with writing some numbers as power of 10 such as:

$$10^3 = 1000$$
$$10^2 = 100$$
$$10^1 = 10$$

We now know that $10^0 = 1$ and we can now also write decimal numbers as powers of 10 using negative exponents such as:

$$10^{-1} = \frac{1}{10^1} = \frac{1}{10} = 0.1$$
$$10^{-2} = \frac{1}{10^2} = \frac{1}{100} = 0.01$$
$$10^{-3} = \frac{1}{10^3} = \frac{1}{1000} = 0.001$$

Also numbers such as $\sqrt{10}$ and $\sqrt[3]{10}$ can be written as a power of 10 using fractional indices such as $10^{\frac{1}{2}}$ and $10^{\frac{1}{3}}$.

If fact, it is possible to write all numbers in the form 10^x using what is known as logarithms. **A logarithm of a positive number** is the power to which a base of 10 must be raised to produce that number.

In other words, according to this definition a logarithm is an exponent. Let's look at how this works with a couple of examples. For instance, using $10^3 = 1000$ we can say that the logarithm of 1000 in base 10 is 3, which is the power to which 10 has been raised. We write this as $\log 1000 = 3$. Looking at another example, if $10^{-2} = 0.01$, we say that the logarithm of 0.01 is -2, this is the power to which 10 has been raised (see the list above), and we write this as $\log 0.01 = -2$. In general, the logarithm of any positive number less than 1 will be negative.

Notice in these examples that if $\log 1000 = 3$, then we can rewrite this as $\log(10^3) = 3$, and if $\log 0.01 = -2$, then we can write this as $\log(10^{-2}) = -2$. We can generalise this relationship between logarithms and the power of 10 in the following way:

$$\log(10^a) = a$$

This relationship shows that when we raise the base of 10 to the power of any number, a, and then take the logarithm, the result is the exponent, a.

Kathy Brady

Logarithm of a positive number
The power to which a base of 10 must be raised to produce a positive number

EXAMPLE 20

Without using a calculator, find the value of the following.

Solutions

a log100

a Since $100 = 10^2$, we can rewrite this expression as $log10^2$. Using the relationship between logarithms and powers of 10, we can say that $log10^2 = 2$.

$$log100 = 2$$

b log100000

b If $100000 = 10^5$, then we can rewrite this expression as $log10^5$. We have established that $log10^5 = 5$.

$$log100000 = 5$$

c log1

c We have already noted that $10^0 = 1$ and so we can rewrite this expression as $log10^0$. We can then evaluate the expression as $log10^0 = 0$.

$$log1 = 0$$

d log0.001

d Since $0.001 = \frac{1}{1000} = \frac{1}{10^3} = 10^{-3}$, we can rewrite this expression as $log10^{-3}$ and evaluate it as $log10^{-3} = -3$.

$$log0.001 = -3$$

PRACTICE 20

Without using a calculator, find the value of the following.

a log1000   **b** log10   **c** log0.1   **d** log0.00001

So far we have only considered logarithms of numbers that are exact powers of 10. However, we can write any *positive number* as a power of 10. For example, we have already established that log100 = 2 and log1000 = 3, so therefore the log500 must be somewhere between 2 and 3. But how can we find it? This is where a scientific calculator with a LOG button comes in very handy. Using a calculator, we can see that log500 = 2.699, which is as we predicted between 2 and 3. What this means is when 10 is raised to the power of 2.699, the result is 500; that is, $10^{2.699} = 500$.

Let's use a calculator again to find log2, which is equal to 0.301; that is, $10^{0.301} = 2$. So, using the relationship that we have established between logarithms and exponents, we can write this as $log2 = log10^{0.301} = 0.301$.

EXAMPLE 21

Use a calculator to find log2 and write this as a power of 10.

Use this relationship to find the value for log20 and confirm the answer using a calculator.

Solution

Using a calculator, we get log2 = 0.301, so log2 = log$10^{0.301}$.

Now we need to use this relationship to find a value for log20. We know that $20 = 2 \times 10$, so this could be a good starting point:

log20 = log(2×10)

log20 = log$(10^{0.301} \times 10^1)$

log20 = log$(10^{0.301+1})$ Apply the product rule for indices in the brackets

log20 = log$10^{1.301}$

log20 = 1.301   Use the base relationship between logarithms and exponents

Using a calculator, we can confirm that log20 = 1.301.

PRACTICE 21: GUIDED

Use a calculator to find log3 and write this as a power of 10.

Use this relationship to find the value for log300 and confirm the answer using a calculator.

log3 = $\square$ so, log$\square$ = log$10^{0.477}$ and $300 = 3 \times \square$

log300 = log$(\square \times 100)$

log300 = log$(10^{0.477} \times 10^{\square})$

log$\square$ = log$(10^{\square + 2})$

log$\square$ = log$10^{\square}$

log300 = $\square$

Using a calculator, log300 = $\square$.

Now try this one:

Use a calculator to find log5 and write this as a power of 10.

Use this relationship to find the value for log0.5 and confirm the answer using a calculator.

8.6   LAWS OF LOGARITHMS

There are three laws of logarithms that can be used to simplify and manipulate logarithmic expressions.

Product rule for logarithms

Consider finding $\log 2 + \log 3$. We have already determined that $\log 2 = 0.301$ and $\log 3 = 0.477$; therefore, $\log 2 + \log 3 = 0.301 + 0.477 = 0.778$. Now let's consider $\log 6$. Using a calculator, we find that $\log 6 = 0.778$. Therefore, it looks like $\log 2 + \log 3 = \log 6$. What is interesting to note is that $2 \times 3 = 6$, so we could rewrite this logarithmic expression as $\log 2 + \log 3 = \log(2 \times 3)$.

Now let's look at $\log 5 + \log 7$. Using a calculator we find that $\log 5 = 0.699$ and $\log 7 = 0.845$; therefore, $\log 5 + \log 7 = 0.699 + 0.845 = 1.544$. Now let's find the logarithm of their product $5 \times 7 = 35$. Using a calculator, we find $\log 35$ is also 1.544, and so we have again the relationship that $\log 5 + \log 7 = \log(5 \times 7)$. These are two examples of the product rule for logarithms. In general terms, the **product rule for logarithms** is:

$$\log a + \log b = \log(ab)$$

EXAMPLE 22

Write the following as a single logarithm.

Solutions

a $\log 6 + \log 8$

   a $\log 6 + \log 8 = \log(6 \times 8) = \log 48$

b $1 + \log 9$

   b   In this example, we first need to write 1 as a logarithm: $\log 10^1 = 1$. Now we have:
$$\log 10 + \log 9 = \log(10 \times 9) = \log 90$$

c $\log 2 + \log 5 + \log 4$

   c   We need to do this example in two applications of the product rule.
$$\begin{aligned} \log 2 + \log 5 + \log 4 &= \log(2 \times 5) + \log 4 \\ &= \log 10 + \log 4 \\ &= \log(10 \times 4) \\ &= \log 40 \end{aligned}$$

d $\log\frac{2}{3} + \log 6 + \log 9$

   d   This example also needs to be carried out in two applications of the product rule.
$$\begin{aligned} \log\frac{2}{3} + \log 6 + \log 9 &= \log\left(\frac{2}{3} \times 6\right) + \log 9 \\ &= \log 4 + \log 9 \\ &= \log(4 \times 9) \\ &= \log 36 \end{aligned}$$

PRACTICE 22

Write the following as a single logarithm.

a $\log 7 + \log 3$

b $\log 8 + 2$

c $\log 3 + \log 4 + \log 5$

d $\log \frac{5}{4} + \log 12 + \log 2$

Quotient rule for logarithms

The next law of logarithms is the quotient rule. Here we will start by considering $\log 8 - \log 4$. Using a calculator, we find that $\log 8 = 0.903$ and $\log 4 = 0.602$; therefore, $\log 8 - \log 4 = 0.903 - 0.602 = 0.301$. This might be a familiar value as we have seen that $\log 2 = 0.301$. So, it looks like $\log 8 - \log 4 = \log 2$. The other thing to notice here is that $\frac{8}{4} = 2$, so we could rewrite this relationship as $\log 8 - \log 4 = \log\left(\frac{8}{4}\right)$.

Now let's look at $\log 15 - \log 5$. First, using a calculator we find $\log 15 = 1.176$ and $\log 5 = 0.699$, so $\log 15 - \log 5 = 1.176 - 0.699 = 0.477$. Now let's find the logarithm of their quotient $\frac{15}{5} = 3$. This should also be starting to look like a familiar value as we have seen that $\log 3 = 0.477$. So again, it seems that $\log 15 - \log 5 = \log\left(\frac{15}{5}\right)$. These are two examples of the quotient rule for logarithms. In general terms, the **quotient rule for logarithms** is:

$$\log a - \log b = \log\left(\tfrac{a}{b}\right)$$

Quotient rule for logarithms
When the logarithms of two numbers are subtracted, the resultant value is the logarithm of the quotient of the first number divided by the second number. In general form:
$\log a - \log b = \log\left(\tfrac{a}{b}\right)$

EXAMPLE 23

Write the following as a single logarithm.

a $\log 24 - \log 4$

b $2 - \log 5$

c $\log 2 + \log 6 - \log 3$

Solutions

a $\log 24 - \log 4 = \log\left(\frac{24}{4}\right) = \log 6$

b In this example, we first need to write 2 as a logarithm: $\log 10^2 = 2$. Now we have:

$\log 10^2 - \log 5 = \log 100 - \log 5$
$= \log\left(\frac{100}{5}\right)$
$= \log 20$

c This example needs to be carried out in two steps working left to right. In the first step, we need to use the product rule for logarithms, and then in the second step the quotient rule for logarithms.

$\log 2 + \log 6 - \log 3 = \log(2 \times 6) - \log 3$
$= \log 12 - \log 3$
$= \log\left(\frac{12}{3}\right)$
$= \log 4$

d $\log 72 - \log 9 - \log 4$

d Again in this example, we need to use two steps working left to right.

$$\log 72 - \log 9 - \log 4 = \log\left(\frac{72}{9}\right) - \log 4$$
$$= \log 8 - \log 4$$
$$= \log\left(\frac{8}{4}\right)$$
$$= \log 2$$

PRACTICE 23

Write the following as a single logarithm.

a $\log 40 - \log 5$

b $3 - \log 50$

c $\log\frac{4}{3} + \log 12 - \log 2$

d $\log 60 - \log 5 - \log 3$

EXAMPLE 24: APPLICATION

The Richter scale measures the magnitude, R, of an earthquake with intensity, I, and is calculated by the formula:

$$R = \log I$$

The 1960 Chilean earthquake recorded a magnitude of 9.5 and the 2011 earthquake off the coast of Japan recorded a magnitude of 9.0. What was the difference in intensity between these earthquakes?

Solution

We do not know the intensity of either of these earthquakes, so we could label the intensity of the Chilean earthquake I_C and the intensity of the Japanese earthquake I_J.

We do know the magnitude of each of these earthquakes and so can form two formulae as follows:

$$9.5 = \log I_C \text{ (Chile)}$$
$$9.0 = \log I_J \text{ (Japan)}$$

To find the difference in the magnitudes we need to subtract:

$$9.5 - 9.0 = \log I_C - \log I_J$$

On the right-hand side of this equation, we can apply the quotient rule for logarithms:

$$0.5 = \log\left(\frac{I_C}{I_J}\right)$$

Now using the fundamental relationship between exponents and logarithms we can say:

$$0.5 = \log 10^{0.5} = \log\left(\frac{I_C}{I_J}\right)$$

Therefore, $\frac{I_C}{I_J} = 10^{0.5}$.

Using a calculator, we find that $10^{0.5} = 3.16$ and therefore we can say that the Chilean earthquake was 3.16 times more intense than the Japanese earthquake.

PRACTICE 24: APPLICATION

The most damaging earthquake in Australia occurred in Newcastle, NSW, in 1989. This earthquake had a magnitude of 5.9. However, in Meekering, WA, an earthquake with a greater magnitude of 6.5 occurred in 1968. What was the difference in intensity of these earthquakes?

Power rule for logarithms

The third law of logarithms is the power rule.

Let's start with this by considering $2\log3$. If $\log3 = 0.477$, then $2\log3 = 2 \times 0.477 = 0.954$. Now let's look at a slightly different expression: $\log3^2$. In this expression, we could start with $3^2 = 9$, so we now have $\log9$. Using a calculator, we can find $\log3^2 = \log9 = 0.954$, which is equal to the value for $2\log3$. So, it looks like $2\log3 = \log3^2$.

Now let's look at $3\log5$. If $\log5 = 0.699$, then $3\log5 = 3 \times 0.699 = 2.097$. Now we will consider the alternative expression $\log5^3$. If $5^3 = 125$, then $\log5^3 = \log125 = 2.097$. So, $3\log5 = \log5^3$. These are two examples of the **power rule for logarithms**. In general terms, the power for logarithms is:

$$n\log a = \log a^n$$

EXAMPLE 25

Use the power rule for logarithms to determine the unknown values in these expressions.

a $5\log6 = \log6^{\square}$

b $\square\log4 = \log\square^3$

c $4\log\square = \log16$

Solutions

a In this example, on the LHS $\log6$ is multiplied by 5; therefore, 5 becomes the power on the RHS, which is the unknown value.

b On the LHS, we do not know the value by which $\log4$ is being multiplied. However, we can see on the RHS that the power is 3; therefore, the unknown value on the LHS must be 3. The second unknown value in this expression is on the RHS, which must be equal to 4 to reflect the LHS.

c In this example, we do not known the value of which the logarithm is being taken on the LHS. Also the power seems to be missing on the RHS. What we do know is the logarithm on the LHS is multiplied by 4, so this value needs to become the power on the RHS. This means we need to rewrite 16 as a value raised to the power of 4. This value must be 2 as $2 \times 2 \times 2 \times 2 = 2^4 = 16$. This means that the RHS is $\log2^4$, and to reflect this the unknown value on the LHS is 2.

> **Power rule for logarithms**
> When the logarithm of a value is multiplied by another value, the result is the logarithms of the first value raised to the power of the second value. In general form:
> $n\log a = \log a^n$

Kathy Brady

Use the power rule for logarithms to determine the unknown values in these expressions.

a $\square\log 3 = \log 3^8$ b $6\log\square = \log 5^\square$ c $3\log\square = \log 27$

We can combine all three laws of logarithms to simplify logarithmic expressions.

EXAMPLE 26

Write the following as a single logarithm.

a $5\log 2 - 2\log 2$

b $2\log 5 + 3\log 2$

c $\log 64 - 4\log 2 + \log 5$

Solutions

a   In this example, the first term in the expression is simplified using the power rule as $5\log 2 = \log 2^5 = \log 32$. The second term in the expression also simplifies using the power rule as $2\log 2 = \log 2^2 = \log 4$. So now the whole expression has become $\log 32 - \log 4$. Now we can use the quotient rule to simplify further:

$$\log 32 - \log 4 = \log\left(\frac{32}{4}\right) = \log 8$$

b   The first term here is simplifying using the power rule as $2\log 5 = \log 5^2 = \log 25$. The second term also simplifies using the power rule as $3\log 2 = \log 2^3 = \log 8$. Now the expression has become $\log 25 + \log 8$. Now we can use the product rule to complete the simplification:

$$\log 25 + \log 8 = \log(25 \times 8) = \log 200$$

c   We can carry out the simplification in this example working from left to right. So starting with $\log 64 - 4\log 2$, we need to use the power rule to simplify $4\log 2$ as $4\log 2 = \log 2^4 = \log 16$. Now we can apply the quotient rule:

$$\log 64 - \log 16 = \log\left(\frac{64}{16}\right) = \log 4$$

Now the expression has become $\log 64 - 4\log 2 + \log 5 = \log 4 + \log 5$. Finally, we use the product rule to write as a single logarithm:

$$\log 4 + \log 5 = \log(4 \times 5) = \log 20$$

d $\frac{1}{2}\log121 + 2\log3$

d The first term will be simplified using the power rule $\frac{1}{2}\log121 = \log121^{\frac{1}{2}}$. Previously in this chapter, we have seen that the exponent $\frac{1}{2}$ indicates that we need to take the square root of the base. Therefore, $\log121^{\frac{1}{2}} = \log\left(\sqrt{121}\right) = \log11$. Now the second term also needs to be simplified using the power rule $2\log3 = \log3^2 = \log9$. Now the expression has become:

$$\frac{1}{2}\log121 + 2\log3 = \log11 + \log9$$

Now we use the product rule to write as a single logarithm:

$$\log11 + \log9 = \log(11 \times 9) = \log99$$

e $2 + \frac{1}{3}\log27 - \log5$

e We carry out the simplification in this example by working from left to right. We start with $2 + \frac{1}{3}\log27$. The first step that needs to be taken is to write 2 as a logarithm: $2 = \log10^2 = \log100$. Now we have $\log100 + \frac{1}{3}\log27$. Next we need to apply the power rule so that $\frac{1}{3}\log27 = \log27^{\frac{1}{3}}$. We have seen earlier in this chapter that the exponent $\frac{1}{3}$ indicates that we need to take the cube root of the base. Therefore, $\log27^{\frac{1}{3}} = \log\left(\sqrt[3]{27}\right) = \log3$. This means that the first two terms have now become $2 + \frac{1}{3}\log27 = \log100 + \log3$. We can simplify these using the product rule as $\log100 + \log3 = \log(100 \times 3) = \log300$. Now the expression has become:

$$2 + \frac{1}{3}\log27 - \log5 = \log300 - \log5$$

Finally, we can use the quotient rule to write the expression as a single logarithm:

$$\log300 - \log5 = \log\frac{300}{5} = \log60$$

PRACTICE 26

Write the following as a single logarithm.

a $3\log4 - 4\log2$

b $3\log5 + 2\log2$

c $\log72 - 2\log3 + \log6$

d $\frac{1}{2}\log49 + \log6$

e $1 + \frac{1}{3}\log64 - 3\log2$

Kathy Brady

8.7   CHAPTER SUMMARY

SKILL OR CONCEPT	DEFINITION OR DESCRIPTION	EXAMPLES
Base	A value that is multiplied by itself.	Exponent Base $\rightarrow 4^8 \leftarrow$ Exponent
Exponent	A value that indicates how many times to carry out a repeat multiplication of the base.	$3 \times 3 \times 3 \times 3 = 3^4$ $0.2 \times 0.2 \times 0.2 \times 0.2 \times 0.2 = 0.2^5$
Base of 1	Whenever 1 is a base, no matter what the exponent, the resultant multiplication will always be equal to 1.	$1^{17} = 1$ $1^{-8} = 1$
Exponent of 1	Any value raised to the power of 1 equals itself.	$14^1 = 14$ $(5.6)^1 = 5.6$ $\left(\dfrac{3}{8}\right)^1 = \dfrac{3}{8}$
Exponent of 0	Any non-zero value, raised to the power of 0 will always be equal to 1.	$23^0 = 1$ $(9.2)^0 = 1$ $\left(\dfrac{4}{5}\right)^0 = 1$
Product rule for exponents	When two numbers with the *same base* are multiplied, the resultant product is obtained by adding their exponents. $a^x \times a^y = a^{x+y}$	$6^2 \times 6^5 = 6^{2+5} = 6^7$ $-4^3 \times -4^1 = -4^{3+1} = -4^4$
Quotient rule for exponents	When two numbers with the *same base* are divided, the resultant quotient is obtained by subtracting the exponent in the denominator from the exponent in the numerator. $\dfrac{a^x}{a^y} = a^{x-y}$	$\dfrac{8^7}{8^2} = 8^{7-2} = 8^5$ $\dfrac{0.2^6}{0.2^4} = 0.2^{6-4} = 0.2^2$
Power rule for exponents	When a value in exponential form is raised to another power, the exponent in the resultant expression is found by finding the product of the two original exponents. $(a^x)^y = a^{xy}$	$(8^3)^4 = 8^{12}$ $(-6^5)^2 = -6^{10}$
Negative exponents	A negative exponent is equal to the reciprocal of the same expression with a positive exponent. $a^{-x} = \dfrac{1}{a^x}$	$4^{-5} = \dfrac{1}{4^5}$ $(-3)^{-6} = \dfrac{1}{(-3)^6}$
Fractional exponent of $\dfrac{1}{2}$	The fractional exponent $\dfrac{1}{2}$ represents the square root of the base: $a^{\frac{1}{2}} = \sqrt{a}$	$25^{\frac{1}{2}} = \sqrt{25} = 5$
Cube root	The cube root of a number is a special value that when multiplied three times results in that number. Written as: $\sqrt[3]{a}$	$\sqrt[3]{64} = 4$
Fractional exponent of $\dfrac{1}{3}$	The fractional exponent $\dfrac{1}{3}$ represents the square root of the base: $a^{\frac{1}{3}} = \sqrt[3]{a}$	$8^{\frac{1}{3}} = \sqrt[3]{8} = 2$
Scientific notation	Numbers written using scientific notation have three components: the co-efficient, the base and the exponent.	$7.81 \times 10^{12} \leftarrow$ Exponent Co-efficient Base

SKILL OR CONCEPT	DEFINITION OR DESCRIPTION	EXAMPLES
Rules for scientific notation	The co-efficient must be greater than or equal to 1 but less than 10. The base is always 10. The exponent is a positive or a negative whole number.	
Converting standard numbers to scientific notation	With standard numbers, move the decimal point to where the resultant number is greater than or equal to 1 but less than 10. This is the co-efficient. The co-efficient is multiplied by 10 raised to the power of the number of decimal places moved. If the original standard number is greater than or equal to 1, move the decimal point to the left and the exponent will be positive. If the original standard number is less than 1, move the decimal point to the right and the exponent will be negative.	$460\,000$ $460\,000 = 46 \times 10^{5}$ 0.000000137 $0.000000137 = 1.37 \times 10^{-7}$
Converting scientific notation to standard numbers	Examine the exponent: If it is positive, the standard number will be greater than one and the decimal point will move *to the right*. If it is negative, the standard number will be less than one and the decimal point will move *to the left*. Move the decimal point the number of places indicated by the exponent.	6.910000 $6.91 \times 10^{6} = 6\,910\,000$ $0.00003.5$ $3.5 \times 10^{-5} = 0.000035$
Calculations using scientific notation	Group the numbers and the values written using index notation. Separately multiply or divide the numbers and the terms in index notation. Use the product rule or the quotient rule for exponents when handling the terms in index notation. Adjust the result to create the correct format for scientific notation.	$5.6 \times 10^{-5} \times 7.1 \times 10^{8}$ $= 5.6 \times 7.1 \times 10^{-5} \times 10^{8}$ $= 39.76 \times 10^{-5+8}$ $= 39.76 \times 10^{3}$ $= 3.976 \times 10^{1} \times 10^{3}$ $= 3.976 \times 10^{4}$ $\dfrac{5.2 \times 10^{-3}}{4.9 \times 10^{7}}$ $= \dfrac{5.2}{4.9} \times \dfrac{10^{-3}}{10^{7}}$ $= 1.06 \times 10^{-3-7}$ $= 1.06 \times 10^{-10}$
Logarithm	The logarithm of any *positive number* is the power to which a base of 10 must be raised to produce that number.	$\log 100 = 2$ $\log 2 = 0.301$
Relationship between exponents and logarithms	When we raise a base of 10 to the power of any number, a, and then take the logarithm, the result is the exponent, a. $\log(10^{a}) = a$	$\log 10^{2} = 2$ $\log 10^{0.301} = 0.301$
Product rule for logarithms	When logarithms of two numbers are added, the result is the logarithm of the product of these numbers: $\log a + \log b = \log(ab)$	$\log 7 + \log 9$ $= \log(7 \times 9)$ $= \log 63$

Kathy Brady

SKILL OR CONCEPT	DEFINITION OR DESCRIPTION	EXAMPLES
Quotient rule for logarithms	When logarithms of two numbers are subtracted, the result is the logarithm of the quotient of the first number divided by the second number: $\log a - \log b = \log \frac{a}{b}$	$\log 60 - \log 5$ $= \log\left(\frac{60}{5}\right)$ $= \log 12$
Power rule for logarithms	When a logarithm of a value is multiplied by another value, the result is the logarithm of the first value raised to the power of the second value: $n\log a = \log a^n$	$6\log 2 = \log 2^6 = \log 64$

8.8   REVIEW QUESTIONS

A SKILLS

1   Write using index notation.

   a $6 \times 6 \times 6 \times 6 \times 6 \times 6$ **b** $\frac{1}{3} \times \frac{1}{3} \times \frac{1}{3} \times \frac{1}{3}$

   c $(-0.75) \times (-0.75) \times (-0.75)$

2   Determine the value of the following.

   a 4^4 **b** $\left(\frac{2}{3}\right)^3$ **c** $(-1.5)^3$

3   Simplify the following using the laws of exponents if possible.

   a $4^3 \times 4^6$ **b** $(-7)^5 \times (-7)$ **c** $\dfrac{(-3)^5}{(-3)^2}$

   d $\dfrac{11^2}{7^2}$ **e** $(8^3)^2$ **f** $((-5)^4)^0$

4   Write the following with positive exponents.

   a 6^{-2} **b** $7^3 \times 7^{-5}$ **c** $\dfrac{5^4}{5^8}$ **d** $(4^4)^{-2}$

5   Without using a calculator, evaluate the following.

   a $49^{\frac{1}{2}}$ **b** $1000^{\frac{1}{3}}$ **c** $144^{-\frac{1}{2}}$ **d** $64^{-\frac{1}{3}}$

6   Fully simplify the following, writing the answer with a positive exponent.

   a $\left(6^{-\frac{1}{2}}\right)^8$ **b** $7^{\frac{1}{2}} \times 7^{-1}$ **c** $\dfrac{3^{-\frac{1}{3}}}{3^{\frac{2}{3}}}$

7   Write using scientific notation.

   a $371\,000\,000$ **b** 0.00000854

8   Write as a number in standard form.

   a 4.81×10^5 **b** 1.095×10^{-6}

9   Evaluate the following writing the answer in scientific notation.

   a $(7.3 \times 10^4) \times (5.8 \times 10^{-7})$ **b** $(2.9 \times 10^{-5}) \times (8.1 \times 10^8)$

   c $\dfrac{8.4 \times 10^6}{1.6 \times 10^3}$ **d** $\dfrac{2.4 \times 10^4}{6.4 \times 10^{-2}}$

10 Without using a calculator, find the value of the following.

   a $\log 10000$ **b** $\log 1$ **c** $\log 0.000001$

11 Use a calculator to find $\log 7$ and write this as a power of 10.

   Use this relationship to find the value for $\log 0.07$ and confirm the answer using a calculator.

12 Write as a single logarithm.

    a $\log 7 + \log 3$ b $3 + \log 4$

    c $\log\frac{3}{4} + \log 8 + \log\frac{1}{6}$ d $\log 63 - \log 7$

    e $2 - \log 4$ f $\log 108 - \log 9 - \log 4$

    g $4\log 5$ h $\frac{1}{2}\log 81$

13 Write as a single logarithm.

    a $2\log 8 - 3\log 2$ b $3\log 2 + 2\log 3$

    c $\log 144 - 4\log 2 + \log 5$ d $\frac{1}{2}\log 81 + \log 6 - 2\log 3$

    e $2 - 2\log 5 + \frac{1}{3}\log 125$

B APPLICATIONS

1   The diameter of an atom varies according to its size, but can be approximated to 10^{-8} centimetres.

    a Given this measurement, how many atoms would need to be lined up to measure a length of 1 centimetre?

    b If the volume of a cube, with side length L, is calculated using the formula $V = L^3$. How many atoms would be enclosed in a cube with side length 1 centimetre? Write the answer as a power of 10.

2   Kepler's Third Law describes the motion of planets around the Sun and is given by the formula:

$$P^2 = d^3$$

where P is the period of the planet (the time taken for the planet to orbit the Sun) and d is the distance of the planet from the Sun (regardless of the units).

    a Show that by raising each side of this equation to the power of $\frac{1}{2}$ this formula can be written as $P = d^{\frac{3}{2}}$.

    b Now use the product rule for exponents to show that this re-arrangement can be written as $P = d \times d^{\frac{1}{2}}$.

    c Use this new arrangement to find the value of P when $d = 400$.

    d Using a similar approach to the steps taken in parts (a) and (b), show that the formula can also be written as $d = P \times P^{-\frac{1}{3}}$.

    e Use this arrangement to find the value of d when $P = 125$.

3 In 2006, the National Aeronautics and Space Administration (NASA) launched the New Horizons spacecraft for an exploration mission to Pluto. This spacecraft travelled at a speed of 58 000 km/hour and the distance it travelled to complete the fly-by of Pluto was 4.76 billion kilometres. (Source: www.nasa.gov)

a If the spacecraft travelled at 58 000 km/hour, how many kilometres did it travel in one day? Write this value in scientific notation.

b Write the distance that New Horizons travelled to Pluto in scientific notation.

c Using the values written in scientific notation, divide the total distance travelled to Pluto by the distance travelled in one day to determine how many days the spacecraft took to reach Pluto. Convert the number of days to years (to one decimal place).

d If the spacecraft was launched in July 2006, in the month of which year did it fly-by Pluto?

4 Coulomb's law calculates the magnitude of electric forces between two charged points. The formula for Coulomb's law is:

$$F = \frac{(9 \times 10^9) \times Q_1 \times Q_2}{r^2}$$

where Q_1 and Q_2 are the charges on the two points and r is the distance between the points.

a Substitute the values $Q_1 = 4 \times 10^{-9}$, $Q_2 = 3 \times 10^{-9}$ and $r = 2$ into the formula for F.

b Show that the numerator in the formula will be equal to 1.08×10^{-7} once the multiplications have been carried out.

c Show that $F = 2.7 \times 10^{-8}$ once the division by the denominator has been carried out.

5 A decibel (dB) is a measure that describes the magnitude, or loudness, of a sound. The formula used to calculate decibels is:

$$dB = 10 \times \log\frac{I}{10^{-12}}$$

where I is the intensity of the sound.

a The intensity of the spoken human voice is $I = 10^{-6}$. Substitute this into the formula for decibels and then use the quotient rule for exponents to simplify the RHS.

b Now, use the fundamental relationship between exponents and logarithms to complete the calculation to determine the number of decibels in the human voice.

c The intensity of sound in the front row of a rock concert is $I = 10^{-1}$. Repeat the steps in part (a) and part (b) to calculate the decibel measurement.

d It is also possible to find the difference in the intensities of two sounds using the following formula:

$$dB_2 - dB_1 = 10\log\left(\frac{I_2}{I_1}\right)$$

A vacuum cleaner measures 80 dB (dB_2) and a whisper measures 20 dB (dB_1). What is the difference in the intensities of these sounds? (Hint: This question can be solved in much the same way as finding the difference in the intensities of earthquakes in Example 24).

Kathy Brady

6 In chemistry, the pH of a solution is a measure of its acidity or alkalinity. The pH values are measured on a scale from 0–14. Values less than 7 are acidic and values greater than 7 are alkaline. If a particular solution has a pH of 7, then it is regarded as neutral. The pH of a solution is calculated using the formula:

$$pH = -\log[\,H^+]$$

For the purposes of this question we will not be concerned with the technical meaning of $[H^+]$. Rather we will concentrate on using the formula and interpreting the results.

a The value of $[H^+]$ of a particular solution $[H^+] = 10^{-4}$. By substituting this value into the formula and applying the power law for logarithms, calculate the pH of this solution? Is the solution acidic or alkaline?

b The value of $[H^+]$ of another solution is $[H^+] = 10^{-12}$. What is the pH of the solution? Is it acidic or alkaline?

9
ALGEBRA FUNDAMENTALS

CHAPTER CONTENT

» What is algebra?
» Algebraic expressions
» Evaluating algebraic expressions
» Like terms

» Simplifying algebraic expressions by collecting like terms
» Simplifying algebraic expressions by expanding brackets

CHAPTER OBJECTIVES

» Generalise relationships using algebraic expressions
» Write algebraic expressions using correct notation and conventions
» Translate words into algebraic expressions
» Evaluate algebraic expressions using substitution

» Recognise like terms in algebraic expressions
» Simplify algebraic expressions by collecting like terms
» Simplify algebraic expressions by expanding brackets using the distributive law

RACETRACK RELATIONSHIPS

The Bathurst 1000 is a 1000-kilometre touring car race held annually on the Mount Panorama Circuit in Bathurst, New South Wales. In the 2015, the event winner was Craig Lowndes (with his partner Steve Richards). Lowndes average speed over the event was 159.5 kilometres per hour. How could we use this information, together with algebra, to calculate how long it took him to complete the course? (Source: v8supercars.com.au)

A special relationship exists between the variables: distance, speed and time. This relationship is commonly written using the algebraic formula:

$$\text{speed} = \frac{\text{distance}}{\text{time}}$$

This formula can also be rearranged to show alternative relationships between speed, distance and time, such as:

$$\text{time} = \frac{\text{distance}}{\text{speed}}$$

Using this arrangement, we can now calculate how long Lowndes took to complete the course. Given that we know that the distance is 1000 kilometres and that his speed was 159.5 kilometres per hour, we can now calculate his time as: $\frac{1000}{159.9}$, which is 6.25 hours, or 6 hours and 15 minutes.

9.1   WHAT IS ALGEBRA?

Algebra is a mathematical language that translates ordinary words into mathematical expressions by using letters and symbols to represent values. Algebra is an area of mathematics that is interested in relationships, and algebra is used to make general statements about these relationships. As a result, algebra has many practical applications that enable a wide range of problems and situations to be solved efficiently.

EXAMPLE 1

What is the relationship between the number of tea towels and the number of pegs? What is the maximum number of tea towels that can be hung using 45 pegs?

Solution

We can see that each tea towel requires two pegs. This means that the number of pegs required will always be double the number of tea towels. This could be written as the relationship:

$$\text{pegs} = 2 \times \text{tea towels}$$

If the number of pegs is always twice the number of tea towels, then an even number of pegs will be required to hang out any given number of tea towels. This means that if we have 45 pegs it will only be possible to use up to 44 of them to hang out tea towels. Therefore, if the number of pegs is double the number of tea towels, the maximum number of tea towels that can be hung with 45 pegs is 22.

PRACTICE 1

What is the relationship between the number of T-shirts in this picture and the number of pegs required to hang them? Write this relationship as a mathematical sentence. What is the maximum number of T-shirts that could be hung with 17 pegs?

EXAMPLE 2

Welcome to Wheel Wonderland, a shop that sells bicycles and tricycles. On the day that you visit Wheel Wonderland, you notice that there are 61 wheels in the shop. If there are 8 bicycles, how many tricycles must there be in the shop?

Solution

The 8 bicycles in the shop will have a total of 16 wheels. This means that the remaining wheels are all on tricycles, and the number of remaining wheels is $61 - 16 = 45$. If each tricycle has 3 wheels, we can write the relationship:

$$3 \times \text{number of tricycles} = 45$$

To find the number of tricycles we need to divide the number of wheels by 3. This means that there are 15 tricycles in the shop.

PRACTICE 2

Next to Wheel Wonderland is another store, Legs Galore. This store sells three-legged bar stools and four-legged coffee tables. When you pop into Legs Galore, you notice that there are a total of 59 legs on the furniture items. If there are 9 bar stools, how many coffee tables must there be in the store?

Kathy Brady

EXAMPLE 3

The next time you visit Wheel Wonderland you find that they now sell bicycles and go-carts. The bicycles and go-carts each have one seat, and you count that there are 21 seats and 54 wheels. How many bicycles and go-carts are in the shop?

Solution

Because there are 21 seats altogether the total of bicycles and go-carts must be equal to 21. This could be written as:

$$\text{bicycles} + \text{go-carts} = 21$$

Now each bicycle has 2 wheels and each go-cart has 4 wheels. So, if we multiply the number of bicycles in the store by 2, and the number of go-carts in the store by 4, the total will be 54 wheels. This could be written as:

$$2 \times \text{bicycles} + 4 \times \text{go-carts} = 54$$

But how does this help to work out how many bicycles and go-carts are in the shop? Well, what we have done is to establish some relationships that might be useful. For instance, what if we guessed that there were 10 bicycles? This would mean that there would need to be 11 go-carts, as the total of bicycles and go-carts must always be 21. Now the 10 bicycles will have 20 wheels and the 11 go-carts will have 44 wheels. This will give a total of 64 wheels. Unfortunately, this is not a good guess because we need to get a total of 54 wheels. But it does give us a hint as to what our next try might be. It appears that we need to reduce the number of go-carts to bring down the total number of wheels.

Let's try 8 go-carts, which means there must be 13 bicycles. There will then be 32 go-carts wheels and 26 bicycles wheels, a total of 58. Still too many! So let's try 6 go-carts. Therefore, there will be 15 bicycles and the total number of wheels will be 24 on go-carts and 30 on bicycles. This gives 54 wheels, which exactly what we want!

PRACTICE 3

When you drop into your local coffee shop there are two people ahead of you at the counter. One of them orders one coffee and 2 cakes and pays $23; the other orders 5 coffees and 1 cake and pays $34. You just want a coffee and a cake. How much will you pay?

In each of these questions the relationships between two or more quantities or values gave the clues to the solution. But to get these solutions we did not use very efficient techniques. What we needed to use was algebra, which is the ideal tool to apply relationships in an efficient way to find solutions to simple or complex situations.

9.2   ALGEBRAIC EXPRESSIONS

In the same block of shops as Wheel Wonderland and Legs Galore is Mandy's Market. Mandy sells fruit and vegetables and she displays her prices on the board at the front of the store:

Apples $1
Bananas $2
Mangoes $3
Tomatoes 50c
Carrots 20c
Potatoes 10c

Looking at Mandy's price list we could write a mathematical expression to represent the cost of an apple and a banana as apple + banana = 3. And the mathematical expression to represent the cost of a mango and a potato would be mango + potato = 3.10.

But these mathematical expressions are fairly wordy, and mathematicians are by nature lazy! Why write a word when you could write just one letter? For example, the expression for the cost of an apple and a banana could be abbreviated to $a + b = 3$.

Now let's think about the cost of 3 tomatoes. We could write the expression for this as $t + t + t$, where the price of a tomato is represented by the letter t. But even this expression is too long for lazy mathematicians! Think of $t + t + t$ as 3 lots of t, or t times by 3 and so in algebra we shorten $t + t + t$ to $3t$. Notice that in the expression $3t$ the multiplication sign has been omitted, as it is implied, and we have written $3t$ (for 3 tomatoes), not $t3$. The reason that the multiplication sign is omitted is that is could be confused with the letter x, which is quite often used in algebra.

Kathy Brady

Variables and constants

In the **algebraic expression** $3t$, the t is known as a **variable**. This is because the price of Mandy's tomatoes may change. The number 3, on the other hand, is known as a **constant**. This is because the value of 3 will always remain the same.

A convention in mathematics is an agreed way of doing something. In algebra, there are a number of conventions that are associated using variables and constants when writing products. These conventions are to always write the constant before the variable in a product, and for a product involving more than one variable always write the variables in alphabetic order.

Algebraic expression
A mathematical sentence that combines variables, constants and arithmetic operations

Variable
A letter that is used to represent an unknown value or a value that can change

Constant
A value that is known and does not change

Making connections
To revise squared numbers and the notation for squaring, check Chapter 3.

Making connections
Check Chapter 5 to revise the representations of fractions, including the concept of fractions as division.

EXAMPLE 4

Write the following expressions using correct algebraic conventions.

Solutions

a $a \times 3$

    a This is a product of a variable and a constant, so we omit the multiplication sign. The multiplication is implied by the absence of a sign. The convention requires that the constant 3 be written before the variable a. So, the expression is written as $3a$.

b $n \times m \times p$

    b This is a product of more than one variable. The multiplication sign is omitted and the variables are written in alphabetic order as mnp.

c $y \times y$

    c In this product, y is being multiplied by itself or squared. The notation that is used for squaring is a power of 2. So, the expression is written as y^2.

d $c \div d$

    d Another convention adopted in algebra involves writing divisions. The division sign $\div$ is not used, rather the algebraic notation for division draws upon the concept of fractions as a way to represent division. Therefore, this division is written as $\frac{c}{d}$.

e $(a + b) \div c$

    e The BEDMAS order of operations still applies in algebraic expressions. This means that in this expression the whole of the content of the brackets $a + b$ is being divided by c. Therefore, this expression is written as $\frac{a + b}{c}$.

Translating words to algebraic expressions

So far, we have practised writing algebraic expressions using the correct conventions. Now we are going to look at how to translate words into algebraic expression. This skill is important to be able to work on worded questions or in real-life applications. When translating words into algebraic expressions, it is important to identify the key words or phrases that describe particular mathematical operations. The following table provides a range of examples.

OPERATION	KEY WORD/PHRASE
Addition	• plus • more than • the sum of • increased by • added to
Subtraction	• minus • less than • the difference of • less • decreased by • subtracted from
Multiplication	• times • the product of • multiplied by • of
Division	• the quotient of • divided by
Powers	• the square of; squared • the cube of; cubed

EXAMPLE 5

Translate the following sets of words into algebraic expressions.

a 16 more than m

b The product of b and 5

Solutions

By checking against the table above, the required mathematical operation can be identified to give the following solutions.

a $m + 16$

b $5b$. Notice the correct use of the algebraic convention to record the constant before the variable in a product.

c *a* decreased by 4	**c** *a* −4
d *n* divided by 10	**d** $\frac{n}{10}$. Notice the correct use of the algebraic notation for division.
e Double *g*	**e** 2*g*. The word double means to multiply by two.
f *a* squared	**f** a^2

PRACTICE 5

Translate the following sets of words into algebraic expressions.

a The sum of *p* and *q*   **b** *x* times *y*   **c** 3 less than *n*

d Triple *y*   **e** The cube of *g*

EXAMPLE 6: APPLICATION

If *w* lotto winners equally share the jackpot prize, what is each winner's share if the jackpot is $50 000?

Solution

This is a worded question that describes what could be a real-life situation. In this example, it is important to identify the key word that indicates a mathematical operation, and this is the word 'share'. Whilst the word 'share' cannot be found in the table above, we have previously described the concept of division as equal sharing in Chapter 2. This means the required operation is division and the algebraic expression to describe this situation is $\frac{50\,000}{w}$.

PRACTICE 6: APPLICATION

A lecture theatre is divided into *x* sections. Each section has *y* rows and each row has 6 seats. How many seats are in the lecture theatre?

9.3   EVALUATING ALGEBRAIC EXPRESSIONS

Suppose there are 140 days in the university academic year. How could we write an algebraic expression to describe how many days a student would be on campus if they were off campus for *d* days? That would be:

$$140 - d \text{ days}$$

And then we could use this expression to calculate how many days a student is off campus for any particular number of *d* days, by replacing *d* with that number. For example:

If the student was off campus for 28 days, it would be 140 − 28.

This process is known as evaluating an algebraic expression. When evaluating an expression, you substitute a specific value for each variable and then perform the mathematical operations to get a numerical solution. Therefore, the steps involved in evaluating an algebraic expression are:

1   substitute given value for each variable
2   carry out the calculation.

EXAMPLE 7

Evaluate each algebraic expression for the given value.

Solutions

a $d + 12$, for $d = 7$

a   In the expression, we need to substitute the d with the given value of 7 and calculate: $7 + 12 = 19$

b $c - 9$, for $c = 15$

b   Substituting the c with 15: $15 - 9 = 6$

c $w - 9.6$, for $w = 10$

c   Algebraic expressions do not always contain whole number constants, as in this example.
Substituting w with 10 and evaluating: $10 - 9.6 = 0.4$

d $v + 3\frac{1}{2}$, for $v = \frac{3}{8}$

d   This example involves the use of fractions.
We start by substituting $\frac{3}{8}$ for v: $\frac{3}{8} + 3\frac{1}{2}$
The addition can then be rewritten: $3 + \frac{3}{8} + \frac{1}{2}$
We now need to write the fractions in this addition with 8 as a common denominator: $3 + \frac{3}{8} + \frac{4}{8}$
Now we can complete the addition: $3\frac{7}{8}$

e $1.3t$, for $t = 5$

e   The lack of a mathematical operator in the algebraic expression implies multiplication. Once the substitution has been carried out, the multiplication sign can be used for clarity: $1.3 \times 5 = 6.5$

f $\frac{x}{0.2}$, for $x = 1.8$

f   This expression indicates a division. We could substitute $x = 1.8$ into this expression and then treat it as a fraction: $\frac{1.8}{0.2}$
If we move the decimal point one place to the right in both the denominator and the numerator, we get: $\frac{18}{2} = 9$

Making connections
Check Chapter 5 to revise the addition of fractions.

PRACTICE 7

Evaluate each algebraic expression for the given value.

a $8 - x$, for $x = 3.5$

b $1.5y$, for $y = 0.2$

c $\frac{1}{6}a$, for $a = 2\frac{1}{2}$

d $\frac{m}{0.5}$, for $m = 3.5$

Kathy Brady

Algebraic expressions need not necessarily be limited to containing only one variable. Expressions can contain multiple variables. To evaluate these expressions care needs to be taken to ensure the correct value for each variable has been substituted before carrying out the calculation.

EXAMPLE 8

If $a = 3$, $b = 2$, $c = 5$ and $d = 6$, find the value of the following expressions.

Solutions

a $a + b$

a Substituting $a = 3$ and $b = 2$: $3 + 2 = 5$

b $3c - d$

b Substituting $c = 5$ and $d = 6$, and noting that c is multiplied by $3 : 3 \times 5 + 6$

   The correct order of operations must be used so that the multiplication is carried out first, and then the addition: $15 + 6 = 21$

c $2(d - b)$

c Because the 2 written on the outside of the brackets is associated with no given mathematical operator, multiplication is implied. However, this expression contains brackets, so the order of operation requires that we evaluate the contents of the brackets first, before multiplying by the 2. Substituting $b = 2$ and $d = 6 : 3(6 - 2) = 3 \times 4 = 12$

d $4a^2$

d In this expression a is squared, which means it is multiplied by itself, and then that value is multiplied by 4. Substituting $a = 3 : 4 \times 3 \times 3 = 36$

e $\dfrac{a - b}{c}$

e This expression involves a division, but the order of operations also needs to be considered as well. The numerator in this expression is in itself another algebraic expression that needs to be evaluated first before dividing by the denominator value. It's like there is a set of invisible brackets around the numerator. So substituting $a = 3$, $b = 2$, $c = 5$: $\dfrac{3 - 2}{5} = \dfrac{1}{5}$

PRACTICE 8

If $m = 4$, $n = 2$, $g = 7$ and $h = 5$, find the value of the following expressions.

a $3m + n$ 　　　　　 b $2g - 3h$ 　　　　　 c $3(n + h)$

d $m^2 + g^2$ 　　　　 e $\dfrac{g + h}{m + n}$

All of the examples we have looked at so far have involved the substitution of positive numbers. However, we do not necessarily need to be limited to the use of positive numbers, as in real life situations negative numbers may also need to be used.

*Making
connections*
*To review
mathematical
operations
that involve
using positive
and negative
directed
numbers
check back to
Chapter 4.*

EXAMPLE 9

If $p = 2$, $q = 5$ and $r = -3$, find the value of the following expressions.

Solutions

a $-4 - p$

a   Substituting $p = 2$: $-4 - 2 = -6$

b $2qr$

b   This expression implies multiplication, so for $q = 5$ and $r = -3$: $2 \times 5 \times (-3) = -30$

c $-p + r^2$

c   The BEDMAS order of operations needs to be applied here to handle the exponent (the square) before the addition. Substituting $p = 2$ and $r = -3$: $-2 + (-3)^2 = -2 + 9 = 7$

d $\dfrac{q - r}{p + q}$

d   The order of operations also needs to be considered in this division. Both the numerator and denominator are sub-algebraic expressions (with invisible brackets) that need to be evaluated before the division can be carried out.

So for $p = 2$, $q = 5$ and $r = -3$: $\dfrac{5 - (-3)}{2 + 5} = \dfrac{8}{7}$. It would be quite acceptable to leave this as a top-heavy common fraction.

PRACTICE 9

If $a = 4$, $b = -3$ and $c = 5$, find the value of the following expressions.

a abc

b $b^2 - c^2$

c $2(a - c)$

d $\dfrac{c - b}{-a}$

CRITICAL THINKING

In Australian Rules football games, teams can kick either goals or behinds. Each goal is worth six points and each behind is worth one point. The team's final score (in points) is calculated according to the expression $6g + b$, where g is the number of goals kicked and b is the number of behinds kicked.

a   If $g = 5$ and $b = 7$, what would the score be?

b   If the score was 31 points, what could be the values of g and b? Try to think of all possible combinations.

c   If $b = 11$, and the score is a two-digit number, what possible values could g take?

Kathy Brady

Sometimes it is necessary to make successive substitutions of different values into the same algebraic expression. In this situation, recording the results is often completed in what is known as a table of values.

EXAMPLE 10

Complete the following table of values.

x	−2	−1	0	1	2
$-3x + 2$					

Solution

To complete the table, one-by-one substitute the values −2, −1, 0, 1 and 2 into the expression $3x + 2$.

For $x = -2 : 3 \times (-2) + 2 = -4$

For $x = -1 : 3 \times (-1) + 2 = -1$

For $x = 0 : 3 \times 0 + 2 = 2$

For $x = 1 : 3 \times 1 + 2 = 5$

For $x = 2 : 3 \times 2 + 2 = 8$

The completed table of values would be:

x	−2	−1	0	1	2
$3x + 2$	−4	−1	2	5	8

PRACTICE 10

Complete the following table of values.

x	−2	−1	0	1	2
$-2x + 1$					

Making connections
In Chapter 10, you will discover how to plot graphs using the information contained in a table of values.

9.4   LIKE TERMS

Earlier in this chapter, we defined an algebraic expression as a mathematical sentence that combines variables, constants and arithmetic operations, where a variable is a letter that is used to represent an unknown value that can change and a constant is a value that is known and does not change. In this section, we use these basic definitions to expand further the language of algebraic expressions.

Each component of an algebraic expression is known as a **term**. For example, the algebraic expression $2x^3 + 3y - 5ab + 4$ has four terms that are $2x^3$, $3y$, $5ab$ and 4.

Terms can have **coefficients**. For example, in the terms given above the coefficient of $2x^3$ is 2. This means that x^3 is being multiplied by 2. Likewise, the coefficient of $3y$ is 3, and the coefficient of $5ab$ is 5, meaning that ab is multiplied by 5.

Particular terms in an algebraic expression are known as **like terms**. These are terms that contain exactly the same variables raised to exactly the same power. Like terms can have different coefficients, but this is the only difference. For example, $3x$, x and $-2x$ are like terms; $2x^2$, $-5x^2$ and $\frac{1}{2}x^2$ are another set of like terms, as is xy^2, $3xy^2$ and $-2xy^2$.

On the other hand, $2a$ and $2b$ are not like terms because the variables are not exactly the same. Neither are x and x^2, nor pq^2 and p^2q, because even though the variables are the same they are not raised to the same powers.

EXAMPLE 11

Name the coefficient of each term in the following algebraic expressions, and identify whether terms are like or unlike.

a $2a + 5c$
b $m - 4n$
c $5x + 7$
d $4g^2 + 2g$
e $2a^2b - 3a^2b$
f $5p^2q + pq^2$

Solutions

	ALGEBRAIC EXPRESSIONS	COEFFICIENTS	LIKE OR UNLIKE	REASON
a	$2a + 5c$	2, 5	Unlike	Variables not exactly the same
b	$m - 4n$	1, −4	Like	Exactly the same variable
c	$5x + 7$	5	Unlike	Combination of a variable and a constant
d	$4g^2 + 2g$	4, 2	Unlike	Variable not raised to exactly the same power
e	$2a^2b - 3a^2b$		Like	Exactly the same variables raised to exactly the same powers
f	$5p^2q + pq^2$	5, 1	Unlike	Variables not raised to exactly the same powers

PRACTICE 11

Name the coefficient of each term in the following algebraic expressions, and identify whether terms are like or unlike.

a $3k - 2k$
b $7y + 4xy$
c $4q^2r - qr$
d $-3x^2y^2 + 4x^2y^2$
e $5ac - 7ac + 3$

Term
A component of an algebraic expression that is a constant, a variable or the product of constants and variables

Coefficient
A number that is used to multiply a variable or variables in an algebraic term

Like terms
When two or more algebraic terms contain exactly the same variable or variables, which are raised to exactly the same powers

Kathy Brady

CRITICAL THINKING

A particular algebraic expression contains 26 terms.

It starts: $a + 2b + 4c + 8d + 16e + \ldots$

a   What is the next term?

b   What is the coefficient of m?

c   What is the coefficient in the last term?

d   Which variable has 1024 as its coefficient?

9.5   SIMPLIFYING ALGEBRAIC EXPRESSIONS BY COLLECTING LIKE TERMS

Once like terms have been identified, they can be collected, or combined, in order to simplify algebraic expressions. This diagram illustrates how and why this is possible:

| 3 lots of 5 | + | 2 lots of 5 | = | 5 lots of 5 |
| 3×5 | + | 2×5 | = | $(3 + 2) \times 5$ |

OR

| 3 lots of 5 | – | 2 lots of 5 | = | 1 lot of 5 |
| 3×5 | | 2×5 | = | $(3 - 2) \times 5$ |

Another way of thinking about this is:

3 apples + 2 apples = 5 apples   or   3 apples − 2 apples = 1 apple

In algebraic terms, this could be written as:

$3a + 2a = (3 + 2)a = 5a$ or $3a - 2a = (3 - 2)a = 1a = a$

This is the process that is known as **collecting or combining like terms**.

Collecting like terms
The process of simplifying an algebraic expression by adding, or subtracting, the coefficients of like terms

EXAMPLE 12

Simplify the following algebraic expressions by collecting the like terms.

Solutions

a $2x + x$

b $a + 3 + a + 7$

c $5y - 3y$

d $3x - 8x$

e $9b - 2 + 7b - 4$

f $2ab + 3ab$

g $3b^2 + b^2$

h $x^2 + 2x - x^2 + 5x$

a We can combine $2x + x$ by adding the coefficients. Note here that the x term has an unwritten coefficient of 1.

$$2x + x = (2 + 1)x = 3x$$

b There are two sets of like terms in the expression $a + 3 + a + 7$: the a terms and the constant terms. We need to combine these sets of like terms separately. Again note here that the a terms have unwritten coefficients of 1.

$$a + 3 + a + 7 = (1 + 1)a + 3 + 7 = 2a + 10$$

c We can combine $5y - 3y$ by subtracting the coefficients.

$$5y - 3y = (5 - 3)y = 2y$$

d We can combine $3x - 8x$ by subtracting the coefficients. To carry out the subtraction we needed to use the subtraction rules that apply to positive and negative numbers.

$$3x - 8x = (3 - 8)x = -5x$$

e There are two sets of like terms in this expression, the b terms and the constant terms, which need to be combined separately.

$$9b - 2 + 7b - 4 = (9 + 7)b - 2 - 4 = 16b - 6$$

f The terms $2ab + 3ab$ are like because they contain the same variables. This means that they can be combined by adding the coefficients.

$$2ab + 3ab = (2 + 3)ab = 5ab$$

g The terms $3b^2 + b^2$ are like because they contain the same variable raised to the same power. To combine we add the coefficients.

$$3b^2 + b^2 = (3 + 1)b^2 = 4b^2$$

h This expression also contains two sets of like terms, the x terms and the x^2 terms. This means that we need to combine these sets of like terms separately. In this case, because the terms look similar, we need to do this with care.

$$x^2 + 2x - x^2 + 5x = (1 - 1)x^2 + (2 + 5)x$$
$$= 0x^2 + 7x$$
$$= 7x$$

Notice that when the coefficients of the x^2 term are subtracted, the resultant coefficient is 0, which eliminates the x^2 in the final answer.

PRACTICE 12

Simplify the following algebraic expressions by collecting the like terms.

a $x + 3x$

b $3a + 2 + a + 4$

c $7y - 8y$

d $18c + 5 - 11c - 4$

e $3xy + 4xy$

f $a - 2b + 3a + b$

g $2p^2 - p^2$

h $b^2 + 2ab + b^2 - 5ab$

EXAMPLE 13: APPLICATION

Find the perimeter of this triangle, writing the answer as simply as possible:

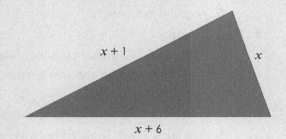

Solution

The perimeter is the distance around the boundary of a two-dimensional figure.

In this figure the perimeter of the triangle will be the sum of the lengths of each of the sides: $x + 1 + x + x + 6$

The algebraic expression that has been formed can be simplified by collecting like terms; that is, the x terms and the constant terms.

$$x + 1 + x + x + 6 = (1 + 1 + 1)x + 1 + 6 = 3x + 6$$

Notice that in this example we collected three like terms (the x terms). There is no limit to the number of like terms that can be collected when simplifying expressions.

Making connections
The concept of perimeter and how to calculate it will be covered fully in Chapter 12.

PRACTICE 13: APPLICATION

Find the perimeter of this rectangle, writing the answer as simply as possible.

$2x + 1$

$3a$

9.6   SIMPLIFYING ALGEBRAIC EXPRESSIONS BY EXPANDING BRACKETS

When algebraic expressions involve the use of brackets, quite often removing the brackets can help to simplify the expression.

To begin to understand what we need to do to remove brackets in algebraic expressions let's consider the algebraic expression $4(a + b)$. If we were to evaluate this expression for the values $a = 7$ and $b = 5$, we would have:

$$4(7 + 5) = 4(12), \text{ which can be rewritten as } 4 \times 12 = 48$$

Now let's consider another algebraic expression $4a + 4b$. If we were to evaluate this expression for the same values $a = 7$ and $b = 5$, we would then have:

$$4 \times 7 + 4 \times 5 = 28 + 20 = 48$$

It appears from these two examples that the expressions $4(a + b)$ and $4a + 4b$ are equivalent because when we substituted the same values for a and b, the resulting answers were equal.

That the expressions $4(a + b)$ and $4a + 4b$ are equivalent is a demonstration of what is known as the distributive law. The distributive law states that when the sum of two values is multiplied by a third value, it is equal to the sum of each of the two values being individually multiplied by the third value. In this case, $4(7 + 5) = 4 \times 7 + 4 \times 5$. In general terms, for any three values a, b and c:

$$a(b + c) = a \times b + a \times c$$

or

$$a(b - c) = a \times b - a \times c$$

It is this feature of the distributive law that we use when removing the brackets in an algebraic expression. We simply remove the brackets and multiply each of the values inside the brackets by the value outside the brackets, keeping sure to maintain any addition or subtraction signs that are included in the bracket. This is known as expanding the brackets.

Distributive law
In general terms, for any three values a, b and c:
$a(b + c) =$
$a \times b + a \times c$
or
$a(b - c) =$
$a \times b - a \times c$

Kathy Brady

EXAMPLE 14

Expand the brackets in the following expressions.

Solutions

a $3(a + 2)$

a To expand the brackets we need to multiply everything inside the brackets by the 3 outside of the brackets, maintaining the addition sign as we do so.
$$3(a + 2) = 3 \times a + 3 \times 2 = 3a + 6$$

b $7(b - 3)$

b To expand the brackets we need to multiply everything inside the brackets by the 7 outside of the brackets, maintaining the subtraction sign as we do so.
$$7(b - 3) = 7 \times b - 7 \times 3 = 7b - 21$$

c $-2(x + 5)$

c To expand the brackets we need to multiply everything inside the brackets by the -2 outside of the brackets, maintaining the addition sign as we do so.
$$-2(x + 5) = -2 \times x + (-2) \times 5 = -2x + (-10) = -2x - 10$$

Whenever the value on the outside of the brackets is a negative number, care must be taken with the signs once the brackets are removed. Note in this example the use of some temporary brackets to indicate that the addition of a negative number, which is the same as subtracting.

d $-3(y - 4)$

d To expand the brackets we need to multiply everything inside the brackets by the -3 outside of the brackets, maintaining the subtraction sign as we do so.
$$-3(y - 4) = -3 \times y - (-3) \times 4 = -3y - (-12) = -3y + 12$$

Here temporary brackets have been placed around the -3 to indicate the subtraction of a negative number, which is the same as adding.

PRACTICE 14: GUIDED

Expand the brackets in the following expressions.

a $4(b + 5) = 4 \times \square + \square \times 5 = \square b \square 20$ b $-2(x - 3) = \square \times x - (-2) \times \square = -2x \square 6$

Now try these:

c $6(a - 2)$ d $-4(p + 3)$

The multiplier outside of the brackets does not necessarily need to be a constant value; it can also be a variable, as in the following examples.

EXAMPLE 15

Expand the brackets in the following expressions.

Solutions

a $a(a + 1)$

a To expand the brackets we need to multiply everything inside the brackets by the a outside of the brackets maintaining the addition sign as we do so.

$$a(a + 1) = a \times a + a \times 1 = a^2 + a$$

Notice here that when a is multiplied by itself, we use the power of 2 or squaring notation, and that when a variable is multiplied by 1 we do not include the 1 as a coefficient.

b $2p(3 - p)$

b To expand the brackets we need to multiply everything inside the brackets by the $2p$ outside of the brackets maintaining the subtraction sign as we do so.

$$2p(3 - p) = 2p \times 3 - 2p \times p = 6p - 2p^2$$

Here, when $2p$ is multiplied by 3, we need to be aware that $2p$ is actually $2 \times p$, and so we have altogether $2 \times p \times 3$, which equals $6p$.

c $3x(y + 1)$

c To expand the brackets we need to multiply everything inside the brackets by the $3x$ outside of the brackets maintaining the addition sign as we do so.

$$3x(y + 1) = 3x \times y + 3x \times 1 = 3xy + 3x$$

Notice here that when $3x$ is multiplied by y that the $3x$ is actually $3 \times x$, and so we have altogether $3 \times x \times y$, which is written as $3xy$ according to the algebraic conventions for multiplication.

d $2c(c - d)$

d To expand the brackets we need to multiply everything inside the brackets by the $2c$ outside of the brackets maintaining the subtraction sign as we do so.

$$2c(c - d) = 2c \times c - 2c \times d = 2c^2 - 2cd$$

Here is another example how to multiply three algebraic values and then write the product using the correct algebraic conventions.

PRACTICE 15: GUIDED

Expand the brackets in the following expressions.

a $x(x - 2) = x \times \square - \square \times 2 = x^\square - \square x$ **b** $3f(g + 5) = \square \times g + 3f \times \square = 3\square g + \square f$

Now try these:

c $2y(y - 3)$ **d** $3a(b + 2a)$

Kathy Brady

9.7   CHAPTER SUMMARY

SKILL OR CONCEPT	DEFINITION OR DESCRIPTION	EXAMPLES
Variable	A variable is a letter that is used to represent an unknown value, or a value that can change.	a, g, x
Constant	A constant is a value that is known and does not change.	2, 16, 150
Algebraic expression	An algebraic expression is a mathematical sentence that combines variables, constants and arithmetic operations.	$2a + 3b - c + 7$ $x^2 + 4x - 5$
Multiplication convention	When writing a product of two or more constants and variables, the multiplication sign is omitted, and the expression is written commencing with the constant followed by the variables in alphabetic order.	$3t$ $5ab$ xyz
Division convention	In algebra, the division sign ($\div$) is not used, rather the division is written in fraction form.	$\dfrac{x}{6}, \dfrac{p}{q}, \dfrac{a+4}{b-5}$
Evaluating an algebraic expression	Substitute a specific value for each variable, and then perform the required mathematical operations to get a numerical solution.	Evaluate $2x + 3$, for $x = 4.5$ $2 \times 4.5 + 3 = 9 + 3 = 12$
Algebraic term	A term is a component of an algebraic expression that is either a constant, a variable or the product of a constant and variables.	$5, w, 4x, ab, 3cd, 7p^2$
Coefficient	A coefficient is a number that is used to multiply a variable, or variables, in an algebraic term.	$2x, 8ab$
Like terms	Two or more algebraic terms are known as like terms if they contain exactly the same variable or variables, which are raised to exactly the same powers.	x and $2x$ $3ab$ and $-2ab$ $p^2q, 5p^2q, -p^2q$ x and $2x$ $3ab$ and $-2ab$ $p^2q, 5p^2q, -p^2q$
Collecting or combining like terms	Collecting, or combining, like terms is the process of simplifying an algebraic expression by adding, or subtracting, the coefficients of like terms.	$x + 2x = 3x$ $3ab - 2ab = ab$ $p^2q + 5p^2q - 2p^2q = 4p^2q$
Distributive law	For any three values a, b and c: $a(b + c) = a \times b + a \times c$ or $a(b - c) = a \times b - a \times c$	$3(5 + 2) = 3 \times 5 + 3 \times 2 = 21$ $3(5 - 2) = 3 \times 5 - 3 \times 2 = 9$
Expanding brackets	Use the distributive law to remove the brackets in an algebraic expression by multiplying each of the values inside the brackets by the value outside the brackets, maintaining the addition or subtraction signs that are included inside the brackets.	$3(x + 4) = 3 \times x + 3 \times 4$ $= 3x + 12$ $a(2b - c) = 2ab - ac$ $2d(d + 3e) = 2d^2 + 6de$

9.8   REVIEW QUESTIONS

A SKILLS

1   Write the following expressions using correct algebraic conventions.

   a $x + x - (y + y + y) + 2$ b $5 - (e + e) + h + h + h$

   c $a \div (b \times 5)$ d $x \times 3 - y \div 4$

2   Translate the following sets of words into algebraic expressions using correct algebraic conventions.

   a   8 more than p b   The product of y, z, x and 4

   c   One quarter of a is subtracted from b d   The quotient of c and $d + 3$

3   If $a = 3$, $b = 4$, $c = 2$ and $d = 5$, find the value of the following expressions.

   a $2a + b$ b $3(d - c)$ c $\dfrac{4d}{a + c}$ d $5c^2d$

4   If $p = 3$, $q = -2$, $r = -1$ and $s = 6$, find the value of the following expressions.

   a $(p + r)^2$ b qrs c $\dfrac{p + q}{r}$ d $\dfrac{q + r}{p}$

5   Complete the following table of values.

x	−2	−1	0	1	2
$-4x + 3$					

6   Simplify the following algebraic expressions by collecting the like terms.

   a $6d + 9 - 4d - 2$ b $10x + 5 - 9x - 3x$

   c $11a + 2a^2 - 9a + 5a^2$ d $xy + x^2y - 4xy + xy^2$

7   Expand the brackets in the following expressions.

   a $5(t - 2)$ b $-3(p - 4)$ c $2a(b + c)$ d $y(3y + 2z)$

B   APPLICATIONS

1   An international mobile phone call costs 25 cents to connect and then 40 cents per minute. The cost of making the call could, therefore, be written as $25 + 40m$, where m is the length of the call in minutes. Complete the table of values to calculate the cost of making a call for the given number of minutes.

m	5	10	15	20	25
$25 + 40m$					

2   The length of one side of a triangle is x cm. The second side is 5 cm longer than the first and the third side is 3 cm shorter than double the length of the first side.

   a   Write an expression for the length of each side of the triangle.

   b   Now write an expression, in simplest terms, for the perimeter of the triangle.

Kathy Brady

3   A small business owner decides to invest in some advertising to expand their customer base. She concludes that both television and radio advertising would be effective. The cost of a television advertisement is $400 and the cost of a radio advertisement is $200. Because radio advertising is less expensive, the business owner decides to pay for 50 more advertisements on the radio than on the television. If the business owner pays for t television advertisements, answer the following questions.

   a   Write an expression for the cost of the television advertisements.

   b   In terms of t, how many radio advertisements does she pay for?

   c   Write and expand an expression for the cost of the radio advertisements.

   d   Now write an expression for the total advertising costs.

   e   Simplify this expression by expanding brackets and collecting like terms.

4   The surface area of a cylinder is the total area of the circular top and base and the curved side. The formula for the surface area of a cylinder is given as $S = 2\pi r(h + r)$, where π (pi) is a constant value, r is the radius, and h is the height of the cylinder. Write this formula with the use of brackets.

5   In a laboratory class, there are m male and f female students. In a particular semester, the male students each break 7 test tubes and the female students each break 4 test tubes. Test tubes costs $2 each.

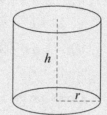

   a   Write an expression for the total number of test tubes broken by the class.

   b   Write and expand an expression for the total cost of the test tubes broken in the semester.

   c   The students also broke one flask each during the semester. Each flask costs $4. Write and expand an expression for the total cost of the flasks broken.

   d   Finally, write and simplify an expression for the total cost of all of the equipment broken.

   e   If there were 10 male students and 12 female students in the class, use this expression to calculate the total cost of the breakages.

10

GRAPHS AND CHARTS

CHAPTER OBJECTIVES

》 Discern between categorical, numerical, discrete and continuous data

》 Select the graph or chart most appropriate to display a particular data type

》 Draw graphs and charts

》 Read and interpret graphs and charts

》 Use graphs and charts in applied contexts

》 Identify the features of the coordinate plane

》 Plot points using a table of values on the coordinate plane to draw a line graph

LONDON 1854

In 1854, a severe outbreak of cholera occurred near Broad Street in the Soho district of London. This outbreak occurred because the Soho district did not have a sewerage system at that time, which caused contamination of the water supply from the River Thames. This outbreak is well known because of the investigations of local physician Dr John Snow that lead to the conclusion that it was the contaminated water, and not the air, that spread the disease.

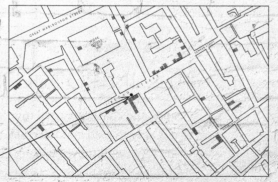

PUMP

(Source: www.mediainstitute.edu)

Dr Snow was sceptical of the dominant theory of the era that cholera was transmitted through the air. Rather he began an investigation of the water supply by plotting the number of cholera cases near a public water pump on Broad Street. In this original plot by Dr Snow, we can see the pump located on the corner of Broad Street and the clusters of cholera cases in each of the nearby dwellings.

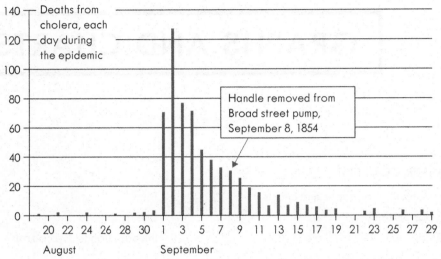

(Source: www.safedrinkingwater.com)

Dr Snow's plot led him to the conclusion that the source of the outbreak was the Broad Street pump. As a result of this investigation, Dr Snow was able to convince the local authorities to remove the handle from the Broad Street pump in order to curtail the spread of the disease. The major outbreak of cholera in Soho commenced on 31 August 1854. Over the next three days, 127 people on or near Broad Street died. By 10 September, 500 people had died, and by the end of the outbreak 616 people were dead.

A graph of the daily deaths from cholera in the Broad Street area that identifies when the handle was removed from the pump appears to confirm Dr Snow's conclusion because the number of deaths quickly declined in the days afterwards. The Broad Street cholera outbreak is a dramatic example of how graphs and charts can be used effectively to address serious issues of public concern. (Source: choleraandthethames.co.uk)

10.1 INTRODUCTION TO GRAPHS AND CHARTS

Graphs and charts are used to summarise information, or data, in a straightforward way that is easy to read and interpret. The beauty of graphs and charts is that they present a visual representation of what could be quite extensive and complex numerical information. Additionally, graphs and charts can effectively illustrate trends and cycles in data, and enable interpretations that involve comparisons.

The terms graphs and charts are frequently used interchangeably. However, there are differences between graphs and charts that are associated with how the data is compiled and represented. Graphs are used to present exact numerical information to illustrate changes or trends in the data over time. They usually feature a left-hand and bottom axis and typically involve the plotting of points that are joined by a line. Charts, on the other hand, are used when the data involves the use of categories and are very effective in visually examining the composition of the sub-categories or comparing categories. Commonly used charts are pictograms, pie charts and bar charts.

Whilst there are differences between graphs and charts, they also have many features in common. In the figure below, the major features have been highlighted.

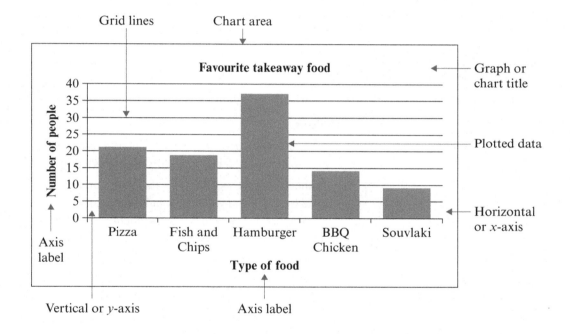

The choice of the best type of graph or chart to use will depend upon the type of data to be displayed. In the next section, we will review the different types of data before proceeding to explore how data can be represented in a range of graphs and charts.

10.2 TYPES OF DATA

Broadly speaking, data is the information that is presented and represented in graphs and charts and there are two main types of data: numerical and categorical. **Numerical data** is counted or measured and recorded as numbers. **Categorical data** is classified or sorted into groups or categories.

Numerical data
Values or observations that can be counted or measured and recorded as numbers

Categorical data
Observations that are classified or sorted into groups or categories

Kathy Brady

Examples of information that is collected as numerical data might be the weight of a baby or how many children there are in a family. Numerical data can be further classified in two different ways: discrete or continuous.

Because **discrete numerical data** involves exact number values it usually results from counting occurrences or observations, for example, how many push-ups you can do or the number of cars in a household. Alternatively, **continuous numerical data** represents values with infinite possibilities such as weight or height. Continuous numerical data usually results from taking measurements and this data is recorded in a reasonable range to make it measurable, for example, recording height in centimetres to one decimal place.

On the other hand, categorical data does not involve observing and recording numerical information rather descriptive information is recorded. Examples of categorical data are eye colour, favourite football team or voting intention.

Discrete data
Numerical
data that
involves exact
number values

Continuous data
Numerical data
that can occur in
an infinite range

EXAMPLE 1

Classify the follow data as discrete numerical, continuous numerical or categorical.

Solutions

a Annual income

a Discrete numerical because annual income is recorded as an exact number; that is, dollars and cents.

b Foot length

b Continuous numerical because the length of a foot could be recorded within an infinite range. However, the measurement would be made within a reasonable range, for instance, in centimetres to two decimal places in order to be able to record it.

c Favourite ice-cream flavour

c Categorical because favourite ice-cream flavour would be recorded in groups such as vanilla, chocolate or strawberry.

d Number of people in a household

d Discrete numerical because people can only be counted in whole numbers.

e Travel method to university

e Categorical as travel method would be recorded in groups such as car, bus or bicycle.

f Travel distance to university

f Continuous numerical because the travel distance could be recorded within an infinite range. However, the measurement would be made within a reasonable range, for instance, in kilometres to two decimal places in order to be able to record it.

PRACTICE 1

Classify the follow data as discrete numerical, continuous numerical or categorical.

a   Price of petrol   b   How far a footballer can kick

c   Number in an endangered species   d   Month of birth

e   Braking reaction time   f   Blood group

The type of graph or chart that is selected to represent a particular data set will depend upon the type of data being used. Regardless of the type of data it is important to remember that if the graph or chart you are presenting is based on data from another source, then you should acknowledge the source of the original data either within the accompanying text or somewhere within the chart area.

10.3 PICTOGRAPH CHARTS AND PIE CHARTS

Pictograph charts

A pictograph chart is a visual method of displaying discrete numerical data that uses related images or symbols to represent the information. It includes a key or scale that indicates the value of each symbol. For example, a pictograph chart showing the sale of cars may use a car symbol to represent 100 cars sold. The advantage of using a pictograph chart is that the data can be quickly interpreted and understood. However, pictograph charts are limited by how the single symbol can be divided to represent a portion of the whole. For example, in a pictograph chart illustrating car sales the reader would not know exactly how many car sales were represent by a small fraction of the whole car symbol.

EXAMPLE 2

The pictograph chart overpage shows the gold medals won by the top eight countries at the 2016 Rio Olympics (Source: rio2016.com). Use the pictograph chart to answer the following questions.

a   How many gold
    medals did the
    USA win?

Solutions

a   The USA has 12 gold medal symbols and each symbol
    represents 10 gold medals. We can also see that
    there is a small portion of the gold medal symbol
    illustrated as well. We could estimate that this small
    section is equal to one medal. Therefore, using the
    information in the pictograph chart the USA won
    $(12 \times 10) + 1 = 121$ gold medals.

b How many more gold medals did Great Britain win than France?

b Great Britain has 6 full symbols, and one partial symbol that clearly represents more than half. We could estimate that this section of the symbol is equal to 7 medals. Therefore, from the pictograph chart we could conclude that Great Britain won $(6 \times 10) + 7 = 67$ gold medals.

France has 4 full symbols and a partial symbol that we could estimate is equal to 2 medals, so France's total is $(4 \times 10) + 2 = 42$ gold medals.

Therefore, Great Britain won $67 - 42 = 25$ more medals than France.

c Among these eight countries, what was the lowest number of gold medals won?

c Among these countries, Korea won the lowest number of gold medals, which we can see from the pictograph is equal to 21.

d Between these eight countries, what was the average number of gold medals won?

d The average is calculated by finding the sum of all of the gold medals won and then dividing by the number of countries in this set of data.

Moving from Korea at the top of the pictograph to USA at the bottom we get:

$21 + 42 + 41 + 42 + 56 + 70 + 67 + 121 = 460$

Therefore, the average number of gold medals is $460 \div 8 = 57.5$.

2016 Rio Olympics gold medal tally

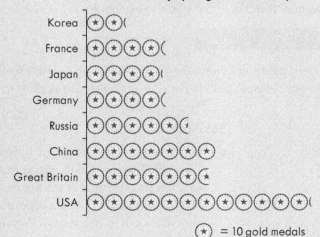

PRACTICE 2

The number of goals scored by the top eight goal kickers in the 2016 Australian Football League season is illustrated in the pictograph chart below (Source: footywire.com). Use this chart to determine the following.

a   How many goals did the top goal kicker in the season score?

b   How many more goals did Eddie Betts score than Tom Hawkins?

c   What was the average number of goals kicked by the top eight players?

Pie charts

A pie graph is a visual method of representing categorical data. Specifically, a pie chart illustrates the categories or parts that go to making up a whole. For example, with regard to the whole that is your income, the different sectors on a pie chart could show how much you spend on rent, food, petrol, socialising or any other category. A pie graph shows the percentage that each category comprises of the whole, so it is essentially a circle divided into 100 equal sections. Each category is represented on the pie graph as a sector that corresponds to, and is labelled with, the percentage that it comprises.

Pie charts are an effective way of comparing categories, especially if each section is a different colour so that readers can easily discern between them. However, if there are too many categories represented on a pie chart, it is difficult for the eye to distinguish between the relative sizes of the different sectors and the chart is difficult to interpret.

EXAMPLE 3

In 2013–2014, there were approximately 11 000 children in Australia who were living in out-of-home care because, for various reasons, they were unable to live with their families. The pie chart below shows the percentage of children in out-of-home care by age (Source: aifs. gov.au). Use this pie chart to answer the following questions.

Making Connections
Revisit Chapter 6 to review how to calculate a percentage.

Solutions

a   Which age group had the greatest number of children?

a   The age group with the greatest percentage of out-of-home care was 1–4 years with 25%.

b   How many children in this age group were in out-of-home care?

b   If there were approximately 11 000 children in out-of-home care, then 25% of this total is $\frac{25}{100} \times \frac{11\,000}{1} = 2750$ children

c   What percentage of children in out-of-home care were aged less than 1 year? How many children in this age group were in out-of-home care?

c   19% of children in out-of-home care were aged less than one. This means that $\frac{19}{100} \times \frac{11\,000}{1} = 2090$ children in this age group were in out-of-home care.

d   What percentage of children in out-of-home care were aged 9 or less?

d   To determine the percentage of children aged 9 or less we need to add the percentages in the sectors that are labelled 5–9, 1–4 and < 1 year. Therefore, the percentage will be 24 + 25 + 19 = 68%.

Age of Australian children in out-of-home care 2013–2014

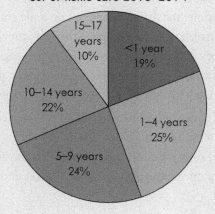

PRACTICE 3

This pie chart presents the percentage of various family types in Australia in 2016. In 2016, there were approximately 6 700 000 families in Australia (Source: aifs.gov.au). Use this pie chart to answer the following questions.

a   What percentage of Australian families comprised one female parent? How many families does this represent?

b   What percentage of families included a couple? How many families does this represent?

c   What percentage of families had children? How many families does this represent?

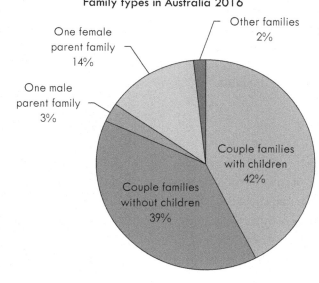

Family types in Australia 2016

To draw a pie chart we could use a technological tool such as a spreadsheet. However, pie charts are also easy to draw using a circle template with 100 equally spaced tick marks, as shown.

We can use template to draw in and shade the sectors that represent the percentages for each category.

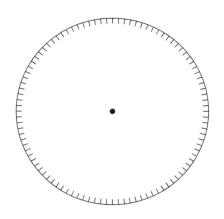

EXAMPLE 4

This table shows the production of various broad acre crops (recorded in tonnes), as a percentage of total production, that were grown in Australia in the year ending 30 June 2015 (Source: abs.gov. au). Draw a pie chart to represent this information.

CROP	PER CENT
Wheat	32%
Sugar cane	44%
Canola	5%
Barley	12%
Sorghum	3%
Other	2%
Oats	2%

Solution

Using the circle template, we draw a line from the centre to the tick mark at the top. Starting with the 32% for wheat we count around 32 tick marks and draw another line to the centre. This sector is then shaded and labelled.

To draw the next sector for sugar cane, which is 44%, we start at the last line we drew for the wheat sector and count around 44 tick marks, and then draw another line to the centre. This sector is then shaded and labelled.

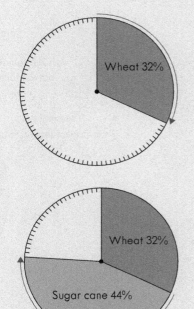

We continue in the same way until the percentages for all of the categories have been counted around, shaded and labelled and we arrive back at the first line that we drew. Don't forget to include a title. The finished pie chart will look like this:

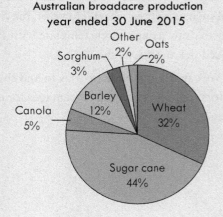

Australian broadacre production year ended 30 June 2015

PRACTICE 4

The following table details the populations of each continent as a percentage of the world's population in 2016 (Source: wikipedia.org). Use this information and the template that has been provided to draw a pie chart to represent this data.

CONTINENT	PER CENT
Asia	60%
Africa	16%
Europe	10%
North America	8%
South America	5%
Australia and Oceania	1%

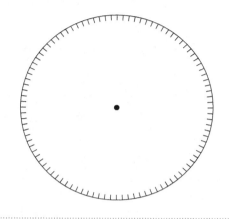

10.4 BAR CHARTS

Bar charts are used to represent categorical data. They are perhaps the most commonly used types of graphs or charts because they are simple to create and are easy to read and interpret. Bar graphs are used to compare the size of different categories, to highlight the breakdown of sub-categories that make up a category or to track changes over a period of time.

An important characteristic of bar charts is that they have a number of variations that can be used to suit a range of applications or contexts. These include vertical bar charts, horizontal bar charts, side-by-side bar charts and stacked bar charts. The common feature of all of these bar chart variations is that the lengths of the bars in the chart are proportional to the size of the category they represent. In this section, we will investigate the features and applications of each of these types of bar charts, examine various bar charts in order to interpret the information that they represent and in some examples discover how to draw the chart.

Vertical bar charts

In a vertical bar chart, the bars are by definition, vertical. Here is an example highlighting its features:

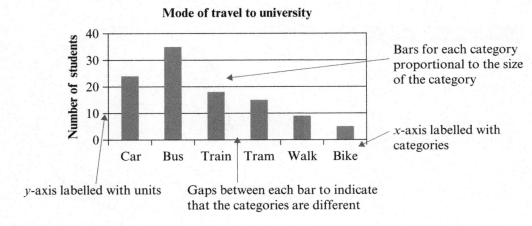

Mode of travel to university

Bars for each category proportional to the size of the category

x-axis labelled with categories

y-axis labelled with units

Gaps between each bar to indicate that the categories are different

Making Connections
Histograms will be covered more fully in Chapter 15.

The horizontal axis, which is known as the *x*-axis, represents the different categories. The vertical axis, known as the *y*-axis, has a numerical scale that indicates the units by which the categories have been tallied. In order to emphasise that the categories are different, a gap is left between the bars on the *x*-axis. A vertical bar chart is somewhat similar to a histogram, which is another type of chart. Histograms, however, do not have gaps between the bars because they represent continuous numeric data, rather than categorical data. Therefore, this is another reason that vertical bar charts have gaps between their bars, as this ensures that the bar chart does not get confused with a histogram.

EXAMPLE 5

The vertical bar chart overpage shows Australian new cars sales for the top eight makes of car in January 2016 (Source: caradvice.com.au). Use this chart to answer the following questions.

a Estimate how many cars were sold of the top-selling make.

b Estimate how many more Mazdas were sold than Fords.

c Which make of car sold approximately 5000 vehicles?

Solutions

a The top-selling make was Toyota and we can estimate that 12 200 were sold.

b From the chart, we can estimate that 10 000 Mazdas were sold and 5500 Fords were sold. Therefore, there were 4500 more Mazdas sold than Fords.

c From the chart, we can conclude that Mitsubishi sold approximately 5000 cars.

d Estimate to the nearest 100 the total number of new cars sold across the top eight makes of car.

d To find the total of the new car sales we need to add the sales for each make, approximated to the nearest 100. From left to right on the chart, this will be:

12 500 + 10 000 + 7000 + 6900 + 5500 + 5500 + 5000 + 4300 = 56 700 cars

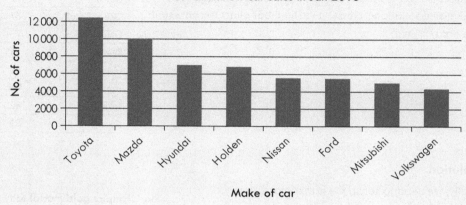

Australian new car sales in Jan 2016

PRACTICE 5

The vertical bar chart below shows the median weekly cost for childcare in Australian states for 2015 (Source: abc.net.au). Use this chart to answer the following questions.

a What was the lowest median cost?

b What is the difference between the median costs in the ACT and in the NT?

c In which state was the median cost approximately $390?

d What was the average median weekly cost of childcare (approximated to the nearest $10) across all of the states in 2015?

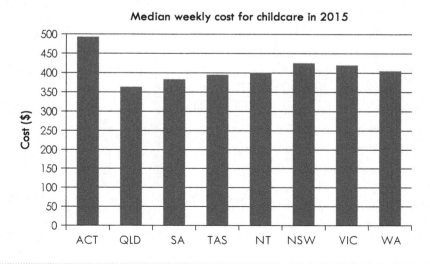

Median weekly cost for childcare in 2015

To draw a vertical bar chart we could use a technological tool such as a spreadsheet. However, these types of charts are easy to draw using grid paper.

EXAMPLE 6

The table details the number of gold medals won by the top eight countries at the 2016 Rio Olympics (Source: rio2016.com). Use this information to draw a vertical bar chart to represent the data.

COUNTRY	MEDALS
USA	121
Great Britain	67
China	70
Russia	56
Germany	42
Japan	41
France	42
Korea	21

Solution

Firstly, we need to set up the horizontal x-axis by labels of the countries using eight equally spaced intervals. Next we need to scale the vertical y-axis. This is done by looking over the data and noting the maximum and minimum values. Using these values, we need to decide how to mark off the y-axis. In this case, the maximum is 121 and the minimum is 21, so marking the y-axis at every 20 would be suitable. Finally, we add a chart title and a label on the y-axis. The chart (shown here) is now ready to draw in the bars.

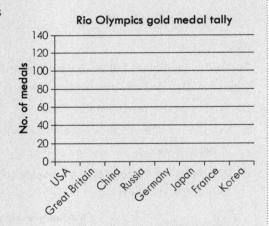

To draw in the bars, start on the x-axis a bit to the left of the USA label, draw a straight-line, parallel to the y-axis, upwards making a close estimate of where 121 lies. Draw across the top of the bar and then down to meet the x-axis again. Do not forget to leave a gap between each bar as you draw them. Keep drawing in the bars trying to keep all of the sides as parallel as possible. Here is what your completed vertical bar chart should look like.

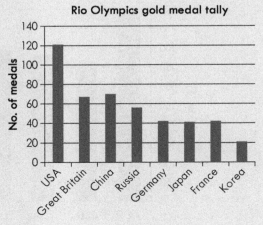

PRACTICE 6

The table details the number of moons orbiting each of the planets in our solar system (Source: solarsystem.nasa.gov). Draw a vertical bar chart to represent this information.

PLANET	MOONS
Earth	1
Mars	2
Jupiter	67
Saturn	62
Uranus	27
Neptune	14
Pluto	5

Horizontal bar charts

A horizontal bar chart is very similar to a vertical bar chart, the difference being that the bars are drawn outward from the y-axis and are therefore, by definition, horizontal. This means that the y-axis is labelled with the different categories, and the x-axis has a numerical scale that indicates the units by which the categories have been tallied. Like a vertical bar chart, a horizontal bar chart also has a gap between each bar. Horizontal bar charts are an effective way of presenting data where the different categories have long titles that are difficult to label below a vertical bar, or where there are a large number of categories that would not fit easily across a page in a vertical bar chart.

EXAMPLE 7

The horizontal bar chart below displays the number of kilojoules in 100 gram serves of popular chocolate snacks (Source: foodstandards.gov.au). Use this chart to answer the following questions.

Solutions

a Estimate how many kilojoules in a chocolate biscuit.

a There are approximately 1880 kilojoules in a chocolate biscuit.

b What is the approximate difference in the number of kilojoules when comparing the snack with greatest number and to the snack with the least?

b The snack with the greatest number of kilojoules is the chocolate bar with approximately 2300, and the snack with the least is the reduced fat chocolate custard with approximately 410. The approximate difference between these is 1900 kilojoules.

c Which snack has approximately 500 kilojoules?

c Regular chocolate custard has approximately 500 kilojoules.

Kathy Brady

d If you were to eat a serve of reduced fat chocolate custard topped with a serve of regular chocolate ice-cream and a chocolate biscuit, approximately how many kilojoules (to the nearest 10) would you consume?

d There is approximately 410 kilojoules in the reduced fat chocolate custard, 790 in the regular chocolate ice-cream and 1880 in the chocolate biscuit. Therefore, you would consume 410 + 790 + 1880 = 3080 kilojoules.

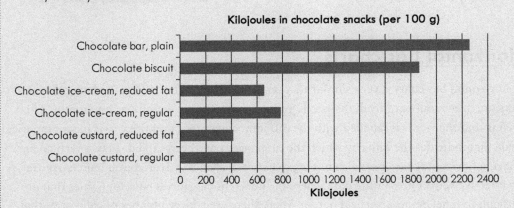

Kilojoules in chocolate snacks (per 100 g)

PRACTICE 7

The horizontal bar chart below shows the number of Group 1 races won by Australia's top eight all-time leading jockeys (Source: australianracingrecords.com.au). Use this chart to answer the following questions.

a Estimate how many races Jim Cassidy won.

b How many more races did Damien Oliver win when compared to Neville Sellwood?

c Who has won 90 races?

d What is the average number of Group 1 races won by these jockeys?

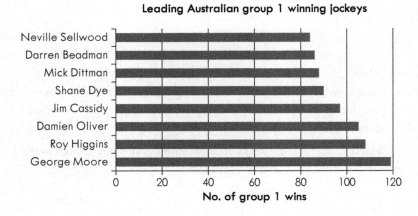

Leading Australian group 1 winning jockeys

Side-by-side bar charts

A side-by-side bar chart can be used to display data with categories that are divided into two or more sub-categories. They can also be used to display changes in categories with the passing of time. A separate bar represents each sub-category and these are usually shaded differently to distinguish between them. A legend is provided to indicate which sub-category the various shadings represent. Care needs to be taken when using side-by-side bar charts not to include too many sub-categories, as the charts will become too complex to read and interpret. Side-by-side bar charts can be presented either horizontally or vertically depending upon the data being used.

EXAMPLE 8

The side-by-side bar chart below presents the long-term mean maximum and minimum temperatures recorded in Alice Springs (Source: bom.gov.au). Use this chart to answer the following questions.

a   What is the highest mean maximum temperature for Alice Springs, and what is the lowest mean minimum temperature?

b   In which month does the smallest difference between the mean maximum and the mean minimum occur? What is this difference?

c   In which month does the greatest difference between the mean maximum and the mean minimum occur? What is this difference?

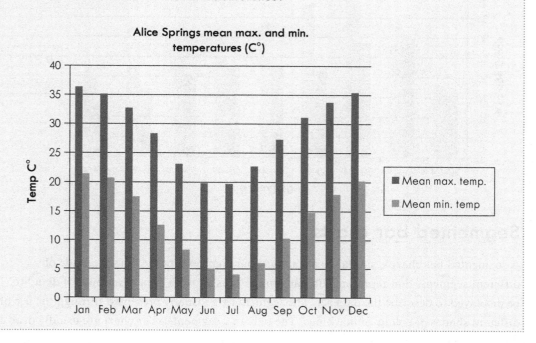

Alice Springs mean max. and min. temperatures (C°)

Solutions

a   The highest mean maximum occurs in January, which is 36°C. The lowest mean minimum occurs in July, which is 4°C.

b   By comparing the differences in length of the bars that show mean maximum and mean minimum, we can see that the smallest difference appears to occur in February. This difference is approximately $35 - 20.5 = 14.5$°C.

c   Again comparing the differences in the lengths of the bars, we can see the greatest different occurs in July. This difference is approximately $19.5 - 4 = 15.5$°C.

PRACTICE 8

The side-by-side chart below shows the percentage of Australians with Bachelor degree or above qualifications comparing various age groups in 2005 and 2015 (Source: abs.gov.au). Use this chart to answer the following questions.

a   In which age groups in 2005 and 2015 did the highest percentage hold a Bachelors degrees or above?

b   In which age group did the percentage holding a Bachelors degree or above increase the most between 2005 and 2015? By how much did it increase?

c   Over the age range 20–44, what was the average percentage holding a Bachelors degree or above in 2005 and 2015?

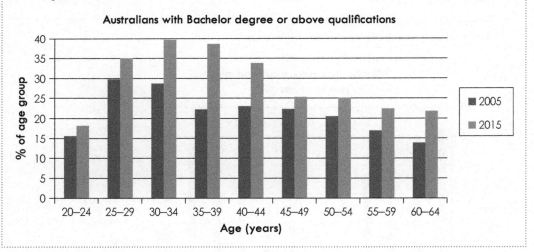

Australians with Bachelor degree or above qualifications

Segmented bar charts

A segmented bar chart is similar to a standard bar chart except the bars are made of different segments that represent sub-categories. This allows a greater amount of detail to be displayed to describe the data set. The segments are visually presented through the use of different shades of colour or markings. The bars in a segmented bar chart are usually drawn the same length because the focus is on the percentage contribution each sub-category makes rather than the size of the category as a whole.

EXAMPLE 9

The segmented bar chart below provides a break down of the various forms of physical activity undertaken in a week across a range of age groups in 2011–2012 (Source: abs.gov.au). Use this chart to answer the following questions.

a Which age group participated in the highest proportion of vigorous activity? What percentage of activity time in this age group was spent in vigorous activity?

b Which age group participated in the highest proportion of moderate activity? What percentage of activity time in this age group was spent in moderate activity?

c Which age group participated in the highest proportion of walking for fitness and recreation? What percentage of activity time in this age group was spent walking for fitness and recreation?

d Which age group participated in the highest proportion of walking for transport? What percentage of activity time in this age group was spent walking for transport?

e In the 45–54 years age group, what proportion of time was spent participating in the four forms of physical activity?

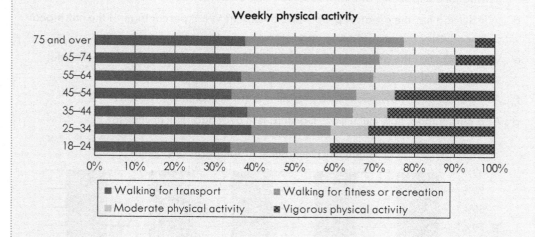

Weekly physical activity

Solutions

a The 18–24 years age group participated in the highest proportion of vigorous activity with approximately 42% of their activity time being spent in this way.

b From the chart, it appears the 65–74 years age group participated in the highest proportion of moderate physical activity. This is represented by the segment that spans from 72% to 91%, which amounts to 19% of their activity time.

c It appears that the 75 years and over age group participated in the highest proportion of walking for fitness or recreation. This is represented by the segment that spans from 37% to 77%, which amounts to 40% of their activity time.

Kathy Brady

d The 25–34 years age group participated in the highest proportion of walking for transport with approximately 38% of their activity time being spent in this way.

e The 45–54 years age group spent 34% of their activity time walking for transport, 31% of their activity time walking for fitness and recreation, 10% of their activity time participating in moderate activity, and 25% of their time participating in vigorous activity.

PRACTICE 9

The segmented bar chart below displays the nutritional composition of 100 grams of each of components in a ham, cheese, tomato and avocado sandwich made with wholemeal bread (Source: foodstandards.gov.au). Use this chart to answer the following questions.

a Which food has the greatest proportion of moisture? What percentage of the nutritional components of this food is made up of moisture?

b Which food has the greatest proportion of protein? What percentage of the nutritional components of this food is made up of protein?

c Which food has the greatest proportion of dietary fibre? What percentage of the nutritional components of this food is made up of dietary fibre?

d Which food has the greatest proportion of starch? What percentage of the nutritional components of this food is made up of starch?

e Which food has the greatest proportion of fat? What percentage of the nutritional components of this food is made up of fat?

f As a percentage, how much more protein is there in ham than in bread?

g As a percentage, how much more moisture is in avocado than in cheese?

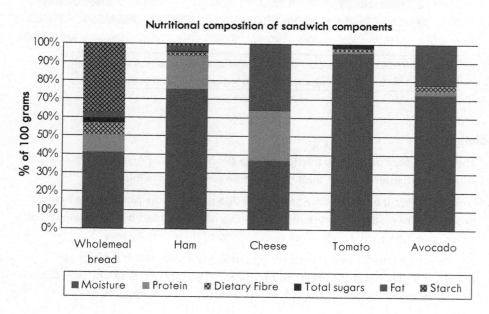

Nutritional composition of sandwich components

10.5 LINE GRAPHS

A line graph is used to represent the changes in a single variable with the passing of time or the relationship between two variables. For this reason, a line graph makes use of discrete numerical data. The key distinguishing feature of a line graph is that it connects the data points with straight-line segments. Without the straight-line, it would be difficult to read and interpret the graph to find patterns in relationships or the passing of time.

Reading and interpreting line graphs

Line graphs are particularly useful for identifying patterns and trends in the data such as seasonal effects, large-scale positive or negative change or turning points in change. When using a line graph, the x-axis represents the independent or non-measured variable (for example, years) whilst the y-axis has a scale and indicates the measurement of the dependent or changing variable (for example, growth in an investment).

    Here is an example of a line graph highlighting its features:

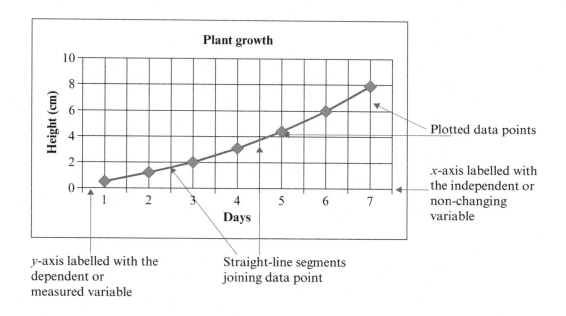

Making Connections Dependent and independent variables will be covered more fully in Chapter 14.

EXAMPLE 10

The line graph below presents information about the per capita consumption of wine (in litres of pure alcohol) in Australia from 1944 to 2009 (Source: abs.gov.au). Use this graph to answer the following questions.

Solutions

a　How many litres of pure alcohol from wine was consumed in 1989–1990?

a　Approximately 24 litres of pure alcohol were consumed in 1989–90.

b　In which annual period was approximately 26 litres of pure alcohol consumed?

b　We can see that approximately 26 litres were consumed in 1999–2000.

c　What was the increase in the consumption of pure alcohol in the twenty-year period 1964–65 to 1984–85?

c　In 1964–65 approximately 7 litres were consumed and in 1984–85 approximately 27.5 litres were consumed. This represents a 20.5 litre increase in consumption.

d　If one standard drink contains 12.5 mL of pure alcohol, how many standard drinks of wine were consumed per capita in 2008–2009?

d　In 2008–2009 approximately 29 litres of pure alcohol from wine was consumed. This converts to 29 000 mL. If there are 12.5 mL in one standard drink, then the per capita consumption of wine measured as standard drinks is:

$$29\,000 \div 12.5 = 2320 \text{ standard drinks}$$

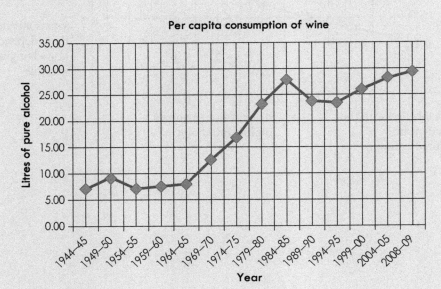

Per capita consumption of wine

PRACTICE 10

The line graph below displays the annual loss of Amazon rainforest (in square kilometres) from 1988 to 2015 (Source: rainforests.mongabay.com). Use this graph to answer the following questions.

a   What was the annual loss of rainforest in 1997?

b   In which year was approximately 11 000 square kilometres of rainforest lost?

c   What was the difference between the area of rainforest lost in 2004 compared to the area lost in 2007?

d   What was the average area of rainforest lost in the period 1994 to 1997?

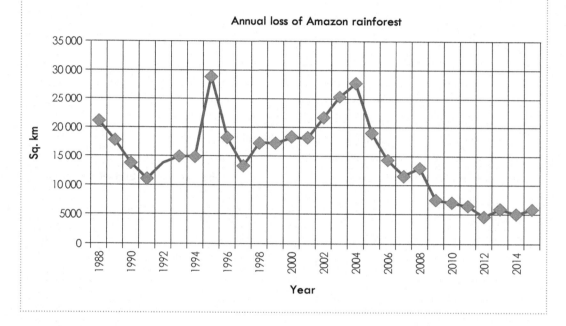

Annual loss of Amazon rainforest

Drawing line graphs

To draw a line graph we could use a technological tool such as a spreadsheet. However, these line graphs can also be easily drawn using grid paper. To draw a line graph we need the data that will be presented listed in a table of values.

EXAMPLE 11

The data in the table presents the number of goals kicked in AFL games each season by former West Coast and Carlton champion Chris Judd (Source: footywire. com). Draw a line graph to represent this data. What observations can we make from this graph?

YEAR	GOALS
2002	21
2003	29
2004	24
2005	15
2006	29
2007	20
2008	15
2009	12
2010	14
2011	14
2012	13
2013	11
2014	7
2015	4

Solution

To draw the line graph, the first thing that needs to be done is to divide both the x-axis and the y-axis using an appropriate scale. There are 14 years that need to be placed on the x-axis, so the length of the axis needs to be marked off with 14 evenly spaced tick marks. Next we need to examine the range of values that will be included on the y-axis, in this case these values range from 4 to 29. So on the y-axis, 0 will be where the axis meets the x-axis and the maximum value of the y-axis could be marked at 30. The axis is then divided using evenly spaced tick marks; in this case, a tick mark at every 5 would be sufficient. Finally, the axis labels and a title should be added.

Once the axes have been set up, they should look like the axes to the right. Now we need to plot the number of goals scored in each year by moving across from the respective value on the y-axis to placing a point above the tick mark for the relevant year. To do this it is best to work systematically by plotting the points moving from left to right across the x-axis. This corresponds to moving down the table of values. Finally, a straight-line segment is drawn to connect one point to another.

Your completed line graph should be similar to the graph to the right. We can see from this graph that the number of goals that Chris Judd kicked each season declined significantly after 2008.

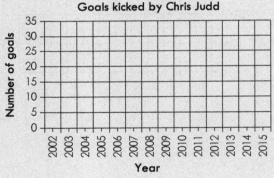

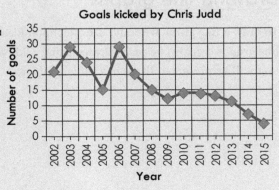

PRACTICE 11

The table below presents the annual number of divorces in Australia from 1970 to 2010 to the nearest 100 (Source: aifs.gov.au). Draw a line graph to represent this data. What observations can be made from this graph?

YEAR	NO. OF DIVORCES
1970	12 200
1975	24 300
1980	39 300
1985	39 800
1990	42 600
1995	49 700
2000	49 900
2005	52 000
2010	50 000

Line of best fit

Sometimes it is not possible to use the data in a table of values to create a neat line graph. This happens because the data has a scattered appearance and if this is the case we say that the data is displayed in a scatter plot.

Here is an example of a scatter plot where the respective weights and heights of a group of people have been plotted.

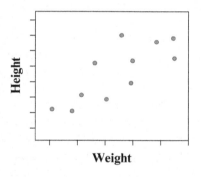

When the data takes the form of a scatter plot, it is still useful, however, to attempt to represent it with a straight line that best represents the trends in the data. This straight line is known as a **line of best fit**.

The graph to the right illustrates how the line of best fit for the weight versus height scatter plot could be drawn. The features of a line of best fit are that it should roughly go through the centre of the points and that the points should be evenly distributed around the line. This is achieved by checking that, of the points that do not lie on the line, approximately half lie above it and half lie below it.

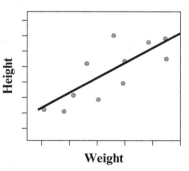

Line of best fit
A straight line
that represents
the data in a
scatter plot

A transparent ruler is useful when drawing a line of best fit because you can easily check the number of points above and below the line. Alternatively, another tip is to position the line of best fit by using a single piece of spaghetti.

EXAMPLE 12

The data table to the left presents the fat content and number of kilojoules in 100 grams of popular snack foods (Source: foodstandards.gov.au). A scatter plot of this data is shown below. Use this scatter plot to draw in a line of best fit.

SNACK FOOD	FAT	KJ
Muesli bar, plain	12.1	1445
Corn chip, regular	26.2	2043
Popcorn, air-popped	4.1	1426
Potato crisp, regular	33.9	2160
Prawn cracker	26.8	2190
Pretzel	7.2	1578
Water cracker	9.3	1697
Rice cracker	3.4	1660
Grissini stick	18.5	2000

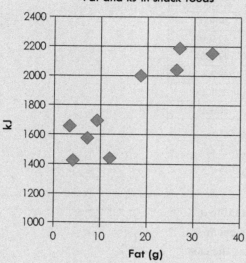

Fat and kJ in snack foods

Solution

A suitable line of best fit is shown here. The things to check are whether the line lies as much as possible in the centre of the points, and of the points that are not on the line, there are as many above the line as below the line.

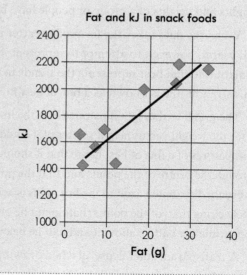

Fat and kJ in snack foods

PRACTICE 12

The weights and heights of a group of women are presented in the data table. Below are axes that have been prepared to plot this data.

WEIGHT (KG)	HEIGHT (CM)
53	148
46	152
53	156
69	160
50	164
67	168
80	172
95	176
75	180
61	183

a   On the prepared axes, plot the points provided in the table.

b   Use this scatter plot to draw in a line of best fit.

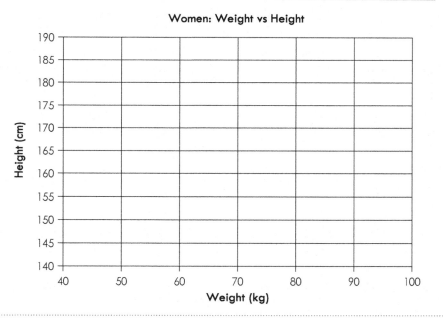

10.6 INTRODUCTION TO THE COORDINATE PLANE

In the final section of this chapter, we will introduce the coordinate plane and the use of coordinate geometry. The coordinate plane and coordinate geometry provides us with a way to describe the position of any point on the plane using what is known as an ordered pair, and to use these points to draw graphs. The French mathematician Renee Descartes (1596–1650) was the first to describe this technique and for this reason, in his honour, the coordinate plane is often referred to as the Cartesian plane.

Kathy Brady

Making
connections
In the next
chapter,
Chapter 11,
we will explore
further the
properties of
geometric figures.

Already in this chapter we have become familiar with the principle features of the coordinate plane: the horizontal x-axis and the vertical y-axis. On the coordinate plane, these axes continue infinitely in each direction. This is because a plane is defined as a two-dimensional surface made up of an infinite number of points.

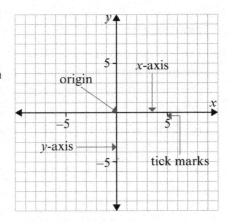

On this coordinate plane, we can see that the horizontal axis is labelled as the x-axis and the arrowheads on each end indicate that the axis continues infinitely in each direction.

The vertical axis is labelled as the y-axis and the arrowhead on each end also indicates that this axis continues infinitely in each direction.

The x-axis and the y-axis intersect at what is known as the origin. At the origin, the x-axis value and the y-axis value are both equal to 0. All the values on either axis are measured from that point.

On the x-axis, the values to the right of the origin are positive, and to the left are negative. As you move right from the origin on the x-axis, the positive values become larger, and as you move left from the origin on the x-axis, the values become smaller. On the y-axis, the values above the origin are positive, and those below are negative. As you move upwards from the origin on the y-axis, the positive values become larger, and as you move downwards from the origin on the y-axis the values become smaller.

Along each axis small tick marks with numbers are included. These are the labels that indicate the scale on the axis. In the diagram, the tick marks are positioned every 5 units; however, this can be altered to suit the range of values being graphed.

Plotting points on the coordinate plane

Ordered pair
A pair of
numbers that
describe the
position of
any point on
the coordinate
plane, written in
the form (x, y)

Every point on the coordinate plane can be described and identified by a unique pair of numbers written in the form (x, y). The first number in this pair is called the x-coordinate and this number indicates how far along the x-axis you need to move. The second number is called the y-coordinate and this represents the vertical distance that you need to move on the y-axis. This pair of numbers is known as an **ordered pair**.

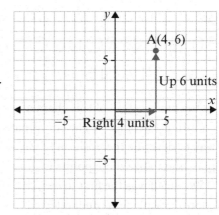

Consider, for example, the ordered pair (4, 6). To find the location of this point, we start at the origin and move along the x-axis 4 units, then we need to move vertically upwards 6 units. A point is then plotted and labelled with the coordinates of the ordered pair. In this case, at that point $x = 4$ and $y = 6$. Often we also label the point with a capital letter. This is helpful when we are plotting more than one point.

The sign of the numbers written in an ordered pair is important. A positive number indicates that you must either move to the right or upwards. A negative number indicates that you must either move to the left or downwards.

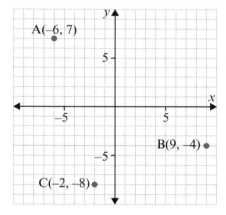

Any point on the x-axis has a y-coordinate of 0. This means that the points $(-7, 0)$ and $(6, 0)$ are both on the x-axis as shown in the graph below. Any point on the y-axis has an x-coordinate of 0. So in this graph, we can see that $(0, 3)$ and $(0, -5)$ are both on the y-axis. The coordinates of the origin are $(0, 0)$ and we use the letter O to label this point as seen on this graph.

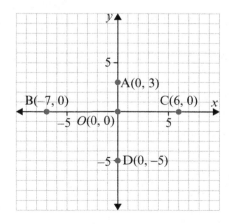

Kathy Brady

EXAMPLE 13

Plot the following points on a coordinate plane.

a A (4, 7) | **b** B (−2, 5) | **c** C (8, 0)
d D (6, −3) | **e** E (−1, −8) | **f** F (0, −6)

Solutions

The points are plotted and
labelled as shown:

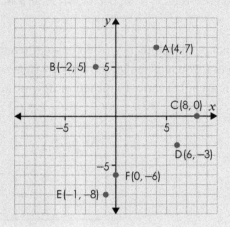

PRACTICE 13

Plot the following points on a coordinate plane.

a J (−6, 2) | **b** K (0, 7) | **c** L (1, 9)
d M (3, −8) | **e** P (0, 4) | **f** Q (−7, −2)

Quadrants

The *x*- and *y*-axes divide the coordinate plane into
four areas known as quadrants. The first quadrant
is on the top right-hand side, above the *x*-axis and
to the right of the *y*-axis. We call this Quadrant
1. In Quadrant 1, both the *x* values and *y* values
are positive. Moving in an anticlockwise direction,
the quadrant that is to the left of the *y*-axis and
above the *x*-axis is Quadrant 2. In Quadrant 2, the
x-values are negative and the *y*-values are positive.
Below the *x*-axis and to the left of the *y*-axis is
Quadrant 3 where both the *x*- and *y*-values are
negative. Finally, to the right of the *y*-axis and below
the *x*-axis is Quadrant 4. In Quadrant 4, the *x*-values are
positive and the *y*-values are negative. The quadrants are shown in this graph.

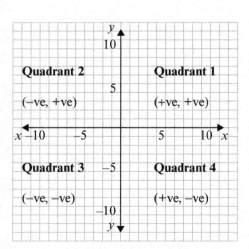

EXAMPLE 14

In which quadrants are the following points located?

		Solutions
a	A (−6, 7)	a Quadrant 2
b	B (4, 1)	b Quadrant 1
c	C (2, −5)	c Quadrant 4
d	D (−3, −8)	d Quadrant 3

PRACTICE 14

In which quadrants are the following points located?

a   M (9, −2)   b   N (−7, −1)

c   P (8, 3)   d   Q (−6, 5)

Table of values

We can also plot points on the coordinate plane using information that is given in a table of values. A table of values is usually used to present the coordinate pairs for a set of points that are plotted and then joined to form a line graph.

EXAMPLE 15

Plot the points given in this table of values and then join the points to form a line graph.

x	−5	−3	0	2	4
y	−9	−5	1	5	9

Solution

Rather than the coordinate pairs being written within brackets, here they are paired on a table of values. So, for instance, the first coordinate pair on the table is (−5, −9). We plot the points given on a table of values in the same way as if they were written within brackets, that is moving right or left to find the x value and then up or down to find the corresponding y value. The points given in the table of values would be plotted as:

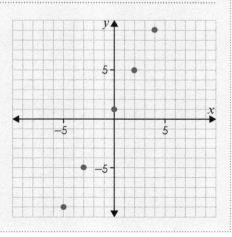

We can now join these points to form a line graph as shown:

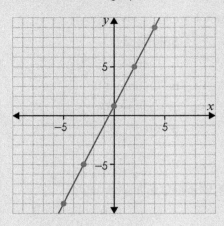

PRACTICE 15

Plot the points given in this table of values and then join to points to form a line graph.

x	−4	−2	0	3	6
y	11	7	3	−3	9

10.7 CHAPTER SUMMARY

SKILL OR CONCEPT	DEFINITION OR DESCRIPTION	EXAMPLES
Graph	Used to present exact numerical information to illustrate changes or trends in the data over time. Usually features a left-hand and bottom axis and involves plotting points that are joined by a line.	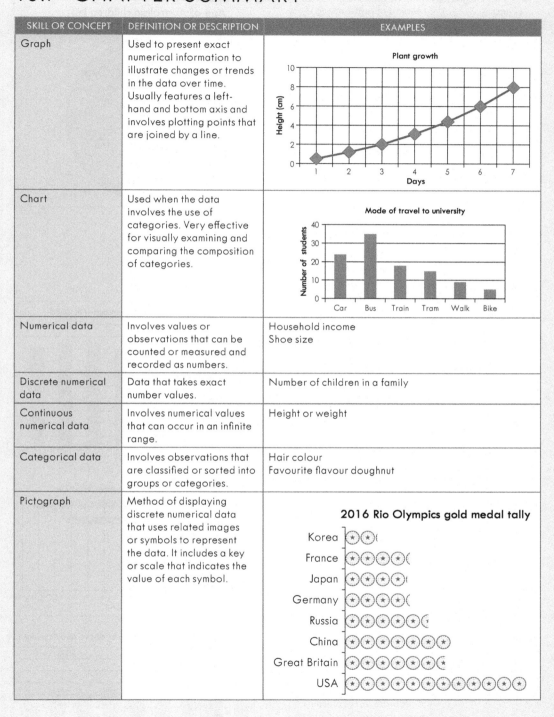
Chart	Used when the data involves the use of categories. Very effective for visually examining and comparing the composition of categories.	
Numerical data	Involves values or observations that can be counted or measured and recorded as numbers.	Household income Shoe size
Discrete numerical data	Data that takes exact number values.	Number of children in a family
Continuous numerical data	Involves numerical values that can occur in an infinite range.	Height or weight
Categorical data	Involves observations that are classified or sorted into groups or categories.	Hair colour Favourite flavour doughnut
Pictograph	Method of displaying discrete numerical data that uses related images or symbols to represent the data. It includes a key or scale that indicates the value of each symbol.	

Kathy Brady

SKILL OR CONCEPT	DEFINITION OR DESCRIPTION	EXAMPLES
Pie chart	Method of representing categorical data. Specifically shows the percentage that each category comprises of the whole, which is represented as a sector that corresponds to, and is labelled with, the percentage that it comprises.	**Age of Australian children in out-of-home care 2013–2014** 15–17 years 10% <1 year 19% 10–14 years 22% 1–4 years 25% 5–9 years 24%
Bar chart	Used to represent categorical data, to compare the size of different categories, to highlight the breakdown of sub-categories that make up a category or to track changes over a period of time.	
Vertical bar chart	Consists of vertical bars, a horizontal axis that represents the different categories and a vertical axis with a numerical scale to indicate the units used to tally the categories. A gap is left between the bars on the x-axis.	**Median weekly cost for childcare in 2015** Cost ($) vertical axis: 0, 50, 100, 150, 200, 250, 300, 350, 400, 450, 500 Categories: ACT, QLD, SA, TAS, NT, NSW, VIC, WA
Horizontal bar chart	Horizontal bars are drawn out from the y-axis, which is labelled with the different categories. The x-axis has a numerical scale to indicate the units used to tally the categories.	**Kilojoules in chocolate snacks (per 100 g)** Chocolate bar, plain Chocolate biscuit Chocolate ice-cream, reduced fat Chocolate ice-cream, regular Chocolate custard, reduced fat Chocolate custard, regular Kilojoules: 0, 200, 400, 600, 800, 1000, 1200, 1400, 1600, 1800, 2000, 2200, 2400
Side-by-side bar chart	Used to display data with categories divided into two or more sub-categories. Can also be used to display changes in categories over time. A separate bar represents each sub-category. Can be presented either horizontally or vertically.	**Australians with Bachelor degree or above qualifications** ■ 2005 ■ 2015 % of age group: 0, 5, 10, 15, 20, 25, 30, 35, 40 Age (years): 20–24, 25–29, 30–34, 35–39, 40–44, 45–49, 50–54, 55–59, 60–64

SKILL OR CONCEPT	DEFINITION OR DESCRIPTION	EXAMPLES
Segmented bar chart	Bars are made of different segments that represent sub-categories. Segments are visually presented using different colours. Bars are usually drawn the same length to focus on the percentage contribution of each sub-category.	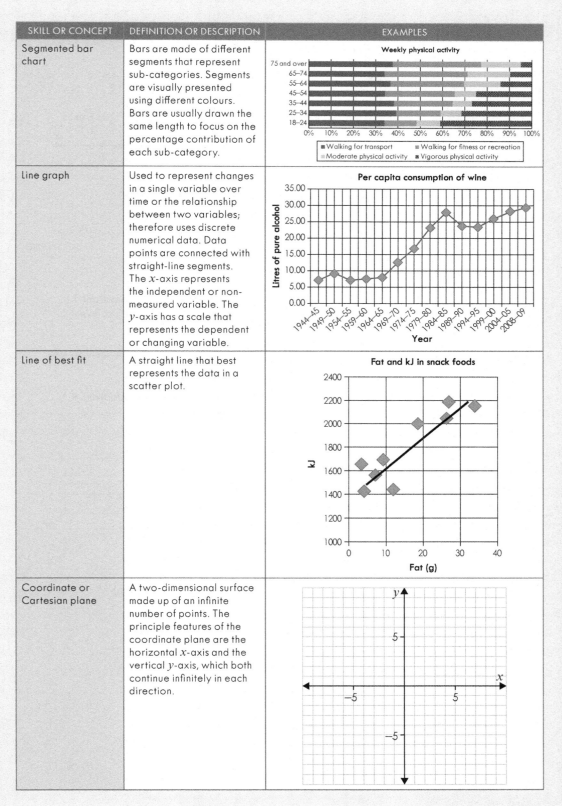
Line graph	Used to represent changes in a single variable over time or the relationship between two variables; therefore uses discrete numerical data. Data points are connected with straight-line segments. The x-axis represents the independent or non-measured variable. The y-axis has a scale that represents the dependent or changing variable.	
Line of best fit	A straight line that best represents the data in a scatter plot.	
Coordinate or Cartesian plane	A two-dimensional surface made up of an infinite number of points. The principle features of the coordinate plane are the horizontal x-axis and the vertical y-axis, which both continue infinitely in each direction.	

Kathy Brady

SKILL OR CONCEPT	DEFINITION OR DESCRIPTION	EXAMPLES
Ordered pair	Describes the position of any point on the coordinate plane. It is written in the form (x, y).	A (4, 6) Up 6 units Right 4 units −5 5 −5
Quadrants	The x- and y-axes divide the coordinate plane into quadrants.	Quadrant 2 (−ve, +ve) Quadrant 1 (+ve, +ve) −5 5 Quadrant 3 (−ve, −ve) Quadrant 4 (+ve, −ve)
Table of values	Used to present the coordinate pairs for a set of points that are plotted and then joined to form a line graph.	x: −5, −3, 0, 2, 4 / y: −9, −5, 1, 5, 9

x	−5	−3	0	2	4
y	−9	−5	1	5	9

10.8 REVIEW QUESTIONS

A SKILLS

1   Fill in the blank spaces by selecting the most appropriate word or phrase from this list.

categorical	Quadrant 2	segmented	vertical
Cartesian plane	line graph	discrete	line of best fit
length	origin	charts	proportional
scatter plot	horizontal	numerical	percentage
Quadrant 4	pictograph	graphs	continuous

a   Data that involves the use of categories is best represented in a _____.

b   In _____, the x-coordinates are negative and the
    y-coordinates are positive.

c   A _____ is a straight-line that represents the
    data in a _____.

d   _____ data involves observations that are classified or
    sorted into groups or categories.

e   The bars in a _____ bar chart are usually drawn the same length to
    focus on the percentage contribution each sub-category makes to the whole.

f   The _____ axis is called the y-axis.

g   A _____ displays discrete numerical data that uses related
    images or symbols to represent the information.

h   The coordinate plane is also referred to as the _____.

i   In a bar chart, the _____ of the bars are _____ to the
    size of the category they represent.

j   In _____, the x-coordinates are positive and the
    y-coordinates are negative.

k   _____ are used to present exact numerical information
    or trends in the data over time.

l   A pie graph shows the _____ that each category comprises of the whole.

m   The point where the x- and y-axes intersect is called the _____.

n   Data that involves values or observations that can be counted or measured and
    recorded as numbers is known as _____ data.

o   A _____ graph makes use of _____
    numerical data.

Kathy Brady

p   In the coordinate plane, the x-axis is the _____ axis.

q   Data that involves numerical values that can occur in an infinite range is known as _____ numerical data.

2   Plot the following points on a coordinate plane.

A   (3, 6)   B   (−5, 1)   C   (0, −7)

D   (8, −2)   E   (−6, −5)   F   (9, 0)

3   In which quadrants are the following points located?

a   M (−7, −4)   b   N (6, 1)

c   P (2, −8)   d   Q (−3, 9)

4   Plot the points given in this table of values and then join the points to form a line graph.

x	−5	−2	0	1	3
y	9	6	4	3	1

B   APPLICATIONS

1   The pictograph chart below shows the total number of medals won by the Australian team at the recent summer Olympic games (Source: wikipedia.com). Use the pictograph chart to answer the following questions.

a   How many medals did Australia win in 2004?

b   How many more medals did Australia win in 1996 than in 2012?

c   What was the highest number of medals won, and in which year?

d   Between these six Olympic Games, what was the average number of medals that Australia won?

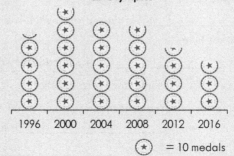

Total medals won by Australia at Olympics

1996   2000   2004   2008   2012   2016

(★) = 10 medals

2   This table shows the production of various fruit crops as a
percentage of total production that were grown in Australia
in the year ended 30 June 2014 (Source: abs.gov.au). Using
the template below, draw a pie chart to represent this
information, and then answer the question that follows.

CROP	PER CENT
Apples	20%
Pears	7%
Bananas	19%
Oranges	26%
Mandarins	7%
Pineapples	7%
Other	14%

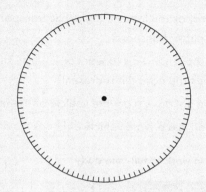

If the total production of fruit was approximately 1 400 000 kilograms, how many more
kilograms of oranges were produced than mandarins?

3   The side-by-side bar chart below presents the long-term mean maximum temperatures recorded
in Perth and Hobart (Source: bom.gov.au). Use this chart to answer the following questions.

a   What is the highest mean maximum temperature in each of these cities, and what is the
lowest mean maximum temperature?

b   In which month does the smallest difference between the mean maximum temperatures in
Perth and Hobart occur? What is this difference?

c   In which month does the greatest difference between the mean maximum temperatures
occur? What is this difference?

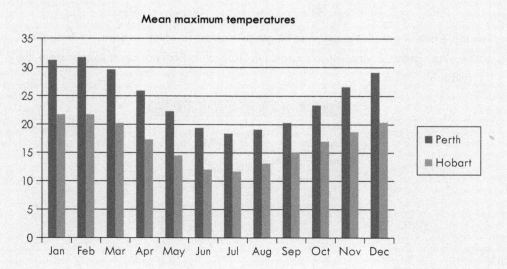

Mean maximum temperatures

4   The segmented bar chart below provides a break down of the main form of transport used
    to travel to work or full-time study in a range of Australian states in the years 2009 to 2012.
    (Source: abs.gov.au). Use this chart to answer the following questions.

    a   In which state did the highest proportion travel as a driver in a private vehicle? What
        percentage of commuters in this state drove a private vehicle?

    b   In which state did the highest proportion travel using public transport? What percentage
        of commuters in this state used public transport?

    c   In which state did the highest proportion walk to work or full-time study? What
        percentage of commuters in this state does this represent?

    d   In the WA, what percentage did not drive a private vehicle or use public transport?

    e   In SA, what percentage travelled in a private vehicle either as the driver or a passenger?

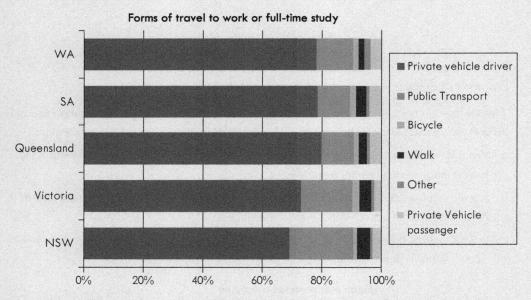

Forms of travel to work or full-time study

5   The table below details the average height (in metres) to which a variety of Australian
    native plants grow (Source: npsp.sa.gov.au). Draw a vertical bar chart to represent this
    information.

PLANT	AV. HEIGHT (M)
Golden wattle	6.5
Blue gum	25
Drooping sheoak	8
Christmas bush	3.5
Rock correa	1

6 The table below presents the number of tourists from China that visited Australia 2006 to 2016 to the nearest 1000 (Source: tra.gov.au). Draw a line graph to represent this data.

YEAR	NO. OF TOURISTS
2006	123000
2007	147000
2008	164000
2009	156000
2010	150000
2011	206000
2012	266000
2013	333000
2014	376000
2015	450000
2016	605000

7 The data table shows the number of games won by the teams in the 2016 AFL competition together with the number of points each team scored in the season to the nearest 10 (Source: footywire.com). Below are axes that have been prepared to plot this data.

a On the prepared axes, plot the points provided in the table.

b Use this scatter plot to draw in a line of best fit.

GAMES WON	POINTS SCORED
17	2220
17	2240
17	2130
16	2380
16	2480
16	2180
15	1860
12	1960
12	1950
10	2060
10	1940
9	1910
8	1710
7	1570
6	1780
4	1570
3	1770
3	1440

Teams: Games won vs Points scored

11

GEOMETRY

CHAPTER OBJECTIVES

» Identify properties of points, lines, planes and angles
» Find unknown angles using angle properties
» Classify two-dimensional figures using properties of angles and sides
» Find unknown angles and sides using properties of two-dimensional figures
» Classify three-dimensional figures using properties of faces, edges and vertices
» Apply Pythagoras' rule to find unknown sides in right-angled triangles
» Apply trigonometric ratios to find unknown sides in right-angled triangles

GEOMETRY AND ARCHITECTURE

The term architecture refers to all that is associated with the design and construction of buildings. Architecture begins with geometry, because geometric principles and properties are the key to designing buildings that are both strong and stable, and are pleasing to the eye.

Triangles, one of the simplest and most easily recognised geometric figures, are frequently used in architecture. Examples come from as far back as the pyramids of Egypt. A triangle is

stable and strong, so it is commonly used to design rafters that will support the tiles or roofing material. Additionally, the sloping side of triangles allows rainwater to run off the roof.

Rectangles are also well-recognised geometric figures that are usually evident in a building's floors, walls, windows and doors. Additionally, the rectangle is the most commonly used shape in multi-storey building design, such as apartment blocks or skyscrapers. This is because the construction of such a building is both cheaper and faster.

Not all geometric figures used in architecture have straight sides. For example, a circle is also a common feature used in arches or doorways. Domes are also a traditional architectural feature in the roofs of religious buildings, and are adapted in the design of sports stadiums. The advantage of a dome roof is that the weight of the dome is supported by its rim. This is particularly advantageous for sports stadiums as there cannot be any columns on the field.

Kathy Brady

11.1   WHAT IS GEOMETRY AND WHY IS IT IMPORTANT?

Geometry is one of the original, classic branches of mathematics. The word 'geometry' comes from Greek and means 'measurement of the Earth'. Ancient peoples developed the principles of geometry from their own observations as a practical tool for construction projects. Thus, geometry was originally about measurement and shape. Whilst these are still two important aspects in the study of geometry, the discipline has expanded over time to consider many more theoretical and analytical dimensions associated with studying lines, angles, figures and space.

In our contemporary world, geometry has numerous applications such as in surveying, navigation and space travel. Additionally, geometry is integral to computer imaging, for example, in the medical field with the use of CT scans or MRIs, or in recreational pursuits such as animation or gaming. In short, geometry is important because it helps us comprehend the spatial relationships associated with lines, angles, shapes and solids that we encounter in our everyday world.

11.2 POINTS, LINES, PLANES AND ANGLES

Points, lines, planes and angles are the foundational building blocks for all geometric concepts and processes.

Defining points, lines and planes

All things geometric start with **points**. A point has no dimensions; it is an exact location in space. We can identify points by labelling them with a single letter such as point *A*, point *B* and point *C* in the first figure.

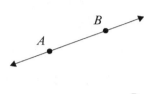

From points, we can build **lines**. A line has only one dimension, length. We highlight that the line extends infinitely in either direction by the arrowheads on each end. Lines can be identified and labelled using two points that lie on the line. In the second figure, we can label the line either $\overleftrightarrow{AB}$ or $\overleftrightarrow{BA}$.

When we talk about a line, we are usually referring to what is known in geometry as a **line segment**. A line segment has a fixed length because it has two distinct endpoints. Line segments can also be identified and labelled using two points that lie on the segment. In the third figure, we can label the line segments as either $\overline{AB}$ or $\overline{BA}$.

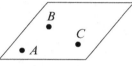

We can also build **planes** from a collection of points. A plane has two dimensions, length and width, but no thickness. Because we need to draw a plane with edges, it is important to remember that it actually extends infinitely in all directions. We need at least three points in order to be able to form and label a plane. In the fourth figure, we can use the label plane *ABC*.

Point
An exact location in space

Line
A collection of points in a straight path that extends infinitely

Line segment
A part of a line that is enclosed by two distinct endpoints

Plane
A collection of points that form a flat two-dimensional surface that extends infinitely in all directions

---

EXAMPLE 1

Draw a diagram to represent the following.

a $\overline{XY}$

Solutions

a This diagram requires a line segment that has *X* and *Y* as the endpoints. It does not matter how long we draw this line segment or in what direction it lies. Nor does it matter which end the points *X* and *Y* are labelled. This means that there can be an infinite number of ways this diagram can be drawn. One way is shown.

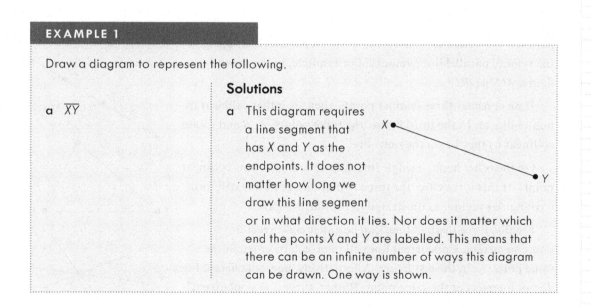

b $\overleftrightarrow{CD}$

b This diagram requires an infinite line that contains the points C and D. Again it does not matter where the points are illustrated on the line or the direction of the line. One possible diagram is shown.

PRACTICE 1

Draw a diagram to represent the following.

a $\overleftrightarrow{PQ}$ **b** $\overline{EF}$

Relationships between points and lines

A number of important relationships exist between points and lines. Firstly, if you have a point and a line, then the point will either lie on the line or it will not. Next, if you have two distinct points, there is only one line that can pass through those points.

Now, if you have two distinct lines they can possibly meet at a point, such as in the first figure.

On the other hand, the lines may never meet at all. If this is the case, then the lines are referred to as being **parallel lines**. Parallel lines are always the same distance apart over their entire length; they will never meet. Diagrams of parallel lines are always drawn with arrowheads to indicate that they are parallel. We write the relationship between parallel lines using the symbol parallel line symbol ∥. For example, in the second figure, $MN \parallel OP$.

If we consider three distinct points, they are either **collinear** or non-collinear. In the third figure, the three points, A, B and C, are collinear as they lie on the same line.

On the other hand, a single line may not connect non-collinear points. If this is the case, the three non-collinear points will form a triangular region, as illustrated in the fourth figure.

Finally, three distinct lines can be either **concurrent** or non-concurrent. Concurrent lines all intersect, or cross, at the same point, as in the last figure. Alternatively, non-concurrent lines do not intersect at the same point. Rather, three non-concurrent lines will enclose a triangular region.

Parallel lines
Two, or more, lines in the same plane that never meet

Collinear points
Three, or more, points that lie on the same line

Concurrent lines
Three, or more, lines that all intersect, or cross, at the same point

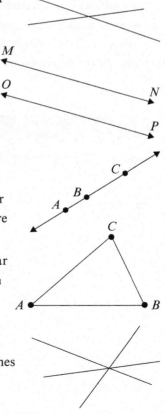

EXAMPLE 2

Sketch and label diagrams to illustrate the following.

Solutions

a *CD*‖ *EF*

a Remember that parallel lines must be drawn with arrowheads to indicate that they are parallel.

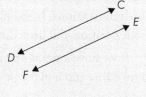

b Three collinear points *XYZ*

b

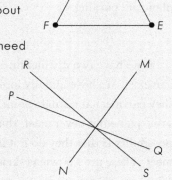

c Three non-collinear points *DEF*

c Remember that three non-collinear points form a triangular region.

d Three distinct concurrent lines

d In this example, we are not given any information about the labels for the three concurrent lines. So we need to select our own. Firstly, draw the lines all passing through the same point, and then select the distinct letters that will label the points at each end of the lines.

e Three distinct non-concurrent lines

e The distinct non-concurrent lines do not intersect at the same point instead they enclose a triangular region as in the diagram for part (c). In this diagram, the three distinct lines are, in no particular order, $\overline{EF}$, $\overline{FD}$ and $\overline{DE}$.

PRACTICE 2

Sketch and label diagrams to illustrate the following.

a *UV* ‖ *WX* ‖ *YZ*

b Four collinear points *QRST*

c Four distinct concurrent lines

d Three distinct non-concurrent lines $\overline{LM}$, $\overline{KL}$ and $\overline{MK}$

Points, lines and planes in space

So far we have only considered the relationships between points and lines in two dimensions; that is, on a plane. Here we will look at the relationship between points, lines and planes in three dimensions. In geometry, this is frequently referred to as **'space'** where objects can have both position and direction.

If we have a point and a plane, the point either lies on the plane or it does not. Now if we have a line and a plane in space, three possibilities can exist:

The line lies completely on the plane.

The line passes through the plane at a point.

The line never meets the plane. This means that the line and the plane are parallel.

If we have two distinct lines in space, again three possibilities can exist. We have already looked at two of these. The first is that they can meet at a single point. The second is that they lie on the same plane but never meet; that is, they are parallel. The third possibility is that they do not lie on the same plane and they never meet. These are known as **skew lines**.

When two lines lie in the same plane, they must either intersect at a point or be parallel. Skew lines do neither; therefore, they must lie on two different planes. This means that skew lines can only exist in a three-dimensional space.

Finally, we will consider the relationship between two distinct planes in space.

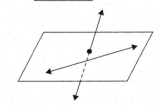

The first possibility is that the planes will intersect in a line.

The second possibility is that the planes will never intersect. This means that the planes are parallel.

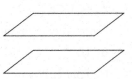

CRITICAL THINKING

Look around a room. How do the planes of the front wall and the ceiling intersect? Now, how do the planes that are the front wall, a side wall and the ceiling intersect? What can you say about the planes that are the front wall and the back wall?

Hold a book up by the spine. How does the plane of each page intersect with the spine? What can you say about how many planes in space might pass through a given line?

EXAMPLE 3

Sketch and label diagrams to illustrate the following.

Solutions

a $\overleftrightarrow{AB}$ lies on a plane

a

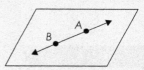

b $\overleftrightarrow{AB}$ intersects a plane at A

b The point of intersection between a line and a plane is a point that lies both on the line and on the plane.

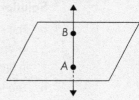

c Skew lines $\overleftrightarrow{AB}$ and $\overleftrightarrow{CD}$

c Skew lines are a bit tricky to sketch. Try to create the illusion that they are on totally different planes, as per the definition.

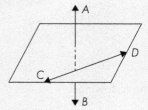

PRACTICE 3

Sketch and label diagrams to illustrate the following.

a $\overleftrightarrow{AB}$ never meets a plane

b Skew lines $\overleftrightarrow{MN}$ and $\overleftrightarrow{OP}$

c Two planes intersect at $\overleftrightarrow{AB}$

Kathy Brady

Angles

Finally, in this section we introduce one of the most important concepts in geometry: **angles**. Like all of the geometrical concepts that have already been introduced, angles are constructed when the basic blocks of points and lines come together to form a **ray**. A ray is only part of a line that has one endpoint and continues infinitely in the other direction.

Unlike lines, we need to be quite specific in naming and labelling rays. The endpoint letter is always named first. The ray in the first figure would be labelled $\overrightarrow{AB}$.

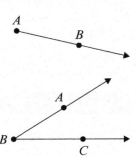

An angle is formed when two rays meet at a common endpoint. This endpoint is known as the vertex. In the second figure, the vertex, is the point B. We can name and label this angle one of two ways: $\angle ABC$ or $\angle CBA$. It is important that the vertex letter is always the middle letter.

EXAMPLE 4

Sketch and label the following.

a $\overrightarrow{PQ}$

b $\angle XYZ$

Solutions

a Remember that the letter on the fixed endpoint of the ray is always named first in the label.

b The letter at the vertex is always the middle letter in the label. It does not matter which of the other letters you use to label each arm of the angle, either way would be correct.

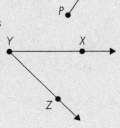

PRACTICE 4

Sketch and label the following.

a $\overrightarrow{XA}$

b $\angle MKL$

c The angle formed by $\overrightarrow{ST}$ and $\overrightarrow{SR}$

11.3 NAMING, CLASSIFYING AND CALCULATING ANGLES

How angles are measured

When a ray is rotated around its endpoint back to its original position, a revolution has occurred. A revolution can be divided into 360 equal parts that are known as **degrees**. The symbol for degrees is °. So, a full revolution is equal to 360°, and 1° is $\frac{1}{360}$ of a revolution. The size of an angle is the amount of rotation the ray makes from its starting position to its finishing position. This means that we can measure the size of any angle using degrees.

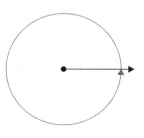

Degree
The unit of measure for the size of an angle

Naming angles according to size

Angles can be named and classified in a variety of ways. For the remainder of this section, we will look at each of these in turn, commencing with how angles are named and classified according to their size.

Straight-line angle

One half of a revolution forms a straight line, which is known as a *straight-line angle*. This angle is equal to 180°, which is half the number of degrees in a full revolution. In the first figure, $\angle PQR = 180°$.

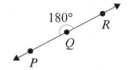

Right angle

One quarter of a revolution measures one quarter of 360°, which is 90°. This angle is called a *right angle*. The small square drawn where the two rays meet is the symbol used to indicate a right angle. In the second figure, $\angle XYZ = 90°$. Notice that X and Z are not labelling any particular points. Rather, the labels are placed at the ends of each ray. This is the convention usually used when labelling angles.

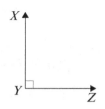

Acute angle

An angle that is less than 90° is called an *acute angle*. In the third figure, $\angle PQR$ is an acute angle. Notice the small curve in the diagram that is used to indicate the angle that is being referred to.

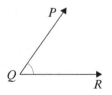

Kathy Brady

Obtuse angle

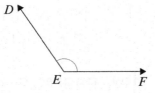

An angle that is greater than 90° but less than 180° is known as an *obtuse angle*. In the first figure, ∠*DEF* is an obtuse angle.

Reflex angle

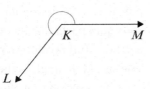

An angle that is greater than 180° but less than 360° is called a *reflex angle*. In the second figure, the curve that indicates the angle of reference is important as it shows that we are referring to the reflex angle ∠*LKM*. The ∠*LKM* could also be an obtuse angle if a small curve was positioned inside of the two rays.

EXAMPLE 5: APPLICATION

What is the size of the angle between the hour hand and the minute hand at 4 p.m.? What type of angle is this?

Solution

A clock face represents a 360° full revolution that is divided into 12 equal sections for each of the hour markers. Therefore, the number of degrees in each of these sections will be $\frac{360}{12} = 30°$. At 4 p.m. the angle between the hour and minute hands will be $4 \times 30° = 120°$. This is an obtuse angle.

PRACTICE 5: APPLICATION

a   What is the size of the angle between the hour and the minute hand at 7 a.m.? What type of angle is this?

b   What is the size of the angle between the hour and the minute hand at 4:30 a.m.? What type of angle is this?

Complementary and supplementary angles

Complementary and supplementary angles are named and classified according to the relationships that can exist between two separate angles.

Complementary angles are two angles that add to 90°. In the figure, the small square at the vertex of the angle tells us that this is a right angle that measures 90°. This means that ∠*a* + ∠*b* = 90°. Therefore, ∠*a* and ∠*b* are complementary.

Complementary
angles
Two angles that
add up to 90°

Supplementary angles are two angles that add to 180°. In the figure, the small semi-circle on the line tells us that this is a half revolution or straight-line angle that measures 180°. This means that $\angle a + \angle b = 180°$. Therefore, $\angle a$ and $\angle b$ are supplementary.

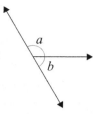

EXAMPLE 6

If possible, find the complement and supplement of:

Solutions

a   32°

a   The complement of any angle is the size of a second angle that will produce an angle sum of 90°. Therefore, the complement of 32° is $90° - 32° = 58°$. This means that 32° and 58° are complementary angles.

The supplement of any angle is the size of a second angle that will produce an angle sum of 180°. Therefore, the supplement of 32° is $180° - 32° = 148°$. This means that 32° and 148° are supplementary angles.

b   105°

b   Since 105° is greater than 90° it does not have a complement. The supplement of 105° will be $180° - 105° = 75°$.

PRACTICE 6

If possible, find the complement and supplement of:

a   89°   **b**   98°

Angles at a point

When two or more lines intersect at one point, at least two angles are created at that point. See some examples below:

One particular type of angle classification results from the intersection of two lines at one point. These are called **vertically opposite angles**.

When we refer to vertically opposite angles, the word 'vertical' means that the angles share the same vertex, rather than the usual 'up and down' meaning of the word. In the figure, $\angle a$ and $\angle c$ are vertically opposite, and $\angle b$ and $\angle d$ are also vertically opposite. A very important property of vertically opposite angles is that they are *equal* in size. This means that $\angle a = \angle c$ and $\angle b = \angle d$.

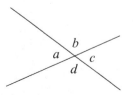

Kathy Brady

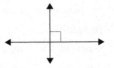

There is a special case of vertically opposite angles that occurs when two lines intersect at right angles, or 90°. In this circumstance, we say that the lines are **perpendicular lines**.

When one line intersects another line at right angles, or 90°, each of the fours angles formed at the point of intersection is 90°.

We can use logical reasoning and the rules associated with angles at a point to calculate the unknown size of angles in geometric diagrams. Some other useful rules to remember when working with angles at a point are that:

- angles in a *revolution* add up to 360°
- angles in a *straight line* add up to 180°.

When rules or reasons are used to calculate the size of angles, both the rule and the angle they apply to must be noted. This means that sometimes a diagram will need to be labelled to identify the various angles that are being referred to in the explanation.

Perpendicular lines
Lines that are at right angles to each other

EXAMPLE 7

Calculate the size of the unknown angle in the following diagrams.

a

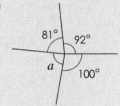

b

c

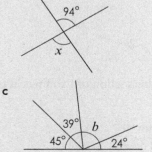

Solutions

a   In this example, the figure represents angles in a revolution. This means that the sum of the angles is 360°. We are given the size of three of the angles but do not know the size of ∠*a*. So, first we add up the given angles: 81° + 92° + 100° = 273°. We can now calculate the size of ∠*a* by finding the difference between 360° and 273°: ∠*a* = 360° − 273° = 87°. (We could check this answer by adding up all of the angles to make sure they equal 360°.)

b   In this example, the figure represents vertically opposite angles, which are equal. Therefore, ∠*x* = 94°.

c   In this example, the figure represents angles around a straight-line angle. This means that the sum of these angles will be 180°. We are given the size of three of the angles, but do not know the size of ∠*b*. So, first we add up the given angles: 45° + 39° + 24° = 108°. We can now calculate the size of ∠*b* by finding the difference between 180° and 108°: ∠*b* = 180° − 108° = 72°. (We could check this answer by adding up the angles to make sure they equal 180°.)

d

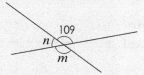

d In this example, some extra labelling of the diagram is useful to identify some of its features. In the figure below, the two lines intersect to create vertically opposite angles, which are equal, so $\angle m = 109°$. To find $\angle n$, the line AB is a straight-line angle. This means that the sum of 109° and $\angle n$ will be 180°, so $\angle n = 180° - 109° = 71°$

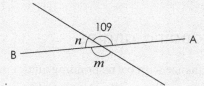

e

e In this example, the figure represents angles in a revolution. We are given the size of one of the angles, 64°. It might appear that there is insufficient information in the diagram to find the size of $\angle x$, but the small square that indicates a right angle is telling us that the angle at that point is 90°. So, the sum of the given angles is $64° + 90° = 154°$. We can now calculate the size of $\angle x$ by finding the difference between 360° and 154°: $\angle x = 360° - 154° = 206°$. (We could check this answer by adding up the angles to make sure they equal 360°.)

PRACTICE 7

Calculate the size of the unknown angle in the following diagrams.

a

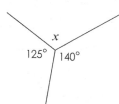

b

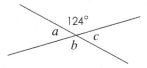

c

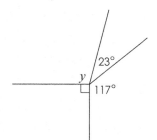

d

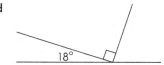

Angles associated with parallel lines

Another way that angles can be named and classified is associated with the situation when two, or more, parallel lines are intersected by a third line. Here is an example:

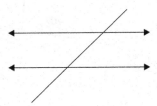

The third line that intersects with the parallel lines is called a *transversal*. Each type of angle that is created in this situation has a special name.

Corresponding angles

We will commence with corresponding angles. *Corresponding angles* are located in corresponding positions with relation to the parallel lines. Corresponding angles are always *equal*. So as shown in the figure:

$\angle 1 = \angle 3$

$\angle 2 = \angle 4$

$\angle 5 = \angle 7$

$\angle 6 = \angle 8$

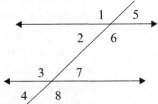

Alternate angles

Alternate angles are located on opposite sides of the transversal, and are inside the two parallel lines. There are two different ways that alternate angles can be paired and they both can be recognised by the Z shape that they form. Alternate angles are always *equal*. So as shown in the figure:

$\angle \blacksquare = \angle \blacksquare$

$\angle \blacksquare = \angle \blacksquare$

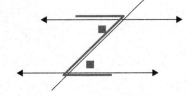
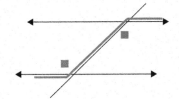

Co-interior angles

Finally, *co-interior angles* lie inside the two parallel lines, but on the same side of the transversal. Co-interior angles are *supplementary*. So as shown in the figure: $\angle a + \angle b = 180°$.

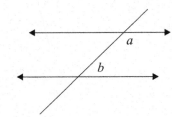

We can use the properties of angles associated with parallel lines to calculate the size of unknown angles in geometric diagrams.

EXAMPLE 8

Calculate the size of the unknown angles in the following diagrams.

Solutions

a

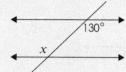

a In the figure, $\angle x$ and 130° are alternate angles. We can see this from the Z-shape that they form. Because alternate angles are always equal $\angle x = 130°$.

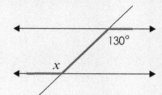

b
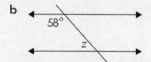

b In this example, the angles shown are co-interior angles. This is because they lie inside the parallel lines and on the same side of the transversal. Co-interior angles are supplementary, so they add up to 180°. This means that $\angle z = 180° - 58° = 122°$.

c
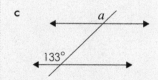

c In this example, the angles shown are corresponding angles because they are located in corresponding positions in relation to the parallel lines. Corresponding angles are always equal, so $\angle a = 133°$.

d
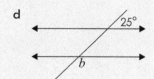

d This is an example of how labelling is required to help identify the angles that are being referred to in the explanation. In the figure below, $\angle ABC = \angle BDE$ because they are corresponding angles. Therefore, $\angle BDE = 25°$. Now $\angle BDE$ and $\angle b$ join together to create a straight-line angle around point D on the line AF. This means that the sum of $\angle BDE$ and $\angle b$ will be 180°. Since $\angle BDE = 25°$, $\angle b = 180° - 25° = 155°$.

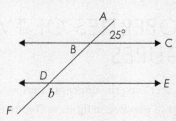

Kathy Brady

e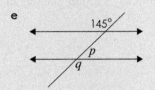

e Some additional labelling has been used to assist with the explanation. In the figure below, the angle that measures 145° is vertically opposite to $\angle ABC$, therefore $\angle ABC = 145°$. Now $\angle ABC$ and $\angle p$ are co-interior angles. This means that the sum of $\angle ABC$ and $\angle p$ is 180°. Since $\angle ABC = 145°$, $\angle p = 180° - 145° = 35°$. Finally, $\angle ABC$ and $\angle q$ are corresponding angles. Therefore, $\angle ABC = \angle q = 145°$.

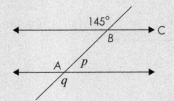

PRACTICE 8

Calculate the size of the unknown angles in the following diagrams.

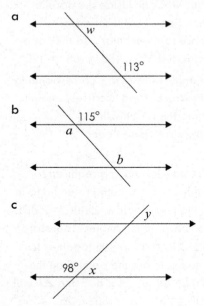

a

b

c

11.4 PROPERTIES OF TWO-DIMENSIONAL FIGURES

In the next two sections, the concepts that we have explored so far will be used to define and describe a range of geometric figures. Geometric figures can be either two-dimensional or three-dimensional.

Polygons

We will focus on two-dimensional figures in this section, commencing with **polygons**.

A polygon is a simple, closed figure that is made up of straight, line segments. In this definition, 'simple' refers to the property that the lines that make up the polygon do not cross. Here are some examples of polygons that are closed figures formed from straight lines that do not cross.

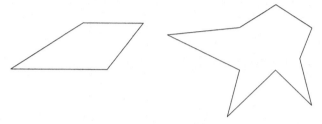

These shapes are examples of figures that are not polygons.

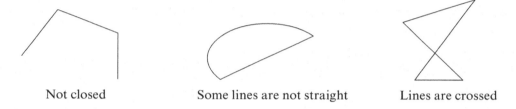

Not closed   Some lines are not straight   Lines are crossed

Naming and classifying polygons

Polygons are named and classified according to the number of sides and the number of angles that they have. Here are some commonly known polygons.

NAME OF POLYGON	NUMBER OF SIDES AND ANGLES	EXAMPLE
Triangle	3	
Quadrilateral	4	
Pentagon	5	
Hexagon	6	
Octagon	8	

Kathy Brady

Using this naming system, if a polygon has sides of equal length and angles of equal size, then it is known as a *regular polygon*. For example, a pentagon with sides of equal length and angles of equal size is known as a regular pentagon.

Polygons are also classified according to whether they are convex or concave.

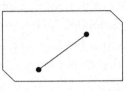

In a *convex* polygon, a line segment joining any two points inside the figure also lies completely inside the figure, as shown in the first figure.

If some part of the line joining the two points inside the figure lies outside of the figure, then it is a *concave* polygon, as shown in the second figure.

Triangles

A triangle is the simplest of all polygons. It has three sides and three angles. Triangles are named and classified according to the length of their sides and/or the size of their angles.

Firstly, an *equilateral triangle* has three equal length sides and three equal angles. Therefore, in the first figure, $\overline{AB} = \overline{BC} = \overline{CA}$ and $\angle ABC = \angle BCA = \angle CAB$. The notation of a small dash on each side in the figure is used to indicate that the sides are equal in length.

In an isosceles triangle, two of the sides are equal in length and the base angles are equal in size. In the second figure, $\overline{XY} = \overline{XZ}$ and $\angle XYZ = \angle XZY$.

A *scalene triangle* has no sides that are equal in length and no angles that are equal in size, as shown in the third figure.

A *right-angled triangle* has one angle that measures 90°. As we have already noted, an angle of 90° is commonly referred to as a right angle and a small square symbol is used to represent the right angle where it is located, as shown in the fourth figure.

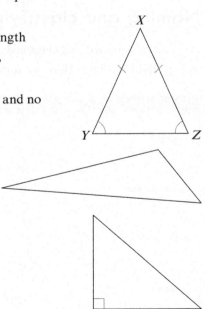

EXAMPLE 9

Sketch and label an isosceles triangle with an obtuse angle.

Solution

In this sketch, the obtuse angle is evident at the top of the triangle and the equal length sides have been indicated by the small dash, that shows $\overline{PQ} = \overline{PR}$.

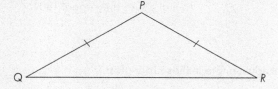

PRACTICE 9

Sketch and label an isosceles right-angled triangle.

Angle sum of a triangle

An extremely important property of triangles is that for all triangles no matter what type, the *sum of each of the angles of the triangle will always be equal to* 180°.

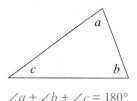

$$\angle a + \angle b + \angle c = 180°$$

We can use this property to determine the size of unknown angles in given triangles.

EXAMPLE 10

Find the size of the unknown angle in these triangles.

a
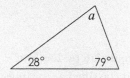

Solutions

a   The sum of the angles in this triangle is equal to 180°. We are given the size of two of the angles but do not know the size of $\angle a$. We commence by working out the total number of degrees in the given angles: $28° + 79° = 107°$. We can now calculate the size of $\angle a$ by finding the difference between 180° and 107°: $\angle a = 180° - 107° = 73°$. (We could check this answer by adding up all of the angles to make sure they equal 180°.)

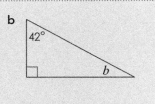

b In this example, we are given information about the size of two of the angles, one is 42° and the other is 90° as indicated by the small square symbol. We commence by finding the sum of these two angles: 42° + 90° = 132°. Now ∠b is the difference between 180° and 132°: ∠b = 180° − 132° = 48°. (We could check this answer by adding up the angles to make sure they equal 180°)

PRACTICE 10

Find the size of the unknown angle in these triangles.

a

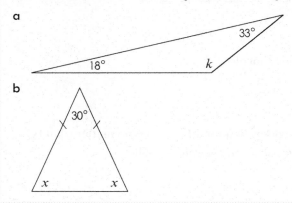

b

CRITICAL THINKING

If the sum of the angles in any triangle is equal to 180°, and in an equilateral triangle each angle is the same size. What is the measure of each angle in an equilateral triangle?

Again, if the sum of the angles in any triangle is equal to 180°, and in a right-angled triangle one of the angles measures 90°, what is the sum of the other two angles in a right-angled triangle?

Quadrilaterals

Parallelogram
A quadrilateral that has opposite sides parallel and equal in length, and opposite angles equal in size

Quadrilaterals are also a well-recognised family of polygons. A quadrilateral has four sides and four angles. Like triangles, quadrilaterals are named and classified according to the length of their sides and/or the size of their angles. We will start by looking at the set of quadrilaterals with two pairs of parallel sides, which are known as **parallelograms**. A parallelogram is a quadrilateral that has opposite sides parallel and equal in length, also the opposite angles are equal in size.

The figure to the right shows a *parallelogram*, the arrowheads and double arrowheads indicate the opposite pairs of parallel sides; that is, $\overline{AB} \parallel \overline{DC}$ and $\overline{AD} \parallel \overline{BC}$. The opposite sides are also equal in length, which means $\overline{AB} = \overline{DC}$ and $\overline{AD} = \overline{BC}$. Also because opposite angles are equal in size, $\angle A = \angle C$ and $\angle B = \angle D$.

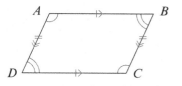

While this is the most commonly recognised form of a parallelogram, there are actually three other special types of parallelograms: rectangle, square and rhombus.

The first figure is a *rectangle*. It has two pairs of opposite sides that are parallel and equal in length, and opposite angles that are equal. By definition, it is a parallelogram. The particular features of a rectangle are all of its angles are equal, each measuring 90°. The small square symbol in each corner indicates that each angle is a right angle.

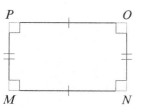

The second figure is a *square*, which is another special type of parallelogram. It has two pairs of opposite sides that are parallel and equal in length, and opposite angles that are equal. The particular properties of a square are that its sides are equal in length, and its angles are also equal. We can see that the angles are all right angles measuring 90°. The small dashes indicate that all of the sides are equal in length.

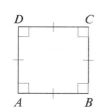

The third figure is a *rhombus,* which is also a special type of parallelogram. It has two pairs of opposite sides that are parallel and equal in length, and opposite angles that are equal. The particular feature of a rhombus is that its sides are all equal in length.

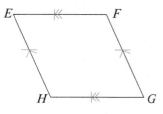

The parallelogram group are not the only types of quadrilaterals. A **trapezium** is another quadrilateral that has only one pair of parallel sides.

The fourth figure shows a *trapezium* as it is usually drawn with the base and the upper sides parallel. The arrowheads on the figure indicate that $\overline{AB} \parallel \overline{CD}$.

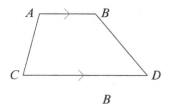

The final member of the quadrilateral family is a **kite**. This is a quadrilateral that has pairs of adjacent sides equal in length.

The fifth figure shows a *kite* in which we can see that there are two pairs of adjacent sides that are equal in length; that is, $\overline{AB} = \overline{CB}$ and $\overline{AD} = \overline{CD}$. We can also see in this diagram that each pair of adjacent sides has a different length. A final property of a kite is that the angles where each pair of sides meet are also equal in length; that is, $\angle A = \angle C$.

Trapezium
A quadrilateral that has *only* one pair of parallel sides

Kite
A quadrilateral with pairs of adjacent sides equal in length

Kathy Brady

Now that the members of the quadrilateral family have been introduced we can link them together on the quadrilateral family tree.

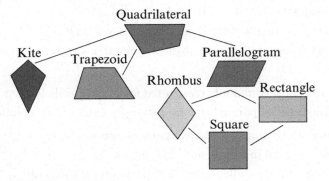

Sketch and label the following figures.

Solutions

a   An isosceles trapezium

a   An isosceles trapezium takes on some of the properties that are associated with isosceles triangles. Firstly, in all trapeziums the base and upper sides are parallel, as indicated in this diagram. Then, like an isosceles triangle the slanting sides are equal in length, and the base angles are equal in size. All of these features have been labelled in this diagram.

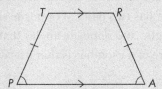

b   A kite with one obtuse angle

b   In the kite shown on page 307, the horizontally opposite angles are both obtuse. This question calls for a kite with just one obtuse angle, which will need to be drawn either at the top, as per the figure below, or at the bottom.

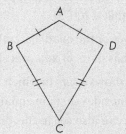

Exterior and interior angles of polygons

When we talk about the angles in a polygon, we are referring to the angle that is formed where two adjacent sides meet. These angles are on the inside of the figure, and are therefore known as **interior angles**.

> Interior angle of a polygon
> An angle that is formed within a polygon where two adjacent sides meet

As the number of sides of a polygon increases, the number of angles will also increase, because in all polygons the number of angles equals the number of sides. We have already seen that in a triangle the sum of the angles is 180°. We can use this property to determine the sum of the interior angles in any polygon. Let's start with a quadrilateral.

This quadrilateral has been divided into two triangles.

The sum of the angles in one of the triangles is $\times + \times + \times = 180°$.

The sum of the angles in the other triangle is $\bullet + \bullet + \bullet = 180°$.

This means that the sum of all the angles is $180° + 180° = 360°$.

So, in general, the:

Sum of the interior angles in a quadrilateral is 360°

Now try a pentagon:

This pentagon is divided into three triangles.

The sum of the angles in one of the triangles is $\times + \times + \times = 180°$.

The sum of the angles in next triangle is $\bullet + \bullet + \bullet = 180°$.

The sum of the angles in the third triangle is $\checkmark + \checkmark + \checkmark = 180°$.

This means that the sum of all the angles is $180° + 180° + 180° = 540°$.

So, the sum of the interior angles in a pentagon is 540°.

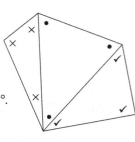

Similarly with a hexagon:

Here the hexagon is divided into four triangles and because the sum of the angles in each triangle is 180°, the sum of the interior angles of the hexagon is $4 \times 180° = 720°$.

To discover a general formula for the sum of the interior angles in a polygon we can look for a pattern using the deductions we have already made:

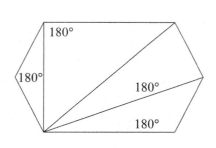

NO. OF SIDES	NO. OF TRIANGLES	ANGLE SUM
3	1	180°
4	2	360°
5	3	540°
6	4	720°

We can see that the number of triangles is two less than the number of sides and the angle sum is the number of triangles × 180°. So, in general, the:

Sum of the interior angles in any polygon with n sides is: $(n − 2) × 180°$

On the other hand, if we are talking about an **exterior angle of a polygon** we are referring to an angle that is, as the term suggests, on the outside of the figure, which is formed by any side of the polygon and the extension of its adjacent side.

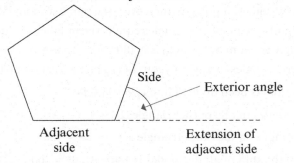

An important relationship exists between the interior and exterior angles in a polygon.

We can see in the figure to the right that adjacent interior and exterior angles lie on the same straight line. Since the angle sum of a straight line is 180°, then the sum of corresponding interior and exterior angles must always equal 180°.

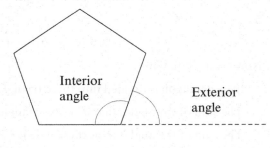

We can use this relationship between interior and exterior angles to establish another important property of exterior angles that is specifically associated with triangles. We will use a specific example to illustrate this property:

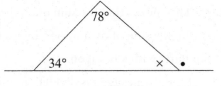

In this triangle, the sum of the given angles is 34° + 78° = 112°. This means that in the triangle, the angle marked x is 180° − 112° = 68°. Now if we look at the straight line made up of × and •, what is the size of •? Given that it is a straight line, • must equal 180° − 68° = 112°. This is an interesting result as we have already calculated this to be the sum of the opposite two angles in the triangle.

The results we found in this example can be used to form the generalisation that:

Any exterior angle in a triangle is equal to the sum of the opposite two interior angles.

This diagram illustrates this property in general terms:

$$c° = a° + b°$$

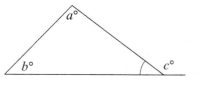

So far in this section, we have explored many properties associated with polygons. These properties can now be applied to find the size of unknown angles in geometric diagrams.

EXAMPLE 12

Find the size of the unknown angles in the following diagrams.

a

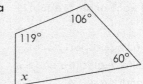

b

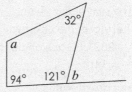

Solutions

a   This is a quadrilateral with an angle sum of 360°. We are given the size of three of the angles in the quadrilateral but do not know the size of $\angle x$. We commence by working out the total number of degrees in the given angles: 119° + 106° + 60° = 285°. We can now calculate the size of $\angle x$ by finding the difference between 360° and 285°: 360° − 285° = 75°. (We could check this answer by adding up the angles to make sure they equal 360°.)

b   This is a quadrilateral that has one exterior angle. We could commence by finding $\angle a$ in the same way as the previous question. Calculate the total number of degrees in the given angles: 32° + 94° + 121° = 247°. We can now calculate the size of $\angle a$ by finding the difference between 360° and 247°: 360° − 247° = 113°. (We could check this answer by adding up the angles to make sure they equal 360°.) Now $\angle b$ is an exterior angle, so we can use the property that adjacent interior and exterior angles have a sum of 180° because they lie on a straight line. Therefore, $\angle b = 180° − 121° = 59°$.

c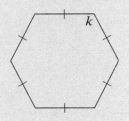

c This is a regular hexagon. We can tell this from the small dashes that show the sides are equal in length. In a regular polygon, the angles are also equal in size. This means that each angle in this hexagon will have the same value as $\angle k$. The angle sum in a hexagon is 720°. Therefore, each angle will be $\frac{720°}{6} = 120°$. So $\angle k = 120°$.

d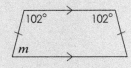

d This is an isosceles trapezium. We can tell this because the small dashes indicate that the slant sides are equal in length. Another property of an isosceles trapezium is that the base angles are equal. This means that both of the base angles will be the same size as $\angle m$. The sum of the given angles is $102° + 102° = 204°$. Therefore, the base angles will total $360° - 204° = 156°$. If each angle is equal in size, then $\angle m = \frac{156°}{2} = 78°$.

e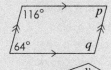

e This is a parallelogram. One key property of a parallelogram is that opposite angles are equal in size. Therefore, $\angle p = 64°$ and $\angle q = 116°$.

f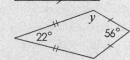

f This is a kite. We can tell this because the small dashed lines indicate that pairs of adjacent sides are equal in length. Another property of a kite is that the angles where each of pairs of adjacent sides meet are also of equal size. This means that the unmarked angle will be equal to $\angle y$. The sum of the given angles is $56° + 22° = 78°$. Therefore, the unknown angles will total $360° - 78° = 282°$. If each angle is equal in size, then $\angle y = \frac{282°}{2} = 141°$.

g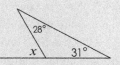

g This is a triangle that has been drawn with an exterior angle. The size of an exterior angle in a triangle is equal to the sum of the opposite two interior angles. Therefore, $\angle y = 28° + 31° = 59°$.

h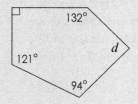

h This is a pentagon. The angle sum in a pentagon is 540°. The total number of degrees in the given angles is $90° + 132° + 121° + 94° = 437°$. We can now calculate the size of $\angle d$ by finding the difference between 540° and 437°: $540° - 437° = 103°$. (We could check this answer by adding up the angles to make sure they equal 540°.)

PRACTICE 12

Find the size of the unknown angles in the following diagrams.

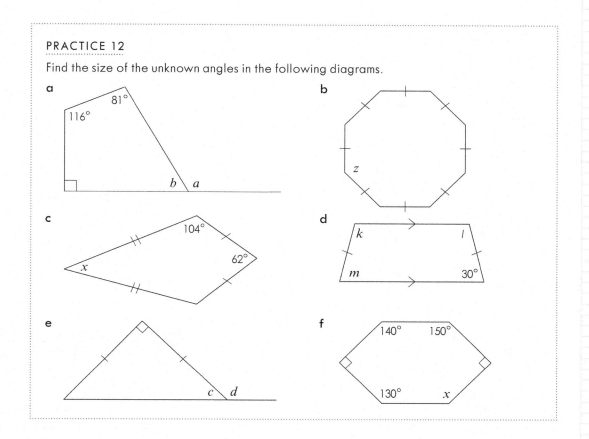

Circles

Finally, in this section, we will look at an important two-dimensional figure in the field of geometry: the **circle**. However, a circle is not considered to be a polygon because it does not fulfil the requirement of having straight-line sides. A circle is a set of points that are all the same distance from a fixed centre point.

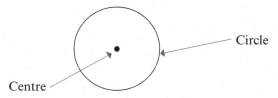

A circle has many geometric features, each of which has a special name. The following table shows the more commonly used features of a circle.

Circle
A set of points that are all the same distance from a fixed centre point

FEATURE	DIAGRAM	DEFINITION
Circumference		The circumference is the boundary or edge of a circle.
Radius		The radius is a line from the centre of the circle to any point on the circumference.
Chord		A chord is a line that joins two points on the circumference of the circle.
Diameter		The diameter is a special chord that passes through the centre point.
Arc		An arc is a continuous section of the circumference of the circle.
Sector		A sector is a section of the circle that is bounded by two radii.
Segment		A segment is a section of a circle that is bounded by a chord and an arc.

We have noted in the section on angles that there are 360° in a full revolution, which is, of course, a circle. Therefore, this means that there are 360° in a circle.

EXAMPLE 13

a Sketch a circle approximately 5 centimetres in diameter.

   i Label the centre of the circle, *C*.

   ii Mark a point, *X*, on the circumference on the circle.

   iii Draw a diameter *XY*.

   iv Shade a sector *YCZ*.

   v Draw a chord *XP*.

   vi Draw a radius *CP*.

b What fraction of a circle are the following sectors?

   i   **ii**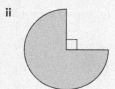

Solutions

a There would be many ways you could draw this diagram to correctly include the required features. Here is one example:

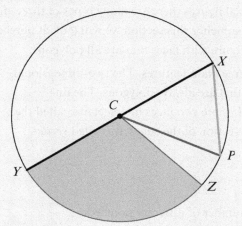

b Given that we know there are 360° in a circle, if we are given the number of degrees in any sector, we can determine what fraction the sector is of the circle.

   i This sector has angle size of 60° so as a fraction of 360° this is $\frac{60°}{360°}$, which is $\frac{1}{6}$.

   ii It seems that there is not enough information in this diagram, but there is. We can see in the unshaded part of the diagram (the section that is not part of the sector), there is a small square symbol that indicates a right angle or 90°. So, if the section that is not in the sector measures 90°, then the measure of the sector must be 360° − 90° = 270°. As a fraction of a circle, this is equal to $\frac{270°}{360°}$, which is $\frac{3}{4}$.

Making Connections
To review how the fractions in these examples have been cancelled take a look back at Chapter 5.

Kathy Brady

PRACTICE 13

a Sketch a circle approximately 5 centimetres in diameter.

i Label the centre of the circle, O.

ii Mark a point, A, on the circumference on the circle.

iii Draw a radius AO.

iv Draw a diameter AOB.

v Shade a sector BOC.

vi Shade a segment CD.

vii Draw a chord BC.

b What fraction of a circle are the following sectors?

i 72°

ii 120°

11.5 PROPERTIES OF THREE-DIMENSIONAL FIGURES

Polyhedron
A solid three-dimensional figure with faces that are all polygons

A three-dimensional geometric figure is called a **polyhedron**. It has length, width and height. Much like two-dimensional figures there are many types of three-dimensional figures with varying properties. To commence this section we will look at polyhedra. A polyhedron is a solid three-dimensional figure with faces that are all polygons.

All polyhedra have three main features. The two-dimensional sides are known as *faces* that are always polygons. The line segments that are formed where two faces intersect are called the *edges*. The points of intersection of the edges are called *vertices*.

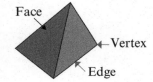

Face — Vertex — Edge

Prisms

Prism
A polyhedron with two opposite end faces that are identical polygons

There are an enormous number of different polyhedra, many of which are quite complex. So, we will limit our review to the more common types, starting with prisms. A **prism** is a polyhedron with two opposite end faces that are identical polygons.

Prisms are named according to the polygon that forms the opposite end faces. Some examples are shown here. As can be seen in top figure, the side faces that are formed are rectangles.

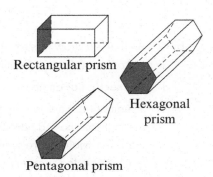

Rectangular prism

Hexagonal prism

Pentagonal prism

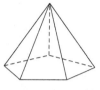
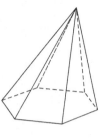

Pyramids

Another polyhedron that is named and classified according to the polygon that forms one of its faces is a **pyramid**. A pyramid has a base that is a polygon and triangular faces that lead up to a single point located above the base called an apex.

A pyramid is named and classified according to the polygon that forms the base. The figures on the right show a square based pyramid and a pentagonal based pyramid. The edges join each of the vertices in the base to the apex and the faces that are formed are triangles.

Pyramids with a base that is a regular polygon can be classified in one of two ways. *Right regular pyramids*, as shown in the figure above, have an apex that is directly above the centre of the base and will have faces that are all isosceles triangles. On the other hand, in an *oblique pyramid* the apex is not directly above the centre of the base and the triangular faces are different shapes. The figure to the right is an oblique hexagonal based pyramid.

Pyramid
A polyhedron with a base that is a polygon and a single point located above the base called the apex

Regular polyhedra

The final group of polyhedra that we will investigate have faces that are all identical, unlike prisms and pyramids. This type of solid is known as **regular polyhedron.** These solids are also known as Platonic solids.

There are only five regular polyhedra. The faces of a tetrahedron, an octahedron and an icosahedron are all equilateral triangles. The faces of a cube are squares and the faces of a dodecahedron are regular pentagons.

Regular polyhedron
A polygon with faces that are all identical regular polygons

Tetrahedron   Cube   Octahedron

Dodecahedron   Icosahedron

EXAMPLE 14

Name and classify the following three-dimensional figures.

Solutions

a
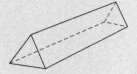

a Triangular prism

b
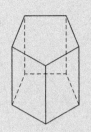

b Pentagonal prism

c
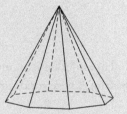

c Right octagonal based pyramid

d
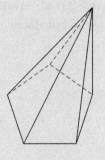

d Oblique pentagonal based pyramid

e
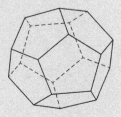

e Dodecahedron

PRACTICE 14

Name and classify the following three-dimensional figures.

a

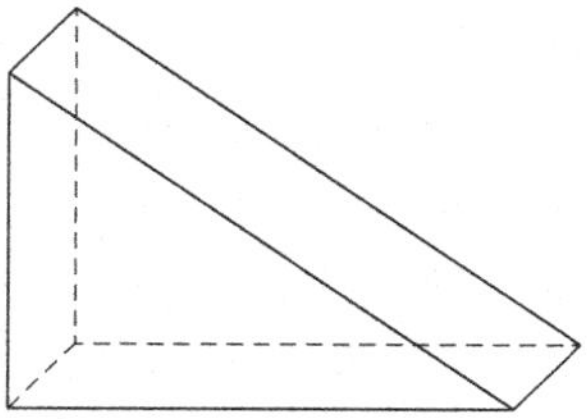

b

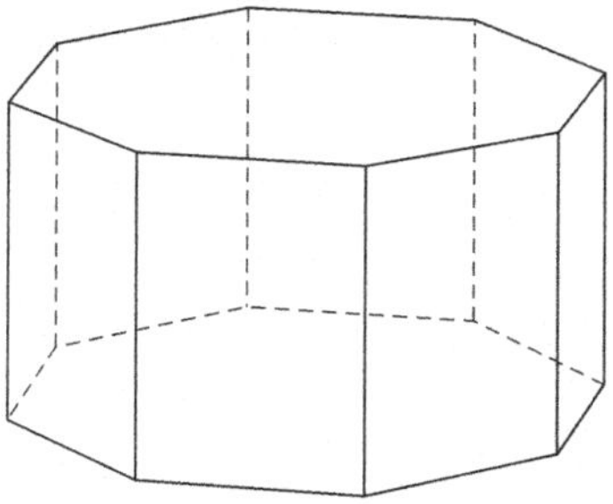

c

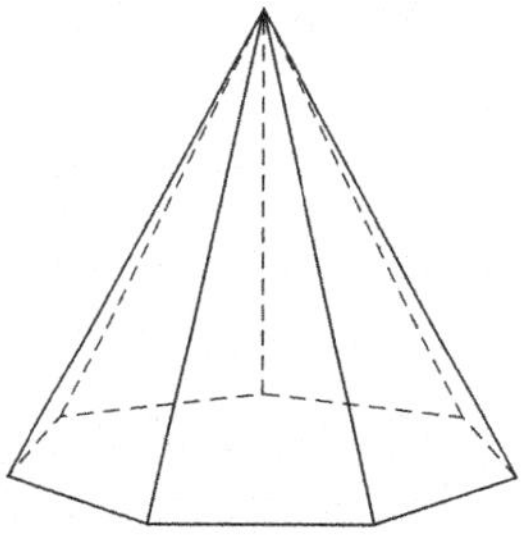

d

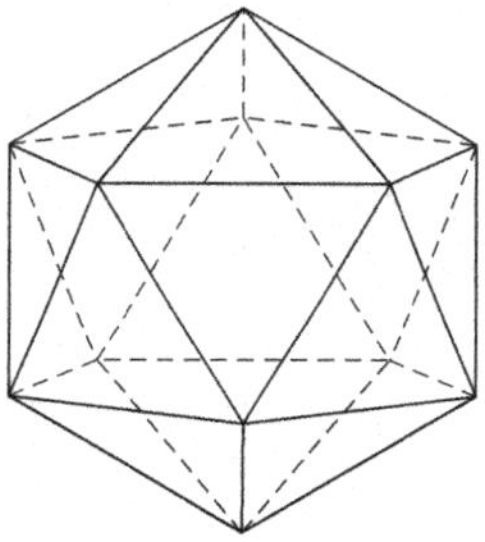

e

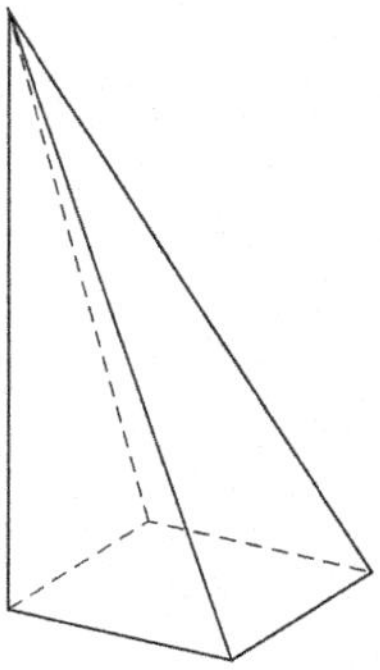

Euler's rule

An interesting relationship exists between the faces, vertices and edges of polyhedra. Here is a table that illustrates this relationship for regular polyhedra.

POLYHEDRON	FACES (F)	VERTICES (V)	EDGES (E)	$F + V$	$E + 2$
Tetrahedron	4	4	6	8	8
Cube	6	8	12	14	14
Octahedron	8	6	12	14	14
Dodecahedron	12	20	30	32	32
Icosahedron	20	12	30	32	32

Kathy Brady

It appears in this table that the number of faces plus the number of vertices is equal to the number of edges plus 2. Indeed, this relationship can be generalised to $F + V = E + 2$ for all polyhedra, not just regular polyhedra, and is known as *Euler's rule*. Here is an example of how Euler's rule works for an octagonal prism.

$F = 8$

$V = 12$

$E = 18$

$F + V = 8 + 12 = 20$

$E + 2 = 18 + 2 = 20$

Therefore, $F + V = E + 2$

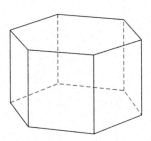

EXAMPLE 15

a Confirm Euler's rule for all of the polyhedra illustrated in Example 14.

b If a polyhedron has 6 faces and 7 vertices, how many edges does it have?

Solutions

a

POLYHEDRON	FACES (F)	VERTICES (V)	EDGES (E)	$F + V$	$E + 2$
Triangular prism	5	6	9	11	11
Pentagonal prism	7	10	15	17	17
Right octagonal based pyramid	9	9	16	18	18
Oblique pentagonal based pyramid	6	6	10	12	12
Dodecahedron	12	20	30	32	32

b If $F + V = E + 2$, and $F = 6$ and $V = 7$, we have $6 + 7 = E + 2$. That is, $13 = E + 2$. Solving this equation by inspection, $E = 11$.

PRACTICE 15

a Confirm Euler's rule for all of the polyhedra illustrated in Practice 14.

b A polyhedron has 9 faces and 21 edges. How many vertices does it have?

Other three-dimensional figures

To conclude this section we will look at some three-dimensional figures that are not classified as polyhedra because they have curved faces and curved edges.

A *cylinder* is a prism-like three-dimensional figure. The opposite end faces are circles, and it has a curved side face.

A *cone* is a pyramid-like three-dimensional figure. It has a circular base and a single vertex that is directly above the centre point of the base. There is a curved side that joins the base to the vertex.

A *sphere* is a perfectly rounded three-dimensional figure, like a ball. The key property of a sphere is that every point on the surface is the same distance from the centre of the sphere.

CRITICAL THINKING

Imagine that you were to open out the curved face of a cylinder. What shaped would it be? Now think about doing this to a cone. What shape would the flattened side be?

If you were to cut a cylinder in half horizontally, what shape would the cut surface be? What if you cut the cylinder in half vertically? Now think about cutting a cone in half horizontally. What shape would the cut surface be? What if you were to cut the cone in half vertically?

11.6 PYTHAGORAS' RULE

In this section, and the next, we will look at some properties that are specifically associated with right-angled triangles. These properties are some of the most commonly used practical applications of geometry. We will begin in this section by exploring Pythagoras' rule, which describes the relationship between the lengths of the sides in a right-angled triangle.

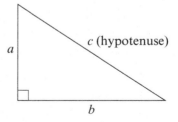

Pythagoras was a Greek mathematician who lived around 500BC. Pythagoras did not actually 'discover' the relationship between the sides in a right-angled triangle; rather he was the first to prove it. Peoples in more ancient civilisations had, through observation, noticed this relationship and had used it extensively in measurement and construction.

In a right-angled triangle two of the sides form the right angle, in this diagram sides a and b. The third side is known as the *hypotenuse*, in the figure above side c. The hypotenuse of a right-angled triangle is opposite the right angle and is the longest side.

Pythagoras' rule involves what happens when we draw a square on each side of a right-angled triangle. According to Pythagoras' rule, the area of the largest square that is drawn on the hypotenuse is equal to the sum of the areas of the squares drawn on the other two sides.

Let's show that this relationship holds true using an example. Here is a right-angled triangle with a square drawn on each of its sides. The length of the hypotenuse is 5 units. The area of the mid-green square drawn on the hypotenuse is 25 square units. The length of the side of the dark green square is 3 units and the area of the square is 9 square units. The length of the side of the light green square is 4 units and the area of the square is 16 square units.

Making Connections
In Chapter 12 square units and the calculation of area will be fully covered.

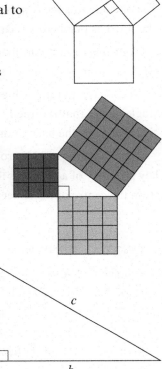

Since 25 = 9 + 16, we have shown that the area of the square on the hypotenuse is equal to the sum of the areas of the squares on the other two sides.

The relationship between the hypotenuse and the other two sides of a right-angled triangle can be generalised as:

In any right-angled triangle, the square of the hypotenuse is equal to the sum of the squares of the other two sides, which is written as:

$$c^2 = a^2 + b^2$$

The power of Pythagoras' rule is that if we know the lengths of two sides in a right-angled triangle we can use the rule to calculate the length of the unknown side.

EXAMPLE 16

Find the length of the unknown side in the following triangles.

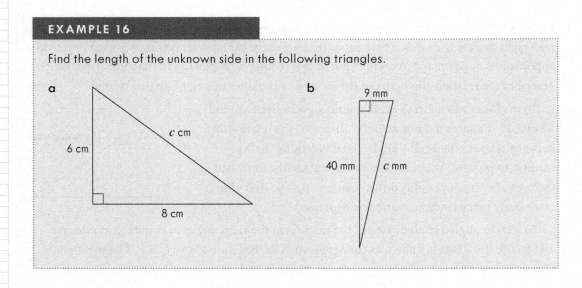

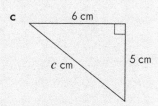

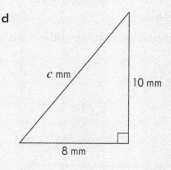

Solutions

In each of these examples the unknown side is the hypotenuse. The lengths of the other two sides are given. Using the formula $c^2 = a^2 + b^2$, it does not matter which of the given sides is labelled a and which is labelled b.

a
$$c^2 = a^2 + b^2$$
$$c^2 = 6^2 + 8^2$$
$$c^2 = 36 + 64$$
$$c^2 = 100$$
$$c = \sqrt{100}$$
$$c = 10 \text{ cm}$$

b
$$c^2 = a^2 + b^2$$
$$c^2 = 9^2 + 40^2$$
$$c^2 = 81 + 1600$$
$$c^2 = 1681$$
$$c = \sqrt{1681}$$
$$c = 41 \text{ mm}$$

Because in examples c and d we are not taking the square root of a square number the answer will not be a whole number. We need to use the $\sqrt{\ }$ button on a scientific calculator to obtain the square root. One decimal place is sufficient.

c
$$c^2 = a^2 + b^2$$
$$c^2 = 5^2 + 6^2$$
$$c^2 = 25 + 36$$
$$c^2 = 61$$
$$c = \sqrt{61}$$
$$c = 7.8 \text{ cm}$$

d
$$c^2 = a^2 + b^2$$
$$c^2 = 8^2 + 10^2$$
$$c^2 = 64 + 100$$
$$c^2 = 164$$
$$c = \sqrt{164}$$
$$c = 12.8 \text{ mm}$$

Making Connections
Revisit Chapter 3 to review squaring numbers and square roots.

Making Connections
To review rounding decimal numbers to a required number of decimal places revisit Chapter 6.

PRACTICE 16

Find the length of the unknown side in the following triangles.

a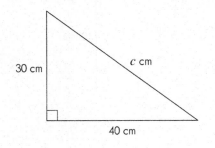

30 cm

c cm

40 cm

b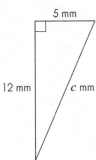

5 mm

12 mm

c mm

c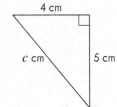

4 cm

c cm

5 cm

d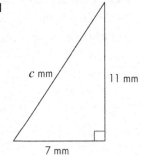

c mm

11 mm

7 mm

EXAMPLE 17: APPLICATIONS

a   A tent is erected and anchored to the ground at the front by a guy rope that runs up to the centre top of the tent. If the guy rope is anchored 1.2 metres from the centre front of the tent and the tent is 1 metre high, how long is the guy rope?

b A delivery truck has a ramp at the back of the truck. If the height from the ground to the where the ramp latches onto the back of the truck is 1.8 metres and the distance along the ground to where the ramp latches onto the truck is 4 metres, how long is the ramp?

c This is a diagram of a wooden frame that will form part of a gabled roof. What length of wood will be required to construct this frame?

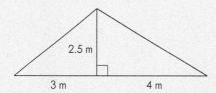

2.5 m

3 m   4 m

Solutions

In application questions such as these it is a good idea to sketch a diagram to represent the situation.

a In this example, a right-angled triangle is formed by the front of the tent, the ground and the guy rope, which is the hypotenuse of the triangle.

$$c^2 = a^2 + b^2$$
$$c^2 = 1^2 + 1.2^2$$
$$c^2 = 1 + 1.44$$
$$c^2 = 2.44$$
$$c = \sqrt{2.44}$$
$$c = 1.6 \text{ m}$$

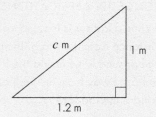

c m   1 m

1.2 m

Kathy Brady

b In this example, a right-angled triangle is formed by the height from the ground to where the ramp latches onto the truck, the distance from where the ramp touches the ground to the back of the truck and the ramp, which is the hypotenuse.

$c^2 = a^2 + b^2$

$c^2 = 1.8^2 + 4^2$

$c^2 = 3.24 + 16$

$c^2 = 19.24$

$c = \sqrt{19.24}$

$c = 4.4$ m

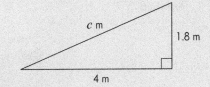

c Labelling of this diagram will assist with making the required explanations. In the right-angled triangle ADC, the side AC is the hypotenuse, so we will commence by calculating the length of AC.

$AC^2 = AD^2 + CD^2$

$AC^2 = 2.5^2 + 4^2$

$AC^2 = 6.25 + 16$

$AC^2 = 22.25$

$AC = \sqrt{22.25}$

$AC = 4.7$ m

Now we will repeat this to find the length of side AB.

$AB^2 = AD^2 + BC^2$

$AB^2 = 2.5^2 + 3^2$

$AB^2 = 6.25 + 9$

$AB^2 = 15.25$

$AB = \sqrt{15.25}$

$AB = 3.9$ m

We now have all of the lengths required to calculate the total length of wood needed. The length of wood is:

$AB + BC + AC + AD = 3.9 + 7 + 4.7 + 2.5 = 18.1$ m

PRACTICE 17: APPLICATIONS

a When a ladder is leaning against a house, the distance from the top of the ladder down to the ground is 4.5 m and the foot of the ladder is 1.5 m away from the wall of the house. How long is the ladder?

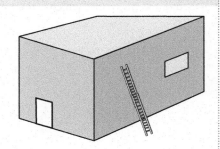

b To walk from one side of a pond to the other, you need to take a path that forms a right angle. One side of this path is 70 m and the other side of the path is 90 m. What would be the distance if you were to swim directly across the pond?

c On a sailing ship, a rope is attached to the top of a mast and then secured on the deck. If the mast is 12 metres high and the rope is secured on the deck 1.6 metres away from the base of the mast, how long is the rope?

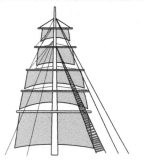

In the examples we have seen so far, the unknown side in each of the triangles has been the hypotenuse. However, we can also use Pythagoras' rule to calculate an unknown length of either of the other two sides as well.

EXAMPLE 18

Find the length of the unknown side in the following triangles.

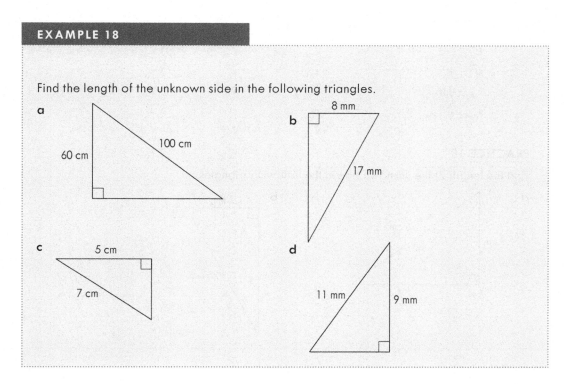

a 60 cm, 100 cm

b 8 mm, 17 mm

c 5 cm, 7 cm

d 11 mm, 9 mm

Making
Connections
*The process of
isolating the
unknown will be
covered fully in
Chapter 14.*

Solutions

In all of these examples, the unknown side could be either the a side or the b side of the triangle, it does not matter which. In these solutions, we have labelled it the b side.

a $a^2 + b^2 = c^2$ This is still Pythagoras' rule – just written on the opposite sides of the = sign.

$a^2 + 60^2 = 100^2$

$a^2 + 3600 = 10000$

$a^2 = 10000 - 3600$ In this step we have subtracted 3600 from each side of the formula to isolate the unknown a on the right-hand side.

$a^2 = 6400$

$a = \sqrt{6400}$

$a = 80 \text{ cm}$

b $a^2 + b^2 = c^2$ **c** $a^2 + b^2 = c^2$

$a^2 + 8^2 = 17^2$ $a^2 + 5^2 = 7^2$

$a^2 + 64 = 289$ $a^2 + 25 = 49$

$a^2 = 289 - 64$ $a^2 = 49 - 25$

$a^2 = 225$ $a^2 = 24$

$a = \sqrt{225}$ $a = \sqrt{24}$

$a = 15 \text{ mm}$ $a = 4.9 \text{ mm}$

d $a^2 + b^2 = c^2$

$a^2 + 9^2 = 11^2$

$a^2 + 81 = 121$

$a^2 = 121 - 81$

$a^2 = 40$

$a = \sqrt{40}$

$a = 6.3 \text{ mm}$

PRACTICE 18

Find the length of the unknown side in the following triangles.

a
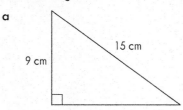
9 cm   15 cm

b
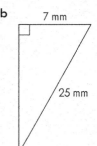
7 mm
25 mm

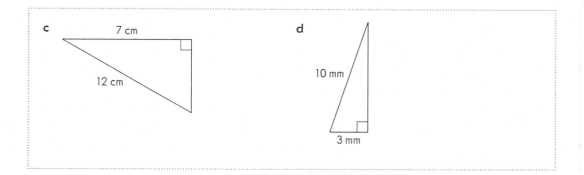

c   7 cm

12 cm

d

10 mm

3 mm

EXAMPLE 19: APPLICATIONS

a The front of a pup tent measures 2.6 m across the base along the ground. The slanting sides of the tent each measure 2.2 m. How high is the tent at its highest point?

b In this diagram of a cone, the height, h, is the perpendicular distance from the top vertex to the centre of the base. The radius, r, is the radius of the circular base. If the slant side of the cone measures 10 cm and the radius measures 4 cm, what is the height of the cone?

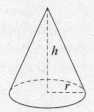

c A shop sign is attached to a wall and supported by a diagonal strut. If the length of the strut is 2 m and it is attached to the wall 1.3 m below the bottom of the sign, how wide is the sign?

Jewellery

Solutions

In application questions such as these it is a good idea to sketch a diagram to represent the situation.

a In this example, two right-angled triangles are formed on the front of the tent. We only need to work with one of them, as they are both the same. To do this we need to halve the base distance to form the b side.

$$a^2 + b^2 = c^2$$
$$a^2 + 1.3^2 = 2.2^2$$
$$a^2 + 1.69 = 4.84$$
$$a^2 = 4.84 - 1.69$$
$$a^2 = 3.15$$
$$a = \sqrt{3.15}$$
$$a = 1.8 \text{ m}$$

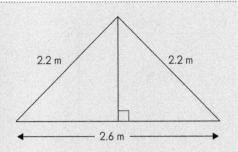

b In this example, the perpendicular height, the radius and the slant side form a right-angled triangle. We will label the radius as side b.

$$a^2 + b^2 = c^2$$
$$a^2 + 4^2 = 10^2$$
$$a^2 + 16 = 100$$
$$a^2 = 100 - 16$$
$$a^2 = 84$$
$$a = \sqrt{84}$$
$$a = 9.2 \text{ cm}$$

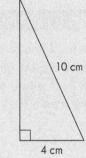

c Here the strut, the wall and the bottom of the sign form a right-angled triangle. The distance down the wall is the b side.

$$a^2 + b^2 = c^2$$
$$a^2 + 1.3^2 = 2^2$$
$$a^2 + 1.69 = 4$$
$$a^2 = 4 - 1.69$$
$$a^2 = 2.31$$
$$a = \sqrt{2.31}$$
$$a = 1.5 \text{ m}$$

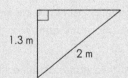

PRACTICE 19: APPLICATIONS

a The size of a flat screen television is usually given in centimetres, which is the diagonal measurement from the top right corner to the bottom left corner. If the size of a particular television is 80 cm, and the width of the television is 60 cm, what is the height of the television?

b In this cone, the height, h, is the

perpendicular distance from the top vertex to the centre of the base. The radius, **r**, is the radius of the circular base. If the slant side of the cone measures 12 cm and the height measures 9 cm, what is the radius of the base?

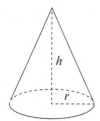

c   A golfer takes two strokes to get onto the green of a particular hole on a golf course. These two shots were at right angles to each other. If the first stroke covered 270 m, and the distance directly from the tee to the green was 440 m, how long was the golfer's second stroke?

11.7 TRIGONOMETRIC RATIOS

The term 'trigonometry' means relating to triangles. In the last section, we explored one aspect of trigonometry: Pythagoras' rule. We discovered that Pythagoras' rule is very useful in finding the lengths of unknown sides in right-angled triangles if we know the lengths of two of the sides. However, sometimes we do not have this much information about side lengths. On the other hand, we may know something about the size of the angles in the right-angled triangle. If this is the case, we can use what is known as trigonometric ratios to find the length of unknown sides.

When using trigonometric ratios, we need to use a different way of naming the sides of the triangle, rather than using the *a*, *b* and *c* sides as we have done using Pythagoras' rule. This is because when using trigonometric ratios we also need to include the information that is provided regarding angles in the right-angled triangle.

Here is a right-angled triangle. In this triangle, there is one angle that has been identified and labelled. When using trigonometric ratios, the convention is to label the identified angle using the Greek letter θ (pronounced 'theta'). The sides that form the right angle are then labelled with reference to the angle θ. The 'opposite' side is directly opposite θ and the 'adjacent' side is next to the angle θ. The hypotenuse is the side that is directly opposite the right angle.

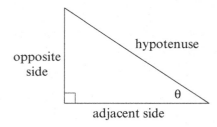

Early Greek and Indian mathematicians noticed that, regardless of the length of their sides, a special relationship existed between the sides of right-angled triangles that had the same size angles. This relationship is that in right-angled triangles with the same size angles when one side of a triangle is divided by another side of that triangle (this is the ratio part), regardless of the side length, the ratio will be the same.

Making Connections
To review the meaning of ratios revisit Chapter 7.

There are three different divisions, or ratios, that can exist in a right-angled triangle. These are the trigonometric ratios that can be generalised using the triangle labelling illustrated above. Each of these trigonometric ratios has a special name: sine, cosine and tangent. However, we usually abbreviate these names to sin, cos and tan. The trigonometric ratios are all written with reference to the labelled angle θ.

$$\sin\theta = \frac{\text{opposite side}}{\text{hypotenuse}} \qquad \cos\theta = \frac{\text{adjacent side}}{\text{hypotenuse}} \qquad \tan\theta = \frac{\text{opposite side}}{\text{adjacent side}}$$

EXAMPLE 20

Calculate the three trigonometric ratios for the given angle in this right-angled triangle.

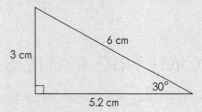

Solution

The labelled angle in the right-angled triangle measures 30°. The side opposite this angle is 3 cm, the side adjacent to this angle is 5.2 cm, and the length of the hypotenuse is 6 cm. The three trigonometric ratios are:

$$\sin\theta = \frac{\text{opposite side}}{\text{hypotenuse}} \qquad \cos\theta = \frac{\text{adjacent side}}{\text{hypotenuse}} \qquad \tan\theta = \frac{\text{opposite side}}{\text{adjacent side}}$$

Using the information in this triangle:

$$\sin 30° = \frac{3}{6} = 0.5 \qquad \cos 30° = \frac{5.2}{6} = 0.866 \qquad \tan 30° = \frac{3}{5.2} = 0.577$$

To carry out these divisions you will need to use your scientific calculator. The decimal values would usually be rounded to three or four decimal places. You can check the answers above on your scientific calculator by using the sin, cos and tan buttons.

PRACTICE 20

Calculate the three trigonometric ratios for the given angle in this right-angled triangle, and check your answers using the sin, cos and tan buttons on a scientific calculator.

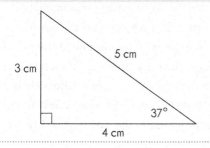

When we use Pythagoras' rule to find the unknown side length in a right-angled triangle, we need to know the lengths of the other two sides. The power of the trigonometric ratios is that to find an unknown side length in a right-angled triangle we only need to know the length of one other side, as long as we are given one of the angles.

EXAMPLE 21

Calculate the length of the side labelled x in the following triangles.

a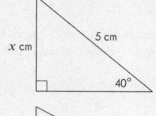

b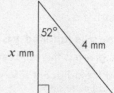

c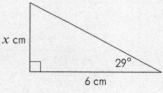

Solutions

When completing examples such as these it is always useful to label the sides of the right-angled triangle with reference to the named angle. Using O for opposite, A for adjacent and H for hypotenuse is a handy abbreviation.

a   In this example, the side labelled x is opposite the identified angle of 40° and the side length that is given is the hypotenuse. The trigonometric ratio involving the opposite side and the hypotenuse is sin.

$$\sin \theta = \frac{\text{opposite side}}{\text{hypotenuse}}$$

$\sin 40° = \frac{x}{5}$ In this step, we have multiplied each side of the formula by 5 to isolate the unknown x on the right-hand side.

$5 \times \sin 40° = x$

$5 \times 0.6428 = x$ This step is carried out using a scientific calculator to find the value of sin 40°.

$3.2 = x$

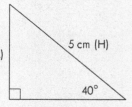

The side marked x is 3.2 cm.

Making Connections
The process of isolating the unknown by multiplying or dividing, and then balancing equations will be covered fully in Chapter 14.

b In this example, the side labelled x is adjacent to the identified angle of 52° and the side length that is given is the hypotenuse. The trigonometric ratio involving the adjacent side and the hypotenuse is cos.

$$\cos\theta = \frac{\text{adjacent side}}{\text{hypotenuse}}$$
$$\cos 52° = \frac{x}{4}$$
$$4 \times \cos 52° = x$$
$$4 \times 0.616 = x$$
$$2.5 = x$$

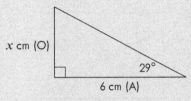

The side marked x is 2.5 mm.

c In this example, the side labelled x is opposite to the identified angle of 29° and the side length that is given is the adjacent side. The trigonometric ratio involving the opposite side and the adjacent side is tan.

$$\tan\theta = \frac{\text{opposite side}}{\text{adjacent side}}$$
$$\tan 29° = \frac{x}{6}$$
$$6 \times \tan 29° = x$$
$$6 \times 0.5543 = x$$
$$3.3 = x$$

The side marked x is 3.3 cm.

PRACTICE 21

Calculate the length of the side labelled x in the following triangles.

a

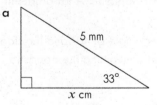

b

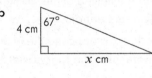

c

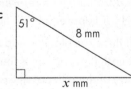

EXAMPLE 22: APPLICATIONS

a   The front of an A-frame house shaped is like an isosceles triangle. If the base angles are 70° and the slanted roof on each side measures 5.1 metres from the ground to the top vertex, what is the vertical height of the house from the ground at the centre to the top vertex?

b   A Boeing 747-400 takes off with an angle of 10° to the horizontal ground. (Source: bangaloreaviation.com) After it has travelled 1100 m over a horizontal distance, what is the plane's vertical height above the ground?

c   A life buoy attached to a rope is thrown from a boat to rescue a person who has fallen overboard. If the rope is 75 m long and makes an angle of 25° with the water, how far away from the boat is the person?

Solutions

In application questions such as these it is a good idea to sketch a diagram to represent the situation.

a   In this example, the side labelled x is opposite the identified angle of 70° and the side length that is given is the hypotenuse. The trigonometric ratio involving the opposite side and the hypotenuse is sin.

$$\sin\theta = \frac{\text{opposite side}}{\text{hypotenuse}}$$
$$\sin 70° = \frac{x}{5.1}$$
$$5.1 \times \sin 70° = x$$
$$5.1 \times 0.94 = x$$
$$4.8 = x$$

The height of the house from the centre of the base at the front to the vertex is 4.8 metres.

b   In this example, the side labelled x is opposite to the identified angle of 10° and the side length that is given is the adjacent side. The trigonometric ratio involving the opposite side and the adjacent side is tan.

$$\tan \theta = \frac{\text{opposite side}}{\text{adjacent side}}$$

$$\tan 10° = \frac{x}{1100}$$

$$1100 \times \tan 10° = x$$

$$1100 \times 0.1763 = x$$

$$194 = x$$

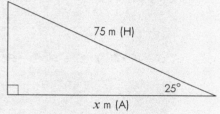

The plane is 194 metres off the ground.

c   In this example, the side labelled x is adjacent to the identified angle of 25° and the side length that is given is the hypotenuse. The trigonometric ratio involving the adjacent side and the hypotenuse is cos.

$$\cos \theta = \frac{\text{adjacent side}}{\text{hypotenuse}}$$

$$\cos 25° = \frac{x}{75}$$

$$75 \times \cos 25° = x$$

$$75 \times 0.9063 = x$$

$$68 = x$$

The person is 68 metres away from the boat.

PRACTICE 22: APPLICATIONS

a   At most airports, the planes will make their final descent at 3° to the runway. (Source: brighthub.com/science/aviation) If a plane's horizontal distance is 40 km from the airport when it commences its descent, how high is it vertically?

b When a kite-board rider launches, the ropes attaching the kite to the rider are at 50°
to the sea. If the ropes are 80 m long, how far above the sea is the kite?

c Tournament water ski rules specify that the length of the ski jump ramp must be
6.8 metres (Source: iwsf.com). If the ramp forms a 15° angle with the water, what is the
length of the ramp that is in contact with the water?

11.8 CHAPTER SUMMARY

SKILL OR CONCEPT	DEFINITION OR DESCRIPTION	EXAMPLES
Point	An exact location in space.	• P
Line	A collection of points in a straight path that extend infinitely.	A ⟷ B
Line segment	A part of a line that is enclosed by two distinct endpoints.	A •——• B
Plane	A collection of points that form a two-dimensional surface that extends infinitely in all directions.	B, A, C
Parallel lines	Two, or more, lines in the same plane that never meet.	⟷ ⟷
Collinear points	Three, or more, points that lie on the same line.	A B C
Concurrent lines	Three, or more, lines that all intersect, or cross, at the same point.	
Space	An infinite three-dimensional region in which objects can have both position and direction.	
Skew lines	Two lines in three-dimensional space that do not intersect and are not parallel.	
Ray	A part of a line that has one endpoint and continues infinitely in the other direction.	A B
Angle	Two rays that meet at a common endpoint, known as the vertex.	

SKILL OR CONCEPT	DEFINITION OR DESCRIPTION	EXAMPLES
Degrees	The unit of measure for the size of an angle. The symbol for degrees is °.	
Revolution	When a ray is rotated around its endpoint until it returns to its original position, which is equal to 360°.	
Straight-line angle	Half of a revolution, which is equal to 180°.	
Right angle	One quarter of a revolution measuring 90°.	
Acute angle	An angle that is less than 90°.	
Obtuse angle	An angle that is greater than 90° but less than 180°.	
Reflex angle	An angle that is greater than 180° but less than 360°.	
Complementary angles	Two angles that add up to 90°.	

SKILL OR CONCEPT	DEFINITION OR DESCRIPTION	EXAMPLES
Supplementary angles	Two angles that add up to 180°.	
Vertically opposite angles	Equally sized angles that are opposite each other when two lines intersect.	
Perpendicular lines	Lines that are at right angles to each other.	
Corresponding angles	Equally sized angles that are located in corresponding positions when a transversal intersects with parallel lines.	
Alternate angles	Equally sized angles that are located on opposite sides of the transversal, and inside the two parallel lines.	
Co-interior angles	Supplementary angles that are on the same side of the transversal and inside the two parallel lines.	
Polygon	A simple closed figure made up of straight, line segments.	
Triangle	A polygon with three sides and three angles.	
Equilateral triangle	A triangle with three sides equal in length and three equal angles.	

SKILL OR CONCEPT	DEFINITION OR DESCRIPTION	EXAMPLES
Isosceles triangle	A triangle with two sides equal in length and two equal base angles.	
Scalene triangle	A triangle that has no sides equal in length, and no equal sized angles.	
Right-angled triangle	A triangle that has one angle that measures 90°.	
Sum of angles in a triangle	The sum of the angles in any triangle will always be equal to 180°.	 $a + b + c = 180$ degrees
Quadrilateral	A polygon with four sides and four angles.	
Parallelogram	A quadrilateral with opposite sides parallel and equal in length, and equally sized opposite angles.	
Rectangle	A parallelogram with all four angles equally sized at 90°.	

SKILL OR CONCEPT	DEFINITION OR DESCRIPTION	EXAMPLES
Square	A parallelogram with all four sides equal in length and all four angles equally sized at 90°.	
Rhombus	A parallelogram with parallel opposite sides, with all sides equal in length, and equally sized opposite angles.	
Trapezium	A quadrilateral that has only one pair of parallel sides.	
Kite	A quadrilateral with pairs of adjacent sides equal in length.	
Interior angle	An angle formed within a polygon where two adjacent sides meet.	
Sum of interior angles	The sum of the interior angles in any polygon with n sides is: $$(n-2) \times 180°$$	For a octagon (8 sides) the sum of interior angles $= (8-2) \times 180° = 1080°$
Exterior angle	Angle that lies on the outside of a polygon, and is formed by any side of the polygon and the extension of its adjacent side.	
Exterior angle of a triangle	Any exterior angle in a triangle is equal to the sum of the opposite two interior angles.	

Kathy Brady

SKILL OR CONCEPT	DEFINITION OR DESCRIPTION	EXAMPLES
Circle	A set of points that are all the same distance from a fixed centre point.	
Circumference	The boundary or edge of a circle.	Circumference
Radius	A line from the centre of a circle to any point on its circumference.	Radius
Chord	A line that joins two points on the circumference of the circle.	
Diameter	A special chord that passes through the centre point of a circle.	
Arc	A continuous section of the circumference of the circle.	
Sector	A section of the circle that is bounded by two radii.	Sector

SKILL OR CONCEPT	DEFINITION OR DESCRIPTION	EXAMPLES
Segment	A section of a circle that is bounded by a chord and an arc.	
Polyhedron	A solid three-dimensional figure with faces that are all polygons.	
Prism	A polyhedron with two opposite end faces that are identical polygons.	
Pyramid	A polyhedron with base that is a polygon with a vertex located above the base called the apex.	
Euler's rule	In any polyhedron, the number of faces plus the number of vertices is equal to the number of edges plus 2, generalised to: $$F + V = E + 2$$	In a square based pyramid $F = 5$, $V = 5$, $E = 8$ so $F + V = E + 2$ is confirmed as $5 + 5 = 8 + 2$
Cylinder	Three-dimensional figure with circular opposite end faces and a curved side face.	
Cone	Three-dimensional figure with a circular base, a single vertex directly above the base, and a curved side face.	
Sphere	A rounded three-dimensional figure. Every point on its surface is the same distance from its centre.	

Kathy Brady

SKILL OR CONCEPT	DEFINITION OR DESCRIPTION	EXAMPLES
Pythagoras' rule	In any right-angled triangle, the square of the hypotenuse is equal to the sum of the squares of the other two sides, written as: $$c^2 = a^2 + b^2$$	A triangle with vertex A at top, side 12 on the left, hypotenuse c, right angle at C, base 5 to B.
Sine	A trigonometric ratio usually abbreviated to sin. In a right-angled triangle with an identified angle θ: $$\sin\theta = \frac{\text{opposite side}}{\text{hypotenuse}}$$	A right-angled triangle with hypotenuse 10, opposite side x, angle $30°$.
Cosine	A trigonometric ratio usually abbreviated to cos. In a right-angled triangle with an identified angle θ: $$\cos\theta = \frac{\text{adjacent side}}{\text{hypotenuse}}$$	A right-angled triangle with hypotenuse 20, angle $60°$, adjacent side x.
Tangent	A trigonometric ratio usually abbreviated to tan. In a right-angled triangle with an identified angle θ: $$\tan\theta = \frac{\text{opposite}}{\text{adjacent}}$$	A right-angled triangle with angle $35°$, opposite side x, adjacent side 10.

11.9   REVIEW QUESTIONS

A SKILLS

1   Sketch and label diagrams to represent the following.

   a $AB \parallel CD$

   b   Four collinear points

   c   Perpendicular lines $\overleftrightarrow{EF}$ and $\overleftrightarrow{MN}$

   d   Ray XY

   e $\angle PQR$

   f   Reflex $\angle XYZ$

   g   An obtuse angled triangle

   h   A concave quadrilateral

2   Sketch a circle approximately 5 centimetres in diameter.

   a   Label the centre of the circle, C.

   b   Mark a point, P, on the circumference on the circle.

   c   Draw a radius CR.

   d   Draw a diameter PCQ.

   e   Shade a sector PCR.

   f   Draw a chord SQ.

3   If possible, find the complement and the supplement of: 59° and 126°.

4   Calculate the size of the unknown angle in the following diagrams.

   a 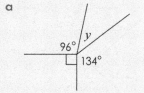   b 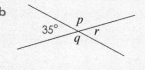   c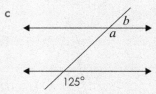

   d 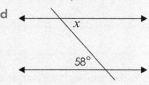   e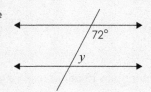

5 Calculate the size of the unknown angle in the following diagrams.

a

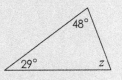

b

c

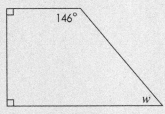

d

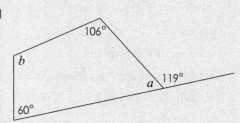

e

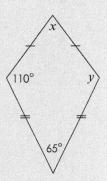

f

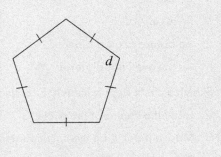

6 Classify and name the following figures.

a

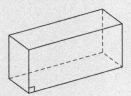

b

c

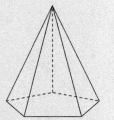

d

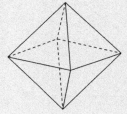

7 If a pentagonal prism has 7 faces and 15 edges, use Euler's rule to determine the number of vertices.

8   Find the length of the unknown sides in the following triangles.

a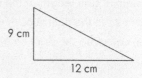
9 cm
12 cm

b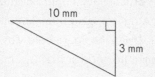
10 mm
3 mm

c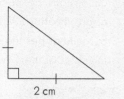
2 cm

9   Find the length of the unknown sides in the following triangles.

a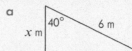
40°
6 m
x m

b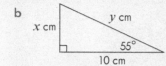
y cm
x cm
55°
10 cm

c
7 mm
x mm
18°

B   APPLICATIONS

1   Safety regulations in South Australia stipulate that the
maximum height of a straight playground slide should be
2.5 metres and the maximum incline of the slide should be
40° at the base of the slide (Source: recsport.sa.gov.au).
For a playground slide with these dimensions:

a   how far away from the ladder does the slide touch
the ground (correct to one decimal place)?

b   what is the length of the slide (correct to one decimal
place)?

2   The Australian Standards Council regulations for the
slopes of wheelchair access ramps into buildings sets the angle of incline where the ramp
touches the ground at 4°. (Source: standards.org.au). For a ramp that slopes to a height of
0.8 metres:

a   what horizontal distance along the ground is required to build the ramp (1 d.p.)?

b   what is the length of the ramp (correct to one decimal place)?

Kathy Brady

3   The Great Pyramid at Giza is a square-based pyramid with the following
    dimensions: perpendicular vertical height approximately 140 metres, side length of the base
    approximately 230 metres. A right-angled triangle could be formed in the centre of the
    pyramid as per the diagram. Using the figure below as a guide, calculate the slant height of
    the sides of the Great Pyramid (to nearest 10 m).

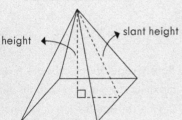

4   A lighthouse emits its signal in a circular motion. The maximum
    distance that the light can be seen is 20 km. A ship is passing
    by the lighthouse parallel to the shoreline. At its closet point,
    the ship is 15 km away from the lighthouse. Using the figure as
    a guide, calculate the total distance the ship travels while it
    is protected by the lighthouse signal (correct to one decimal
    place).

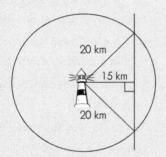

5   A large weight is suspended from a beam by a
    cable. The weight is supported by two wires,
    which are attached to each end of the beam.
    Using the information in the figure calculate:

    a   the length of the cable (correct to two
        decimal places)

    b   the length of the beam (correct to one
        decimal place)

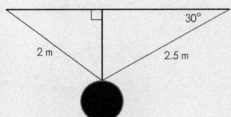

12

MEASUREMENT

CHAPTER OBJECTIVES

» Identify features and properties of metric measurement system

» Convert between metric units of measurement

» Calculate the perimeter of two-dimensional figures

» Calculate the area of two-dimensional figures

» Calculate the volume of three-dimensional figures

» Convert volume to capacity

MEASURING OLYMPIC FACILITIES

Measurement is essential to ensure that the Olympic facilities, such as pools, courts, tracks, and fields, comply with official Olympic regulations.

Olympic pool

The dimensions of an Olympic competition pool comply with standards set down by the international body for swimming. Usually some variation is permitted, provided it falls within a certain range (called the allowed tolerance). The competition pool must be 50.0 metres long with a tolerance of +0.03 metres and −0.00 metres. This means that the pool can be up to 3 centimetres longer than 50 metres, but cannot be even one millimetre shorter. The width must be exactly 25.0 metres. (Source: fina.org)

Olympic stadium

The international body for athletics mandates the dimensions of an Olympic athletics track. The track has an inside running distance of 400 metres. It consists of straight sides, each 84.39 metres long, and curves at each end with a radius of 36.5 metres. There are eight lanes, each 1.22 ± 0.01 metres wide. Tracks must have two independent measurements made, using laser-surveying instruments. These may not differ from each other by more than the allowed tolerances or the track will not be certified for Olympic competition. (Source: iaaf.org)

12.1 METRIC MEASUREMENT SYSTEM

The metric system was developed in France around the time of the French Revolution when scientists of the day recognised the flaws in measuring systems being used at the time that were based on characteristics of the human body, for example, the unit 'feet'. The International System of Units (SI units) is now the metric measurement system that is used by most of our world. The system comprises seven base units and a common set of prefixes that are used to specify multiples or fractions of each unit.

In this chapter, we will consider three of the basic units used in the metric system:

- length: metre (m)
- mass: gram (g)
- capacity: litre (L).

All other units that are used to measure length, mass or capacity are written as multiples or fractions of these units using prefixes that are added to the base unit to indicate how much larger or smaller they are compared to the base unit. Each of these prefixes has a specific symbol that is attached to the symbol for the base unit. For example, the prefix 'kilo' (k) indicates the unit is 1000 times larger than the base unit or the prefix 'milli' (m) indicates the unit is 1000 times smaller, that is 0.001, than the base unit.

This table details the relationship between the base units and some of their commonly used multiples and fractions.

Symbol	Prefix	Number	Name	
G	giga	1 000 000 000	billion	Number of times larger than the base unit
M	mega	1 000 000	million	
k	kilo	1000	thousand	
Base units				
c	centi	0.01	hundredth	Number of times smaller than the base unit
m	milli	0.001	thousandth	
μ	micro	0.000001	millionth	
n	nano	0.000000001	billionth	

12.2 METRIC UNIT CONVERSIONS

The simplest metric unit conversion is to carry out a conversion to base units. To do this we replace the prefix with the required number (using the table above as a reference).

EXAMPLE 1

Solutions

a Convert 1 km to m.

a Replace the prefix 'k' for 'kilo' with the number 1000. Therefore:

  1 km = 1000 m

b Convert 1 mL to L.

b Replace the prefix 'm' for 'milli' with 0.001. Therefore:

  1 mL = 0.001 L

PRACTICE 1

a Convert 1 μg to a base unit.

b Convert 1 ML to a base unit.

Because metric units are all multiples or fractions of each other we can easily convert between different sizes of the same unit by moving the decimal point the required number of places. Tables that set out the relationship between the various units are a helpful guide as to how many jumps of the decimal point will be required to carry out a conversion.

Length

In this table, the relationships between the units of length are shown.

1000 m	100 m	10 m	m	0.1 m	0.01 m	0.001 m
km			m		cm	mm

When moving from a smaller unit to a larger unit, jump the decimal point to the *left* the required number of decimal places. When moving from a larger unit to a smaller unit, jump the decimal point to the *right* the required number of decimal places.

EXAMPLE 2

a   Convert 6.7 km to m.

b   Convert 98 cm to km.

c   Convert 0.78 cm to mm.

Solutions

a   In this conversion, we are moving from a larger unit (km) to a smaller unit (m) by making 3 jumps to the right.

1000 m	100 m	10 m	m	0.1 m	0.01 m	0.001 m
km			m		cm	mm

So, we need to jump the decimal point in 6.7 three places to the right and fill in the empty spaces with zeros.

$$6.700$$

So, 6.7 km = 6700 m

b   In this conversion, we are moving from a smaller unit (cm) to a larger unit (km) by making 5 jumps to the left.

1000 m	100 m	10 m	m	0.1 m	0.01 m	0.001 m
km			m		cm	mm

So, we need to jump the decimal point in 98 (which is to the right of the digit 8) five places to the left and fill in the empty spaces with zeros.

$$0.00098$$

So, 98 cm = 0.00098 km

c   In this conversion, we are moving from a larger unit (cm) to a smaller unit (mm) by moving 1 jump to the right.

1000 m	100 m	10 m	m	0.1 m	0.01 m	0.001 m
km			m		cm	mm

So, we need to jump the decimal point in 0.78 one place to the right.

$$0.78$$

So, 0.78 cm = 7.8 mm

Mass

In this table, the relationships between commonly used mass units are given, and we can use the table to carry out conversions between these units.

1000 g	100 g	10 g	g	0.1 g	0.01 g	0.001 g
kg			g			mg

EXAMPLE 3

a Convert 2400 mg to g.

b Convert 0.007 kg to mg.

Solutions

a In this conversion, we are moving from a smaller unit (mg) to a larger unit (g) by making 3 jumps to the left.

1000 g	100 g	10 g	g	0.1 g	0.01 g	0.001 g
kg			g			mg

So, we need to jump the decimal point in 2400 (which is to the right of the final 0) three places to the left.

2.400

So, 2400 mg = 2.4 g

b In this conversion, we are moving from a larger unit (kg) to a smaller unit (mg) by making 6 jumps to the right.

1000 g	100 g	10 g	g	0.1 g	0.01 g	0.001 g
kg			g			mg

So, we need to jump the decimal point in 0.007 six places to the right and fill in the empty spaces with zeros.

0.007000

So, 0.007 kg = 7000 mg

c   Convert 3 000 000 μg to mg.

c   This conversion involves moving from micrograms (μg) to milligrams (mg). Referring to the metric unit relationship table in Section 12.1, we can see that a microgram is a smaller unit than a milligram. So in this conversion, because we are moving from a smaller unit to a larger unit, making jumps to the left will be required. Although μg does not appear on the conversion table that we have been using, it is possible to modify this table to help us work out how many jumps we need to make.

g	0.1 g	0.01 g	0.001 g	0.0001 g	0.00001 g	0.000001 g
g			mg			μg

Now we can see that 3 jumps of the decimal point in 3 000 000 (which is to the right of the final 0) to the left are needed.

$$3000\underset{\smile}{000},$$

So, 3 000 000 μg = 3000 mg

PRACTICE 3

a   Convert 380 mg to kg.   b   Convert 370 g to mg.   c   Convert 4900 mg to μg.

Capacity

Finally, in this table the relationships between commonly used capacity units are given, and we can use the table to carry out conversions between these units.

1000 L	100 L	10 L	L	0.1 L	0.01 L	0.001 L
kL			L			mL

EXAMPLE 4

Solutions

a   Convert 0.07 kL to L.

a   In this conversion, we are moving from a larger unit (kL) to a smaller unit (L) by making 3 jumps to the right.

1000 L	100 L	10 L	L	0.1 L	0.01 L	0.001 L
kL			L			mL

So, we need to jump the decimal point in 0.07 three jumps to the right and fill the empty space with a zero.

$$0.\underset{\smile}{070}$$

So, 0.07 kL = 70 L

b Convert 0.02 kL to mL.

b In this conversion, we are also moving from a larger unit (kL) to a smaller unit (mL) making 6 jumps to the right.

1000 L	100 L	10 L	L	0.1 L	0.01 L	0.001 L
kL			L			mL

So, we need to jump the decimal point in 0.02 six jumps to the right and fill the empty spaces with zeros.

$$0.\underset{\smile}{020000}$$

So, 0.02 kL = 20 000 mL

c Convert 4500 L to ML.

c This conversion involves moving from litres (L) to megalitres (ML). Referring to the metric unit relationship table in Section 12.1, we can see that a litre is a smaller unit than a megalitre. So in this conversion, because we are moving from a smaller unit to a larger unit, making jumps to the left will be required. Although ML does not appear on the conversion table that we have been using, it is possible to modify this table to help us work out how many jumps we need to make.

1 000 000 L	100 000 L	10 000 L	1000 L	100 L	10 L	L
ML			kL			L

Now we can see that 6 jumps to the left of the decimal point in 4500 (which is to the right of the final 0) are required and the empty spaces to be filled with zeros.

$$0.\underset{\smile}{004500}$$

So, 4500 L = 0.0045 ML

PRACTICE 4

a   Convert 5600 mL to kL.   **b**   Convert 70 mL to L.   **c**   Convert 0.59 ML to L.

EXAMPLE 5: APPLICATION

Antibiotics are commonly prescribed for children as a liquid medication. One form of liquid medication contains 250 mg of the antibiotic per 5 mL of the liquid. A doctor prescribes 400 mg of the antibiotic per day for a child. How much of the liquid medication would the child need to take?

Solution

If there are 250 mg of antibiotic in 5 mL of liquid, then we can deduce that in each 1 mL of liquid there will be 50 mg of antibiotic. The final step is to calculate how many lots of 50 mg are contained in 400 mg, which will be 8 lots. Therefore, the child will need 8 lots of 1 mL, which is 8 mL.

PRACTICE 5: APPLICATION

A particular medication is delivered intravenously. There are 80 mg of the required drug per each 1 mL of liquid in the drip. A doctor prescribes 720 mg for a patient to receive of the drug per day. How much of the solution would need to be administered through the intravenous drip?

12.3 PERIMETER

Perimeter is the length, or distance, around the boundary of a two-dimensional closed figure. Because perimeter is a measurement of length, it is one-dimensional in nature.

Perimeter is calculated by finding the sum of all of the sides of a figure. We will commence by looking at how to calculate the perimeter of some simple closed figures.

> **Perimeter**
> The length, or distance, around the boundary or edge of a two-dimensional closed figure

EXAMPLE 6

Find the perimeter of the following figures.

a

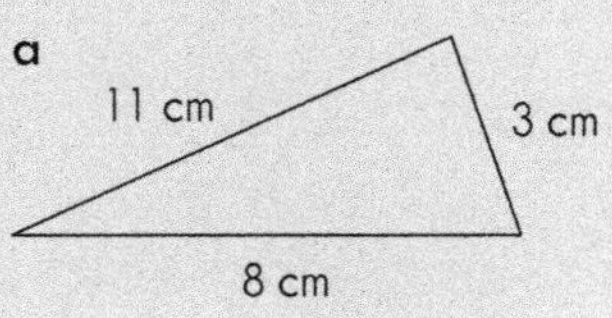

Solutions

a   In this example, the lengths of each side of the triangle are given so the perimeter can be calculated by finding the sum of each side:

Perimeter = 11 + 8 + 3 = 22 cm

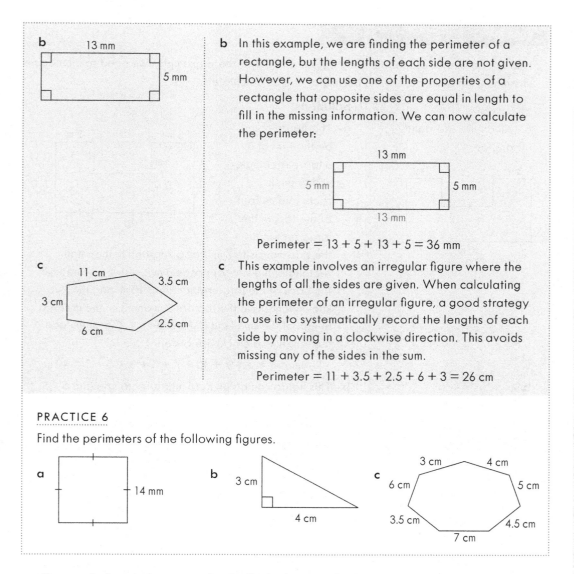

b In this example, we are finding the perimeter of a rectangle, but the lengths of each side are not given. However, we can use one of the properties of a rectangle that opposite sides are equal in length to fill in the missing information. We can now calculate the perimeter:

Perimeter = 13 + 5 + 13 + 5 = 36 mm

c This example involves an irregular figure where the lengths of all the sides are given. When calculating the perimeter of an irregular figure, a good strategy to use is to systematically record the lengths of each side by moving in a clockwise direction. This avoids missing any of the sides in the sum.

Perimeter = 11 + 3.5 + 2.5 + 6 + 3 = 26 cm

Making connections
To review the properties of rectangles and other geometric figures revisit Chapter 11.

PRACTICE 6

Find the perimeters of the following figures.

Geometric figures do not need to be limited to one single type, such as a triangle or a pentagon; rather they can be composite figures. A composite figure is a geometric figure that is made from two or more individual geometric figures. The perimeters of composite figures are, nevertheless, still calculated by finding the sum of all sides.

Kathy Brady

EXAMPLE 7

Find the perimeter of the following figures. In both of these examples we need to fill in some missing information before the perimeter can be calculated.

Solutions

a All angles are right angles.

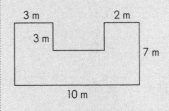

a This figure could be thought of as a large rectangle with a small rectangle cut out of the top. In the large rectangle,

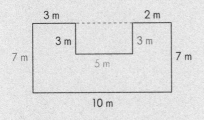

the side opposite 7 m is also 7 m, and in the small rectangle, the side opposite 3 m is also 3 m. Finally, if the base of the large rectangle is 10 m, and the two known lengths in the top are 2 m and 3 m, the missing length of the small cut out will be 5 m. Carefully add the length in a clockwise order:

Perimeter = 2 + 7 + 10 + 7 + 3 + 3 + 5 + 3 = 40 m

b

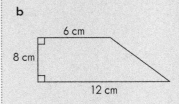

b This figure can be divided into a rectangle and a right-angled triangle. So the side of the rectangle opposite 6 cm is also 6 cm, and the side opposite 8 cm is 8 cm. If the base of the whole figure is 12 cm, then the base of the triangle must then be equal to 6 cm. We need to use Pythagoras' rule to find the length of the hypotenuse.

$c^2 = a^2 + b^2$

$c^2 = 6^2 + 8^2$

$c^2 = 36 + 64$

$c^2 = 100$

$c = 10$ cm

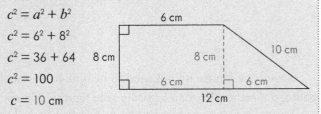

Carefully add the lengths in a clockwise order:

Perimeter = 6 + 10 + 12 + 8 = 36 cm

PRACTICE 7

Find the perimeter of the following figures.

a All angles are right angles.

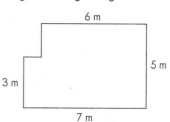

b

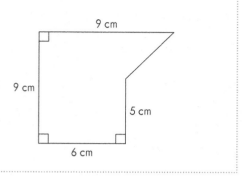

Circumference

Circumference
(measurement)
The distance,
or perimeter,
around the
boundary,
or edge, of
a circle

In Chapter 11, we defined the circumference of a circle as the set of points that are an equal distance from the centre of the circle. The term **circumference**, however, also has a second definition that relates to measurement. It is the distance around the boundary of a circle.

Circles have many fascinating properties. One of these is the special relationship that exists between the circumference and the diameter of a circle. In every circle, no matter what its size, when the length of the circumference is divided by the diameter, the result will be the same constant value. We call this special value π, the Greek letter pi, and write this relationship as $\frac{C}{d} = \pi$. The value of π is the infinite decimal 3.1415926 … which is usually approximated to 3.14, and we will use this value in this chapter.

Since $\frac{C}{d} = \pi$, we can rearrange this to:

$$C = \pi d$$

This equation means that we can calculate the circumference of a circle if we know the diameter.

Making
connections
Chapter 13
details more
fully how to
rearrange
formula.

EXAMPLE 8

Find the perimeters of the following figures.

a

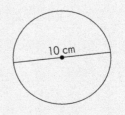

Solutions

a The length of the diameter of this circle is 10 cm. This means we have all the information we need to calculate the circumference. We will use the value $\pi = 3.14$ in this and all other examples that involve circles.

$$C = \pi d$$
$$C = 3.14 \times 10$$
$$= 31.4 \text{ cm}$$

Kathy Brady

b

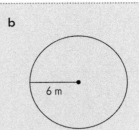

6 m

c

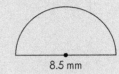

8.5 mm

b In this example, we are given the length of the radius rather than the length of the diameter. The radius of a circle is half the length of the diameter. This means that the diameter of this circle is 12 m.

$$C = \pi d$$
$$C = 3.14 \times 12$$
$$= 37.68 \, m$$

c This example involves finding the perimeter of a semi-circle. We start by calculating the circumference as if it were a full circle with a diameter of 8.5 mm.

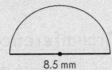

8.5 mm

$$C = \pi d$$
$$C = 3.14 \times 8.5$$
$$= 26.69 \, mm$$

Now we only need half of this circumference—the curve in this diagram. So this will be 26.69 ÷ 2 = 13.35 mm (2 d.p.). But we have not yet fully calculated the whole perimeter of this figure, as we need to add on the base of the semi-circle that is the diameter (the straight line). Therefore:

Perimeter = 13.35 + 8.5 = 21.85 mm

PRACTICE 8

Find the perimeter of the following figures (to two decimal places).

a

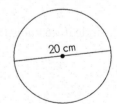

20 cm

b

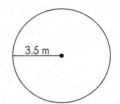

3.5 m

c

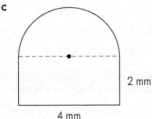

2 mm

4 mm

12.4 AREA

Area is the measurement of the size of a two-dimensional figure. It could be thought of as the amount of space inside the boundary of the figure.

The area of a two-dimensional figure is the number of unit squares that the figure contains, or covers. Therefore, area is generally measured in square units, for example, square centimetres (sq. cm or cm²). However, much larger areas, for instance, expanses of land, are measured in hectares where 1 hectare is $100\,\text{m} \times 100\,\text{m} = 10\,000\,\text{m}^2$

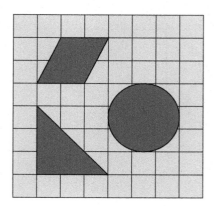

Area
The measurement that describes the size of a two-dimensional figure in terms of square units

Finding the area of rectangles and triangles

The area of a *rectangle* is equal to its length multiplied by its width. The formula for the area of a rectangle is:

$$A = L \times W$$

The area of a *triangle* is equal to half the base multiplied by the perpendicular height. The formula for the area of a triangle is:

$$A = \tfrac{1}{2}(B \times H)$$

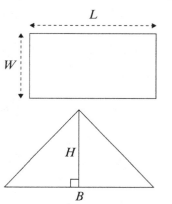

EXAMPLE 9

Find the area of the following figures.

a All angles are right angles.

Solutions

a In this example, the rectangle has length 9 cm and width 7 cm.

$$A = L \times W$$
$$A = 9 \times 7$$
$$= 63 \text{ cm}^2$$

7 cm

9 cm

9 cm

7 cm

Kathy Brady

b All angles are right angles.

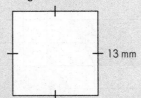

b This is a square, which is a special sort of rectangle. Therefore, we use the same formula as a rectangle, where the length and the width are the same value.

$$A = L \times W$$
$$A = 13 \times 13$$
$$= 169 \text{ mm}^2$$

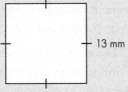

c

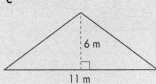

c In this triangle, the perpendicular height is indicated by the dotted line between the base and the apex of the triangle. We know that it is perpendicular by the small square symbol at the base.

$$A = \tfrac{1}{2}(B \times H)$$
$$A = \tfrac{1}{2}(11 \times 6)$$
$$= \tfrac{1}{2} \times 66$$
$$= 33 \text{ m}^2$$

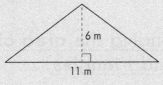

d

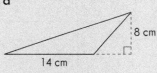

d The perpendicular height of a triangle is not always drawn internally as in this example. In triangles with obtuse angles, such as this one, the perpendicular height is drawn externally meeting an extension of the base line.

$$A = \tfrac{1}{2}(B \times H)$$
$$A = \tfrac{1}{2}(14 \times 8)$$
$$= \tfrac{1}{2} \times 112$$
$$= 56 \text{ cm}^2$$

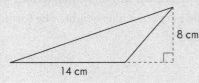

PRACTICE 9

Find the area of the following figures.

a

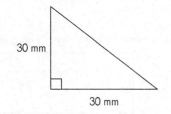

b All angles are right angles.

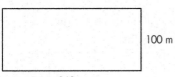

Give your answer in hectares.

c

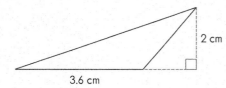

Finding the area of parallelograms and circles

The area of a *parallelogram* is equal to the base multiplied by the perpendicular height. The formula for the area of a parallelogram is:

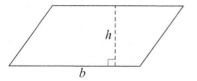

$$A = b \times h$$

To find the area of a circle all we need to know is the length of the radius of the circle. We also need to use the special value π. The area of a circle is equal to the radius squared multiplied by π. The formula for the area of a circle is:

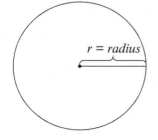

$$A = \pi r^2$$

EXAMPLE 10

Find the area of the following figures.

Solutions

a

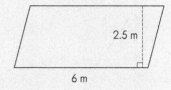

2.5 m

6 m

a In this example, the parallelogram has base length 6 m and perpendicular height of 2.5 m.

$$A = b \times h$$
$$A = 6 \times 2.5$$
$$= 15 \text{ m}^2$$

2.5 m

6 m

b

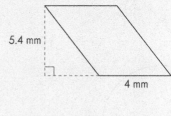

5.4 mm

4 mm

b The perpendicular height of a parallelogram is not always drawn internally, as in this example. It can also be drawn externally meeting an extension of the base line.

$$A = b \times h$$
$$A = 5.4 \times 4$$
$$= 21.6 \text{ mm}^2$$

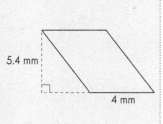

5.4 mm

4 mm

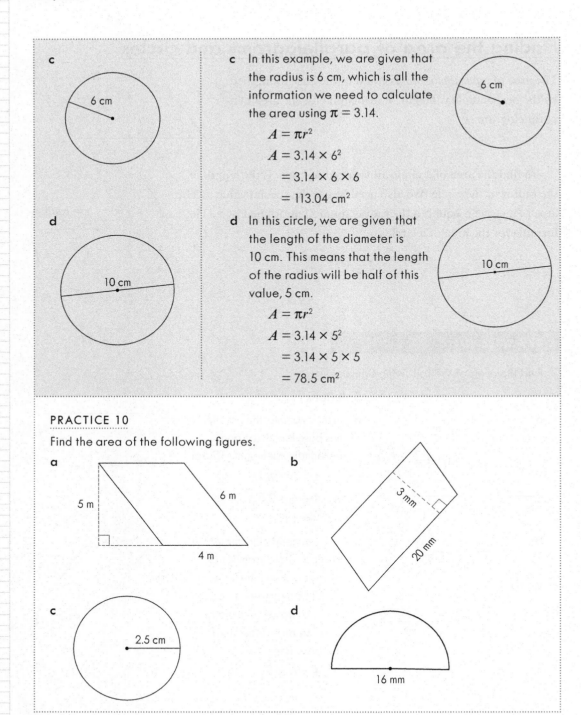

c In this example, we are given that the radius is 6 cm, which is all the information we need to calculate the area using $\pi = 3.14$.

$$A = \pi r^2$$
$$A = 3.14 \times 6^2$$
$$= 3.14 \times 6 \times 6$$
$$= 113.04 \text{ cm}^2$$

d In this circle, we are given that the length of the diameter is 10 cm. This means that the length of the radius will be half of this value, 5 cm.

$$A = \pi r^2$$
$$A = 3.14 \times 5^2$$
$$= 3.14 \times 5 \times 5$$
$$= 78.5 \text{ cm}^2$$

PRACTICE 10

Find the area of the following figures.

a

b

c

d

Finding the area of composite figures

We have seen that composite figures are geometric figures that are made from two or more individual geometric figures. To find the area of composite figures, first the area of each of the component figures is found, and then the sum of these areas is calculated.

EXAMPLE 11

Find the area of the following figures.

a

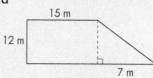

b

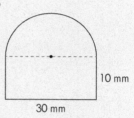

Solutions

a This figure comprises a rectangle and a triangle. The only information that is missing is the perpendicular height of the triangle. This side is common to both the rectangle and the triangle and, therefore, it is 12 m.

$A \text{ rectangle} = L \times W$

$A \text{ rectangle} = 15 \times 12$

$\qquad = 180 \text{ m}^2$

$A \text{ triangle} = \frac{1}{2}(B \times H)$

$A \text{ triangle} = \frac{1}{2}(7 \times 12)$

$\qquad = \frac{1}{2} \times 84$

$\qquad = 42 \text{ m}^2$

Total area = 180 + 42 = 222 m²

b This figure is made up of a rectangle and a semi-circle. The piece of information that appears to be missing is the radius. We can see in the diagram that the diameter of the semi-circle is also one side of the rectangle. Therefore, the diameter of the semi-circle is 30 mm, so the radius will be 15 mm.

$A \text{ rectangle} = L \times W$

$A \text{ rectangle} = 30 \times 10$

$\qquad = 300 \text{ mm}^2$

The area of the semi-circle will be half the area of a full circle with radius 15 mm.

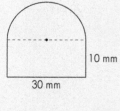

$A \text{ semi-circle} = \frac{1}{2}\pi r^2$

$A \text{ semi-circle} = \frac{1}{2} \times 3.14 \times 15^2$

$\qquad = \frac{1}{2} \times 3.14 \times 15 \times 15$

$\qquad = 353.25 \text{ mm}^2$

Total area = 300 + 353.25

$\qquad = 653.25 \text{ mm}^2$

c

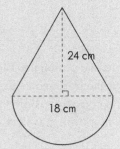

c This is a composite figure made up of a semi-circle and a triangle. The only piece of missing information is the radius. However, we can see in the diagram that the base of the triangle is also the diameter of the semi-circle. Therefore, the radius is 9 cm.

A semi-circle $= \frac{1}{2}\pi r^2$

A semi-circle $= \frac{1}{2} \times 3.14 \times 9^2$

$\qquad = \frac{1}{2} \times 3.14 \times 9 \times 9$

$\qquad = 127.17$ cm^2

A triangle $= \frac{1}{2}(B \times H)$

A triangle $= \frac{1}{2}(18 \times 24)$

$\qquad = \frac{1}{2} \times 432$

$\qquad = 216$ cm^2

Total area $= 127.17 + 216$

$\qquad = 343.17$ cm^2

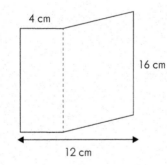

PRACTICE 11

Find the area of the following figures.

a

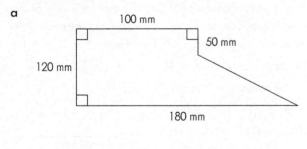

b

c

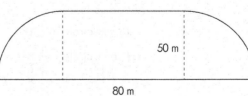

EXAMPLE 12: APPLICATIONS

a The internal grassed area of this running track needs to be re-laid. The grassed area comprises two semi-circles at each end of a central rectangular section. If the diameter of the semi-circles is 60 m and the length of the straight sides is 100 m, what is the total area of grass required?

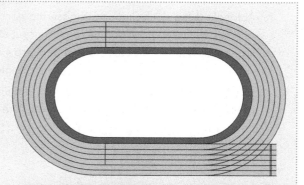

b This swimming pool is 10 m × 3 m. The paving around the pool is 2 m wide. What area of paving has been laid?

Solutions

A good strategy for solving measurement application problems is to draw a geometric diagram that represents the situation.

a The grassed area comprises a rectangle and two semi-circles, which would combine to make a full circle. So, we need to find the area of the rectangle, and the area of a circle with a radius of 30 m (because the diameter is 60 m).

A rectangle $= L \times W$

A rectangle $= 100 \times 60$

$\qquad = 6000$ m^2

A circle $= \pi r^2$

A circle $= 3.14 \times 30^2$

$\qquad = 3.14 \times 30 \times 30$

$\qquad = 2826$ m^2

Total area $= 6000 + 2826$

$\qquad = 8826$ m^2

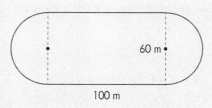

b This example involves a smaller rectangle lying inside a larger rectangle. The smaller rectangle is the swimming pool measuring 10 m × 3 m. The larger rectangle includes both the swimming pool and the paved area. Since the paving is 2 m wide, using the diagram as a guide we can see that an extra 4 m needs to be added to the length and width of the pool to create the larger rectangle, which will be 14 m × 7 m.

If we find the area of larger rectangle and then subtract the area of the smaller rectangle, we would then have the area of the paving that is shaded in the diagram.

A large rectangle $= L \times W$

A large rectangle $= 14 \times 7$

$\qquad = 98$ m^2

A small rectangle $= L \times W$

A small rectangle $= 10 \times 3$

$\qquad = 30$ m^2

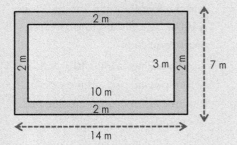

Area of paving = area of large rectangle − area of small rectangle

$A = 98 - 30$

$\qquad = 68$ m^2

PRACTICE 12: APPLICATIONS

a A circular lawn is 10 m in diameter. The paving around it is 1 m wide. What is the area of the paving?

b This is the flag of Brazil. The official dimensions require that the flag measures 20 units × 14 units and that the total area of the rhombus is 87.98 sq. units. Also, the diameter of the circle is 3.5 units. What, in square units, is the area of the flag that surrounds the rhombus? What is the area of the rhombus that surrounds the circle? (Source: Wikipedia.org)

CRITICAL THINKING

Two identical trapeziums can be joined together to form a parallelogram as shown.

How could you use this arrangement to develop a formula for the area of a trapezium?

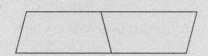

12.5 VOLUME AND CAPACITY

Volume and **capacity** are both associated with the measurement of three-dimensional figures. Whilst these terms are often used interchangeably, they actually describe two different concepts. The volume of a three-dimensional figure is the amount of space that it occupies; and the capacity of a three-dimensional figure is the quantity that it can hold.

Volume is measured in the number of cubic units of space, for example, mm³, cm³ or m³, which a three-dimensional figure occupies. Capacity is the quantity that a three-dimensional space holds and is expressed as a liquid measure, for example, mL or L.

> **Volume**
> The amount of space that a three-dimensional figure occupies
>
> **Capacity**
> The quantity that a three-dimensional container can hold

Units of measurement

When a value is cubed, it is multiplied by itself three times. So when we refer to volume being measured in, for example, cm³, we are thinking about what happens when 1 cm is multiplied by itself three times. That is:

$$1 \text{ cm} \times 1 \text{ cm} \times 1 \text{ cm} = 1 \text{ cm}^3$$

And since 1 cm = 10 mm, we can re-write this cubed expression as:

$$10 \text{ mm} \times 10 \text{ mm} \times 10 \text{ mm} = 1000 \text{ mm}^3$$

Therefore, 1 cm³ = 1000 mm³

Now let's think about what 1 m³ means:

$$1 \text{ m} \times 1 \text{ m} \times 1 \text{ m} = 1 \text{ m}^3$$

And since 1 m = 100 cm, we can re-write this cubed expression as:

$$100 \text{ cm} \times 100 \text{ cm} \times 100 \text{ cm} = 1\,000\,000 \text{ cm}^3$$

Therefore, 1 m³ = 1 000 000 cm³

We can combine these two relationships into a table that will help with converting the units of measurement associated with volume.

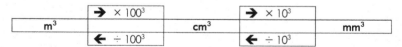

By definition, the capacity of a three-dimensional object is the quantity that it can hold as expressed as a liquid measure such as mL or L. However, the dimensions, for example, the length, width and depth, of a three-dimensional object are measured using linear measurements, such as cm or mm. This means that we need to be able to convert a volume calculated in cubic units to a liquid measure in order to describe the capacity of a three-dimensional object.

The principal relationship that links cubic units to liquid units is:

$$1 \text{ cm}^3 = 1 \text{ mL}$$

Kathy Brady

We can use this key relationship to create some other useful conversions between cubic units and liquid units.

We have already established that $1\,m^3 = 1\,m \times 1\,m \times 1\,m$

And since $1\,m = 100\,cm$, we can re-write this cubed expression as:

$$100\,cm \times 100\,cm \times 100\,cm = 1\,000\,000\,cm^3$$
$$= 1\,000\,000\,mL$$

(because $1\,cm^3 = 1\,mL$)

Now $1\,000\,000\,mL = 1000\,L = 1\,kL$

Therefore, $1\,m^3 = 1\,kL$

EXAMPLE 13

Carry out the following conversions.

Solutions

In some of these examples we will need to refer to this conversion table:

m³	$\times 100^3$ → / ← $\div 100^3$	cm³	$\times 10^3$ → / ← $\div 10^3$	mm³

a $2300\,mm^3$ to cm^3

a   To convert mm^3 to cm^3 we need to divide by 10^3, which is 1000. So we need to jump the decimal point in 2300 (which is to the right of the final 0) three places to the left.

$$2.300,$$

So, $2300\,mm^3 = 2.3\,cm^3$

b $0.035\,m^3$ to cm^3

b   To convert m^3 to cm^3 we need to multiply by 100^3, which is $1\,000\,000$. So we need to jump the decimal point in 0.035 six places to the right and fill the empty spaces with zeros.

$$0.035000,$$

So, $0.035\,m^3 = 35\,000\,cm^3$

c $5600\,cm^3$ to mL

c   Since $1\,mL = 1\,cm^3$, $5600\,cm^3 = 5600\,mL$

d $9000\,L$ to m^3

d   Firstly, $9000\,L = 9\,kL$. And since $1\,kL = 1\,m^3$, $9\,kL$ will equal $9\,m^3$. Therefore, $9000\,L = 9\,m^3$.

e $6.3\,m^3$ to kL

e   Since $1\,m^3 = 1\,kL$, $6.3\,m^3$ will equal $6.3\,kL$.

f $0.75\,m^3$ to L

f   Since $1\,m^3 = 1000\,L$, $0.75\,m^3$ will equal $750\,L$.

g 0.063 m³ to mL

g In this conversion, we need to firstly convert m³ to cm³. To do this we need to jump the decimal point in 0.063 six jumps to the right and fill the empty spaces with zeros.

$$0.063000,$$

So, 0.063 m³ = 63 000 cm³. Now since 1 cm³ = 1 mL, 63 000 cm³ will equal 63 000 mL. So, 0.063 m³ = 63 000 mL.

PRACTICE 13

Carry out the following conversions.

a 4100 cm³ to m³   **b** 0.078 cm³ to mm³   **c** 2900 cm³ to mL

d 5500 cm³ to L   **e** 4.7 kL to m³   **f** 800 L to m³

g 9300 mL to m³

Calculating volume

Many three-dimensional figures, such as prisms or cylinders, have what is known as a **uniform cross- section**, which is created when the figure is cut or sliced parallel to its base or ends. If a three-dimensional figure has a uniform cross-section, then all cross-sections are identical to each other and the base or ends of the figure.

In the first figure, a cylinder, the uniform cross-section that is formed will be a circle that is identical to the base. In the second figure, a triangular prism, the uniform cross-section is a triangle that is identical to the ends. In the third figure, a rectangular prism, the uniform cross-section is a rectangle that is identical to the base.

For any three-dimensional figure with a uniform cross-section, the volume of the figure can be calculated by finding the area of the cross-section and then multiplying the area by the height or length of the figure.

Volume of figure with uniform cross-section = area of cross-section × height (or length)

Uniform cross-section
Cross-sections that are always identical to each other and the base or ends of the figure

EXAMPLE 14

Find the volume of the following figures.

a

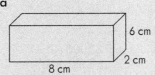

b

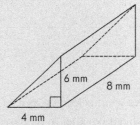

c

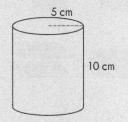

Solutions

a   This figure is a rectangular prism. The uniform cross-section is the rectangle that is shaded with dimensions 6m × 2 m. The length of the figure is 8 m.

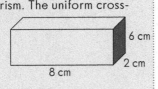

Volume = area of cross-section × length

Volume = area of rectangle × length

Volume = 6 × 2 × 8

       = 96 cm²

b   This figure is a triangular prism. The uniform cross-section is a right-angled triangle with base 4 mm and perpendicular height 6 mm. The length of this figure is 8 mm.

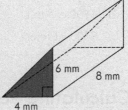

Volume = area of cross-section × length

Volume = area of triangle × length

Volume = $\frac{1}{2}(B \times H)$ × length

Volume = $\frac{1}{2}(4 \times 6)$ × 8

       = $\frac{1}{2}(24)$ × 8

       = 12 × 8

       = 96 mm³

c   This figure is a cylinder. The uniform cross-section of this figure is a circle with radius 5 cm. The height of the figure is 10 cm.

Volume = area of cross-section × length

Volume = area of circle × length

Volume = πr^2 × height

Volume = 3.14 × 5² × 10

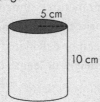

       = 3.14 × 25 × 10

       = 785 cm³

d

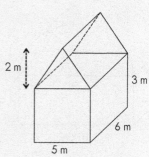

2 m

3 m

6 m

5 m

d This is a composite three-dimensional figure. It comprises a triangular prism sitting on top of a rectangular prism. The uniform cross-section for this figure is a composite two-dimensional figure comprising a rectangle and a triangle. The dimensions of the rectangle are 5 m × 3 m. The base of the triangle is 5 m (as it is the common side with the rectangle) and the perpendicular height is 2 m. We will start by calculating the area of the uniform cross-section.

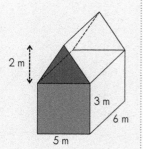

2 m

3 m

6 m

5 m

Area = (area of rectangle) + (area of triangle)

Area = $(L \times W) + (\frac{1}{2}(B \times W))$

Area = $(5 \times 3) + (\frac{1}{2}(5 \times 2))$

$= 15 + \frac{1}{2}(10)$

$= 15 + 5$

$= 20 \text{ m}^2$

Now that we have the area of the cross-section we can find the volume of the figure by multiplying by its length, which is 6 m.

Volume = area of cross-section × length

Volume = 20 × 6

$= 120 \text{ m}^3$

PRACTICE 14

Find the volume of the following figures.

a

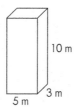

10 m

3 m

5 m

b

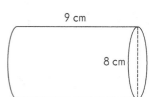

9 cm

8 cm

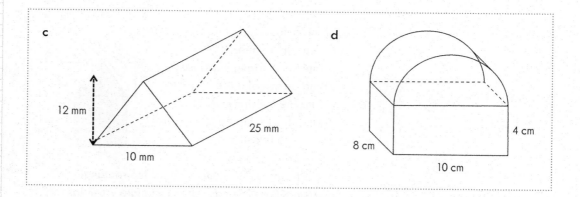

Volume of pyramids, cones and spheres

There are other three-dimensional figures that do not have a uniform cross-section such as pyramids, cones and spheres. We can still calculate the volume of these figures using formulas that apply specifically to each type of figure.

The volume of a cone or pyramid is:

$$\frac{1}{3}(\text{area of the base} \times \text{perpendicular height})$$

The volume of a sphere is:

$$\frac{4}{3}(\pi r^3)$$

EXAMPLE 15

Find the volume of the following figures.

a

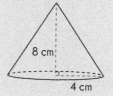

Solutions

a This figure is a cone. The base of the cone is a circle with radius 4 cm and the perpendicular height is 8 cm.

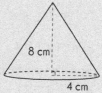

$\text{Volume} = \frac{1}{3}(\text{area of the base} \times \text{perpendicular height})$

$\text{Volume} = \frac{1}{3}(\text{area of circle} \times \text{perpendicular height})$

$\text{Volume} = \frac{1}{3}(\pi r^2 \times \text{perpendicular height})$

$\text{Volume} = \frac{1}{3}(3.14 \times 4^2 \times 8)$

$\quad\quad\quad = \frac{1}{3}(3.14 \times 16 \times 8)$

$\quad\quad\quad = 133.97 \text{ cm}^3 \text{ (2 decimal places)}$

b

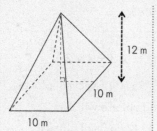

b This figure is a square based pyramid. The square base has 10 m sides and the perpendicular height is 12 m.

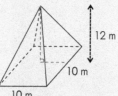

Volume = $\frac{1}{3}$(area of the base × perpendicular height)

Volume = $\frac{1}{3}$(area of square × perpendicular height)

Volume = $\frac{1}{3}$($L \times W \times$ perpendicular height)

Volume = $\frac{1}{3}$(10 × 10 × 12)

$\qquad$ = 400 m³

c

c This figure is a sphere. The radius of the sphere is 5 mm. This is the only piece of information we need to calculate the volume of a sphere.

Volume = $\frac{4}{3}$(πr^3)

Volume = $\frac{4}{3}$(3.14 × 5³)

$\qquad$ = $\frac{4}{3}$(3.14 × 5 × 5 × 5)

$\qquad$ = 523.33 mm³ (2 decimal places)

d

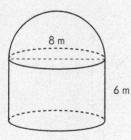

d This is a composite three-dimensional figure. It comprises a half-sphere sitting on top of a cylinder. Both the cylinder and the half-sphere have a diameter of 8 m, which means they have a radius of 4 m. The height of the cylinder is 6 m. The most straightforward way of calculating the volume of this figure is to separately calculate the volume of each component and then find their sum.

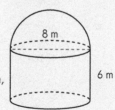

Volume cylinder = area of circular base × base

Volume = πr^2 × height

Volume = 3.14 × 4² × 6

$\qquad$ = 3.14 × 4 × 4 × 6

$\qquad$ = 301.44 m³ (2 decimal places)

To calculate the volume of the half-sphere we will firstly calculate the volume of a full sphere with radius 4 m, and then halve the result.

Kathy Brady

$$\text{Volume sphere} = \frac{4}{3}(\pi r^3)$$

$$\text{Volume sphere} = \frac{4}{3}(3.14 \times 4^3)$$

$$= \frac{4}{3}(3.14 \times 4 \times 4 \times 4)$$

$$= 267.95 \text{ m}^3 \text{ (2 decimal places)}$$

So the volume of the half-sphere will be:

$$267.95 \div 2 = 133.97 \text{ m}^2 \text{ (2 decimal places)}$$

Now we can find the volume of the entire figure by adding the two volumes we have calculated:

$$\text{Volume} = 301.44 + 133.97$$

$$= 435.41 \text{ m}^3 \text{ (2 decimal places)}$$

PRACTICE 15

Find the volume of the following figures.

a

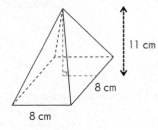

11 cm

8 cm

8 cm

b

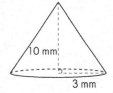

10 mm

3 mm

c

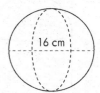

16 cm

d

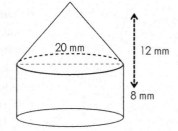

20 mm   12 mm

8 mm

EXAMPLE 16: APPLICATION

This diagram represents the dimensions of a swimming pool. How many kilolitres of water does it contain when it is full?

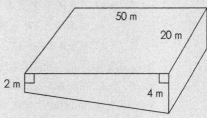

50 m

20 m

2 m

4 m

Solutions

The swimming pool diagram is an example of a three-dimensional figure with a uniform cross-section, which is shaded. This cross-section is a composite figure comprising a rectangle (darker) and a triangle (lighter). The rectangle has the dimensions 2 m × 50 m. Because the deepest side of the pool was given as 4 m, the triangle must therefore have a base of 2 m and its perpendicular height is the same 50 m.

The first step is to calculate the area of the uniform cross-section comprising the rectangle and the triangle.

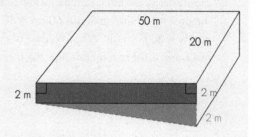

Area of rectangle $= L \times W$

$$= 50 \times 2$$

$$= 100 \text{ m}^2$$

Area of triangle $= \frac{1}{2}(B \times H)$

$$= \frac{1}{2}(2 \times 50)$$

$$= \frac{1}{2}(100)$$

$$= 50 \text{ m}^2$$

Total area of uniform cross-section $= 100 \text{ m}^2 + 50 \text{ m}^2$

$$= 150 \text{ m}^2$$

Now we can calculate the volume of the pool as the area of cross-section × length (or height). But be careful here, the length is not 50 m; rather it is 20 m.

Volume = area of cross-section × length

Volume $= 150 \times 20$

$$= 3000 \text{ m}^3$$

The final step is to convert 3000 m³ into kL. Because 1 m³ = 1 kL, we can determine that the capacity of the pool is 3000 kL.

Kathy Brady

PRACTICE 16: APPLICATIONS

a This propane gas tank is composed of a
   cylinder with a half-sphere at each end. The
   length of the cylindrical section is 8 m and
   the diameter is 4 m. What is the capacity of
   the tank in litres of propane?

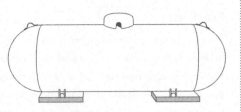

b This concrete planter has solid concrete
   walls that surround an open cavity. On the outside,
   the dimensions of the base are 80 cm × 60 cm and the
   planter is 50 cm high. On the inside of the cavity, the
   dimensions of the base are 60 cm × 40 cm, and the
   cavity is 30 cm high. What volume of concrete, in m³,
   has been used to manufacture the planter?

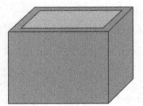

12.6 CHAPTER SUMMARY

SKILL OR CONCEPT	DEFINITION OR DESCRIPTION	EXAMPLES
Length	Distance from one point to another, which is measured using the basic metric unit of metres	6 m 0.2 m
Mass	A measure of how much matter is contained in an object, commonly measured as weight using the basic metric unit of grams	0.4 g 1246 g
Metric prefixes	G (giga) means billion M (mega) means million k (kilo) means thousand c (centi) means hundredth m (milli) means thousandth μ (micro) means millionth n (nano) means billionth	5 GB 4.2 ML 0.6 kg 7.9 cm 3 mL 8 μg 2.6 nm
Metric unit conversions	To move from a smaller unit to a larger unit, jump the decimal point to the *left* the required number of places. To move from a larger unit to a smaller unit, jump the decimal point to the *right* the required number of places.	3600 mg = 3.6 g 79 cm = 0.00079 km 1 300 000 μg = 1300 mg 0.008 kL = 8000 mL 4.5 km = 4500 m
Perimeter	The length, or distance, around the boundary of a two-dimensional closed figure, which is calculated by adding the lengths of each side.	$P = 14 + 2.3 + 15.1$ $P = 31.4$ m
Circumference	The distance around the boundary of a circle or the perimeter of a circle.	
π (Greek letter pi)	When the circumference of any circle is divided by its diameter, $\frac{C}{d} = \pi$, an infinite decimal that is approximated to 3.14	
Calculating circumference	Circumference of a circle is equal to π multiplied by its diameter: $C = \pi d$	$C = \pi d$ $C = 3.14 \times 5$ $= 15.7$ mm
Area	The size of a two-dimensional figure described in terms of square units of length.	

SKILL OR CONCEPT	DEFINITION OR DESCRIPTION	EXAMPLES
Hectare	A unit used to describe the area of large expanses: 1 hectare $= 10000\,m^2$	
Calculating the area of a rectangle	Area of a rectangle is equal to its length multiplied by its width: $A = L \times W$	$A = L \times W$ $A = 13 \times 8$ $\quad = 104\,cm^2$ 8 cm, 13 cm
Calculating the area of a triangle	Area of a triangle is equal to half the base multiplied by the perpendicular height: $A = \frac{1}{2}(B \times H)$	$A = \frac{1}{2}(B \times H)$ $A = \frac{1}{2}(16 \times 7)$ $\quad = \frac{1}{2} \times 112$ $\quad = 56\,m^2$ 7 m, 16 m
Calculating the area of a parallelogram	Area of a parallelogram is equal to the base multiplied by the perpendicular height: $A = b \times h$	$A = b \times h$ $A = 3.6 \times 12$ $\quad = 43.2\,cm^2$ 3.6 cm, 12 cm
Calculating the area of a circle	Area of a circle is equal to the radius squared multiplied by π: $A = \pi r^2$	$A = \pi r^2$ $A = 3.14 \times 1.4^2$ $\quad = 6.15\,mm^2$ (2 d.p.) 1.4 mm
Volume	The amount of space that a three-dimensional figure occupies expressed as cubic units of length.	
Capacity	The quantity that a three-dimensional container can hold expressed as a liquid measure.	
Volume and capacity unit conversions	$1\,cm^3 = 1000\,mm^3$ $1\,m^3 = 1000000\,cm^3$ $1\,cm^3 = 1\,mL$ $1\,m^3 = 1\,kL$	$3\,mL = 3\,cm^3 = 3000\,mm^3$ $7200000\,cm^3 = 7.2\,m^3 = 7.2\,kL$ $6.8\,L = 6800\,mL = 6800\,cm^3$ $4500\,L = 4.5\,kL = 4.5\,m^3$
Uniform cross-section of three-dimensional figure	Created when a three-dimensional figure is cut parallel to the base or the ends of the figure. Resulting cross-sections are all identical to each other and the base or ends.	

SKILL OR CONCEPT	DEFINITION OR DESCRIPTION	EXAMPLES
Calculating the volume of a figure with uniform cross-section	Volume = area of cross-section × height (or length)	$V = A$ of cross-section × length $V = 54 \times 12$ $\quad = 648 \text{ cm}^3$ Area $= 54 \text{ cm}^2$ 12 cm
Calculating the volume of a cone or a pyramid	Volume = $\frac{1}{3}$ × area of the base × perpendicular height	$V = \frac{1}{3}(A \text{ of base} \times \text{height})$ $V = \frac{1}{3}(A \text{ of circle} \times \text{height})$ $V = \frac{1}{3}(\pi r^2 \times \text{height})$ $V = \frac{1}{3}(3.14 \times 5^2 \times 16)$ $\quad = \frac{1}{3}(3.14 \times 5 \times 5 \times 16)$ $\quad = 418.67 \text{ cm}^3$ 16 cm 5 cm
Calculating the volume of a sphere	Volume = $\frac{4}{3}(\pi r^3)$	$V = \frac{4}{3}(\pi r^3)$ $V = \frac{4}{3}(3.14 \times 8^3)$ $\quad = \frac{4}{3}(3.14 \times 8 \times 8 \times 8)$ $\quad = 2143.57 \text{ mm}^3$ 8 mm

Kathy Brady

12.7 REVIEW QUESTIONS

A SKILLS

1   Carry out the following unit conversions.

    a   7400 cm to km   **b**   61 mg to μg

    c   4.9 ML to kL   **d** 1.3 m^3 to cm^3

    e   0.67 mL to cm^3 **f** 5.5 m^3 to kL

2   Find the perimeters of the following figures.

    a

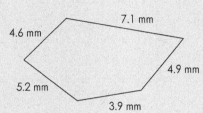

    b

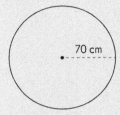

    c

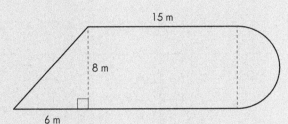

3   Find the area of the following figures.

    a   All angles are right angles.   **b**

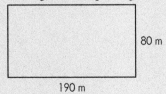

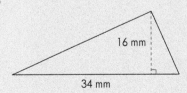

    Give your answer in hectares
    (to two decimal places)

c

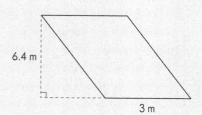

6.4 m

3 m

d

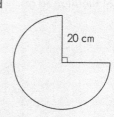

20 cm

e

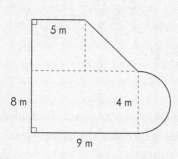

5 m

8 m

4 m

9 m

4 Find the volume of the following figures.

a

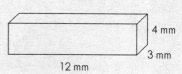

4 mm

3 mm

12 mm

b

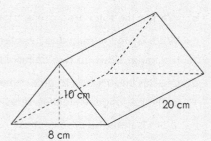

10 cm

20 cm

8 cm

c

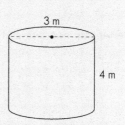

3 m

4 m

d

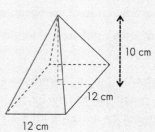

10 cm

12 cm

12 cm

e

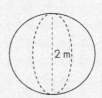

2 m

f

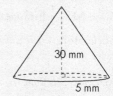

30 mm

5 mm

Kathy Brady

5 Find the volume of the shaded portion of this figure.

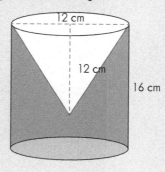

12 cm

12 cm

16 cm

B APPLICATIONS

1 A small camping tent has sides at the front and the back that are isosceles triangles. The base of the triangular sides is 1.5 m and the height at the centre is 1 m. The length of the tent is 2.1 m and it also has an attached base.

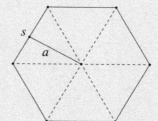

   a How many sides does the tent have altogether, including the base?

   b What is the total area of the triangular sides?

   c What is the length of the hypotenuse on the triangular sides (to two decimal places)?

   d Use this information to determine the dimensions of the rectangular slanting sides of the tent.

   e What is the total area of these rectangular sides (to two decimal places)?

   f What are the dimensions of the base?

   g What is the area of the base (to two decimal places)?

   h What is the total area of fabric that is needed to construct the tent (to two decimal places)?

2 The area of a regular hexagon can be calculated if we know the side length (s), and the length from the centre of a side to the centre of the hexagon (a), which are marked in this diagram. Using these two measurements the formula for calculating the area of a regular hexagon is:

$$A = \tfrac{1}{2}(\text{perimeter} \times a)$$

   a What is the perimeter of a regular hexagon with a side length of 2 metres?

   b If a regular hexagon has 2 metre sides, the length of a will be 1.73 metres (to two decimal places). Use the formula above to calculate the area of such a regular hexagon.

c The hexagon that has been described forms
the uniform-cross section of an above ground
swimming pool. If the depth of the pool is
1.4 metres, what is the volume of the pool in cubic
metres (to two decimal places)?

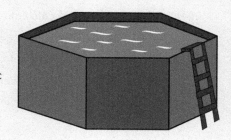

d What is the capacity of the pool in kilolitres (to
two decimal places)?

3 According to the Brazilian National Institute for Space
Research, in 2015 the estimated loss of forest in the
Brazilian Amazon was 5831 km^2 (Source: inpe.br).

a If 1 km = 1000 m, what is 1 km^2 in terms of m^2?

b What is the 2015 loss of Brazilian Amazon forest in
terms of m^2?

c How many hectares of Brazilian Amazon forest
were lost in 2015?

4 The London Eye is a giant Ferris wheel on the South
Bank of the River Thames in London. It has a diameter
of 120 metres.

a When riding on the London Eye, how far does a
passenger capsule travel in one full revolution of
the wheel?

When the London Eye was constructed, sections of
the wheel were floated up the Thames and then the
whole wheel was assembled lying flat on platforms in
the river. Once this had been completed, the wheel
was hydraulically lifted into an upright position.
(Source: wikipedia.com)

b What area did the London Eye cover when it was
completed and ready to be lifted into place?

5 The circumference of a NBA Size 7 competition
basketball is 74.5 cm (Source: spalding.com)

a Substitute this value into the formula $C = \pi d$.

b Now divide both sides of this formula by the value
of π (3.14) to find the diameter of the ball to two
decimal places.

c What is the radius of the ball (to two decimal
places)?

Kathy Brady

d   Use this information to calculate the volume of the ball to the nearest cm^3.

e   How many litres of air (to one decimal place) need to be pumped into the ball to expand it fully? (Hint: first convert cm^3 to mL)

f   What is the relationship between the size of the ball and its capacity?

13

ALGEBRA: SOLVING EQUATIONS

CHAPTER CONTENT

» What are equations and why are they important?

» Isolating the unknown

» The balancing principle

» Solving equations using addition or subtraction

» Solving equations using multiplication or division

» Solving equations using a combination of operations

» Solving equations involving collecting terms and removing brackets

» Formula

CHAPTER OBJECTIVES

» Solve equations by inspection

» Use backtracking techniques to isolate the unknown in an algebraic expresssion

» Isolate the unknown using the balancing principle

» Solve linear equations that involve addition, subtraction, multiplication and division, or a combination of these operations

» Solve linear equations that involve removing brackets and collecting like terms

» Rearrange and evaluate formulas

BIG SHOES TO FILL

It is possible to estimate your Australian shoe size using the length of your foot in millimetres (mm).

The equation that describes the relationship between foot length and shoe size is:

$$S = 0.125l - 22.5$$

where S is the shoe size, and l is the foot length in mm. (Source: sandler.com.au). For a size 10 shoe, the equation would become:

$$10 = 0.125l - 22.5$$

So to find out what foot length would fit size 10 shoes we would need to solve this equation to obtain $l = 260$ mm.

13.1 WHAT ARE EQUATIONS AND WHY ARE THEY IMPORTANT?

Equation
A mathematical sentence that describes the equality of two algebraic expressions, and comprises left and right-hand sides that are separated by an equal sign

The power and importance of **equations** is demonstrated by the extent to which we rely upon them in our everyday life. For example, the computer chips that are integral to all aspects of our daily activities require the application of equations, or the processes that are behind getting successful hits when doing an internet search are also driven by equations. Put simply, equations have changed, and will continue to change, our world.

So, what is an equation? An equation is a mathematical sentence that is composed of two algebraic expressions separated by an equals sign. An equation, therefore, has two sides: the left-hand side (LHS) that is to the left of the = sign and the right-hand side (RHS) that is to the right of the = sign.

Some examples of equations are:

$$5x + 7 = 19$$
$$2(x - 5) = 3(x + 3) - 4$$
$$\frac{3x + 2}{4} = 5(x - 1)$$
$$x^2 - 3x + 5 = 0$$

Making Connections
To further revise the language associated with using algebra revisit Chapter 9.

When talking about and using equations, we use the same language and terminology as when using algebraic expressions. Here is a recap using $x^2 - 3x + 5 = 0$ as an example:

- x is known as a *variable* as it represents an unknown value
- 5 is a *constant*, a known value
- -3 is a *coefficient*, a constant that is used to multiply a variable
- x^2, $-3x$ and 5 are all *terms* that make up the equation

In Chapter 9, we were introduced to the technique of evaluating algebraic expressions by substituting a specific given value for each variable in the expression and then performing mathematical operations to get a numerical answer. However, when using equations we do not evaluate, rather we solve the equation. The **solution of an equation** is the value of the variable that makes the equation true; that is, the LHS will equal the RHS. Finding this value is known as solving the equation.

Solution of an equation
The value of the variable required to make the LHS of the equation equal to the RHS of the equation

There are many techniques that can be used to solve equations, depending upon its complexity. The simplest method is called solving by inspection. This involves looking at the equation and mentally working out the solution.

EXAMPLE 1

Solve the following equations by inspection.

Solutions

a $b + 2 = 5$

a In this equation, we ask what should we add to 2 to get 5.

Solution: $b = 3$

b $13 - x = 7$

b In this equation, we ask what should we subtract from 13 to get 7.

Solution: $x = 6$

c $\frac{x}{5} = 10$

c For this example, we ask what divided by 5 is equal to 10.

Solution: $x = 50$

d $3c = 21$

d In this example, we ask what multiplied by 3 is equal to 21.

Solution: $c = 7$

Notice in each of these solutions, the final answer is written in a form that describes which value of the variable makes the equation true, for example, $b = 3$.

PRACTICE 1

Solve the following equations by inspection.

a $n + 6 = 11$ b $x + 5 = 5$ c $3 + b = 0$

d $12 - a = 2$ e $\frac{x}{3} = 15$ f $4g = -8$

CRITICAL THINKING

Here is what looks like a very complex equation:

$$10 - x^2 + (x - 6) + 2(x - 2) + x(x - 3) + x = 0$$

Try to apply the techniques of expanding brackets and collecting like terms that were covered in Chapter 9 to find the simple solution to this equation.

13.2 ISOLATING THE UNKNOWN

Many equations, however, cannot be solved by inspection because they are more complex. This means that we need to use other algebraic methods to solve equations. Whichever method is used, the goal is to find the value of the unknown that will make the equation true. In order to do this, we need to isolate the unknown on one side of the equal sign, which means that the wanted value will be on the other side.

Building expressions

To be able to isolate the unknown, firstly it is important to understand how algebraic expressions are built up and a flowchart is a useful tool for this analysis. For example, consider how the expression $2x + 3$ has been built up:

$$\boxed{x} \xrightarrow{\times 2} \boxed{2x} \xrightarrow{+3} \boxed{2x + 3}$$

We started with an x, which was then multiplied by 2 to give $2x$, after that a 3 was added to result in $2x + 3$.

EXAMPLE 2

Complete the following flowcharts to show how each expression was built up.

a $\boxed{x} \xrightarrow{\times 5} \boxed{} \xrightarrow{+2} \boxed{}$

b $\boxed{x} \xrightarrow{-1} \boxed{} \xrightarrow{\times 3} \boxed{}$

c $\boxed{x} \xrightarrow{+4} \boxed{} \xrightarrow{\div 3} \boxed{}$

d $\boxed{x} \xrightarrow{\div 2} \boxed{} \xrightarrow{-5} \boxed{}$

Solutions

a $\boxed{x} \xrightarrow{\times 5} \boxed{5x} \xrightarrow{+2} \boxed{5x + 2}$

b   In this example, because all of $x - 1$ is multiplied by 3 in the second step, brackets are used to indicate this.

$$\boxed{x} \xrightarrow{-1} \boxed{x - 1} \xrightarrow{\times 3} \boxed{3(x - 1)}$$

c   In this example, because all of $x + 4$ is divided by 3 in the second step, $x + 4$ is the numerator in the division when written in fraction form.

$$\boxed{x} \xrightarrow{+4} \boxed{x + 4} \xrightarrow{\div 3} \boxed{\dfrac{x + 4}{3}}$$

d   In the first step, x is divided by 2 (expressed as a fraction), so 5 is subtracted from that whole expression in the second step.

$$\boxed{x} \xrightarrow{\div 2} \boxed{\dfrac{x}{2}} \xrightarrow{-5} \boxed{\dfrac{x}{2} - 5}$$

PRACTICE 2

Complete the following flowcharts to show how each expression was built up.

a x $\xrightarrow{\times 4}$ ☐ $\xrightarrow{-3}$ ☐

b x $\xrightarrow{+5}$ ☐ $\xrightarrow{\times -2}$ ☐

c x $\xrightarrow{-1}$ ☐ $\xrightarrow{\div -4}$ ☐

d x $\xrightarrow{\div 3}$ ☐ $\xrightarrow{+7}$ ☐

e x $\xrightarrow{\times 3}$ ☐ $\xrightarrow{-1}$ ☐ $\xrightarrow{\div 4}$ ☐

f x $\xrightarrow{+5}$ ☐ $\xrightarrow{\times 2}$ ☐ $\xrightarrow{-3}$ ☐

EXAMPLE 3

Use a flowchart to show how each expression was built up.

a $4x - 2$ b $3(x - 2)$ c $\dfrac{3x - 4}{2}$

d $\dfrac{x}{3} + 5$ e $\dfrac{x + 6}{3} - 4$

Solutions

In these examples, the flowchart boxes and the order in which the steps occurred when building up the expression have not been provided. Rather those decisions need to be made by analysing the expression. In each flowchart, the first box will be x, so the question that we need to ask is: "What was the first thing that happened to the x?" Then careful decisions need to be made about the order in which the steps occurred.

a The first thing that happened to the x in this example was that it was multiplied by 4. Notice that it would not be correct to say that the first thing to happen to the x was a subtraction of 2, if this were the case the expression would be $4(x - 2)$. Therefore, the flowchart is:

x $\xrightarrow{\times 4}$ $4x$ $\xrightarrow{-2}$ $4x - 2$

Kathy Brady

b In this example, the first thing that happened to the x was a subtraction of 2, then the whole of $x - 2$ was multiplied by 3.

$$\boxed{x} \xrightarrow{\;-2\;} \boxed{x-2} \xrightarrow{\;\times 3\;} \boxed{3(x-2)}$$

c In this example, the first thing that happened to the x was a multiplication by 3, after that 4 was subtracted. The final step was a division of the whole expression by 2.

$$\boxed{x} \xrightarrow{\;\times 3\;} \boxed{3x} \xrightarrow{\;-4\;} \boxed{3x-4} \xrightarrow{\;\div 2\;} \boxed{\dfrac{3x-4}{2}}$$

d In this example, the first thing that happened to the x was a division by 3. Then 5 was added to the whole expression.

$$\boxed{x} \xrightarrow{\;\div 3\;} \boxed{\dfrac{x}{3}} \xrightarrow{\;+5\;} \boxed{\dfrac{x}{3}+5}$$

e In this example, the first thing that happened to the x was an addition of 6, then the next step was a division by 3 of $x + 6$. In the final step, 4 was subtracted from the whole expression.

$$\boxed{x} \xrightarrow{\;+6\;} \boxed{x+6} \xrightarrow{\;\div 3\;} \boxed{\dfrac{x+6}{3}} \xrightarrow{\;-4\;} \boxed{\dfrac{x+6}{3}-4}$$

PRACTICE 3

Use a flowchart to show how each expression was built up.

a $-3x + 1$ **b** $5(x + 6)$ **c** $\dfrac{-2x + 1}{3}$

d $\dfrac{x}{7} - 4$ **e** $\dfrac{2x - 1}{4 + 3}$

Backtracking

Backtracking is the process of undoing an expression in the order that it was built up. Using a flowchart, we work in reverse starting with the final expression on the right-hand side of the flowchart and working back to the beginning variable (we have been using x) on the left-hand side of the flowchart. This process of getting x by itself is known as isolating the unknown.

When we backtrack, we apply opposite operations to undo those that were used when building up the expression. The opposite operations are also known as inverse operations.

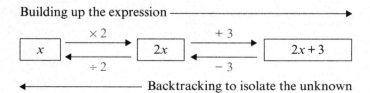

Building up the expression ⟶

$$\boxed{x} \;\underset{\div 2}{\overset{\times 2}{\rightleftarrows}}\; \boxed{2x} \;\underset{-3}{\overset{+3}{\rightleftarrows}}\; \boxed{2x+3}$$

⟵ Backtracking to isolate the unknown

EXAMPLE 4

Backtrack the following expressions working RHS to LHS to isolate the unknown.

a $\boxed{x}$ $\xrightarrow{\times 4}$ $\boxed{4x}$ $\xrightarrow{+3}$ $\boxed{4x+3}$

b $\boxed{x}$ $\xrightarrow{+3}$ $\boxed{x+3}$ $\xrightarrow{\div 4}$ $\boxed{\dfrac{x+3}{4}}$

Solutions

Working from the RHS to the LHS and undoing each building operation by using the reverse operation we get the following.

a $\boxed{x}$ $\overset{\times 4}{\underset{\div 4}{\rightleftarrows}}$ $\boxed{4x}$ $\overset{+3}{\underset{-3}{\rightleftarrows}}$ $\boxed{4x+3}$

b $\boxed{x}$ $\overset{+3}{\underset{-3}{\rightleftarrows}}$ $\boxed{x+3}$ $\overset{\div 4}{\underset{\times 4}{\rightleftarrows}}$ $\boxed{\dfrac{x+3}{4}}$

PRACTICE 4

Backtrack the following expressions working RHS to LHS to isolate the unknown.

a $\boxed{x}$ $\xrightarrow{-2}$ $\boxed{x-2}$ $\xrightarrow{\times 5}$ $\boxed{5(x-2)}$

b $\boxed{x}$ $\xrightarrow{\div 2}$ $\boxed{\dfrac{x}{2}}$ $\xrightarrow{+5}$ $\boxed{\dfrac{x}{2}+5}$

c $\boxed{x}$ $\xrightarrow{-2}$ $\boxed{x-2}$ $\xrightarrow{\times 4}$ $\boxed{4(x-2)}$ $\xrightarrow{+1}$ $\boxed{4(x-2)+1}$

We can skip the building step and instead move straight to using a flowchart with inverse operations to backtrack, or undo, the expression and isolate the unknown.

EXAMPLE 5

Use a flowchart to isolate the unknown in the following expressions.

a $3x-5$ b $\dfrac{x-1}{3}$

c $\dfrac{2x}{3}+1$ d $\dfrac{5x-2}{4}-7$

Kathy Brady

Solutions

a As the expression was built up, the last thing that would have occurred was -5, so this is the first thing to undo using the inverse operation $+5$. The next expression that is being backtracked is $3x$, which means x multiplied by 3. The inverse operation required to undo this is divide by 3. We have now isolated the unknown, x.

$$\boxed{3x - 5} \xrightarrow{+5} \boxed{3x} \xrightarrow{\div 3} \boxed{x}$$

b The question we must ask each time is what was the last thing that occurred as the expression was built up. In this case, it was the divide by 3. So this must be undone first using the inverse operation of multiply by 3. Next, the -1 is undone using the inverse operation $+1$.

$$\boxed{\dfrac{x-1}{3}} \xrightarrow{\times 3} \boxed{x-1} \xrightarrow{+1} \boxed{x}$$

c $\boxed{\dfrac{2x}{3} + 1} \xrightarrow{-1} \boxed{\dfrac{2x}{3}} \xrightarrow{\times 3} \boxed{2x} \xrightarrow{\div 2} \boxed{x}$

d Even though this expression looks quite complicated, undoing it to isolate the unknown, x, just requires carefully analysing the order in which things occurred as it was built up and then backtracking using inverse operations.

$$\boxed{\dfrac{5x-2}{4} - 7} \xrightarrow{+7} \boxed{\dfrac{5x-2}{4}} \xrightarrow{\times 4} \boxed{5x-2} \xrightarrow{+2} \boxed{5x} \xrightarrow{\div 5} \boxed{x}$$

PRACTICE 5

Use a flowchart to isolate the unknown in the following expressions.

a $2x + 1$ b $2 - 4x$ c $\dfrac{3x - 1}{2}$

d $\dfrac{x + 3}{4} + 5$ e $\dfrac{2x + 1}{3} - 2$

13.3 THE BALANCING PRINCIPLE

Backtracking using a flowchart is not the most efficient way to isolate the unknown. An algebraic approach is far more efficient and the key to solving equations algebraically is making sure that we always keep the equation in balance. The equation needs to be in balance because there is an $=$ sign between the LHS and the RHS, and so whatever we do to the equation we must make certain that the LHS is always equal to the RHS. Care needs to be taken never to upset this balance. This means that whenever we do something to one side of the equation, we must do exactly the same thing to the other side of the equation.

EXAMPLE 6

Use the balancing principle to *add*:

a 6 to both sides of $x - 6 = 4$

b 11 to both sides of $2x - 11 = 3$

Use the balancing principle to *subtract*:

c 2 from both sides of $x + 2 = 3$

d 5 from both sides of $2x + 5 = 0$

Solutions

a When we add 6 to both sides of the equation, we get:

$x - 6 + 6 = 4 + 6$

and then by carrying out the additions:

$x = 10$

b Adding 11 to both sides of the equation we get:

$2x - 11 + 11 = 3 + 11$

next carry out the additions:

$2x = 14$

c When we subtract 2 from both sides of the equation, we get:

$x + 2 - 2 = 3 - 2$

and then by carrying out the subtractions:

$x = 1$

d Subtracting 5 from both sides of the equation we get:

$2x + 5 - 5 = 0 - 5$

next carry out the subtractions:

$2x = -5$

PRACTICE 6

Use the balancing principle to *add*:

a 7 to both sides of $x - 7 = 6$

b 13 to both sides of $3x - 13 = 2$

Use the balancing principle to *subtract*:

c 4 from both sides of $x + 4 = 2$

d 9 from both sides of $3x + 9 = 1$

Kathy Brady

Making
Connections
*To review
multiplication of
fractions and use
of cancellation
revisit Chapter 5.*

EXAMPLE 7

Use the balancing principle to *multiply*:

a each side of $3x = 1$ by 3

b each side of $\frac{x}{4} = 2$ by 4

Use the balancing principle to *divide*:

c each side of $3x = 12$ by 3

d each side of $2x + 8 = 10$ by 2

Solutions

a When we multiply each side of the equation by 3, we get:

$$3 \times 3x = 3 \times 1$$

and then by carrying out the multiplications:

$$9x = 3$$

b Multiplying each side of the equation by 4 we get:

$$\frac{x}{4} \times \frac{4}{1} = 2 \times 4$$

next carry out the multiplications:

$$\frac{x}{4} \times \frac{\cancel{4}}{1} = 8$$

which results in:

$$x = 8$$

Notice here that we needed to use the process of cancelling fractions to complete the multiplication.

c When we divide each side of the equation by 3, we get:

$$\frac{3x}{3} = \frac{12}{3}$$

Notice here that we used the algebraic notation for division, writing it in fractional form. Now we carry out the divisions:

$$\frac{3x}{3} = \frac{12}{3}$$

which results in:

$$x = 4$$

d Dividing each side of the equation by 2 we get:

$$\frac{2x + 8}{2} = \frac{10}{2}$$

On the LHS, everything in the numerator must in individually divided by 2, which results in:

$$x + 4 = 5$$

e each side of $4(x + 2) = 20$ by 4

e When we divide each side of the equation by 4, we get:

$$\frac{4(x + 2)}{4} = \frac{20}{4}$$

In this example, the division on the LHS is easier because the 4 outside of the brackets indicates that the contents of the brackets are to be multiplied by 4. So when we divide by 4 on the LHS, what will remain is the bracket contents, which results in:

$$x + 2 = 5$$

PRACTICE 7

Use the balancing principle to *multiply*:

a each side of $2x + 1 = 3$ by 2

b each side of $\frac{x + 2}{3} = 1$ by 3

Use the balancing principle to *divide*:

c each side of $5x = 30$ by 5

d each side of $3x - 9 = 24$ by 3

e each side of $6(x - 1) = 18$ by 6

13.4 SOLVING EQUATIONS USING ADDITION OR SUBTRACTION

As we have defined earlier in this chapter, the solution of an equation is the value of the unknown that is required to make the LHS of the equation equal to the RHS. Solving an equation involves carrying out the processes that are needed to find the solution. When solving an equation, we need to apply backtracking, using inverse operations to isolate the unknown, simultaneously with the balancing principle.

Solving addition equations

In an addition equation, on the LHS a particular value is added to the unknown, for example, $y + 9 = 14$ or $b + 2 = -1$. To solve an addition equation we need to isolate the unknown on the LHS using the inverse operation; that is, by subtracting the value that has been added. Then the balancing principle must be applied by subtracting the same value on the RHS.

Making
Connections
To review
mathematical
operations that
involve using
directed numbers
check back to
Chapter 4.

EXAMPLE 8

Find the solutions of the following equations.

Solutions

a $y + 9 = 14$

a   The inverse operation required to isolate the unknown is to subtract 9, which must occur on both sides of the equation according to the balancing principle.

$$y + 9 = 14$$
$$y + 9 - 9 = 14 - 9 \quad \text{Apply inverse operation, subtract 9, to isolate unknown and balance equation}$$
$$y = 5$$

Notice how the solution to this equation has been set out. Each step in finding the solution is written vertically one under the other, with the = signs aligned vertically. This is the accepted convention for laying out the steps required for solving equations, which we will use for the remainder of this chapter.

b $b + 2 = -1$

b $$b + 2 = -1$$
$$b + 2 - 2 = -1 - 2 \quad \text{Apply inverse operation, subtract 2, to isolate unknown and balance equation}$$
$$b = -3$$

PRACTICE 8

Find the solutions of the following equations.

a $a + 12 = 30$ b $p + 21 = 0$

Solving subtraction equations

In a subtraction equation, on the LHS a particular value is subtracted from the unknown, for example, $t - 4 = -4$ or $m - 6 = 1$. To solve a subtraction equation we need to isolate the unknown on the LHS using the inverse operation; that is, by adding the value that has been subtracted. Then the balancing principle must be applied by adding the same value on the RHS.

EXAMPLE 9

Find the solutions of the following equations.

Solutions

a $t - 4 = -4$

a   The inverse operation required to isolate the unknown is to add 4, which must occur on both sides of the equation according to the balancing principle.

$$t - 4 = -4$$
$$t - 4 + 4 = -4 + 4$$ Apply inverse operation, add 4, to
$$t = 0$$ isolate unknown and balance equation

b $m - 6 = 1$

b $$m - 6 = 1$$
$$m - 6 + 6 = 1 + 6$$ Apply inverse operation, add 6, to
$$m = 7$$ isolate unknown and balance equation

PRACTICE 9

Find the solutions of the following equations.

a $n - 3 = -2$ b $b - 5 = -10$

Translating words into equations

At the beginning of this chapter, we saw examples of the power of equations in many practical applications. In most of these situations, the required equation is not provided, rather it must be developed using given information or observations before it can be solved.

A key skill in this process is being able to translate words into equations. Three steps are required:

- Step 1: Decide what the unknown is and assign a letter to represent it.
- Step 2: Look for words or phrases that indicate which operation(s) are involved; for example, 'decreased by' or 'the difference between' means subtract, and 'more than' or 'increased by' indicates add.
- Step 3: Form the equation around the = sign. Look for words or phrases that indicate =; for example, 'results in', 'is equal to', 'the result is' or simply 'is'.

Making Connections Review Chapter 9 for a more complete list of key words and phrases that indicate various mathematical operations.

EXAMPLE 10

Find the value of the unknown.

a 3 more than n is 11

Solutions

a The phrase 'more than' indicates addition so the LHS of the equation will be $n + 3$. The word 'is' indicates the $=$ sign. So the whole equation will be $n + 3 = 11$. Now we need to solve the equation to find the value of the unknown. This is an addition equation so we will isolate the unknown using the inverse operation, subtraction, and then apply the balancing principle.

$$n + 3 = 11$$
$$n + 3 - 3 = 11 - 3$$
$$n = 8$$

b 6 less than y results in 7

b 'Less than' indicates subtraction. However, unlike addition care needs to be taken when writing subtraction expressions with regard to which value is being subtracted. In this example, 6 less than y, means that 6 is being subtracted from y and is written as $y - 6$. The phrase 'results in' indicates the $=$ sign. The whole equation will be $y - 6 = 7$. This is a subtraction equation so isolate the unknown by using the inverse operation, addition, and then apply the balancing principle. The solution is:

$$y - 6 = 7$$
$$y - 6 + 6 = 7 + 6$$
$$y = 13$$

c When a number is increased by 3.6, the result is 9

c In this example, the unknown has not been assigned a letter to represent it. In these circumstances, we can choose which letter we would like. Let's use 'k'. The phrase 'is increased by' indicates addition, so the LHS of the equation will be $k + 3.6$. 'The result is' indicates the $=$ sign, so the whole equation is $k + 3.6 = 9$, and the solution will be:

$$k + 3.6 = 9$$
$$k + 3.6 - 3.6 = 9 - 3.6$$ Apply inverse operation, subtract 3.6, to isolate unknown
$$k = 5.4$$ and balance equation

Notice here that equations do not necessarily have whole number solutions.

d n decreased by 19 is equal to 35

d 'Decreased by' tells us that we need to subtract 19 from n, and 'is equal to' indicates where the $=$ sign is positioned. So the resultant equation is $n - 19 = 35$ and the solution is:

$$n - 19 = 35$$
$$n - 19 + 19 = 35 + 19$$
$$n = 54$$

Apply inverse operation, add 19, to isolate unknown and balance equation

PRACTICE 10

Find the value of the unknown.

a 37 less than a equals 20

b 14 more than y is 33

c The difference between x and 6 is 4.5

d 5 more than a number is 7.5

EXAMPLE 11: APPLICATION

When a dental hygienist cleaned a patient's teeth, the total bill was $125. This amount was partially covered by medical benefits insurance. If the patient paid $60, use an equation to determine how much of the bill was covered by insurance.

Solution

There are not many words or phrases in this example to provide clues on how to set up the required equation. So we need to analyse the situation for hints. Let's start with determining the unknown value. The question is asking how much of the patient's bill is covered by their insurance, so this is the unknown value. We need represent the unknown with a letter, and the obvious choice is I. What do we know about I? Well, we know that the patient's bill was $125, and that the patient paid the bill with a combination of the amount covered by medical benefits together with $60. This implies an addition of $I + 60$, which totalled $125. So now we can write and solve the equation:

$$I + 60 = 125$$
$$I + 60 - 60 = 125 - 60$$
$$I = 65$$

The insurance covered $65 of the bill.

PRACTICE 11: APPLICATION

A male born in 1882 could expect to live to 47.2 years. By comparison, a male born in 2012 is expected to live 32.9 years longer (Source: worldbank.org). Use an equation to determine the life expectancy of a male born in 2012.

Kathy Brady

13.5 SOLVING EQUATIONS USING MULTIPLICATION OR DIVISION

Solving multiplication equations

In a multiplication equation, on the LHS the unknown is multiplied by a particular value, for example, $5x = 35$ or $4a = -12$. To solve a multiplication equation we need to isolate the unknown on the LHS using the inverse operation; that is, by dividing by the multiplying value. Then the balancing principle must be applied by dividing by the same value on the RHS.

EXAMPLE 12

Find the solutions of the following equations.

a $5x = 35$

b $4a = -12$

Solutions

a $5x = 35$

$\dfrac{5x}{5} = \dfrac{35}{5}$ Apply inverse operation, divide by 5, to isolate unknown and balance equation

$x = 7$

b $4a = -12$

$\dfrac{4a}{4} = \dfrac{-12}{4}$ Apply inverse operation, divide by 4, to isolate unknown and balance equation

$a = -3$

PRACTICE 12

Find the solutions of the following equations.

a $7y = 21$ b $2t = -8$

Solving division equations

In a division equation, on the LHS the unknown is divided by a particular value, for example, $\frac{x}{2} = 8$ or $\frac{y}{5} = -1$. To solve a division equation we need to isolate the unknown on the LHS using the inverse operation; that is, multiplying by the dividing value. Then the balancing principle must be applied by multiplying by the same value on the RHS.

EXAMPLE 13

Find the solutions of the following equations.

a $\dfrac{x}{2} = 8$

Solutions

a $\dfrac{x}{2} = 8$

$\dfrac{x}{2} \times \dfrac{2}{1} = 8 \times 2$ Apply inverse operation, multiply by 2, to isolate unknown and balance equation

$x = 16$

b $\dfrac{y}{5} = -1$

b $\dfrac{y}{5} = -1$

$\dfrac{y}{5} \times \dfrac{5}{1} = -1 \times 5$ Apply inverse operation, multiply by 5, to isolate unknown and balance equation

$y = -5$

PRACTICE 13

Find the solutions of the following equations.

a $\dfrac{a}{7} = 4$ **b** $\dfrac{n}{6} = -9$

Translating words into equations

There are also key words or phrases that indicate either a multiplication or division equation; for example, 'the product of' means multiply and 'the quotient of' means divide. The words or phrases that are used to indicate the = sign are the same as those used in addition or subtraction equations.

EXAMPLE 14

Find the value of the unknown.

Solutions

a The product of 9 and n results in 45

a 'The product' indicates multiplication, so the LHS will be $9n$ (using the algebraic multiplication convention). The phrase 'results in' indicates the position of the = sign. Therefore, the whole equation is $9n = 45$. This is a multiplication equation so we will isolate the unknown using the inverse operation, division, and then apply the balancing principle.

$9n = 45$

$\dfrac{9n}{9} = \dfrac{45}{9}$ Apply inverse operation, divide by 9, to isolate unknown and balance equation

$n = 5$

b One quarter of a number is equal to 12

b Finding 'one quarter of' is the same as dividing by 4. In this example, we can choose our own letter to represent the unknown, so let's use q, which is being divided by 4. Therefore, the LHS of the equation will be $\dfrac{q}{4}$, which is equal to 12, so the whole equation will be $\dfrac{q}{4} = 12$. This is a division equation so we will isolate the unknown using the inverse operation, multiplication, and then apply the balancing principle.

$\dfrac{q}{4} = 12$

$\dfrac{q}{4} \times \dfrac{4}{1} = 12 \times 4$

$q = 48$ Apply inverse operation, multiply by 4, to isolate unknown and balance equation

Kathy Brady

c The result of 3 times m is 27

c The word 'times' is commonly used for multiply and we are told the result of the multiplication of 3 and m is 27. So the equation is $3m = 27$, and the solution is:

$$3m = 27$$
$$\frac{3m}{3} = \frac{27}{3}$$ Apply inverse operation, divide by 3, to isolate unknown and balance equation
$$m = 9$$

d When a number is divided by 100, the quotient is 4

d In this example, we firstly need to select a letter to represent the unknown number. This time let's use p. We are told that p is divided by 100 so the LHS will be $\frac{p}{100}$. The phrase that indicates where to position the $=$ sign is 'the quotient is', so the equation is $\frac{p}{100} = 4$. The solution to find the unknown value is:

$$\frac{p}{100} = 4$$
$$\frac{p}{100} \times \frac{100}{1} = 4 \times 100$$ Apply inverse operation, multiply by 100, to isolate unknown and balance equation
$$p = 400$$

PRACTICE 14

Find the value of the unknown.

a The product of 12 and m is 3

b One third of a number is equal to 21

c When x is divided by 5, the result is 11

d Four times a is equal to 88

EXAMPLE 15: APPLICATION

A student plans to buy a tablet computer in 7 weeks time. If the device costs $420, use an equation to determine how much money the student must save each week in order to buy the tablet computer.

Solution

There are not many words or phrases in this example to provide clues for how to set up the required equation. So we need to analyse the situation for hints. Let's start with determining the unknown value. The question is asking how much money the student must save each week, so this is the unknown value. We need to represent the unknown with a letter, so let's use M. What do we know about M? Well we know that each week, for 7 weeks, the student will be saving M dollars. This means that the total amount that the student will save is 7 lots of M, or algebraically $7M$, which will be equal to the required $420. So now we can write and solve the equation.

$$7M = 420$$
$$\frac{7M}{7} = \frac{420}{7}$$
$$M = 60$$

Making Connections
Check Chapter 12 to review the relationship between the sides and the perimeter of a square

PRACTICE 15: APPLICATION

The perimeter of a square block of land is 280 metres. What is the length of one side of the block?

13.6 SOLVING EQUATIONS USING A COMBINATION OF OPERATIONS

The equations we looked at in the previous two sections involved the use of just one operation in building up the equation, and so just one inverse operation was needed to backtrack the equation and isolate the unknown. In each case, the solution to these equations was obtained in one step. These are the most straightforward sorts of equations. In most circumstances, however, equations are more complex than these simple one-step equations because they require a combination of different inverse operations to isolate the unknown. When this is the case, it can take two, three or even more steps to find the solution. In this section, we will look at how to solve a range of two-step and three-step equations that use a combination of operations.

One type of two-step equation involves the use of both multiplication and addition or subtraction as the equation is built up, for example, $2x + 3 = 1$ or $3x - 4 = 8$. To solve these sorts of equations, we need firstly to examine the equation and determine the order in which it should be undone to isolate the unknown. This is the process that we practised in section 13.2 earlier in this chapter.

EXAMPLE 16

Solve the following equations.

In each case, we need to ask what was the last thing that happened as the equation was built up, because this will be the first thing to be undone using the inverse operation. Then we continue to backtrack the equation in the reverse order that it was built up using the appropriate inverse operations, remembering to always apply the balancing principle, until the unknown has been isolated.

Solutions

a $2x + 3 = 7$

a

$2x + 3 = 7$

$2x + 3 - 3 = 7 - 3$ Apply inverse operation, subtract 3, to isolate unknown and balance equation

$\frac{2x}{2} = \frac{4}{2}$ Apply inverse operation, divide by 2, to isolate unknown and balance equation

$x = 2$

After the solution has been found, check on its accuracy by substituting the value back into the original equation to ensure that the LHS = RHS.

Check: $2(2) + 3 = 4 + 3 = 7$ ✓

b $3x - 4 = 8$

$3x - 4 = 8$	
$3x - 4 + 4 = 8 + 4$	Apply inverse operation, add 4, to isolate unknown and balance equation
$\dfrac{3x}{3} = \dfrac{12}{3}$	Apply inverse operation, divide by 3, to isolate unknown and balance equation
$x = 4$	

Check: $3(4) - 4 = 12 - 4 = 8$ ✓

c $7x + 2 = -12$

$7x + 2 = -12$	
$7x + 2 - 2 = -12 - 2$	Apply inverse operation, subtract 2, to isolate unknown and balance equation
$\dfrac{7x}{7} = \dfrac{-14}{7}$	Apply inverse operation, divide by 7, to isolate unknown and balance equation
$x = -2$	

Check: $7(-2) + 2 = -14 + 2 = -12$ ✓

d $2x - 1 = -5$

$2x - 1 = -5$	
$2x - 1 + 1 = -5 + 1$	Apply inverse operation, add 1, to isolate unknown and balance equation
$\dfrac{2x}{2} = \dfrac{-4}{2}$	Apply inverse operation, divide by 2, to isolate unknown and balance equation
$x = -2$	

Check: $2(-2) - 1 = -4 - 1 = -5$ ✓

e $-5x - 4 = 6$

$-5x - 4 = 6$	
$-5x - 4 + 4 = 6 + 4$	Apply inverse operation, add 4, to isolate unknown and balance equation
$\dfrac{-5x}{-5} = \dfrac{10}{-5}$	Apply inverse operation, divide by −5, to isolate unknown and balance equation
$x = -2$	

Check: $-5(-2) - 4 = 10 - 4 = 6$ ✓

f $-3x + 7 = -5$

$-3x + 7 = -5$	
$-3x + 7 - 7 = -5 - 7$	Apply inverse operation, subtract 7, to isolate unknown and balance equation
$\dfrac{-3x}{-3} = \dfrac{-12}{-3}$	Apply inverse operation, divide by −3 to isolate unknown and balance equation
$x = 4$	

Check: $-3(4) + 7 = -12 + 7 = -5$ ✓

PRACTICE 16

Solve the following equations.

a $3x + 4 = 10$ **b** $4x - 3 = 13$ **c** $5x + 7 = -8$

d $2x - 9 = -11$ **e** $-4x - 1 = 7$ **f** $-5x + 6 = -4$

Another type of two-step equations involves the use of both division and addition or subtraction as the equation is built up, for example, $\frac{x + 2}{3} = 4$ or $\frac{x}{4} - 1 = 2$. As with the previous set of examples, to solve these sorts of equations we need firstly to determine the order in which it should be undone to isolate the unknown.

EXAMPLE 17

Solve the following equations.

a $\dfrac{x + 2}{3} = 1$

b $\dfrac{x - 6}{2} = 4$

c $\dfrac{x + 3}{2} = -1$

d $\dfrac{x - 5}{3} = -2$

e $\dfrac{x}{5} + 3 = 7$

f $\dfrac{x}{2} - 1 = 5$

g $\dfrac{x}{4} + 2 = -1$

Solutions

a $\dfrac{x + 2}{3} = 1$

$\dfrac{x + 2}{3} \times \dfrac{3}{1} = 1 \times 3$ Apply inverse operation, multiply by 3, to isolate unknown and balance equation

$x + 2 - 2 = 3 - 2$ Apply inverse operation, subtract 2, to isolate unknown and balance equation

$x = 1$

Check: $\dfrac{1 + 2}{3} = \dfrac{3}{3} = 1$ ✓

b $\dfrac{x - 6}{2} = 4$

$\dfrac{x - 6}{2} \times \dfrac{2}{1} = 4 \times 2$ Apply inverse operation, multiply by 2, to isolate unknown and balance equation

$x - 6 + 6 = 8 + 6$ Apply inverse operation, add 6, to isolate unknown and balance equation

$x = 14$

Check: $\dfrac{14 - 6}{2} = \dfrac{8}{2} = 4$ ✓

c $\dfrac{x + 3}{2} = -1$

$\dfrac{x + 3}{2} \times \dfrac{2}{1} = -1 \times 2$ Apply inverse operation, multiply by 2, to isolate unknown and balance equation

$x + 3 - 3 = -2 - 3$ Apply inverse operation, subtract 3, to isolate unknown and balance equation

$x = -5$

Check: $\dfrac{-5 + 3}{2} = \dfrac{-2}{2} = -1$ ✓

d $\dfrac{x - 5}{3} = -2$

$\dfrac{x - 5}{3} \times \dfrac{3}{1} = -2 \times 3$ Apply inverse operation, multiply by 3, to isolate unknown and balance equation

$x - 5 + 5 = -6 + 5$ Apply inverse operation, add 5, to isolate unknown and balance equation

$x = -1$

Check: $\dfrac{-1 - 5}{3} = \dfrac{-6}{3} = -2$ ✓

e $\dfrac{x}{5} + 3 = 7$

$\dfrac{x}{5} + 3 - 3 = 7 - 3$ Apply inverse operation, subtract 3, to isolate unknown and balance equation

$\dfrac{x}{5} \times \dfrac{5}{1} = 4 \times 5$ Apply inverse operation, multiply by 5, to isolate unknown and balance equation

$x = 20$

Check: $\dfrac{20}{5} + 3 = 4 + 3 = 7$ ✓

f $\dfrac{x}{2} - 1 = 5$

$\dfrac{x}{2} - 1 + 1 = 5 + 1$ Apply inverse operation, add 1, to isolate unknown and balance equation

$\dfrac{x}{2} \times \dfrac{2}{1} = 6 \times 2$ Apply inverse operation, multiply by 2, to isolate unknown and balance equation

$x = 12$

Check: $\dfrac{12}{2} - 1 = 6 - 1 = 5$ ✓

g $\dfrac{x}{4} + 2 = -1$

$\dfrac{x}{4} + 2 - 2 = -1 - 2$ Apply inverse operation, subtract 2, to isolate unknown and balance equation

Kathy Brady

$$\frac{x}{4} \times \frac{4}{1} = -3 \times 4$$

Apply inverse operation, multiply by 4, to isolate unknown and balance equation

$$x = -12$$

Check: $\frac{-12}{4} + 2 = -3 + 2 = -1$ ✓

h $\frac{x}{3} - 1 = -4$

h $\frac{x}{3} - 1 = -4$

$$\frac{x}{3} - 1 + 1 = -4 + 1$$

Apply inverse operation, add 1, to isolate unknown and balance equation

$$\frac{x}{3} \times \frac{3}{1} = -3 \times 3$$

Apply inverse operation, multiply by 3, to isolate unknown and balance equation

$$x = -9$$

Check: $\frac{-9}{3} - 1 = -3 - 1 = -4$ ✓

PRACTICE 17

Solve the following equations.

a $\frac{x+1}{4} = 2$　　　　**b** $\frac{x-2}{5} = 3$　　　　**c** $\frac{x+4}{3} = -2$

d $\frac{x-3}{2} = -5$　　　　**e** $\frac{x}{2} + 4 = 7$　　　　**f** $\frac{x}{3} - 2 = 4$

g $\frac{x}{5} + 1 = -3$　　　　**h** $\frac{x}{4} - 3 = -1$

EXAMPLE 18: APPLICATION

The perimeter of this figure is 62 cm. Find the value of x.

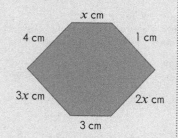

Making Connections
Review collecting like terms in Chapter 9

Solution

The perimeter of a figure is the distance around all of its sides. In this example, we have been given information about the length of each side, so we can add all of these lengths to give us an expression for the perimeter:

$$x + 1 + 2x + 3 + 3x + 4$$

We can use the technique of collecting like terms to simplify this expression for the perimeter to $6x + 8$.

The information provided also tells us that the actual length of the perimeter is 62 cm. We can use this information to create an equation and then solve it to find the value of x.

$$6x + 8 = 62$$
$$6x + 8 - 8 = 62 - 8$$
$$\frac{6x}{6} = \frac{54}{6}$$
$$x = 9$$

PRACTICE 18: APPLICATION

A 13 cm long piece of wire is bent to form a triangle. One side of the triangle measures A cm, the second side is 3 cm longer than A, and the third side is 2 cm less than A. Find the lengths of each side of the triangle.

Some equations that combine operations can take three or more steps to solve. However, this does not make them any harder because the process is always the same. Analyse the equation to determine what was the last thing that happened as the equation was built up, then make this the first thing to be undone using the inverse operation. Then, continue to backtrack the equation in the reverse order that it was built up using the appropriate inverse operations, remembering to always apply the balancing principle, until the unknown has been isolated.

EXAMPLE 19

Solve the following equations.

Solutions

a $\dfrac{2x+1}{7} = 3$

a
$$\dfrac{2x+1}{7} = 3$$

$$\dfrac{2x+1}{7} \times \dfrac{7}{1} = 3 \times 7 \quad \text{Apply inverse operation, multiply by 7, to isolate unknown and balance equation}$$

$$2x + 1 - 1 = 21 - 1 \quad \text{Apply inverse operation, subtract 1, to isolate unknown and balance equation}$$

$$2x = 20$$

$$\dfrac{2x}{2} = \dfrac{20}{2} \quad \text{Apply inverse operation, divide by 2, to isolate unknown and balance equation}$$

$$x = 10$$

Check: $\dfrac{2(10)+1}{7} = \dfrac{20+1}{7} = \dfrac{21}{7} = 3 \checkmark$

b $\dfrac{3x-4}{5} = 4$

b
$$\dfrac{3x-4}{5} = 4$$

$$\dfrac{3x-4}{5} \times \dfrac{5}{1} = 4 \times 5 \quad \text{Apply inverse operation, multiply by 5, to isolate unknown and balance equation}$$

$$3x - 4 + 4 = 20 + 4 \quad \text{Apply inverse operation, add 4, to isolate unknown and balance equation}$$

$$3x = 24$$

$$\dfrac{3x}{3} = \dfrac{24}{3} \quad \text{Apply inverse operation, divide by 3, to isolate unknown and balance equation}$$

$$x = 8$$

Check: $\dfrac{3(8)-4}{5} = \dfrac{24-4}{5} = \dfrac{20}{5} = 4 \checkmark$

PRACTICE 19: GUIDED

Solve the following equation.

$$\dfrac{2x+7}{3} = 9$$

$$\dfrac{2x+7}{3} \times \dfrac{3}{1} = 9 \times \square$$

$$2x + 7 - \square = 27 - 7$$

$$2x = \square$$

$$\dfrac{2x}{\square} = \dfrac{20}{\square}$$

$$x = \square$$

Now try this one:

$$\dfrac{3x+2}{4} = 5$$

13.7 SOLVING EQUATIONS INVOLVING COLLECTING TERMS AND REMOVING BRACKETS

When solving some equations, it may be necessary to expand any brackets that are part of the equation, or collect some like terms, before embarking on the usual process for finding a solution.

Equations with brackets

Making
Connections
*To review the
process for
expanding
brackets revisit
Chapter 9*

Equations with brackets might look like $5(x - 3) = 35$ or $3(2x + 1) = 30$. Before solving these types of equations, we will need to expand the brackets by multiplying everything inside the bracket by the value on the outside.

Once we have expanded the brackets, we need to determine how the equation has been built up. Then backtrack in the reverse order using inverse operations to isolate the unknown, being sure to also apply the balancing principle.

EXAMPLE 20

Solve the following equations.

Solutions

In each case the first step is to expand the brackets on the LHS.

a $2(x + 6) = 14$

a
$$2(x + 6) = 14$$
$$2x + 12 = 14$$
$$2x + 12 - 12 = 14 - 12$$

Expand the brackets and apply the inverse operation, subtract 12, to isolate unknown and balance equation

$$2x = 2$$
$$\frac{2x}{2} = \frac{2}{2}$$
$$x = 1$$

Apply inverse operation, divide by 2, to isolate unknown and balance equation

Check: $2(1 + 6) = 2(7) = 14$ ✓

b $5(x - 3) = 40$

b
$$5(x - 3) = 40$$
$$5x - 15 = 40$$
$$5x - 15 + 15 = 40 + 15$$

Expand the brackets and apply the inverse operation, add 15, to isolate unknown and balance equation

$$5x = 55$$
$$\frac{5x}{5} = \frac{55}{5}$$
$$x = 11$$

Apply inverse operation, divide by 5, to isolate unknown and balance equation

Check: $5(11 - 3) = 5(8) = 40$ ✓

c $2(2x + 1) = 30$

c $2(2x + 1) = 30$

$4x + 2 = 30$

$4x + 2 - 2 = 30 - 2$ Expand the brackets and apply the inverse operation, subtract 2, to isolate the unknown and balance the equation

$4x = 28$

$\dfrac{4x}{4} = \dfrac{28}{4}$ Apply inverse operation, divide by 4, to isolate unknown and balance equation

$x = 7$

Check: $2(2(7) + 1) = 2(14 + 1) = 2(15) = 30$ ✓

d $4(2x - 7) = 20$

d $4(2x - 7) = 20$

$8x - 28 = 20$

$8x - 28 + 28 = 20 + 28$ Expand the brackets and apply the inverse operation, add 28, to isolate unknown and balance equation

$8x = 48$

$\dfrac{8x}{8} = \dfrac{48}{8}$ Apply inverse operation, divide by 8, to isolate unknown and balance equation

$x = 6$

Check: $4(2(6) - 7) = 4(12 - 7) = 4(5) = 20$ ✓

PRACTICE 20

Solve the following equations.

a $3(x + 5) = 18$

b $2(x - 7) = 22$

c $3(3x + 6) = 27$

d $5(2x - 7) = 5$

Equations with like terms

In some equations the unknown can appear more than once, for example, $2(x + 3) - 3x = 5$. Before solving an equation such as this, we first need to expand any brackets that are contained in the equation, and then collect the like terms. Then we will carry out the usual backtracking and balancing processes to isolate the unknown.

EXAMPLE 21

Solve the following equations.

Solutions

In each case, we need to firstly expand the brackets and collect the like terms on the LHS of the equation.

a $3(x - 2) - x = 4$

a $3(x - 2) - x = 4$

$3x - 6 - x = 4$ Expand brackets and collect like x terms

$$2x - 6 = 4$$

Apply inverse operation, add 6, to isolate unknown and balance equation

$$2x - 6 + 6 = 4 + 6$$

$$2x = 10$$

$$\frac{2x}{2} = \frac{10}{2}$$

Apply inverse operation, divide by 2, to isolate unknown and balance equation

$$x = 5$$

Check: $3(5 - 2) - 5 = 3(3) - 5 = 9 - 5 = 4$ ✓

b $5(x - 3) - 3x = -7$

b $5(x - 3) - 3x = -7$

$$5x - 15 - 3x = -7$$

Expand brackets and collect like x terms

$$2x - 15 = -7$$

$$2x - 15 + 15 = -7 + 15$$

Apply inverse operation, add 15, to isolate unknown and balance equation

$$2x = 8$$

$$\frac{2x}{2} = \frac{8}{2}$$

Apply inverse operation, divide by 2, to isolate unknown and balance equation

$$x = 4$$

Check: $5(4 - 3) - 3(4) = 5(1) - 12 = 5 - 12 = -7$ ✓

c $3(x - 2) + 2(x - 4) = 1$

c $3(x - 2) + 2(x - 4) = 1$

$$3x - 6 + 2x - 8 = 1$$

Expand brackets and collect like x terms

$$5x - 14 = 1$$

$$5x - 14 + 14 = 1 + 14$$

Apply inverse operation, add 14, to isolate unknown and balance equation

$$5x = 15$$

$$\frac{5x}{5} = \frac{15}{5}$$

Apply inverse operation, divide by 5, to isolate unknown and balance equation

$$x = 3$$

Check: $3(3 - 2) + 2(3 - 4) = 3(1) + 2(-1) = 3 - 2 = 1$ ✓

d $2(2x + 3) + 3(x - 4) = 15$

d $2(2x + 3) + 3(x - 4) = 15$

$$4x + 6 + 3x - 12 = 15$$

Expand brackets and collect like x terms

$$7x - 6 = 15$$

$$7x - 6 + 6 = 15 + 6$$

Apply inverse operation, add 6, to isolate unknown and balance equation

$$7x = 21$$

$$\frac{7x}{7} = \frac{21}{7}$$

Apply inverse operation, divide by 7, to isolate unknown and balance equation

$$x = 3$$

Check: $2(2(3) + 3) + 3(3 - 4) = 2(6 + 3) + 3(3 - 4)$

$$= 2(9) + 3(-1) = 18 - 3 = 15$$ ✓

PRACTICE 21: GUIDED

a   Solve the following equation.

$$4(x - 2) - 2x = 2$$

$$\square x - \square - 2x = 2$$

$$4x - 8 - \square x = 2$$

$$\square x - 8 = 2$$

$$2x - 8 + \square = 2 + \square$$

$$2x = \square$$

$$\frac{2x}{\square} = \frac{10}{\square}$$

$$x = \square$$

Now try this one:

b $2(3x - 2) - x = 1$

c   Solve the following equation:

$$5(x + 1) + 3(x + 2) = 27$$

$$\square x + 5 + 3x + \square = 27$$

$$8x + \square = 27$$

$$8x + \square - 11 = 27 - \square$$

$$8x = \square$$

$$\frac{8x}{\square} = \frac{16}{\square}$$

$$x = \square$$

Another one to try:

d $3(2x - 5) + 4(x + 1) = 9$

When the unknown is on both sides of the equation

A final type of equation we should consider is when the unknown appears on both the LHS and the RHS of the equation, for example, $3(4x - 3) = 3x + 4$. What should we do in this situation? Well, the first step is the same as in some of the previous examples: to expand any brackets. The next step involves removing the unknown from one side of the equation being careful to apply the balancing principle when we do this. When carrying out this step, the aim is always to have a *positive* co-efficient on the resultant unknown. This means that this unknown may end up on either side of the equals sign. Doing this will simplify any of the final backtracking steps that will be taken in order to isolate the unknown. We then need to gather all of the constants to the side of the equation that is opposite the unknown. Finally, the equation is solved using backtracking and applying the balancing principle. All of these steps are fully explained in the following examples.

EXAMPLE 22

Solve the following equations.

a $2x + 5 = 11 - x$

b $x - 3 = 5x - 7$

c $8 - 2x = 5 + x$

d $7x + 2 = 2x + 17$

e $2(4x + 1) = 2(x + 7)$

Solutions

a As there are no brackets to expand, we can move straight into removing the unknown from one side of the equation. Remember the aim is to have a positive co-efficient on the resultant unknown. We remove the unknown from one side of the equation by adding or subtracting it, just as if it were a constant, and then balancing on the other side of the equation. In this example, we can achieve this by adding x to both sides of the equation as follows.

$$2x + 5 + x = 11 - x + x$$
$$3x + 5 = 11$$

So now we have the unknown on just one side of the equation, the LHS, and it has a positive co-efficient as required. Next we need to gather the constants on the other side of the equation, the RHS. We can do this by subtracting 5 from the LHS and then balancing on the RHS.

$$3x + 5 - 5 = 11 - 5$$
$$3x = 6$$

Now the remaining step is to carry out a division by 3 on both sides of the equation.

$$\frac{3x}{3} = \frac{6}{3}$$
$$x = 2$$

Check: LHS: $2(2) + 5 = 9$; RHS: $11 - 2 = 9$; LHS = RHS ✓

b As there are no brackets to expand, we can move straight onto gathering the unknowns on one side of the equation and the constants on the other. If we were to attempt to bring the unknowns to the LHS of the equation by subtracting $5x$ from the RHS and then balancing, the following would occur.

$$x - 5x - 3 = 5x - 5x - 7$$
$$-4x - 3 = -7$$

As we can now see, there is a negative co-efficient on the unknown, which is not the desired outcome. What we should have done in this case was to bring the unknowns to the RHS of the equation by subtracting x from the LHS and then balancing on the RHS as follows.

$$x - x - 3 = 5x - x - 7$$
$$-3 = 4x - 7$$

Now the unknown has the desired positive co-efficient. It is worth noting that it is not a cause for concern that the unknown is on the RHS of this equation. It is only by convention that equations are usually written with the unknown on the LHS. In reality, either way is mathematically correct because of the presence of the $=$ sign. Next step is to gather the constants on the LHS of the equation by adding 7 to each side.

$$-3 + 7 = 4x - 7 + 7$$
$$4 = 4x$$

Now we can isolate the unknown by dividing each side by 4.

$$\frac{4}{4} = \frac{4x}{4}$$

$1 = x$ or alternatively $x = 1$

Check: LHS: $1 - 3 = -2$; RHS: $5(1) - 7 = 5 - 7 = -2$; LHS = RHS ✓

c $\qquad 8 - 2x = 5 + x$

$8 - 2x + 2x = 5 + x + 2x$ Move the unknown to the RHS to create the desired positive co-efficient

$\qquad 8 = 5 + 3x$

$8 - 5 = 5 - 5 + 3x$ Move the constants to the LHS

$\qquad 3 = 3x$

$\qquad \dfrac{3}{3} = \dfrac{3x}{3}$ Apply inverse operation, divide by 3, to isolate unknown and balance equation

$\qquad 1 = x$ or alternatively $x = 1$

Check: LHS: $8 - 2(1) = 8 - 2 = 6$; RHS: $5 + 1 = 6$; LHS = RHS ✓

d $\qquad 7x + 2 = 2x + 17$

$7x - 2x + 2 = 2x - 2x + 17$ Move the unknown to the LHS to create the desired positive co-efficient

$\qquad 5x + 2 = 17$

$5x + 2 - 2 = 17 - 2$ Move the constants to the RHS

$\qquad 5x = 15$

$\qquad \dfrac{5x}{5} = \dfrac{15}{5}$ Apply inverse operation, divide by 5, to isolate unknown and balance equation

$\qquad x = 3$

Check: LHS: $7(3) + 2 = 21 + 2 = 23$; RHS: $2(3) + 17 = 6 + 17 = 23$; LHS = RHS ✓

e $2(4x + 1) = 2(x + 7)$

$8x + 2 = 2x + 14$ Expand the brackets

$8x - 2x + 2 = 2x - 2x + 14$ Move the unknown to the LHS to create the desired positive co-efficient

$6x + 2 - 2 = 14 - 2$ Move the constants to the RHS

$6x = 12$

$\dfrac{6x}{6} = \dfrac{12}{6}$ Apply inverse operation, divide by 6, to isolate unknown and balance equation

$x = 2$

Check: LHS: $2(4(2) + 1) = 2(8 + 1) = 2(9) = 18$; RHS: $2(2 + 7) = 2(9) = 18$; LHS = RHS ✓

PRACTICE 22: GUIDED

a Solve the following equation.

$8x - 9 = 4x - 1$

$8x - 9 - \square = 4x - 1 - 4x$

$\square x - 9 = -1$

$4x - 9 + \square = -1 + \square$

$4x = \square$

$\dfrac{4x}{\square} = \dfrac{8}{\square}$

$x = \square$

Now try this one:

b $3x + 7 = 11 - x$

e Solve the following equation.

$3(x + 2) = 2(2x + 1)$

$3(x + 2) = 2(2x + 1)$

$\square x + 6 = 4x + \square$

$3x + 6 - \square x = 4x + 2 - 3x$

$6 - \square = \square + 2 - \square$

$\square = x$

f And another one to try:

$2(3x - 2) = 5(x + 1)$

c Solve the following equation.

$3x + 4 = 5x + 2$

$3x + 4 - 3x = 5x + 2 - \square$

$4 = \square x + 2$

$4 - \square = 2x + 2 - \square$

$\square = 2x$

$\dfrac{2}{\square} = \dfrac{2x}{\square}$

$\square = x$

d Another one to try:

$14 - 3x = 2 - x$

13.8 FORMULA

A major application of algebra is in its use of rules to describe real situations, for example, in the sciences, economics, business or technology. These rules are known as formula. Quite often the terms 'equation' and 'formula' are used interchangeably, but they are quite different. An equation contains unknowns that are found by solving the equation. On the other hand, a formula is used to find the value of an unknown.

What is a formula?

A formula describes the relationship between two or more variables. Some formulas that you might have seen or heard of are $A = \frac{1}{2}bh$ (for the area of a triangle), $C = \pi d$ (for the circumference of a circle) or even $E = mc^2$ (Einstein's special theory of relativity). Formulas are *evaluated* by substituting given values for particular variables in the formula, then carrying out the required calculations to find the value of the remaining unknown variable.

The formula for the area of a rectangle is $A = lw$. There are three variables in this formula: A, l and w. Quite often the letters that are selected to name particular variables are the first letter of the word that the variable represents. So, in this case, A represents area, l is the length of the longer pair of sides, and w is the length of the shorter pair of sides, which is often called the width. In this formula, there is no mathematical operator between l and w, which implies, according to the conventions of algebra, that they are multiplied. To calculate the area of a rectangle we need to know both the value of the length and the width.

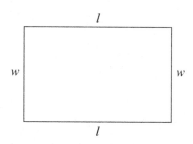

Making Connections
Review how to substitute into and evaluate algebraic expressions in Chapter 9.

The **subject of a formula** is the value that is being determined and the **rule of the formula** indicates how the subject should be calculated. So, A is the subject of the formula, $A = lw$, and lw is the rule.

Subject of a formula
The value that is being determined

Rule in a formula
Indicates how the subject should be calculated

Evaluating formula using substitution

EXAMPLE 23: APPLICATIONS

a   The formula for the perimeter of a rectangle is
$P = 2(l + w)$, where P is the perimeter,
l is the length and w is the width (see below). If
$l = 2.5\,\text{m}$ and $w = 1.5\,\text{m}$, find P.

b   The formula for the circumference of a circle is
$C = \pi d$, where C is the circumference,
d is the diameter of the circle, and π is a constant
value that approximates to 3.14. If $d = 200\,\text{mm}$, find C.

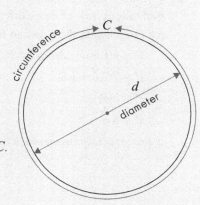

c   The length of the hypotenuse in a right-angled triangle is given by Pythagoras'
rule: $c^2 = a^2 + b^2$, where c is the length of the hypotenuse, and a and b are the lengths
of the other two sides. If $a = 4$ cm and $b = 3$ cm, find c.

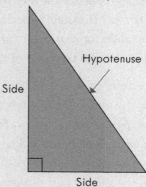

d   The amount of money in a bank account earning simple interest is given by the formula
$A = P(1 + rt)$, where A is the current amount in the account, P is the principal
(the original amount deposited in the account), r is the annual interest rate as a
percentage and t is the time in years that the account has operated.

If a principal of $5000 was deposited in an account 2 years ago earning 6% per
annum, what is the current amount in the account?

e   The formula to convert degrees measured in Fahrenheit to degrees measured
in Celsius is $C = \frac{5}{9}(F - 32)$, where C is degrees in Celsius and F is degrees in
Fahrenheit. If a temperature was recorded at 86 degrees Fahrenheit, what is the
equivalent temperature in Celsius measure?

Solutions

a P is the subject of the formula, which is the value we are finding. $2(l + w)$ is the rule
that tells us how to calculate the subject, P. We do this by substituting the given values
for l and w into the rule and then evaluating.

$P = 2(l + w)$ where $l = 2.5$ m and $w = 1.5$ m

$P = 2(2.5 + 1.5)$

$P = 2(4)$

$P = 8$ m

Important reminder: always take care to provide your solution using any given units.

b In this example, we have one variable in the rule, the length of the diameter, d. The rule also contains the constant value π, and the absence of a mathematical operator implies π and d are multiplied.

$C = \pi d$ where $d = 200$ mm

$C = \pi(200)$

$C = 3.14 \times 200$

$C = 628$ mm

c In this example, c is the subject of the formula, the value that we are finding, and $a^2 + b^2$ is the rule.

$c^2 = a^2 + b^2$ where $a = 4$ cm and $b = 3$ cm

$c^2 = 4^2 + 3^2$

$c^2 = 16 + 9$

$c^2 = 25$

$c = \sqrt{25}$

$c = 5$ cm

d Even though this formula has four variables, and the information looks complicated, it is not difficult as long as care is taken when substituting the values into the rule and then the calculations carried out methodically.

$A = P(1 + rt)$ where $P = \$5000$, $r = 6\%$, $t = 2$ years

$A = 5000\left(1 + \dfrac{6}{100} \times 2\right)$ Notice that the various units are omitted once the values are substituted into the rule, and that the % is written is fraction notation.

$A = 5000\left(1 + \dfrac{12}{100}\right)$ Multiply the fraction by 2

$A = 5000\left(1\dfrac{12}{100}\right)$ Add the whole number 1 and the fraction to form a mixed number

$A = 5000\left(\dfrac{112}{100}\right)$ Convert the mixed number to a common fraction

$A = \dfrac{5000}{1} \times \dfrac{112}{100}$ Remove the brackets and write the calculation in fraction multiplication form

$A = \dfrac{50}{1} \times \dfrac{112}{1}$ Cancel the numerator 5000 and denominator 100 by dividing by 100

$A = \$5600$ Multiply the fractions and include the units in the answer

e This is an example of a formula commonly used in the sciences. The subject of the formula is C, degrees in Celsius, and the rule is $\dfrac{5}{9}(F - 32)$. Given that there is only one variable in the rule it is reasonably straightforward to evaluate.

$C = \dfrac{5}{9}(F - 32)$ where $F = 86$ degrees

$C = \dfrac{5}{9}(86 - 32)$

Making Connections
To review squaring numbers and calculating square roots revisit Chapter 3. Revisit Chapter 11 to review Pythagoras' Rule.

Kathy Brady

$$C = \frac{5}{9}(54)$$

$$C = \frac{5}{9} \times \frac{54}{1}$$ Remove the brackets and write the calculation in fraction multiplication form

$$C = \frac{5}{1} \times \frac{6}{1}$$ Cancel the numerator 9 and denominator 54 by dividing by 9

$$C = 30 \text{ degrees}$$

PRACTICE 23

Evaluate the following formulas for the given values of the variables.

FORMULA	GIVEN VALUES	TO FIND THE SUBJECT
$P = 2(l + w)$	$l = 4.2$ m and $w = 3.6$ m	P
$C = \pi d$	$\pi \cong 3.14$ and $d = 40$ mm	C
$c^2 = a^2 + b^2$	$a = 6$ cm and $b = 8$ cm	c
$A = P(1 + rt)$	$P = \$3000$, $r = 5\%$, $t = 3$ years	A
$C = \frac{5}{9}(F - 32)$	$F = -4$ degrees	C

Rearranging formula

Sometimes the required unknown is not the given subject of the formula; rather it is one of the variables contained in the rule. In this case, we need to re-arrange the formula so that the required unknown becomes the subject of the formula. To do this we use all of the backtracking and undoing strategies that are used when we solve equations.

EXAMPLE 24: APPLICATIONS

a   Rearrange the $P = 2(l + w)$ to make l the subject of the formula.

b   Rearrange $C = \pi d$ to make d the subject of the formula.

c   Rearrange $C = \frac{5}{9}(F - 32)$ to make F the subject of the formula.

d   Rearrange $c^2 = a^2 + b^2$ to make b the subject of the formula.

Solutions

a   In the formula $P = 2(l + w)$, we need to move variables and constants around to achieve the required outcome that just l is on one side of the equals sign, thus becoming the subject of the formula, and the other variables and constants are on the opposite side of the = sign to form the new rule to calculate l. It does not matter which side of the = sign the subject sits. In this case, as the l is on the RHS, to simplify the process we will keep it there, and then move other variables and constants to the LHS.

$$P = 2(l + w)$$

$$P = 2l + 2w \quad \text{Expand the brackets}$$

$$P - 2w = 2l \qquad \text{Remove the variable } 2w \text{ from the RHS by using the inverse operation, subtract. This}$$
is carried out on both sides to balance the equation

$$\frac{P - 2w}{2} = \frac{2l}{2} \qquad \text{Apply inverse operation, divide by 2 and balance, to make } l \text{ the subject of the}$$
formula

$$\frac{P - 2w}{2} = l \text{ or alternatively } l = \frac{P - 2w}{2}$$

l is now the subject of the formula

b Remember that the value π is a constant. So to rearrange this formula to make d the subject all we need to do is to divide both sides by π.

$$C = \pi d$$

$$\frac{C}{\pi} = \frac{\pi d}{\pi}$$

$$\frac{C}{\pi} = d \text{ or alternatively } d = \frac{C}{\pi}$$

c We can see that F is on the RHS of this formula and to keep things as simple as possible we will leave it there, moving the other variables and constants to the LHS. On the RHS of the equation, we would normally start by expanding the brackets; that is, multiplying everything inside the brackets by the constant on the outside. However, in this case, because the constant is a fraction, doing so would not be the simplest approach to use. Rather, we can just undo the whole fraction by multiplying by its reciprocal.

$$C = \frac{5}{9}(F - 32)$$

$$\frac{9}{5} \times C = \frac{9}{5} \times \frac{5}{9}(F - 32) \qquad \text{Multiply both sides by the reciprocal to cancel out the}$$
whole fraction on the RHS and to balance on the LHS

$$\frac{9}{5}C + 32 = F - 32 + 32 \qquad \text{Apply inverse operation, add 32, to both sides to}$$
make F the subject of formula

$$\frac{9}{5}C + 32 = F \text{ or alternatively } F = \frac{9}{5}C + 32$$

d This formula involves squared variables. If we want to undo the squaring operation, we must use the inverse operation of taking the square root.

$$c^2 = a^2 + b^2$$

$$c^2 - a^2 = a^2 + b^2 - a^2 \qquad \text{Remove the variable } a^2 \text{ from the LHS by applying the inverse operation,}$$
subtract, to both sides of the equation to balance.

$$c^2 - a^2 = b^2$$

$$\sqrt{c^2 - a^2} = \sqrt{b^2} \qquad \text{Apply inverse operation, square root, on both sides to make } b \text{ the subject}$$
of formula

$$\sqrt{c^2 - a^2} = b \text{ or alternatively } b = \sqrt{c^2 - a^2}$$

In the final step, note that while the square root undoes the b^2 on the RHS, it does not individually undo the c^2 or a^2 on the LHS. This is because on the LHS the square root is being taken of the whole of the expression $c^2 - a^2$, rather than each of the individual squared components.

PRACTICE 24: APPLICATIONS

a Rearrange the $P = 2(l + w)$ to make w the subject of the formula.

Making Connections Review fractions and their reciprocals in Chapter 5.

b   Rearrange the $c^2 = a^2 + b^2$ to make a the subject of the formula.

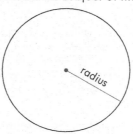

c   The formula for the area of a circle is $A = \pi r^2$, where A is the area and r is the radius of the circle. Rearrange this formula to make r the subject.

d   The distance travelled by a free-falling object is described by the formula $s = \frac{1}{2}gt^2$, where s is the distance travelled, g is the acceleration due to gravity and t is the length of time the object has been falling. This is a commonly used formula in the physical sciences. Rearrange this formula to make g the subject.

Rearranging and evaluating formula

Once the required variable has been made the subject of the formula, the given values for the other variables can be substituted into the new rule, which can then be evaluated to find the value of the unknown variable.

EXAMPLE 25: APPLICATIONS

a   The formula to calculate the amount of simple interest earned is $I = Prt$, where I is the interest earned, P is the principal (the original amount deposited), r is the interest rate and t is the time in years the principal has been accruing interest.

If the amount of interest earned over 2 years is $60 and the principal was $1000, find the interest rate.

b

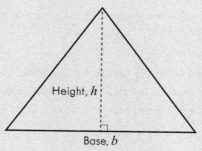

Height, h

Base, b

The formula for the area of a triangle is $A = \frac{bh}{2}$, where A is the area, b is the length of the base, and h is the perpendicular height. If the area of the triangle is 42 cm² and the length of the base is 12 cm, find the height.

c

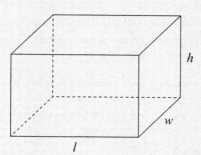

The formula for the volume of a box is $V = lwh$, where V is the volume, l is the length of one set of sides in the base, w is the length of the other set of sides in the base and h is the height of the sides of the box. If the volume of a box is 3 m³, when the width is 6 m and the height is 2 m, find the length of the base of the box.

d   In the physical sciences, formulas are often used to calculate the velocity and acceleration of moving objects. One of these formulas is $v = u + at$, where v is the velocity at the end, u is the velocity at the start, a is the constant acceleration of the object and t is the amount of time the object was moving in seconds. If a particular object had a starting velocity of $u = 0$, a final velocity of $v = 75$ and constant acceleration of $a = 20$, for how long did it travel? (Note in this example we are not going to complicate the question by including specialised units for u, v and a, rather our focus is on rearranging and evaluating the formula.)

Solutions

a   In this example, we have been told the values of I, P and t, and are required to find the value of r, which is not the subject of the formula provided. This means that we need to firstly rearrange the formula to make r the subject. Remember it does not matter in which order multiplication is performed.

$I = Prt$

$I = Ptr$ Change the order of Ptr to help simplify in the next step

$\dfrac{I}{Pt} = \dfrac{Ptr}{Pt}$ Apply inverse operation, division by Pt, to both sides of equation to make r the subject of the formula on the RHS

$\dfrac{I}{Pt} = r$ or alternatively $r = \dfrac{I}{Pt}$

Our first goal has been completed: to make r the subject of the formula. Now we need to find the value of r for the given values of the other variables.

$r = \dfrac{I}{Pt}$ where $I = \$60$, $P = \$1000$ and $t = 2$ years

$r = \dfrac{\cancel{60}}{\cancel{1000} \times 2}$ Cancel the numerator 60 and the denominator 1000 by dividing by 10

$r = \dfrac{\cancel{6}}{100 \times \cancel{2}}$ Cancel the numerator 6 and the denominator 2 by dividing by 2

$r = \dfrac{3}{100}$

The variable, r, represents the interest rate, which is usually written as a percentage. Converting $r = \dfrac{3}{100}$ into a percentage, the result is $r = 3\%$.

Making Connections
To review converting fractions into percentage revisit Chapter 6.

Making
Connections
*Revisit Chapter 12
to review the
concept of area*

b In this example, we have been told the values of A and b and are required to find the value of h, which is not the subject of the formula provided. This means that we need to firstly rearrange the formula to make h the subject.

$$A = \frac{bh}{2}$$

$\frac{2}{1} \times A = \frac{2}{1} \times \frac{bh}{2}$ Apply inverse operation, multiply by 2, to both sides of equation

$$2A = bh$$

$\frac{2A}{b} = \frac{bh}{b}$ Apply inverse operation, divide by b, to both sides of equation to make h the subject of the formula

$\frac{2A}{b} = h$ or alternatively $h = \frac{2A}{b}$

The first step has been completed: h is now the subject of the formula. We now need to find the value of h for the given values of the other variables.

$h = \frac{2A}{b}$ where $A = 42$ cm^2 and $b = 12$ cm

$$h = \frac{2 \times 42}{12}$$

$$h = \frac{84}{12}$$

$$h = 7 \text{ cm}$$

So, the height of the triangle is 7 cm.

Making
Connections
*Review the
concept of volume
in Chapter 12*

c Here we have been told the values of V, w and h and are required to find the value of l, which is not the subject of the formula provided. This means that we need to firstly rearrange the formula to make l the subject.

$$V = lwh$$

$\frac{V}{wh} = \frac{lwh}{wh}$ Apply inverse operation, divide by wh, to both sides of equation to make l the subject of formula

$\frac{V}{wh} = l$ or alternatively $l = \frac{V}{wh}$

Our first goal has been completed: l is now the subject of the formula. We now need to find the value of l for the given values of the other variables.

$l = \frac{V}{wh}$ where $V = 3$ m^2, $w = 6$ m and $h = 2$ m

$$l = \frac{3}{6 \times 2}$$

$$l = \frac{3}{12}$$

$l = \frac{1}{4}$ Cancel the numerator 3 and the denominator 12 by dividing by 3

So we have that the length is $\frac{1}{4}$ m.

d In this example, we have been told the values of u, v and a are required to find the value of t, which is not the subject of the formula provided. This means that we need to firstly rearrange the formula to make t the subject.

$$v = u + at$$

$v - u = u + at - u$ Apply inverse operation, subtract u, from both sides of equation

$\frac{v - u}{a} = \frac{at}{a}$ Apply inverse operation, divide by a, to make t the subject of formula

$\frac{v - u}{a} = t$ or alternatively $t = \frac{v - u}{a}$

The first step has been completed: t is now the subject of the formula. We now need to find the value of t for the given values of the other variables.

$t = \frac{v - u}{a}$ where $u = 0$, $v = 75$ and $a = 20$

$t = \frac{75 - 0}{20}$

$t = \frac{75}{20}$

$t = \frac{15}{4}$ Cancel the numerator 75 and the denominator 20 by dividing 5

So the object was travelling for $\frac{15}{4}$ seconds, or alternatively we could write that as $t = 3.75$ seconds.

Making Connections
To revise converting fractions to decimals revisit Chapter 6

PRACTICE 25: APPLICATIONS

a The formula to calculate the amount of simple interest earned is $I = Prt$.

If the amount of interest earned is \$120 at a rate of 3% and the principal was \$1000, how long did it take for the interest to be earned?

b The formula for the area of a triangle is $A = \frac{bh}{2}$. If the area of the triangle is 60 cm² and the perpendicular height is 20 cm, find the length of the base.

c The formula for the volume of a box is $V = lwh$. If the volume of a box is 54 m² when the length is 3 m and the height is 9 m, find the width of the base of the box.

d A formula used to calculate the velocity and acceleration of moving objects is $v = u + at$, where v is the velocity at the end, u is the velocity at the start, a is the constant acceleration of the object and t is the time the object has been moving. If a particular object had a final velocity of $v = 120$ and constant acceleration of $a = 8$, and has travelled for $t = 9$, find its starting velocity, u. (This question is not going to be complicated by including the specialised units for the variables used in the formula.)

e Another formula used in the physical sciences is $F = ma$, which describes the relationship between an object's mass (m) and the amount of force (F) needed to accelerate it (a). If the amount of force $F = 120$ is need to accelerate the object to $a = 4$, find the mass of the object. (Again no specialist units have been included just so we can focus on the rearrangement and evaluation of the formula.)

13.9 CHAPTER SUMMARY

SKILL OR CONCEPT	DEFINITION OR DESCRIPTION	EXAMPLES
Equation	A mathematical sentence that describes the equality of two algebraic expressions.	$4x - 6 = 15$ $\dfrac{2x + 5}{3} = 3(x + 4)$ $4(x - 1) = 2(x + 5) - 11$ $x^2 - 5x + 1 = 0$
Equation solution	The value of the variable required to make the LHS of the equation equal to the RHS of the equation.	$3 + b = 9; b = 6$ $a - 7 = 4; a = 11$ $\dfrac{x}{4} = 7; x = 28$ $5g = -20; g = -4$
Backtracking	The process of undoing an expression in the order that it was built up.	
Inverse operations	The opposite operations that undo those that were used when building up the expression.	$+ \Leftrightarrow -$ $\times \Leftrightarrow \div$
Balancing principle	Whatever is done to one side of the equation must be done in exactly the same way to the other side of the equation.	$a + 2 = 10$ $a + 2 - 2 = 10 - 2$ $a = 8$
Solving equations	Carrying out the processes that are needed to find the solution by: • expanding brackets • collecting like terms • backtracking using inverse operations to isolate the unknown • simultaneously applying the balancing principle • checking the solution by substituting into the original equation	$2(3x + 4) = 2(x + 2)$ $6x + 8 = 2x + 4$ $6x - 2x + 2x = 2x - 2x + 4$ $4x + 8 - 8 = 4 - 8$ $4x = -4$ $\dfrac{4x}{4} = \dfrac{-4}{4}$ $x = -1$ Check: $2(3(-1) + 4) = 2((-1) + 2)$ $2(-3 + 4) = 2(-1 + 2)$ $2(1) = 2(1)$ $2 = 2\checkmark$
Formula	Describes the relationship between two or more variables.	$P = 2(l + w)$ $C = \pi d$ $c^2 = a^2 + b^2$ $A = P(1 + rt)$ $C = \dfrac{5}{9}(F - 32)$
Subject of a formula	The value that is being determined.	$C = \dfrac{5}{9}(F - 32)$
Formula rule	Indicates how the subject should be calculated.	$C = \dfrac{5}{9}(F - 32)$
Evaluating a formula	Substituting given values for particular variables into the formula, then carrying out the required calculations to find the value of the remaining unknown variable.	$v = u + at$ where $u = 1$, $a = 15$ and $t = 5$ $v = 1 + 15(5)$ $v = 76$
Rearranging a formula	Using backtracking and balancing processes to re-write the formula so that the required unknown becomes the subject.	$A = \pi r^2$, make r the subject $\dfrac{A}{\pi} = r^2$ $\sqrt{\dfrac{A}{\pi}} = \sqrt{r^2}$ $\sqrt{\dfrac{A}{\pi}} = r$

13.10 REVIEW QUESTIONS

A SKILLS

1 Solve by inspection.

    a $\quad m - 5 = 2$ $\qquad\qquad\qquad$ b $\quad 3k = -6$

2 Use a flowchart to isolate the unknown in the following expressions.

    a $\quad \dfrac{2x + 4}{3}$ $\qquad\qquad\qquad$ b $\quad \dfrac{3x - 2}{4} - 1$

3 Solve the following equations.

    a $\quad x + 7 = 11$ $\qquad\qquad\qquad$ b $\quad c - 4 = -9$

    c $\quad 5d = -35$ $\qquad\qquad\qquad$ d $\quad \dfrac{y}{7} = -3$

4 Translate the following phrases into an equation and then solve.

    a $\quad$ 14 less than a number is 3

    b $\quad$ When a number is increased by 8, the result is 12

    c $\quad$ When a number divided by 6, the quotient is 7

    d $\quad$ The product of 7 and a number is 84

5 Solve the following equations.

    a $\quad 4x + 3 = 11$ $\qquad\qquad\qquad$ b $\quad -2x - 5 = 3$

    c $\quad \dfrac{x + 5}{4} = -3$ $\qquad\qquad\qquad$ d $\quad \dfrac{x}{3} - 2 = -5$

6 Solve the following equations.

    a $\quad 4(x - 5) = 24$ $\qquad$ b $\quad 3(2x + 7) = 9$ $\qquad$ c $\quad 2(x - 4) + 5x = 10$

    d $\quad 3(x - 2) + 2(x - 4) = 1$ $\qquad$ e $\quad 5x - 3 = 21 - 3x$ $\qquad$ f $\quad 3(3x - 2) = 2(x + 1)$

7 Evaluate the following formulas for the given values of the variables.

FORMULA	GIVEN VALUES	TO FIND THE SUBJECT
$V = \frac{1}{3}\pi r^2 h$	$\pi \cong 3.14$, $r = 10$ cm and $h = 6$ m	V
$A = P(1 + r)^n$	$P = \$100$, $r = 0.1$ and $n = 2$	A

8 Rearrange $V = \frac{1}{3}\pi r^2 h$ to make:

    a $\quad h$ the subject of the formula

    b $\quad r$ the subject of the formula

9   Rearrange the following formulas to be in terms of the given subject, and then evaluate for the given values.

FORMULA	GIVEN VALUES	TO FIND THE SUBJECT
$V = I^2 R$	$V = 1250$ and $R = 50$ (omitting technical units)	I
$A = P(1 + rt)$	$A = \$200$, $r = 0.1$ and $t = 2$	P

B   APPLICATIONS

1   In Australia, the official soccer field dimensions specify the maximum length and width for a field, with the maximum length being 120 metres. (Source: dsr.wa.gov.au)

   a   If the maximum width of the field is w, write an expression for the perimeter of the field.

   b   According to the official dimensions, the maximum perimeter of a soccer field will be 420 metres. Use this information to algebraically determine the maximum width of a soccer field.

2   Patrice has an average mark of 84 for her first two Economics tests. After she completed the third test, Patrice's average mark increased to 89.

   a   What were the total marks that Patrice earned in the first two tests?

   b   If Patrice achieved m marks in the third test, write an expression for the average mark Patrice earned over three tests.

   c   Use this expression to create an equation for Patrice's average mark over three tests and then solve the equation to find m.

3   In 2015, the statistics for the number of Australian organ donations revealed the following information:

   •   The number of liver donations was 74 more than two times the number of heart donations.

   •   The number of kidney donations was 247 less than ten times the number of heart donations.

   •   There were a total number of 1062 liver, kidney and heart donations.

   (Source: donatelife.gov.au)

   a   If there were h heart donations, write an expression for the number of liver donations.

   b   Write another expression for the number of kidney donations in terms of h heart donations.

   c   Write an expression for the total number of liver, kidney and heart donations in terms of h, then simplify this expression.

   d   Use this expression to create an equation for the total number of liver, kidney and heart donations in 2015.

   e   Solve this equation to find h, the number of heart donations.

   f   Finally, use the value for h to find the number of liver and kidney donations.

4   Ali is preparing to run a marathon. He weighs 80 kg, and for an athlete of this weight his recommended daily calorie intake is as follows:

-   240 more calories should come from carbohydrates than from fats.

-   The number of calories that come from protein should be 40 more than one-fifth of the calories that come from fats.

-   The total number of calories that should come from carbohydrates, protein and fats in total is 4680.

(Source: ausport.gov.au)

a   If the number of calories coming from fats is f, write an expression for the number of calories that should come from carbohydrate.

b   Write another expression for the number of calories that should come from protein in terms of f.

c   Now write an expression for the total number of calories that should come from carbohydrates, protein and fats in terms of f, and simplify this expression.

d   Use this expression to create an equation for the total number of calories from carbohydrates, protein and fats, in terms of f.

e   Solve this equation to find f, the number of calories that should come from fats.

f   Finally, use this value to find the number of calories that should come from carbohydrates and protein.

5   The area of a trapezium is given by the formula $A = \left(\dfrac{a+b}{2}\right) \times h$, where a and b are the lengths of the parallel sides and h is the perpendicular height, as per this diagram.

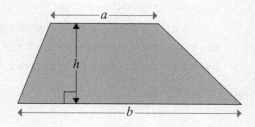

If the length of the base of a trapezium is 10 cm, the perpendicular height is 12 cm, and the area is 96 cm², find the length of a.

6   In her Chemistry practical, Lucy requires 10 litres of a 15% acid solution for a particular experiment. However, she only has on hand a 10% solution and a 30% solution so she needs to mix these solutions to create the required 15% solution. The first thing that Lucy must do is convert all of the percentages to decimals: 10% = 0.1, 30% = 0.3 and 15% = 0.15. Then Lucy will carry out the following steps.

a   If the required amount of the 0.1 solution is x litres, in terms of x, how many litres will be required of the 0.3 solution to create a total of 10 litres?

Kathy Brady

b If there are x litres of the 0.1 solution, then there will be $0.1x$ litres of acid in that solution. Use this relationship to write an expression, in terms of x, for how many litres of acid will be in the 0.3 solution.

c Next write an expression, in terms of x, for the total litres of acid when the 0.1 solution and the 0.3 solution are mixed. Simplify this expression.

d Now calculate how many litres of acid will be in the resulting 10 litres of the 0.15 solution.

e Use this value to write an equation, in terms of x, for the total litres of acid in the mixture.

f Solve this equation to find x, the number of litres of the 0.1 solution that will be required.

g Finally, determine how many litres of the 0.3 solution will be required to make the 10 litres of the 0.15 solution.

7 The Body Mass Index (BMI) calculates the ratio between a person's weight and their height. The BMI is one way of indicating whether a person might be underweight, a healthy weight or overweight. (Source: heartfoundation.org.au)

   The formula to calculate BMI is weight (in kilograms) divided by height (in metres) squared, or $BMI = \dfrac{kg}{h^2}$.

   If a person weighs 75 kg and their height is 1.6 m, what is their BMI (to one decimal place)?

8 In the physical sciences, the Ideal Gas Law describes the relationships that exist in various forms of gases. The formula for the Ideal Gas Law is $PV = nRT$, where P is the pressure the gas is under, V is the volume of the gas, n is the number of moles of the gas, T is the temperature of the gas and R is a constant value known as the gas constant. (For the purposes of this question, some of the specialised units are not included; rather the focus is on rearranging and evaluating the formula).

   20 L of hydrogen is being stored under pressure. The specific conditions of its storage gives rise to the following variable values: $n = 10$, $T = 290$ and $R = 8.314$. Given these conditions, find the pressure under which the hydrogen gas is being stored (to the nearest 100).

14

STRAIGHT-LINE GRAPHS

CHAPTER CONTENT

» Introduction to straight-line graphs

» Drawing straight-line graphs

» Slope of straight-line graphs

» Equations of straight lines in slope–intercept form

» Equations of straight lines in general form

CHAPTER OBJECTIVES

» Use table of values, and axis intercepts to draw straight-line graphs

» Identify the characteristics of slope in straight-line graphs

» Calculate slope

» Identify slope and y-intercept in equations written in slope–intercept form

» Write equations in slope–intercept form given the slope and/or points on the line

» Draw straight-line graphs from equations written in slope–intercept form

» Convert equations in slope–intercept form to general form

IT'S A BIT OF A STRETCH

In laboratory experiments, it is usual to control one variable and see how it affects another variable. Straight-line graphs clearly illustrate the sorts of relationships where one variable directly affects the other. For example, in a physics experiment, a large spring is hung from a hook and various weights are attached to the spring. The investigator Vanessa records the stretched length of the spring each time a set of weights is added and records this information on a table of values. She then plots the data from the table of values to form a straight-line graph.

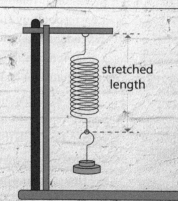
stretched length

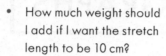

Weight (kg)	1	2	5	10
Stretch length (cm)	4.5	6.5	12.5	22.5

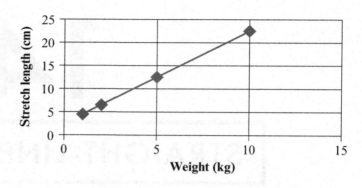

Using the straight-line graph she has drawn, Vanessa can now answer questions such as:

- How much weight should I add if I want the stretch length to be 10 cm?
- What would the stretch length be if I added 15 kg?
- What is the length of the spring with no added weight?
- How can I describe the relationship between the stretch length and the weight added using an equation?

14.1 INTRODUCTION TO STRAIGHT-LINE GRAPHS

Graphing is a diagrammatic method of representing the relationships between quantities or variables. A straight-line graph is a type of graph that illustrates how two variables are in a linear relationship.

When two variables are in a linear relationship, the value of one variable will depend upon the value of the other variable. That is, as one variable changes so will the other variable. By convention, we called these variables x and y. So, if y changes whenever x does, we say that the value of y is dependent on the value of x, and we name y the **dependent variable** and x the **independent variable**.

In linear relationships, any change in the independent variable will always produce a corresponding change in the dependent variable. The linear relationship between the dependent and independent variables is described mathematically using a **linear equation**. The graph of a linear equation is a straight line that comprises all the points with coordinates make the equation true.

One example of a linear relationship is the distance formula, distance = speed × time, which we can abbreviate to $d = s \times t$. If a car is travelling at a constant speed of 80 km/hour, we can establish the linear relationship $d = 80t$ (notice that the × sign is implied here).

This means that distance travelled is the dependent variable, as it will vary directly with the time taken, which is the independent variable. The equation $d = 80t$ is a linear equation that will produce the straight-line graph overpage.

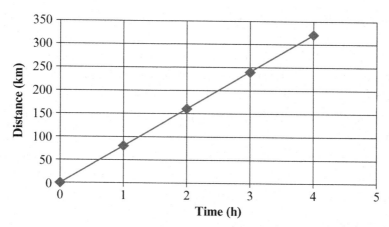

Notice that the *independent* variable, time, is graphed on the *horizontal* axis, and the *dependent* variable, distance, is graphed on the *vertical* axis.

Another example of a linear relationship can be found in the area of geometry, where we determined that the sum of the interior angles of a polygon increases as the number of sides increase. This means that the angle sum is the dependent variable, as it changes according to the number of sides, which is the independent variable. This linear relationship is described by the linear equation, angle sum = $(n - 2) \times 180°$, where n is the number of sides. The graph of this relationship is shown below.

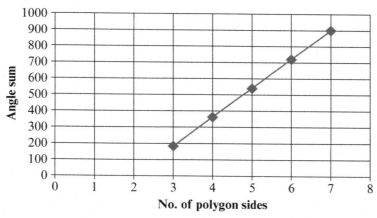

Notice again that the independent variable, sides, is graphed on the horizontal axis and the dependent variable, angle sum, is graphed on the vertical axis.

Making connections
Take another look at Chapter 11 to review angle sum and other properties of polygons.

CRITICAL THINKING

In the graph above, why are no points plotted for the values 0, 1 and 2 on the horizontal axis?

Linear relationships and their straight-line graphs can be found in many practical contexts and areas of study and research. For this reason, they play an important part in how we mathematically represent, model and interpret many situations.

Kathy Brady

Making connections
To review how to plot a graph from a table of values revisit Chapter 10.

14.2 DRAWING STRAIGHT-LINE GRAPHS

There are a number of ways we can draw a straight-line graph, and the method that is selected with be dependent upon the information that is available. In Chapter 10, we discussed how to plot a graph from a given table of values and this is one method that could be used.

In this section, we will present two further methods that can be used to draw a straight-line graph. The first of these techniques also involves using a table of values, but in this method the values in the table need to be calculated in order to draw the graph. When using this method, the linear equation that will generate the straight-line graph is provided. However, the independent variable, x, may or may not be given.

EXAMPLE 1

Draw the graph of $y = 2x + 3$ by first completing this table of values.

x	−2	−1	0	1	2
y					

Solution

To complete the table, one-by-one substitute the x values −2, −1, 0, 1 and 2 into the equation $y = 2x + 3$ to determine the value of y.

For $x = -2$; $y = 2 \times (-2) + 3 = -1$

For $x = -1$; $y = 2 \times (-1) + 3 = 1$

For $x = 0$; $y = 2 \times 0 + 3 = 3$

For $x = 1$; $y = 2 \times 1 + 3 = 5$

For $x = 2$; $y = 2 \times 2 + 3 = 7$

The completed table of values is:

x	−2	−1	0	1	2
y	−1	1	3	5	7

We now plot the coordinate points (−2, −1), (−1, 1), (0, 3), (1, 5) and (2, 7) to form a straight line.

Notice the arrowhead on each end of the line we have graphed. These arrows indicate that the line actually continues indefinitely in each direction.

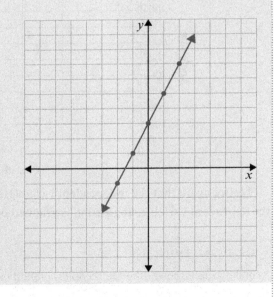

PRACTICE 1

Draw the graph of $y = -3x + 1$ by first completing this table of values.

x	−2	−1	0	1	2
y					

Sometimes the only information that we are given is the linear equation and we need to create a table of values by selecting appropriate values for x to be substituted into the equation. Keep in mind that just two points is all we need to draw a straight line, so our table of values does not need to be extensive. Four or five x values are quite sufficient, and stick to easy to calculate values.

EXAMPLE 2

Draw the graph of $y = -2x - 1$ by first completing a table of values.

Solution

It does not matter which values of x we select to create the coordinate points that will make the equation true, so why not select easy to calculate values such as those we used in the previous example.

To create the table, one-by-one substitute the x values $-2, -1, 0, 1$ and 2 into the equation $y = -2x - 1$ to determine the value of y.

For $x = -2; y = -2 \times (-2) - 1 = 4 - 1 = 3$

For $x = -1; y = -2 \times (-1) - 1 = 2 - 1 = 1$

For $x = 0; y = -2 \times 0 - 1 = -1$

For $x = 1; y = -2 \times 1 - 1 = -2 - 1 = -3$

For $x = 2; y = -2 \times 2 - 1 = -4 - 1 = -5$

Therefore, our table of values is:

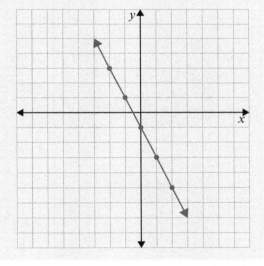

x	-2	-1	0	1	2
y	3	1	-1	-3	-5

We now plot the coordinate points $(-2, 3), (-1, 1), (0, -1), (1, -3)$ and $(2, -5)$ to form a straight line.

PRACTICE 2

Draw the graph of $y = 3x + 4$ by first completing a table of values.

EXAMPLE 3: APPLICATION

In a physics laboratory, students are studying the laws of motion. They toss an object upwards with an initial velocity of 49 m/s. After t seconds, the object will be travelling at a velocity of v m/s according to the equation:

$v = 49 - 9.8t$

Draw the graph of this equation by completing the following table of values.

t	0	5	10
v			

Kathy Brady

Solution

To create the table, one-by-one substitute the t values 0, 5 and 10 into the equation $v = 49 - 9.8t$ to determine the value of v.

For $t = 0$; $v = 49 - 9.8 \times 0 = 49 - 0 = 49$

For $t = 5$; $v = 49 - 9.8 \times 5 = 49 - 49 = 0$

For $t = 10$; $v = 49 - 9.8 \times 10 = 49 - 98 = -49$

Therefore the table of values will be:

t	0	5	10
v	49	0	−49

Note that v is the dependent variable, because it changes whenever t does, so v is graphed on the vertical axis and t, the independent variable, is graphed on the horizontal axis. We also need to choose an appropriate scale for each axis to create this graph.

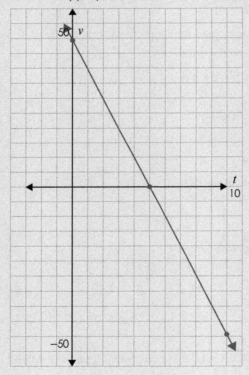

PRACTICE 3: APPLICATION

The students continue to study the laws of motion in another physics laboratory class. This time they are looking at the movement of an object with a constant acceleration of $5 \, \text{m/sec}^2$. If the initial velocity of the object is 10 m/s, after t seconds the object will be travelling at a velocity of v m/s according to the equation:

$$v = 10 + 5t$$

Draw the graph of this equation by completing the following table of values.

t	0	1	2	3	4
v					

Axis intercepts

So far, we have drawn graphs of straight lines from linear equations in the form $y = \ldots$ When using linear equations in this form, it is easy to determine the value of y with each x that is substituted. However, sometimes linear equations can take a different form.

Consider the equation $2x + y = 4$. This is still a linear equation but it is not in the form $y = \ldots$, since the x and y variables are both on the left-hand side of the equation.

Let's try to create a table of values for this equation using the following values for x.

x	−2	0	2	4	6
y					

We will substitute the x values, one-by-one, into the equation $2x + y = 4$ to determine the corresponding y value.

For $x = -2$; $2 \times -2 + y = 4 \rightarrow -4 + y = 4$. This means that $y = 8$.

For $x = 0$; $2 \times 0 + y = 4 \rightarrow 0 + y = 4$. This means that $y = 4$.

For $x = 2$; $2 \times 2 + y = 4 \rightarrow 8 + y = 4$. This means that $y = 0$.

For $x = 4$; $2 \times 4 + y = 4 \rightarrow 8 + y = 4$. This means that $y = -4$.

For $x = 6$; $2 \times 6 + y = 4 \rightarrow 12 + y = 4$. This means that $y = -8$.

The completed table of values will be:

x	−2	0	2	4	6
y	8	4	0	−4	−8

We now plot the coordinate points (−2, 8), (0, 4), (2, 0), (4, −4) and (6, −8) to form a straight line.

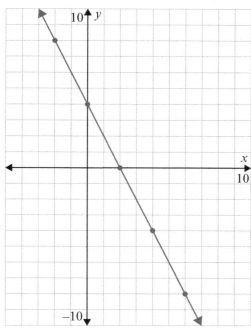

Kathy Brady

y-intercept
The point
where a
graph crosses
the *y*-axis,
when $x = 0$

Chapter 14

Notice on this straight-line graph that the line crosses the *y*-axis at the point where $y = 4$ with the coordinate (0, 4). This point is called the **y-intercept** and occurs when the *x* value is 0. Likewise, the line crosses the *x*-axis at the point $x = 2$ with the coordinate (2, 0). This is called the **x-intercept** and occurs when the *y* value is 0.

As we have already noted, we need just two points to draw a straight-line graph. So if we were to find the *y*-intercept and the *x*-intercept, they would be the only points that we would need to draw the straight-line graph that passes through these points.

A linear equation in the form $2x + y = 4$ is said to be in the general form, and equations in this form make it easy to determine the *x*- and *y*-intercepts, and therefore draw the straight-line graph. (We will present more about the general form of a linear equation in Section 5 of this chapter.)

x-intercept
The point where
a graph crosses
the *x*-axis,
when $y = 0$

EXAMPLE 4

By determining the *x*- and *y*-intercepts, draw the graphs of:

a $x + y = 5$ **b** $3x + 4y = 12$

Solutions

a   The *y*-intercept of this graph will occur when $x = 0$. So, we will substitute this value of *x* into the equation to get $0 + y = 5$. Therefore, the *y*-intercept is the point where $y = 5$, with the coordinate (0, 5). The *x*-intercept of this graph will occur when $y = 0$. So, we will substitute this value of *y* into the equation to get $x + 0 = 5$. Therefore, the *x*-intercept is the point where $x = 5$, with the coordinate (5, 0). We now have the two points we need to draw the graph.

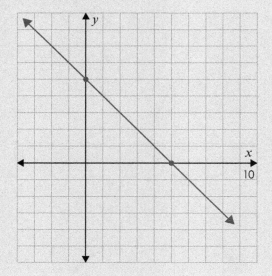

b The y-intercept of $3x + 4y = 12$ will occur when $x = 0$. We substitute this value of x into the equation to get $3 \times 0 + 4y = 12 \rightarrow 0 + 4y = 12$. This means that $y = 3$.

Therefore, the y-intercept is the point where $y = 3$, with the coordinate $(0, 3)$. The x-intercept of this graph will occur when $y = 0$. So, we will substitute this value of y into the equation to get $3x + 4 \times 0 = 12 \rightarrow 3x + 0 = 12$. This means that $x = 4$. Therefore, the x-intercept is the point where $x = 4$, with the coordinate $(4, 0)$. We now have the two points we need to draw the graph.

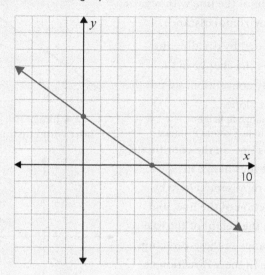

PRACTICE 4

By determining the x- and y-intercepts, draw the graphs of:

a $2x + y = 6$ **b** $5x + 3y = 15$

14.3 SLOPE OF STRAIGHT-LINE GRAPHS

An important feature of a straight-line graph is its **slope**. When considering the slope of a line, we are interested in measuring both its steepness and its direction.

In straight-line graphs, slope is always measured from left to right.

This line rises upwards from left to right so it is said to have a **positive slope**.

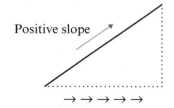

Positive slope

Slope
A number that describes the steepness and direction of a straight-line graph

Positive slope
Occurs when a line goes upwards from left to right

Kathy Brady

Negative slope
Occurs when
a line that
downward from
left to right

This line descends from left to right and so it has a **negative slope**.

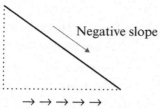

Negative slope

As we have defined, the slope of a straight line describes both its direction and its steepness. The steepness of the slope is determined by looking at the change in the height of the line with regard to the change in the horizontal distance.

The slope is calculated by dividing the change in the height by the change in the horizontal distance. By convention, we use the letter m as the label for slope.

$$m = \frac{\text{change in height}}{\text{change in horizontal distance}}$$

Change in height

Change in horizontal distance

Let's now draw this line on the coordinate plane. We can see that the change in horizontal distance is represented here by a change in the x-direction of the line, and likewise the change in the height of the line is represented by a change in the y-direction. We refer to the change in the x-direction as the x-step, and the change in the y-direction and the y-step. So now we can re-define the slope, m, as:

Change in y-direction

Change in x-direction

$$m = \frac{\text{change in height}}{\text{change in horizontal distance}} = \frac{y\text{-step}}{x\text{-step}}$$

The size of the y-step and the x-step can be calculated if we know just two points on the line, which we refer to as (x_1, y_1) and (x_2, y_2) on the following graph.

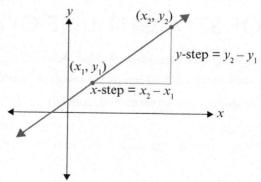

(x_2, y_2)

(x_1, y_1)

$y\text{-step} = y_2 - y_1$

$x\text{-step} = x_2 - x_1$

We can calculate the y-step by subtracting the y value of the first point from the y value of the second point, and similarly for the x-step. So now we can write a formula for m, the slope of a line, given two points on the line (x_1, y_1) and (x_2, y_2):

$$m = \frac{y\text{-step}}{x\text{-step}} = \frac{y_2 - y_1}{x_2 - x_1}$$

EXAMPLE 5

Find the slopes of the following lines.

a

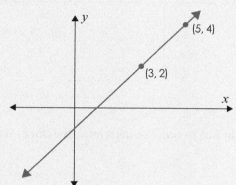

b

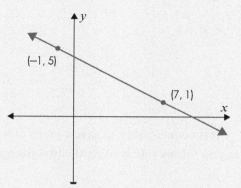

Solutions

a   We have been given two points $(3, 2)$ and $(5, 4)$. Let's call $(3, 2)$ the (x_1, y_1) point and $(5, 4)$ the (x_2, y_2) point. This means that $x_1 = 3$, $y_1 = 2$, $x_2 = 5$ and $y_2 = 4$. Now we need to carefully substitute these values into the formula for the slope to get:

$$m = \frac{y_2 - y_1}{x_2 - x_1} = \frac{4 - 2}{5 - 3} = \frac{2}{2} = 1$$

It does not matter which points are called (x_1, y_1) and (x_2, y_2). The same value for m will result when they are substituted into the formula.

b   Notice, firstly, that this graph goes downward from left to right so it has a negative slope. Using the points on this graph, we will call $(-1, 5)$ the (x_1, y_1) point and $(7, 1)$ the (x_2, y_2) point. Therefore, $x_1 = -1$, $y_1 = 5$, $x_2 = 7$ and $y_2 = 1$. Carefully substituting these values into the formula for the slope, we get:

$$m = \frac{y_2 - y_1}{x_2 - x_1} = \frac{1 - 5}{7 - (-1)} = \frac{-4}{8} = -\frac{1}{2}$$

The resultant slope is a negative number, as predicted.

Find the slopes of the following lines.

a

b

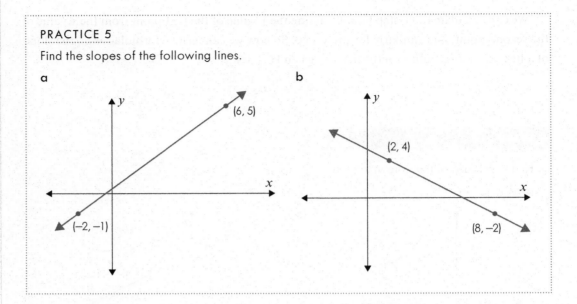

It is not necessary to have a graph of a straight line in order to determine the slope. We can do this by simply using the two given points.

EXAMPLE 6

Find the slope of the line that passes through $(-3, 1)$ and $(4, -6)$.

Solution

Let's call $(-3, 1)$ the (x_1, y_1) point and $(4, -6)$ the (x_2, y_2) point. This means that $x_1 = -3$, $y_1 = 1$, $x_2 = 4$ and $y_2 = -6$. Now we can substitute these values into the formula for the slope to get:

$$m = \frac{y_2 - y_1}{x_2 - x_1} = \frac{-6 - 1}{4 - (-3)} = \frac{-7}{7} = -1$$

PRACTICE 6

Find the slope of the line that passes through $(5, 1)$ and $(-3, -7)$.

Horizontal and vertical lines

Consider the points $(3, 4)$ and $(-2, 4)$. What is the slope of the line that passes through these points? Let's call $(3, 4)$ the (x_1, y_1) point and $(-2, 4)$ the (x_2, y_2) point. This means that $x_1 = 3$, $y_1 = 4$, $x_2 = -2$ and $y_2 = 4$. Substituting these values into the formula for the slope, we get:

$$m = \frac{y_2 - y_1}{x_2 - x_1} = \frac{4 - 4}{-2 - 3} = \frac{0}{-5} = 0$$

Now if these points are plotted on the coordinate plane and joined, the resultant straight line is **horizontal**:

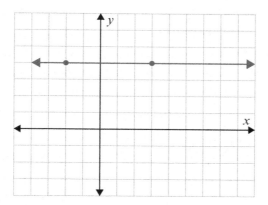

Horizontal line
A straight-line graph that has a slope of 0 and the y values of each point are equal

On this graph, we can see that the y values of all points on this line will be equal to 4. If a straight-line graph has a slope of 0 and y values of each point that are all the same, then the line is horizontal.

Let's now think about another pair of points (3, −5) and (3, 2). What is the slope of the line that passes through these points? Let's call (3, −5) the (x_1, y_1) point and (3, 2) the (x_2, y_2) point. This means that $x_1 = 3$, $y_1 = -5$, $x_2 = 3$ and $y_2 = 2$. Substituting these values into the formula for slope, we get:

Making connections
Review the meaning of division by 0 in Chapter 2.

$$m = \frac{y_2 - y_1}{x_2 - x_1} = \frac{2 - (-5)}{3 - 3} = \frac{7}{0}$$

Any division by 0 is undefined. So the slope of the line joining these points is said to be undefined.

If these two points are plotted on the coordinate plane and joined, the resultant straight line is vertical and we can see that the x value of every point on this line is equal to 3. If a straight-line graph has a slope that is undefined and x values of each point that are all the same, then the line is **vertical**.

Vertical line
A straight line that has a slope that is undefined and the x values of each point are equal

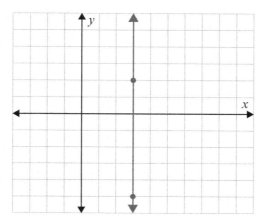

Making
connections
Revisit
Chapter 11
to review the
definition
and other
characteristics
of parallel lines.

Parallel lines
Two straight lines
that have slopes
that are equal

Parallel and perpendicular lines

We can use information about the slope of two different lines to determine whether they might be either parallel or perpendicular.

By definition, **parallel lines** are two or more lines that extend endlessly and never meet. This means, by implication, that if two lines are parallel they will have the same slope. That is, if the slope of one line is m_1 and the slope of the other line is m_2, then $m_1 = m_2$.

EXAMPLE 7

Determine whether this pair of lines is parallel.

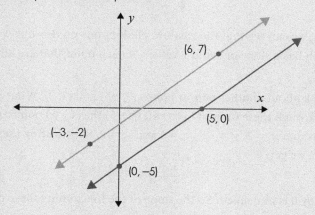

Solution

These lines might appear to be parallel but we need to calculate their slopes in order to determine whether they are or not. Consider the light-coloured line first. Let's call $(-3, -2)$ the (x_1, y_1) point and $(6, 7)$ the (x_2, y_2) point. This means that $x_1 = -3$, $y_1 = -2$, $x_2 = 6$ and $y_2 = 7$. Substituting these values into the formula for the slope, we get:

$$m = \frac{y_2 - y_1}{x_2 - x_1} = \frac{7 - (-2)}{6 - (-3)} = \frac{9}{9} = 1$$

Next, calculate the slope of the dark-coloured line. Call $(0, -5)$ the (x_1, y_1) point and $(5, 0)$ the (x_2, y_2) point. This means that $x_1 = 0$, $y_1 = -5$, $x_2 = 5$ and $y_2 = 0$. Substituting these values into the formula for the slope, we get:

$$m = \frac{y_2 - y_1}{x_2 - x_1} = \frac{0 - (-5)}{5 - 0} = \frac{5}{5} = 1$$

Since the slopes of the two lines are equal, we can conclude that the lines are parallel.

PRACTICE 7

Determine whether this pair of lines is parallel.

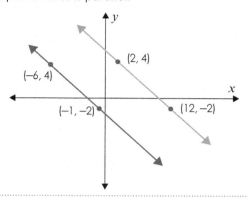

EXAMPLE 8

The points $(-4, -3)$ and $(-2, 1)$ lie on one straight line. The points $(-1, -5)$ and $(7, 2)$ lie on another straight line. Determine whether these lines are parallel.

Solution

Let's call the first straight line, Line 1, which has a slope of m_1. The points $(-4, -3)$ and $(-2, 1)$ lie on Line 1. We can calculate m_1 as follows:

$$m_1 = \frac{y_2 - y_1}{x_2 - x_1} = \frac{1 - (-3)}{-2 - (-4)} = \frac{1+3}{-2+4} = \frac{4}{2} = 2$$

Next, we calculate the slope of Line 2 with a slope of m_2. The points $(-1, -5)$ and $(7, 2)$ lie on Line 2. We can calculate m_2 as follows:

$$m_2 = \frac{y_2 - y_1}{x_2 - x_1} = \frac{2 - (-5)}{7 - (-1)} = \frac{2+5}{7+1} = \frac{7}{8}$$

Since $m_1 \neq m_2$, the lines are not parallel.

PRACTICE 8

The points $(1, 4)$ and $(13, -2)$ lie on one straight line. The points $(-5, 4)$ and $(-1, 2)$ lie on another straight line. Determine whether these lines are parallel.

Perpendicular lines are at right angles to each other. Two lines are perpendicular if the product of their slopes is -1. So in the diagram below, if the slope of the light-coloured line m_1 and the slope of the dark-coloured line is m_2, then the lines will be perpendicular if $m_1 \times m_2 = -1$.

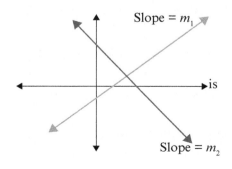

Perpendicular lines

Two straight lines that have the product of their slopes equal to -1

Kathy Brady

EXAMPLE 9

Show that these two lines are perpendicular.

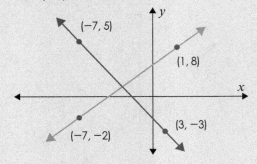

Solution

Even though these lines appear to be perpendicular we need to calculate their slopes in order to determine whether they are or not. Let's start with finding the slope of the light-coloured line. Let's call $(-7, -2)$ the (x_1, y_1) point and $(1, 8)$ the (x_2, y_2) point. This means that $x_1 = -7$, $y_1 = -2$, $x_2 = 1$ and $y_2 = 8$. Substituting these values into the formula for the slope, we get:

$$m_1 = \frac{y_2 - y_1}{x_2 - x_1} = \frac{8 - (-2)}{1 - (-7)} = \frac{8 + 2}{1 + 7} = \frac{10}{8} = \frac{5}{4}$$

Next, calculate the slope of the dark-coloured line. Call $(-7, 5)$ the (x_1, y_1) point and $(3, -3)$ the (x_2, y_2) point. This means that $x_1 = -7$, $y_1 = 5$, $x_2 = 3$ and $y_2 = -3$. Substituting these values into the formula for the slope, we get:

$$m_2 = \frac{y_2 - y_1}{x_2 - x_1} = \frac{-3 - 5}{3 - (-7)} = \frac{-8}{10} = -\frac{4}{5}$$

Now we need to find the product of the slopes of these two lines:

$$\frac{5}{4} \times -\frac{4}{5} = -\frac{20}{20} = -1$$

Since the product equals -1, we can conclude that the lines are perpendicular.

PRACTICE 9

Show that these two lines are perpendicular.

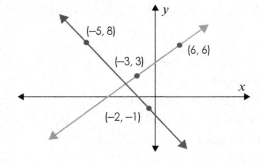

EXAMPLE 10

Line A passes through the points $(2, 6)$ and $(10, 9)$

Line B passes through the points $(-1, -5)$ and $(5, -15)$

Line C passes through the points $(1, -7)$ and $(4, -12)$

Line D passes through the points $(-1, 0)$ and $(2, -8)$

Determine which pair of lines is parallel and which pair of lines is perpendicular.

Solution

The first step is to calculate the slopes of each line.

The slope of Line A that passes through $(2, 6)$ and $(10, 9)$ is:

$$m_A = \frac{y_2 - y_1}{x_2 - x_1} = \frac{9 - 6}{10 - 2} = \frac{3}{8}$$

The slope of Line B that passes through $(-1, -5)$ and $(5, -15)$ is:

$$m_B = \frac{y_2 - y_1}{x_2 - x_1} = \frac{-15 - (-5)}{5 - (-1)} = \frac{-15 + 5}{5 + 1} = \frac{-10}{6} = -\frac{5}{3}$$

The slope of Line C that passes through $(1, -7)$ and $(4, -12)$ is:

$$m_C = \frac{y_2 - y_1}{x_2 - x_1} = \frac{-12 - (-7)}{4 - 1} = \frac{-12 + 7}{3} = -\frac{5}{3}$$

The slope of Line D that passes through $(-1, 0)$ and $(2, -8)$ is:

$$m_D = \frac{y_2 - y_1}{x_2 - x_1} = \frac{-8 - 0}{2 - (-1)} = -\frac{8}{3}$$

Since $m_B = m_C$, Line B is parallel to Line C.

Looking at $m_A = \frac{3}{8}$ and $m_D = -\frac{8}{3}$, we can see that $\frac{3}{8} \times -\frac{8}{3} = -\frac{24}{24} = -1$, so Line A and Line D are perpendicular.

PRACTICE 10

Line E passes through the points $(2, -11)$ and $(-3, 4)$

Line F passes through the points $(-6, -7)$ and $(5, -2)$

Line G passes through the points $(1, 9)$ and $(10, 12)$

Line H passes through the points $(8, -4)$ and $(3, 7)$

Determine which pair of lines is parallel and which pair of lines is perpendicular.

Drawing straight-line graphs using slope

We have already seen how to draw straight-line graphs from a table of values or using their axis intercepts. In this section, we will look at how to draw a straight-line graph given its slope and one point on the line.

EXAMPLE 11

Draw the graphs of the following lines.

a The slope of the line is 2 and it passes through the point (2, 3).

b The slope of the line is −3 and it passes through the point (−1, 2).

· ·

Solutions

a Recall that we described the slope of a line as $\frac{y\text{-step}}{x\text{-step}}$. In this example, the slope is 2, which can be written as $\frac{2}{1}$. Therefore, we have:

$$\frac{y\text{-step}}{x\text{-step}} = \frac{2}{1}$$

The first step in drawing the graph is to plot the given point (2, 3).

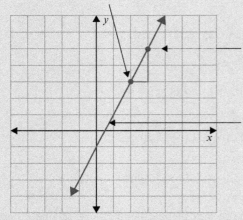

Then starting at that point we move 1 unit to the right for the x-step and 2 units upwards for the y-step to arrive at the point (3, 5).

We now have the two points we need to draw the straight line.

b In this example, the slope of the line is negative, which means that it goes downward as we move from left to right. The steps taken are the same as above, being careful to take the *x*- and *y*-steps in the correct direction to create a negative slope. Therefore, we need to have:

$$\frac{y\text{-step}}{x\text{-step}} = \frac{-3}{1}$$

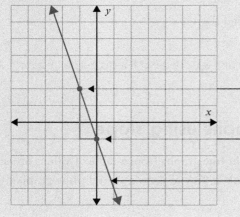

The first step is to plot the given point (−1, 2).

Then starting at that point we move 1 unit to the right for the x-step and 2 units downwards for the y-step to arrive at the point (0, −1).

Now we have the two points needed to draw the line.

14.4 EQUATIONS OF STRAIGHT LINES IN SLOPE–INTERCEPT FORM

The slope–intercept form is the most common way that a straight-line graph can be described using a linear equation. In fact, in Section 14.2 we have already seen a number of linear equations written in slope–intercept form without even realising it! These are equations such as $y = 2x + 3$ or $y = -3x + 1$. As the name implies, the **slope–intercept form of a linear equation** provides information about both the slope of the line and an axis-intercept, specifically, the y-axis intercept. For example, in the equation $y = 2x + 3$, the coefficient of x, the number 2, represents the slope of the graph and the constant value, the number 3, is the y-axis intercept of the graph. The slope–intercept form of a line is written in the form $y = mx + c$, where m is the slope and c is the y-axis intercept at the point $(0, c)$.

Slope–intercept form of a linear equation
A linear equation that is written as $y = mx + c$, where m is the slope and c is the y-axis intercept at the point $(0, c)$

EXAMPLE 12

Complete the following table providing the slope and y-axis intercept for each equation.

EQUATION	SLOPE	y-AXIS INTERCEPT
$y = 3x + 2$		
$y = -5x - 1$		
$y = \frac{1}{3}x + 4$		
$y = -4x$		
$y = 2$		

Solutions

In each case, we need to identify the coefficient of x, which is the slope, and the constant value, the y-axis intercept at $(0, c)$.

EQUATION	SLOPE	y-AXIS INTERCEPT
$y = 3x + 2$	3	(0, 2)
$y = -5x - 1$	−5	(0, −1)
$y = \frac{1}{3}x + 4$	$\frac{1}{3}$	(0, 4)
$y = -4x$	−4	(0, 0)
$y = 2$	0	(0, 2)

Kathy Brady

Note that in $y = -4x$ the absence of the constant term implies that it is 0. The equation could have been written as $y = -4x + 0$. In $y = 2$, the absence of an x term implies that the coefficient of x is 0; that is, the equation could have been written as $y = 0x + 2$.

PRACTICE 12

Complete the following table providing the slope and y-axis intercept for each equation.

EQUATION	SLOPE	y-AXIS INTERCEPT
$y = -4x + \frac{3}{2}$		
$y = 2x - 7$		
$y = -\frac{1}{2}x - 1$		
$y = 5$		
$y = -3x$		

If we are provided with information about the slope of a graph and its y-axis intercept, we can then write the equation of the line using the slope–intercept form.

EXAMPLE 13

For the given slope and y-axis intercept write the equation of the line.

SLOPE	y-AXIS INTERCEPT	EQUATION
4	$(0, -2)$	
-2	$(0, \frac{2}{3})$	
$-\frac{3}{4}$	$(0, -5)$	
6	$(0, 0)$	
0	$(0, -3)$	

Solutions

In each case, the coefficient of x in the equation will be the slope and the constant value in the coordinate $(0, c)$ will be the value of the y-axis intercept, c.

SLOPE	y-AXIS INTERCEPT	EQUATION
4	$(0, -2)$	$y = 4x - 2$
-2	$(0, \frac{2}{3})$	$y = -2x + \frac{2}{3}$
$-\frac{3}{4}$	$(0, -5)$	$y = -\frac{3}{4}x - 5$
6	$(0, 0)$	$y = 6x$
0	$(0, -3)$	$y = -3$

Note that when the y-axis intercept is $(0, 0)$, the value of $c = 0$ and the equation could have been written as $y = 6x + 0$ but by convention we do not write the addition of 0. When the slope is 0, the equation could have been written as $y = 0x - 3$, but we do not normally write a variable with a coefficient of 0.

PRACTICE 13

For the given slope and y-axis intercept write the equation of the line.

SLOPE	y-AXIS INTERCEPT	EQUATION
0.4	(0, 1)	
-2	(0, 3.6)	
$\frac{3}{2}$	$(0, \frac{5}{2})$	
0	(0, 0)	

Point–intercept form of a linear equation

We are not always provided with information about either the slope, the y-axis intercept or both. For example, we may be provided the slope and another point on the line. In this case, it is still possible to determine the equation of the line in slope–intercept form by first writing the equation in what is known as the **point–intercept form**. This is when the equation of a line is written as $y - y_1 = m(x - x_1)$, where m is the slope of the line and (x_1, y_1) is a point on the line.

Point–intercept form of a linear equation

A linear equation that is written in the form $y - y_1 = m(x - x_1)$, where m is the slope of the line and (x_1, y_1) is a point on the line

EXAMPLE 14

Find the equation, in slope–intercept form, of the following lines.

a   The slope of the line is 2 and it passes through the point (2, 3).

b   The slope of the line is -3 and it passes through the point $(-1, -2)$.

Solutions

a   In this example, $m = 2$ and the point (2, 3) is (x_1, y_1). We can now substitute these values into the point–intercept equation.

$$y - y_1 = m(x - x_1)$$
$$y - 3 = 2(x - 2)$$

We could leave the equation in this form. However, it is often more useful to convert it to slope–intercept form. To do this we use the algebraic strategies to isolate y on the LHS of the equation.

$$y - 3 = 2(x - 2)$$
$$y - 3 = 2x - 4 \qquad \text{The brackets have been expanded on the RHS}$$
$$y - 3 + 3 = 2x - 4 + 3 \qquad \text{y is isolated on the LHS by adding 3 to each side}$$
$$y = 2x - 1 \qquad \qquad \text{The equation is now in slope–intercept form.}$$

Making connections
To review the algebraic strategies of backtracking and balancing to isolate the unknown revisit Chapter 13.

b In this example, $m = -3$ and the point $(-1, -2)$ is (x_1, y_1). We can now substitute these values into the point–intercept equation.

$$y - y_1 = m(x - x_1)$$
$$y - (-2) = -3(x - (-1))$$
$$y + 2 = -3(x + 1)$$

Again we will convert this equation to slope–intercept form.

$$y + 2 = -3(x + 1)$$
$$y + 2 = -3x - 3$$
$$y + 2 - 2 = -3x - 3 - 2 \qquad \text{The brackets have been expanded on the RHS}$$
$$y = -3x - 5 \qquad \text{y is isolated on the LHS by subtracting 2 from each side}$$

The equation is now in slope–intercept form.

PRACTICE 14: GUIDED

The slope of the line is 3 and it passes through the point $(-4, 5)$. Find its equation in slope–intercept form.

$$y - 5 = \square\,(x - (-4))$$
$$y - \square = 3(x + \square)$$
$$y - \square = \square x + 12$$
$$y - 5 + \square = 3x + 12 + \square$$
$$y = 3x + \square$$

Now try this one:

The slope of the line is -2 and it passes through the point $(6, -1)$. Find its equation in slope–intercept form.

We have now established that to graph a line all we require are the coordinates of two points on the line. We can also find the equation of a line in slope–intercept form if we are given two points.

EXAMPLE 15

Find the equation of the line in slope–intercept form of:

a the line that passes through $(4, 6)$ and $(8, 22)$

b the line that passes through $(-3, 3)$ and $(2, -7)$

Solutions

a The first step is to find the slope of the line joining the two points (4, 6) and (8, 22) using:

$$m = \frac{y_2 - y_1}{x_2 - x_1} = \frac{22 - 6}{8 - 4} = \frac{16}{4} = 4$$

Now we proceed as in the previous example by selecting one of the given points. Let's use (4, 6). We can now substitute these values into the point–intercept equation.

$$y - y_1 = m(x - x_1)$$
$$y - 6 = 4(x - 4)$$

Now convert this equation to slope–intercept form.

$$y - 6 = 4(x - 4)$$
$$y - 6 = 4x - 16$$
$$y - 6 + 6 = 4x - 16 + 6$$
$$y = 4x - 10$$

The equation is now in slope–intercept form.

b To find the slope of the line joining (−3, 3) and (2, −7):

$$m = \frac{y_2 - y_1}{x_2 - x_1} = \frac{-7 - 3}{2 - (-3)} = \frac{-10}{5} = -2$$

Select one of the points. Let's use (−3, 3) and substitute these values into the point–intercept equation.

$$y - 3 = -2(x - (-3))$$
$$y - 3 = -2(x + 3)$$

Now convert this equation to slope–intercept form.

$$y - 3 = -2(x + 3)$$
$$y - 3 = -2x - 6$$
$$y - 3 + 3 = -2x - 6 + 3$$
$$y = -2x - 3$$

The equation is now in slope–intercept form.

PRACTICE 15

Find the equation of the line in slope–intercept form of:

a the line that passes through (4, 11) and (−2, −1)

b the line that passes through (2, −5) and (−4, 7)

Kathy Brady

CRITICAL THINKING

It does not matter which of the two given points you select for substitution into the point–intercept equation. Why?

DRAWING STRAIGHT-LINE GRAPHS USING SLOPE–INTERCEPT FORM

In the previous section, we looked at how to draw a graph of a straight line given the slope and one point on the line. We used exactly the same procedure when drawing the graph of a line in slope–intercept form $y = mx + c$. The steps required are:

*   plot the y-axis intercept $(0, c)$
*   using the slope, m, locate a second point on the line
*   draw the line joining these two points.

EXAMPLE 16

Draw the graphs of the following lines.

a $y = 4x + 2$

b $y = -\frac{5}{3}x + 3$

Solutions

a   In this example, the coefficient of x is 4, which is the slope. This number is positive so the line will go upwards as we move from left to right. This can be written as $\frac{4}{1}$.
Therefore, we have $\frac{y\text{-step}}{x\text{-step}} = \frac{4}{1}$.

The first step is to plot the constant value 2 as the y-axis intercept, $(0, 2)$.

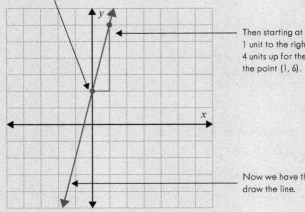

Then starting at that point we move 1 unit to the right for the x-step and 4 units up for the y-step to arrive at the point $(1, 6)$.

Now we have the two points needed to draw the line.

b In this example, the coefficient of x is $-\frac{5}{3}$, so slope of the line is negative, which means that it goes downward as we move from left to right. Therefore, we have $\frac{y\text{-step}}{x\text{-step}} = \frac{-5}{3}$.

The first step is to plot the constant value 3 as the y-axis intercept, $(0, 3)$.

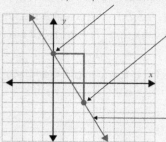

Then starting at that point we move 3 units to the right for the x-step and 5 units downward for the y-step to arrive at the point $(3, -2)$.

Now we have the two points needed to draw the line.

PRACTICE 16

Draw the graphs of the following lines.

a $y = \frac{3}{4}x - 5$ **b** $y = -3x + 1$

EXAMPLE 17: APPLICATION

When catching a taxi, the fare is divided into two parts. The first part is known as flag fall and this is an initial charge incurred when the passenger gets into the taxi before the driver starts driving. Once the trip is underway, a fixed charge per kilometre is also made. The total fare is the sum of these two fares. For a particular taxi trip, the passenger pays a flag fall of \$4 and the charge per kilometre is \$2.50.

a Write an equation that describes the fare paid, F, in terms of the distance travelled, d.

b Draw a graph of this relationship.

c Calculate the slope of this graph. In terms of the taxi fare, what is the significance of the slope?

d With regard to the taxi fare, what is the significance of the vertical axis intercept?

e Use the graph to estimate the taxi fare for a 15 km trip and then confirm this algebraically.

Solution

This example brings together many of the aspects that have been covered in the previous sections.

a   The amount paid per kilometre is $2.50. So if the trip is d kilometres, that part of the fare will be $\$2.50 \times d$, or written algebraically $2.5d$. To this we add the flag fall of $4. This means that the equation that describes the fare, F, in terms of the distance travelled, d, is $F = 2.5d + 4$.

b   To draw the graph we will select some values for d to calculate the fare, F. Let's use $d = 0, 10$ and 20.

When $d = 0, F = 2.5 \times 0 + 4 = 4$

When $d = 10, F = 2.5 \times 10 + 4 = 25 + 4 = 29$

When $d = 20, F = 2.5 \times 20 + 4 = 50 + 4 = 54$

We now have three coordinate pairs: $(0, 4)$, $(10, 29)$, $(20, 54)$ that we can plot on the coordinate plane.

Note that F is the dependent variable, because it changes whenever d does, so F is graphed on the vertical axis and d, the independent variable, is graphed on the horizontal axis. We also need to choose an appropriate scale for each axis to create the graph.

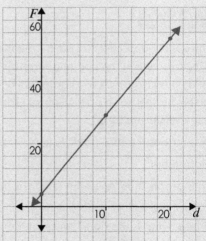

We can restrict our graph to the first quadrant on the coordinate plane because the values of d and F will always be positive.

c   Using the points $(0, 4)$ and $(10, 29)$, we can use the slope formula to calculate the slope by substituting d and F for x and y in our original formula.

$$m = \frac{F_2 - F_1}{d_2 - d_1} = \frac{29 - 4}{10 - 0} = \frac{25}{10} = 2.5$$

We could have also determined the slope from the equation that we established in (a), as it is the coefficient of d in that equation. The significance of the slope in this situation is that the fare increases by $2.50 for every kilometre travelled.

d The fare, F, is graphed on the vertical axis. The vertical axis intercept is the point $(0, 4)$, which means that the fare paid is $4 before any kilometres have been travelled.

e From the graph, we can estimate the fare for 15 kilometres to be $41.

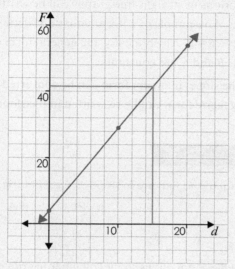

We can calculate this exactly by substituting $d = 15$ into our equation for the fare:

$F = 2.5d + 4$

$F = 2.5 \times 15 + 4$

$F = 37.5 + 4$

$F = 41.5$

So the exact fare is $41.50.

PRACTICE 17: APPLICATION

A home line telephone plan has a fixed monthly line rental charge of $24, and calls are billed on top of this at 30 cents each.

a Write an equation that describes the monthly phone bill, B, in terms of the number of calls made, C.

b Draw a graph of this relationship.

c Calculate the slope of this graph. In terms of the monthly bill, what is the significance of the slope?

d With regard to the monthly bill, what is the significance of the vertical axis intercept?

e Use the graph to estimate the monthly bill if 25 calls are made, and then confirm this algebraically.

14.5 EQUATIONS OF STRAIGHT LINES IN GENERAL FORM

General form of a
linear equation
A linear equation is
written in the form
$Ax + By + C = 0$,
where A, B and C
are constants

Another way that the equations of straight-line graphs can be written is using what is known as the general form. An example of a linear equation written in the **general form of a linear equation** is $3x + 4y + 6 = 0$. Algebraically, the general form of a linear equation is $Ax + By + C = 0$, where A, B and C are constants.

Equations written in the slope–intercept form can be converted to the general form, and likewise equations in the general form can be re-written in the slope–intercept form.

EXAMPLE 18

Convert the following equations that are in slope–intercept form into the general form.

a $y = 3x - 7$ 　　　　　　　　　　　　　b $2y = -5x + 4$

Making connections
To review the
algebraic strategies
of backtracking
and balancing to
isolate the unknown
revisit Chapter 13.

Solutions

We need to re-arrange these equations so that all the terms are on the LHS using the same backtracking and balancing techniques that we use when solving linear equations.

a

$$y = 3x - 7$$
$$y + 7 = 3x - 7 + 7 \quad \text{Undo the } -7 \text{ on the RHS by adding 7 to each side}$$
$$y + 7 = 3x$$
$$y + 7 - 3x = 3x - 3x \quad \text{Undo the } 3x \text{ on the RHS by subtracting } 3x \text{ from each side}$$
$$y + 7 - 3x = 0$$
$$-3x + y + 7 = 0 \quad \text{Re-arrange the terms on the LHS so that they are in the form } Ax + By + C = 0$$

b

$$2y = -5x + 4$$
$$2y - 4 = -5x + 4 - 4 \quad \text{Undo the } +4 \text{ on the RHS by subtracting 4 from each side}$$
$$2y - 4 = -5x$$
$$2y - 4 = -5x + 5x \quad \text{Undo the } -5x \text{ on the RHS by adding } 5x \text{ to each side}$$
$$2y - 4 + 5x = 0$$
$$5x + 2y - 4 = 0 \quad \text{Re-arrange the terms on the LHS so that they are in the form } Ax + By + C = 0$$

PRACTICE 18

Convert the following equations that are in slope–intercept form into the general form.

a $y = -4x + 6$ 　　　　　　　　　　　b $-5y = -x + 9$

EXAMPLE 19

Convert the following equations that are in general form into slope–intercept form.

a $5x + y - 6 = 0$

b $-4x - 3y + 12 = 0$

Solutions

We need to re-arrange these equations so that the x term and the constant term are on the RHS using backtracking and balancing techniques.

a

$$5x + y - 6 = 0$$

$$5x - 5x + y - 6 = 0 - 5x \qquad \text{Undo the } 5x \text{ on the LHS by subtracting } 5x \text{ from each side}$$

$$y - 6 = -5x$$

$$y - 6 + 6 = -5x + 6 \qquad \text{Undo the } -6 \text{ on the LHS by adding 6 to each side}$$

$$y = -5x + 6$$

The equation is now in slope–intercept form.

b

$$-4x - 3y + 12 = 0$$

$$-4x + 4x - 3y + 12 = 0 \qquad \text{Undo the } -4x \text{ on the LHS by adding } 4x \text{ to each side}$$

$$-3y + 12 = 4x$$

$$-3y + 12 - 12 = 4x - 12 \qquad \text{Undo the } +12 \text{ on the LHS by subtracting 12 from each side}$$

$$-3y = 4x - 12$$

$$\frac{-3y}{-3} = \frac{4}{-3}x - \frac{12}{-3} \qquad \text{Undo the multiplication by } -3 \text{ on the LHS by dividing each term by } -3$$

$$y = -\frac{4}{3}x + 4$$

This equation is now in slope–intercept form.

PRACTICE 19

Convert the following equations that are in general form into slope–intercept form.

a $-3x + y + 5 = 0$

b $2x + 5y - 10 = 0$

Kathy Brady

14.6 CHAPTER SUMMARY

SKILL OR CONCEPT	DEFINITION OR DESCRIPTION	EXAMPLES
Dependent variable	Variable in a linear relationship with a value that is determined by the other variable (the independent variable).	The growth of a plant is *dependent* upon the amount of fertiliser applied, which is the *independent* variable.
Independent variable	Variable in a linear relationship that can be any chosen value.	
Linear equation	An equation involving two variables that produces a straight line when it is graphed.	$y = 4x + 7$ $3x - 2y - 5 = 0$
Graphing from a table of values	Create a table of values by selecting appropriate values for x to be substituted in the equation. Plot the points to draw the straight line.	$y = 2x + 1$ | x	−1	0	1	2
y	−1	1	3	5	
y-intercept	The point where a graph crosses the y-axis, when $x = 0$.	
x-intercept	The point where a graph crosses the x-axis, when $y = 0$.	
Graphing using axis intercepts	Use an equation in the general form to determine the x- and y-intercepts to draw the graph.	$2x + y = 8$ When $x = 0, y = 8$ When $y = 0, x = 4$
Slope	A value, m that describes the steepness and direction of a straight line.	
Positive slope	Goes upwards from left to right.	Positive slope

SKILL OR CONCEPT	DEFINITION OR DESCRIPTION	EXAMPLES
Negative slope	Goes downwards from left to right.	Negative slope
Calculating slope	For two points on a line (x_1, y_1) and (x_2, y_2), $m = \dfrac{y\text{-step}}{x\text{-step}} = \dfrac{y_2 - y_1}{x_2 - x_1}$	For $(2, -4)$ and $(-1, 5)$ $m = \dfrac{y\text{-step}}{x\text{-step}}$ $= \dfrac{5 - (-4)}{-1 - 2}$ $= \dfrac{9}{-3}$ $= -3$
Horizontal line	If the slope of a straight line is 0, then the line is horizontal, and the y values of each point are equal.	
Vertical line	If the slope of a straight line is undefined, then the line is vertical, and the x values of each point are equal.	
Parallel lines	Two straight lines are parallel if their slopes are equal. If the slope of one line is m_1 and the slope of the other line is m_2, then $m_1 = m_2$.	 Slope of light-coloured line $m_1 = \dfrac{y_2 - y_1}{x_2 - x_1}$ $= \dfrac{5 - (-2)}{9 - (-5)}$ $= \dfrac{7}{14}$ $= \dfrac{1}{2}$ Slope of dark-coloured line $m_2 = \dfrac{y_2 - y_1}{x_2 - x_1}$ $= \dfrac{0 - (-4)}{8 - 0}$ $= \dfrac{4}{8}$ $= \dfrac{1}{2}$ $m_1 = m_2$

Kathy Brady

SKILL OR CONCEPT	DEFINITION OR DESCRIPTION	EXAMPLES
Perpendicular lines	Two straight lines are perpendicular if the product of their slopes is equal to -1. If the slope of one line is m_1 and the slope of the other line is m_2, then $m_1 \times m_2 = -1$.	 Slope of light-coloured line $$m_1 = \frac{y_2 - y_1}{x_2 - x_1}$$ $$= \frac{5 - (-2)}{1 - (-5)}$$ $$= \frac{7}{6}$$ Slope of dark-coloured line $$m_2 = \frac{y_2 - y_1}{x_2 - x_1}$$ $$= \frac{4 - (-2)}{-6 - 1}$$ $$= \frac{6}{-7}$$ $$m_1 \times m_2 = \frac{7}{6} \times \frac{6}{-7} = -1$$
Slope–intercept form	A linear equation written as $y = mx + c$, where m is the slope and c is the y-axis intercept at the point $(0, c)$.	$y = 6x - 5$ Slope $= m = 6$ y-intercept $= c = -5$
Point–intercept form	When the equation of a line written as $y - y_1 = m(x - x_1)$, where m is the slope of the line and (x_1, y_1) is a point on the line.	The slope of the line is 3 and it passes through $(4, -1)$ in point-intercept form: $y - (-1) = 3(x - 4)$
Converting point–intercept to slope–intercept form	Use algebraic strategies of undoing and backtracking to isolate the unknown, y, on the LHS of the equation	$y - (-1) = 3(x - 4)$ $y + 1 = 3x - 12$ $y + 1 - 1 = 3x - 12 - 1$ $y = 3x - 13$
Drawing lines from slope–intercept form	• Write the slope, m, as $\frac{y\text{-step}}{x\text{-step}}$ • Plot, c, the y-axis intercept • Starting at c, draw in the x-step and the y-step to create another point • Join the points with a straight line	Draw the graph of $y = \frac{2}{3}x + 3$ $$\frac{y\text{-step}}{x\text{-step}} = \frac{2}{3}$$ y-axis intercept $= 3$
General form	A linear equation is written as $Ax + By + C = 0$, where A, B and C are constants.	
Converting general form to slope–intercept form	Use algebraic strategies of undoing and backtracking to isolate the unknown, y, on the LHS of the equation	$4x - 5y - 15 = 0$ $4x - 4x - 5y - 15 = 0 - 4x$ $-5y - 15 + 15 = -4x + 15$ $-5y = -4x + 15$ $\frac{-5}{-5}y = \frac{-4}{-5}x + \frac{15}{-5}$ $y = \frac{4}{5}x - 3$

14.7 REVIEW QUESTIONS

A SKILLS

1 Draw the graph of $y = -2x + 5$ by first completing a table of values.

2 By determining the x- and y-intercepts, draw the graph of $4x + 5y = 20$.

3 State whether the slopes of the following graphs are positive, negative, zero or undefined.

a
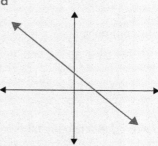

b
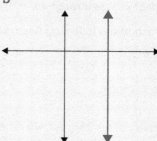

c
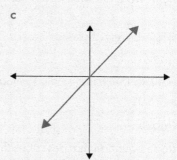

d
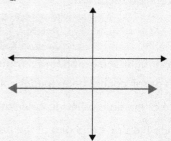

4 Calculate the slopes of the following.

a
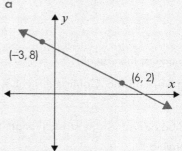

b A line that passes through $(2, -5)$ and $(-4, 1)$.

5 Determine which pairs of lines are parallel and which pairs of lines are perpendicular.

a $y = \frac{1}{2}x + 2$ b $2x + y - 2 = 0$

c $x - 2y - 4 = 0$ d $y = -2x - 1$

Kathy Brady

6 For the given slope and y-axis intercept, write the equation of the line.

SLOPE	y-AXIS INTERCEPT	EQUATION
-0.8	$(0, 2)$	
5	$(0, 2.5)$	
$\frac{5}{6}$	$(0, \frac{3}{2})$	
undefined	$(0, 0)$	

7 Find the equation, in slope–intercept form, of the following lines.

a The slope of the line is -5 and it passes through the point $(-1, 6)$.

b The line that passes through $(6, -11)$ and $(4, -3)$.

8 Draw the graph of the following lines using the slope and y-intercept only.

a $y = -5x + 3$ **b** $y = \frac{3}{4}x - 2$ **c** $2x + 3y - 3 = 0$

B   APPLICATIONS

1 At the beginning of this chapter, we read about Vanessa who was conducting an experiment measuring the stretched length of a spring when various weights were added. Here is the data that Vanessa recorded.

Weight (kg)	1	2	5	10
Stretch length (cm)	4.5	6.5	12.5	22.5

a Draw a straight-line graph for this data, extending the graph to touch the y-axis. Explain why the graph cannot be extended into the negative quadrants.

b Use two of the data points to calculate the slope of this graph.

c Substitute the slope and one of the points into a point–slope equation that describes the relationship between the stretch length, S, in terms of the weight added, W.

d Re-arrange the point–slope equation into slope–intercept form.

e What is the value of c, in slope–intercept equation?

f Does the graph cross the y-axis at this value?

g With regard to Vanessa's experiment, what is the meaning of this value?

h Use your graph to determine the stretch length if 15 kg were added. Confirm this value by substituting into the formula.

i Use your graph to determine how much weight should be added for the stretch length to be 10 cm. Confirm this value by substituting into the formula.

2   The equation that is used to convert degrees in Celsius to degrees in Fahrenheit is:

$$F = \tfrac{9}{5}C + 32$$

a   What are the dependent and independent variables in this equation?

b   Against which axis will each of these variables be graphed?

c   What is the slope of the graph of this equation? What is the vertical axis intercept?

d   Choose an appropriate scale to draw a straight-line graph of this equation.

e   Explain why the graph can be extended into the negative quadrants.

f   With regard to this relationship, what is the meaning of the value where the graph crosses the vertical axis?

g   At what value does the graph cross the horizontal axis? What is the meaning of this value?

h   Use the graph to determine degrees in Fahrenheit equivalent to 10°C. Confirm this value algebraically.

i   Use the graph to determine the degrees in Celsius equivalent to 20°F. Confirm this value algebraically.

15

STATISTICS

CHAPTER OBJECTIVES

- » Understand what statistics is and why it is useful
- » Calculate mean, median and mode and know when to use which one
- » Calculate range, interquartile range and standard deviation and understand the information they present

---

LOOKING FOR A REAL RELATIONSHIP

Statistics helps us work out whether what seems like a relationship between two variables is a real relationship, or just coincidence, or even perhaps that two variables are connected by a third variable. For example, people with low socio-economic status are more likely to die from lung cancer than people with greater wealth. Does this mean that having less money causes lung cancer?

No! Rather, people from lower socio-economic backgrounds are more likely to smoke cigarettes than those from wealthier backgrounds. It is smoking cigarettes that causes cancer, not socio-economic status. Statistical analysis enables the relevant data to be analysed to find out whether a relationship between two variables, in this case, socio-economic background and dying from lung cancer, is causal. The statistical analysis shows that being from a low socio-economic background does not cause lung cancer; smoking causes lung cancer.

15.1 WHAT IS STATISTICS?

According to the Oxford English Dictionary (Source: oed.com) statistics involves the collection, arrangement, analysis and interpretation of numerical data. Descriptive statistics refers to the analysis of patterns in a particular data collection. Inferential statistics refers to analysis that allows us to infer or deduce patterns in a whole data set based on analysis of a representative sample.

Statistical analysis often involves large quantities of data. It allows us to find information that would otherwise be hidden. It allows us to predict future trends from existing data. Statistics helps us answer questions like:

- How does a streaming media service know what movies you will be interested in?
- Does mobile phone usage increase your risk of brain cancer?
- Has the result of a sporting contest been fixed ahead of time?
- What is causing the increase of allergies in children in some countries like Australia?

EXAMPLE 1: APPLICATION

Here is the end of the 2015 season ladder and associated data (Source: footywire.com) for the Australian Football League (AFL). What interesting questions might a detailed statistical analysis of this data enable us to answer?

RANKING	OVERALL SEASON	NUMBER OF KICKS	NUMBER OF HANDBALLS	NUMBER OF DISPOSALS
1	Fremantle	Hawthorn	Sydney Swans	Hawthorn
2	West Coast Eagles	West Coast Eagles	Hawthorn	Sydney Swans
3	Hawthorn	Fremantle	West Coast Eagles	West Coast Eagles
4	Sydney Swans	Sydney Swans	Fremantle	Fremantle
5	Richmond	North Melbourne	North Melbourne	North Melbourne
6	Western Bulldogs	Adelaide Crows	Western Bulldogs	Western Bulldogs
7	Adelaide Crows	Richmond	Essendon	Richmond
8	North Melbourne	Western Bulldogs	GWS Giants	Essendon
9	Port Adelaide	Essendon	Collingwood	Adelaide Crows
10	Geelong Cats	Collingwood	Port Adelaide	Collingwood
11	GWS Giants	St Kilda	Geelong Cats	GWS Giants
12	Collingwood	Port Adelaide	Brisbane Lions	Port Adelaide
13	Melbourne	GWS Giants	St Kilda	St Kilda
14	St Kilda	Brisbane Lions	Richmond	Brisbane Lions
15	Essendon	Carlton	Adelaide Crows	Geelong Cats
16	Gold Coast Suns	Gold Coast Suns	Melbourne	Carlton
17	Brisbane Lions	Geelong Cats	Carlton	Melbourne
18	Carlton	Melbourne	Gold Coast Suns	Gold Coast Suns

Solution

Detailed statistical analysis would allow us to see whether the apparent relationship between the number of kicks and position on the AFL ladder is a 'real' relationship or simply a matter of chance. The same questions could be answered for position on the ladder and number of handballs, and position on the ladder and number of disposals. If there turns out to be a relationship between all of the statistics and position on the ladder, further analysis could also tell us which statistic—kicks, handballs, or disposals— has the most influence on ladder position. These questions are not the only interesting ones that might be answered; you may be able to think of others.

PRACTICE 1: APPLICATION

Lilly manages a pizza shop. She is considering two venues for a possible new location. Venue A is larger, next to a liquor store, with a few businesses around the area, and a gym a couple of streets away. Venue B is smaller, with a few businesses next door and a high school only a block away. Which venue would you choose if you were Lilly? What data would you need in order to decide?

CRITICAL THINKING

The GINI index is a measure of disparity of wealth in a given country (Source: Investopedia.com). The GINI index ranges from 0 to 1. A GINI index of 0 indicates that every person in a country has equal wealth, whereas an index of 1 indicates that one person has all the wealth of a given country. Recent GINI index values for several countries include 0.394 (United States, 2014), 0.281 (Sweden, 2013), and 0.465 (Chile, 2013) (Source: www.stats.oecd.org). What other variables might the GINI index be related to?

Attempting to find out the answer to a question using statistics requires the following steps or stages:

1. working out how to obtain the data you need
2. collecting the data
3. presenting and analysing the data
4. interpreting the data and drawing conclusions.

The next section focuses on Step 1: working out how to obtain the data that you need. The remaining sections in this chapter cover aspects of Steps 2, 3 and 4 above.

15.2 SAMPLING TECHNIQUES

One way of obtaining the data you need for a statistical investigation is to involve directly each person or unit in your **population**.

Examples of populations include all of a person's red blood cells, all of the university graduates in Australia or all of the volcanoes in the southern hemisphere. Conducting a statistical investigation by directly involving every unit in a whole population is a census. However, often it is impractical to investigate every unit in a population; it may be too time-consuming or expensive. If this is the case, a **sample** of the population is used for investigative purposes.

For example, it is virtually impossible to measure the level of haemoglobin in every blood cell in a person's body, so a sample is taken, and the level of haemoglobin in every cell is estimated using the sample. A rough guide is that for a population of size n a sample size of $\sqrt{n}$ will be large enough.

In order to accurately represent the whole population, a sample must be a representative sample, and it must accurately represent the variability in the population. For example, if studying the ethnic backgrounds of students, a collection of students from only one university may not accurately represent the backgrounds of students from all of the universities in a country. This means that the sample must be both **random** and **independent**.

For example, to choose a random sample of a group of people, we would put each person's name in a hat and pull out a sample of names, or at university, since each student has an ID number a random number generator on a computer can be used to select students randomly. On the other hand, a sample is not random if only people who are friends of the person taking the sample are given surveys to fill in. A sample is not independent if one person is given five copies of a survey—one for themselves and one to distribute to four other people.

In reality, it can be difficult to generate random and independent samples so other methods are used and checks put in place to make sure the sample approximates a random sample. Five other sampling methods that are commonly used are stratified random sampling, quota sampling, systematic sampling, judgement sampling and convenience sampling.

Stratified random sampling

Random sampling can still result in a biased or non-typical sample; for example, in a sample of 40 university students, it would be possible for all students to be first-year students. To overcome this, a sample can be stratified, or divided into categories, and a sample from each stratum (or category) is randomly collected. The size of the sample in each stratum is based on the size of the stratum in the population.

Tiffany Winn

Population
A group of events, objects, results or individuals, all of which share some unifying characteristic

Sample
A portion of a population that represents the whole population

Random sample
A sample where all members of the population have an equal chance of selection

Independent sample
A sample where the choice of one member does not influence the choice of another

Quota sampling

In a quota sample, the sample is stratified but responses are collected using any convenient method rather than randomly. For example, patrons of a café might be approached until each stratum has been filled.

Systematic sampling

In systematic sampling, the sample selection occurs in an organised way. For example, every fortieth student in an alphabetical list of students is chosen.

Judgement sampling

Judgement sampling involves the use of a sample that the data collector believes accurately represents the population. For example, selecting a tutorial group from a topic that you believe most accurately represents the students in the topic.

Convenience sampling

Finally, convenience sampling as the name suggests simply involves selecting a sample in the most convenient way possible. For example, the first forty people you see in a café are selected for your survey on the food from the café.

EXAMPLE 2

Which of the following represents simple random sampling?

A   Selecting volunteers from a particular topic

B   Listing people according to their country of origin and choosing a proportion from within each country of origin at random

C   Using a computer program to generate random numbers to guide the selection of participants

D   Selecting schools randomly and then taking everyone within the school as the sample

....

Solutions

The third option (C) is the only one that is a simple random sampling technique. The second option (B) is an example of stratified random sampling. The first option (A) is an example of judgement sampling.

PRACTICE 2

Which of the following is an example of stratified random sampling?

A Using a random number generator to choose which bank to open an account with

B Splitting university students into groups according to their country of origin and taking a random sample with the proviso that each country of origin group has a certain number of members, according to the percentage of their group in the whole population

C Pulling names out of a hat at random

D Selecting schools randomly and then taking everyone within the school as the sample

EXAMPLE 3: APPLICATION

A questionnaire was distributed to find out which skills Australian university graduates had developed while at university that they believed were relevant for their current employment.

In this example, the population is:

A the skills set typically acquired at university

B Australian university students

C a survey

D Australian university graduates

Solution

Option (D), Australian university graduates, is the population. Option (B) would be incorrect because the survey was conducted to find out about skills of 'Australian university graduates'.

PRACTICE 3: APPLICATION

A researcher investigated the effectiveness of a given learning program to prepare 100 fifth grade students from five different schools for the NAPLAN test. What is the sample?

A Fifth grade students

B Five different schools

C Australian fifth grade students

D 100 fifth grade students from five different schools

15.3 MEASURES OF CENTRAL TENDENCY

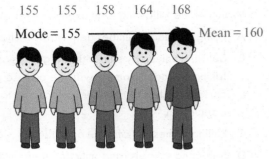

155   155   158   164   168

Mode = 155 ——————— Mean = 160

Median = 158

If Jordan was manufacturing clothing and needed some kind of measure of the middle or average value of the height of males, to guide his clothing sizes, what measure would he use?

There are three commonly used **measures of central tendency**: mean, median and mode. Each is appropriate in different situations. We will explore each measure and once we have done that, we will be able to say which measure is most appropriate for our clothing manufacturer.

Mean

The **mean** is the average of a collection of numbers. To calculate the mean add up all the values in the set and divide by the total number of values. The mean is represented by either $\bar{x}$ or the Greek letter μ.

In maths language, the mean is written as follows:

$$\bar{x} = \frac{\sum_{i=1}^{n} x_i}{n}$$

The line over the x means average

Sum of data values (Greek letter $\sum$ means 'sum of')

The symbol n is used for the total number of data values

While this may initially appear confusing, it is important to remember that this is just technical language for describing how to calculate the mean: add up all the values and divide by the number of values. The Greek symbol $\sum$ means the 'sum of'. The small notations following $\sum$, which in the figure are $i = 1$ at the bottom and n at the top, together with the x_i, tell us that we want to apply the 'sum of' (add up) to all of the values that we have, from the first value (1) up to the last value (n). Each number that we have is given a symbol: $x_1, x_2, x_3,$ and so on, up to the last value x_n. For example, if there are eight numbers to take the average of, then $n = 8$ and the last value is x_8. Once we have added up all the values, we divide by the number of values (n). For example, if there are eight numbers to take the average of, then $n = 8$ and we divide by 8.

EXAMPLE 4: APPLICATIONS

a   In a group of ten males, the heights of each person in centimetres are as follows: 165, 180, 162, 190, 186, 183, 177, 167, 165 and 170. What is the average height of this group of males?

b   If the maximum temperature over the past five days has been 27°, 31°, 34°, 32° and 29°, what is the average (mean) maximum temperature?

Solutions

a   First, add up all the values:

$$165 + 180 + 162 + 190 + 186 + 183 + 177 + 167 + 165 + 170 = 1745$$

Then, divide by the total number of values to get the mean:

$$\text{mean } (\mu) = \frac{1745}{10} = 174.5 \text{ cm}$$

The average height of the group of males is 174.5 cm.

b   First, add up all the values:

$$27 + 31 + 34 + 32 + 29 = 153$$

Then, divide by the total number of values to get the mean:

$$\text{mean } (\mu) = \frac{153}{5} = 30.6°C$$

The average maximum temperature over the five days is 30.6°C.

PRACTICE 4: APPLICATIONS

a   In a dozen eggs, the weight in grams of each egg is: 57, 53, 52, 61, 50, 62, 51, 52, 60, 62, 53 and 50.

What is the average weight of these eggs? If eggs are sold by the dozen in cartons as: large (minimum weight 600 g); extra-large (minimum weight 700 g); and jumbo (minimum weight 800 g), in which of these categories will this dozen eggs fall?

b   Eight students sit a quiz and receive the following percentages: 72, 54, 40, 88, 60, 65, 55 and 70. What is the average percentage received for the quiz?

Median

The **median** is the middle value of a data set, when the data is in order from smallest to largest. When there are an odd number of values in the set, the median is the exact middle value. For an even number of data values, the median is the mean of the two middle values.

Median
The middle value in a data set

EXAMPLE 5: APPLICATION

Five houses sell for the following values: $2 235 000, $425 000, $332 400, $254 300 and $537 600. What is the median sale price?

Solution

First, arrange the values in order from smallest to largest:

$254 300	$332 400	$425 000	$537 600	$2 235 000

Then, cross off values in pairs, one from each end, until there are only one or two values left:

First step:	~~$254 300~~	$332 400	$425 000	$537 600	~~$2 235 000~~
Second step:	~~$254 300~~	~~$332 400~~	$425 000	~~$537 600~~	~~$2 235 000~~

↑ Median

Since there is only one value left, this value is the median. The median sale price is $425 000.

PRACTICE 5: APPLICATION

A baker makes up bags of doughnuts but varies the number of doughnuts in each bag, adding extra doughnuts if the doughnuts appear smaller. The number of doughnuts in each bag is: 12, 10, 9, 10, 12, 11, 11, 12 and 10. What is the median number of doughnuts per bag?

EXAMPLE 6: APPLICATION

The weekly salaries of eight employees at a coffee shop are $190, $140, $220, $90, $180, $140, $110 and $200. What is the median salary?

Solution

First, arrange the values in order from smallest to largest:

$90	$110	$140	$140	$180	$190	$200	$220

Then, cross off values in pairs, one from each end, until there are only one or two values left. This time there is an even number of values so there are two values left:

First step:	~~$90~~	$110	$140	$140	$180	$190	$200	~~$220~~
Second step:	~~$90~~	~~$110~~	$140	$140	$180	$190	~~$200~~	~~$220~~
Third step:	~~$90~~	~~$110~~	~~$140~~	$140	$180	~~$190~~	~~$200~~	~~$220~~

Since there are two values left, the median is the mean of those two values.

$$\text{Median} = \frac{140 + 180}{2} = \$160$$

The median salary is $160 per week.

PRACTICE 6: APPLICATION

Anna is advertising her investment agency. She has recorded how much her last 10 customers have made from their $1000 investments. The amounts her last ten customers have made are as follows: 90, 88, 68, 101, 300, 124, 500, 90, 79 and 60. Which average (mean or median) should she use on her website to attract more customers?

Mode

The **mode** is often used as the measure of central tendency with non-numerical data and other situations where it does not make sense to calculate a mean or a median. The mode is the most frequently occurring value in a data set. If more than one value occurs most frequently, then each of these values is known as a modal value and the set of data is said to be multimodal. Specifically, a data set with two modal values is bi-modal.

Mode
The most frequently occurring value in a data set

EXAMPLE 7: APPLICATION

In a class of fifteen overseas students, country of origin is as follows: China, South Korea, India, United States, United Kingdom, China, China, Malaysia, India, China, India, China, Brazil, Malaysia and South Korea.

Solution

Count the number of people from each country and arrange the data in a table.

COUNTRY OF ORIGIN	COUNT
Brazil	1
China	5
India	3
Malaysia	2
South Korea	2
United Kingdom	1
United States	1

Since the country with the greatest number of students represents the mode, the mode is China with 5 students.

PRACTICE 7: APPLICATION

When considering the twelve months in a calendar year, what is the modal value for the number of days in a month?

EXAMPLE 8: APPLICATION

A department store has tracked their sales of women's t-shirts by size over the past six months. Which size represents the mode for t-shirts sold? Why would the department store be interested in this information? If a mean value were calculated using the count column in the table, what would this number represent? Would it make sense to calculate this mean value?

SIZE	COUNT
Extra small (XS)	72
Small (S)	123
Medium (M)	123
Large (L)	64
Extra large (XL)	33

Solution

Size small (S) and size medium (M) both have the most t-shirts sold so they are both modal values; the distribution of sizes is bi-modal. A department store could use this information in buying stock for future months. A mean value would represent the average number of t-shirts sold per size (across all sizes) in the store. It would not make sense to calculate this mean value because it is unlikely to be useful; knowing the individual values per size is more helpful from the point of view of managing stock in the store.

PRACTICE 8: APPLICATION

A company is analysing feedback from their customers as to their service quality. Customers have replied to a survey question with a number from 1 to 4 where 1 = excellent, 2 = good, 3 = average, and 4 = poor. Responses from thirty customers are as follows: 4, 3, 3, 2, 4, 1, 3, 2, 1, 2, 2, 2, 1, 1, 2, 3, 2, 3, 2, 3, 2, 2, 2, 1, 1, 4, 3, 2, 1 and 3. What is the modal rating for service from these thirty customers?

15.4 MEASURES OF VARIABILITY

Many times, summarising data by using a middle value does not give enough information. For example, consider the following sets of numbers:

   4, 4, 4, 4, 4   4, 3, 5, 5, 3   7, 4, 4, 2, 3

Each of these sets of numbers has the same average, 4, yet the spread of numbers included in each of the sets is quite different.

For Jordan, a clothing manufacturer, just knowing the average height of Australian males, 175 cm, does not give him enough information to make his clothing. If we were to look at a pair of jeans, the length of an inseam on a pair of pants is, on average, 45% of the person's overall height. Since 45% of 175 cm is 78.75 cm, Jordan knows that on average, the length of the inseam of a pair of men's jeans will need to be 78.75 cm. However, in using just the average value Jordan cannot tell whether, for example, three-quarters of all males are within a centimetre or two of 175 cm, or within three to four centimetres, or within seven to eight centimetres. Knowing this information will change the range of different sizes of jeans that Jordan makes, and the quantity he makes in each size. Therefore, both the average value and a measure of variability are important values for summarising data. Three common measures of variability are the range, interquartile range and standard deviation. In this section, we will explore each measure and then we will be able to say which measure is most appropriate for our clothing manufacturer.

Range

The **range** is the simplest and most general measure of variability. It is calculated by subtracting the smallest data value from the largest data value.

Range
The difference between largest and smallest values in a set of data

EXAMPLE 9

A class of seven students completed a quiz and received the following scores:

63, 77, 52, 89, 90, 46 and 70

What is the range of scores?

Solution

The lowest quiz score is 46 and the highest is 90, so the range of quiz scores is $90 - 46 = 44$.

PRACTICE 9

A class of eight students completed a quiz and received the following scores:

64, 87, 62, 49, 90, 56, 69 and 70

What is the range of scores?

EXAMPLE 10: APPLICATION

In a hospital, each patient's pulse rate is taken three times a day. Patient Tom's pulse is 72, 76 and 74, while patient Bill's is 72, 91 and 59. What is the pulse range for each patient? What is the mean for each patient?

Solution

The lowest value for Tom's pulse is 72 and the highest 76, so the range is $76 - 72 = 4$. The lowest value for Bill's pulse is 59 and the highest 91, so the range is $91 - 59 = 32$. The average (mean) for Tom's pulse is $\frac{72 + 76 + 74}{3} = 74$. The average (mean) for Bill's pulse is $\frac{72 + 91 + 59}{3} = 74$. Knowing the range as well as the average tells us that Bill's pulse has been a lot more variable than Tom's. This may indicate some health issues for Bill that Tom does not have, despite both their average pulses being the same.

Tiffany Winn

Interquartile range

Interquartile range (IQR)
The difference between the upper (Q3) and the lower quartile (Q1) of a data set

The second measure of variability that we will consider is the **interquartile range**. A quartile is one of three points that divides a collection of data into four equal groups. To understand what a quartile is, think of dividing your data into quarters; the quartiles are the boundary points between the groups. We label the lower quartile Q1, the middle quartile Q2 and the upper quartile Q3.

Consider the data set 170, 165, 149, 165, 155, 157, 156, 199, 167, 176, 168 and 170. If we arrange this set in order from lowest to highest, we get:

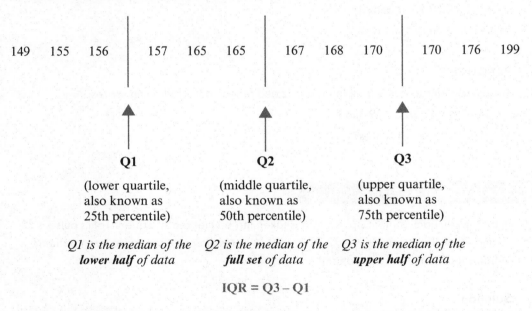

149   155   156 | 157   165   165 | 167   168   170 | 170   176   199

Q1
(lower quartile, also known as 25th percentile)

Q2
(middle quartile, also known as 50th percentile)

Q3
(upper quartile, also known as 75th percentile)

*Q1 is the median of the **lower half** of data*   *Q2 is the median of the **full set** of data*   *Q3 is the median of the **upper half** of data*

IQR = Q3 − Q1

To calculate the IQR in the example above, note that the middle line represents Q2, the median of the full set of data. The line to the left represent Q1, the median of the bottom half of the data, and the line to the right Q3, the median of the top half of the data.

$Q2 = \frac{165 + 167}{2} = 166$ (remember that where a median falls between two data values it is equal to the mean of those two values)

$Q1 = \frac{156 + 157}{2} = 156.5$ $Q3 = \frac{170 + 170}{2} = 170$

$IQR = Q3 - Q1$

$= 170 - 156.5$

$= 13.5$

The IQR provides a measure of variability that, unlike the range, is not skewed by extreme values. For example, in the set above, the range would be $199 - 149 = 50$ and this is much larger than it would otherwise be because of the value 199 that is more than 20 units greater than any other value in the data. The IQR is much smaller at 13.5 because it is the range of data between the two solid black lines, so it excludes any extreme values at the bottom or top of the collection of data values. It gives a more accurate picture of the overall variability in most of the data.

EXAMPLE 11: APPLICATION

A class of nine students completed a quiz and received the following scores:

63, 77, 52, 89, 90, 46, 30, 64 and 70

What is the IQR of the data? How does it compare to the range?

Solution

First, arrange the data in order and divide it into quartiles:

| 30 | 46 | 52 | 63 | 64 | 70 | 77 | 89 | 90 |

Q1   Q2   Q3

Since there are an odd number of data values, Q2 falls exactly on the middle data point ($Q2 = 64$). Since Q2 falls exactly on a data point rather than between two data points, the Q2 value (64) is excluded from the collection of values used to calculate Q1 and Q3.

Q1 is the middle value of 30, 46, 52 and 63, $Q1 = \frac{46 + 52}{2} = 49$

Q3 is the middle value of 70, 77, 89 and 90, $Q3 = \frac{77 + 89}{2} = 83$

$IQR = Q3 - Q1$

$= 83 - 49$

$= 34$

The range of the data is $90 - 30 = 60$, which is much bigger than the IQR.

PRACTICE 11: APPLICATIOIN

The birth lengths of a group of babies are as follows (in cm):

    50, 47, 52, 48, 50, 51, 54, 49 and 50

What is the IQR for this collection of values? How does it compare to the range?

EXAMPLE 12: APPLICATION

A class of eight students completed a quiz and received the following scores:

    63, 77, 52, 89, 46, 30, 64 and 70

What is the IQR of the data? How does it compare to the range?

Solution

First, arrange the data in order and divide it into quartiles:

Since there is an even number of data values, Q2 falls in between the middle two data points $Q2 = \frac{63 + 64}{2} = 63.5$. Since Q2 falls in between the middle two data points rather than exactly on a middle data point, the two middle values (63 and 64) are included in the data used to calculate Q1 and Q3.

Q1 is the middle value of 30, 46, 52 and 63, which makes $Q1 = \frac{46 + 52}{2} = 49$.

Q3 is the middle value of 64, 70, 77 and 89, which makes $Q3 = \frac{70 + 77}{2} = 73.5$.

$$IQR = Q3 - Q1$$
$$= 73.5 - 49$$
$$= 24.5$$

The range of the data is $89 - 30 = 59$, which is much bigger than the IQR.

PRACTICE 12: APPLICATION

The recorded daily maximum temperatures, in Celsius, for Adelaide over 12 days in January are as follows:

    27, 30, 38, 36, 35, 29, 29, 33, 38, 40, 36 and 32

What is the IQR for this collection of values? How does it compare to the range?

EXAMPLE 13: APPLICATION

The recorded daily maximum temperatures, in Celsius, for Adelaide over one week in July are as follows:

17, 19, 18, 16, 15, 19 and 16

What is the IQR for this collection of values? How does it compare to the range?

Solution

Since there are an odd number of data values, Q2 falls exactly on the middle data point (Q2 = 17). Since Q2 falls exactly on a data point rather than between two data points,

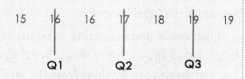

the Q2 value (17) is excluded from the collection of values used to calculate Q1 and Q3. Q1 is the middle value of 15, 16 and 16, which means Q1 falls exactly on 16 (Q1 = 16). Similarly, Q3 is the middle value of 18, 19 and 19, so Q3 = 19.

$$IQR = Q3 - Q1$$
$$= 19 - 16$$
$$= 3$$

The range of the data is 19 − 15 = 4, which is not much larger than the IQR, but as there are no extreme values in this data set, the difference between the IQR and range is not as marked as in the other example questions.

PRACTICE 13: APPLICATION

Ten houses have sold in Bedford Park in the last week for the following prices:

$410 000, $389 000, $476 000, $512 000, $350 000, $960 000 $432 000, $543 000, $529 000 and $499 000

What is the IQR for these selling prices? How does it compare to the range?

Standard deviation

Another commonly used measure of variability in a data set is the **standard deviation**. While the IQR tells you about the middle 50% of the data, the standard deviation tells you about the spread of all the data points. Since the IQR and standard deviation measure different things, they can be different for the same set of data. For example, consider these two data sets:

Set A = 1, 1, 1, 1, 1, 1, 1
Set B = 1, 1, 1, 1, 1, 1, 1000

Standard
deviation
A measure of
the average
distance of *all*
data points from
the mean and
is represented
by the Greek
letter σ

The IQR for both data sets is $1 - 1 = 0$, whereas the standard deviation for the first set is zero (all values are at the mean value) and for the second set is approximately 350. The IQR is based on two data points: the medians of the top and bottom halves of the data. It captures the variability in the middle 50% of data. On the other hand, the standard deviation is based on all the data points and captures variance in all the data. For this reason, the standard deviation is sensitive to extreme values, and this needs to be considered when deciding if the standard deviation is an appropriate measure of variability. The standard deviation is especially useful for data that has a normal distribution, which we will discuss in Section 15.6 of this chapter.

One way of understanding standard deviation is to think about shots at a target. Each target below shows twelve shots (in dark green) at a target. The green circle indicates the average distance of all shots from the target. In both cases, the average (mean) distance of all shots from the target is the same. However, for the target on the left, there are a number of shots very close to the target and others quite some distance away. In contrast, for the target on the right, there are fewer shots very close to the target and the shots that are furthest from the target are not as far away as the target on the left. This means there is less variation in the distance from the target on the right, and there is a smaller standard deviation for the shots on the right target than for those on the left.

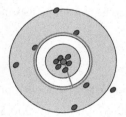 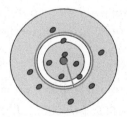

Trying to calculate a measure of the average distance from the mean is tricky, because values that are less than or more than the mean can cancel out each other. For example, if the mean is 6, and there are two values at 10 (4 more than the mean) and another two values at 2 (4 less than the mean), then the distance from the mean for those four values will be $+4$, $+4$, -4 and -4. Adding these together gives zero: $4 + 4 - 4 - 4 = 0$. For this reason, the steps to calculate an average distance from the mean needed for the standard deviation are a bit trickier than just adding and subtracting numbers.

There are five steps required to calculate the standard deviation for a set of data. Here we demonstrate these steps for two different data sets.

Data set A = 3, 4, 2, 5, 6

1   Calculate the mean of the data:

$$\text{Mean } (\mu) = \frac{3 + 4 + 2 + 5 + 6}{5}$$
$$= \frac{20}{5}$$
$$= 4$$

2   Calculate the square of the distance from the mean for each data value:

$(3 - 4)^2 = (-1)^2 = 1$

$(4 - 4)^2 = 0^2 = 0$

$(2 - 4)^2 = (-2)^2 = 4$

$(5 - 4)^2 = 1^2 = 1$

$(6 - 4)^2 = 2^2 = 4$

3   Sum the squares calculated in Step 2:

$1 + 0 + 4 + 1 + 4 = 10$

4   Divide the result from Step 3 by the number of data values: $\frac{10}{5} = 2$

5   Take the square root of the result in Step 4 to get the standard deviation (you will need a calculator for this step):

Standard deviation $(\sigma) = \sqrt{2} = 1.41$ (2 d.p.)

This means that for the set of data values 3, 4, 2, 5 and 6, the average distance of all points from the mean is 1.41 units.

Now consider the data set:

B = 3, 4, 4, 5, 4

1   Calculate mean of the data set:

$$\text{Mean } (\mu) = \frac{3 + 4 + 4 + 5 + 4}{5}$$
$$= \frac{20}{5}$$
$$= 4$$

2   Calculate the square of the distance from the mean for each data value:

$(3 - 4)^2 = (-1)^2 = 1$

$(4 - 4)^2 = 0^2 = 0$

$(4 - 4)^2 = 0^2 = 0$

$(5 - 4)^2 = 1^2 = 1$

$(4 - 4)^2 = 0^2 = 0$

3   Sum the squares: $1 + 0 + 0 + 1 + 0 = 2$

4   Divide by number of data values: $\frac{2}{5} = 0.4$

5   Take square root: $\sqrt{0.4} = 0.63$ (2 d.p.)

Standard deviation $(\sigma) = 0.63$

EXAMPLE 14

Consider the data sets A = 4, 3, 5, 3, 5 and B = 4, 5, 4, 3, 4. Which data set would you expect to have the greater standard deviation and why?

Solution

You would expect the first data set to have a greater standard deviation because while the mean of each data set is 4, the first data set has more values further away from the mean than the second data set.

PRACTICE 14

Consider the data sets {7, 4, 3, 5, 2, 3, 5, 9, 8, 3} and {2, 4, 5, 4, 6, 5, 3, 4, 9, 4}. Which data set would you expect to have the greater standard deviation and why?

EXAMPLE 15

Calculate the standard deviation of the data set 3, 4, 5, 1, 7.

Solution

1   Calculate the mean of the data:

$$\text{Mean } (\mu) = \frac{3 + 4 + 5 + 1 + 7}{5}$$
$$= \frac{20}{5}$$
$$= 4$$

2   Calculate the square of the distance from the mean for each data value:

$(3 - 4)^2 = (-1)^2 = 1$

$(4 - 4)^2 = 0^2 = 0$

$(5 - 4)^2 = 1^2 = 1$

$(1 - 4)^3 = (-3)^2 = 9$

$(7 - 4)^2 = 3^2 = 9$

3   Sum the squares calculated in Step 2:

$1 + 0 + 1 + 9 + 9 = 20$

4   Divide the result from Step 3 by the number of data values:

$$\frac{20}{5} = 4$$

5   Take the square root of the result in Step 4 to get the standard deviation:

Standard deviation $(\sigma) = \sqrt{4} = 2$

This means that for the set of data values 3, 4, 5, 1 and 7, the average distance of all points from the mean is 2 units.

PRACTICE 15: GUIDED

Calculate the standard deviation of the data set 4, 6, 5, 1, 8.

1 Calculate the mean of the data:

$$\text{Mean } (\mu) = \frac{4 + 6 + 5 + 1 + 8}{5}$$

$$= \frac{\square}{5}$$

$$= 4.8$$

2 Calculate the square of the distance from the mean for each data value:

$$(4 - 4.8)^2 = (-0.8)^2 = 0.64$$

$$(6 - \square)^2 = (1.2)^2 = 1.44$$

$$(\square - 4.8)^2 = (0.2)^2 = \square$$

$$(1 - 4.8)^2 = (\square)^2 = 14.44$$

$$(\square - \square)^2 = (3.2)^2 = \square$$

3 Sum the squares calculated in Step 2:

$$\square + \square + 0.04 + \square + 10.24 = \square$$

4 Divide the result from Step 3 by the number of data values:

$$\frac{26.8}{\square} = 5.36$$

5 Take the square root of the result in Step 4 to get the standard deviation:

$$\text{Standard deviation } (\sigma) = \sqrt{\square} = 2.32 \text{ (2 d.p.)}$$

Now try this one:

Calculate the standard deviation of the data set 3, 5, 2, 1, 6.

EXAMPLE 16: APPLICATION

Two tutorial classes sat the same exam. Each tutorial class had ten students. Scores, out of 10, were as follows:

Class 1: 9, 4, 8, 4, 6, 6, 6, 5, 5 and 7

Class 2: 3, 7, 6, 7, 4, 4, 8, 8, 8 and 5

Which class had the greater variability in scores, as measured by the standard deviation?

Solution

First, calculate the standard deviation for Class 1:

1 Calculate the mean of the data:

$$\text{Mean } (\mu) = \frac{9 + 4 + 8 + 4 + 6 + 6 + 6 + 5 + 5 + 7}{10}$$

$$= \frac{60}{10}$$

$$= 6$$

2   Calculate the square of the distance from the mean for each data value:

$(9 - 6)^2 = 3^2 = 9$

$(4 - 6)^2 = (-2)^2 = 4$

$(8 - 6)^2 = 2^2 = 4$

$(4 - 6)^2 = (-2)^2 = 4$

$(6 - 6)^2 = 0^2 = 0$

$(6 - 6)^2 = 0^2 = 0$

$(6 - 6)^2 = 0^2 = 0$

$(5 - 6)^2 = (-1)^2 = 1$

$(5 - 6)^2 = (-1)^2 = 1$

$(7 - 6)^2 = 1^2 = 1$

3   Sum the squares calculated in Step 2:

$9 + 4 + 4 + 4 + 0 + 0 + 0 + 1 + 1 + 1 = 24$

4   Divide the result from Step 3 by the number of data values:

$\frac{24}{10} = 2.4$

5   Take the square root of the result in Step 4 to get the standard deviation:

Standard deviation $(\sigma)^2 = \sqrt{2.4} = 1.55$ (2 d.p.)

Next, calculate the standard deviation for Class 2:

1   Calculate the mean of the data:

Mean $(\mu) = \frac{3 + 7 + 6 + 7 + 4 + 4 + 8 + 8 + 8 + 5}{10} = \frac{60}{10} = 6$

2   Calculate the square of the distance from the mean for each data value:

$(3 - 6)^2 = (-3)^2 = 9$

$(7 - 6)^2 = 1^2 = 1$

$(6 - 6)^2 = 0^2 = 0$

$(7 - 6)^2 = 1^2 = 1$

$(4 - 6)^2 = (-2)^2 = 4$

$(4 - 6)^2 = (-2)^2 = 4$

$(8 - 6)^2 = 2^2 = 4$

$(8 - 6)^2 = 2^2 = 4$

$(8 - 6)^2 = 2^2 = 4$

$(5 - 6)^2 = (-1)^2 = 1$

3   Sum the squares calculated in Step 2:

$9 + 1 + 0 + 1 + 4 + 4 + 4 + 4 + 4 + 1 = 32$

Divide the result from Step 3 by the number of data values: $\frac{32}{10} = 3.2$

4   Take the square root of the result in Step 4 to get the standard deviation:

  Standard deviation $(\sigma) = 3.2 = 1.79$ (2 d.p.)

Overall, the average distance of all exam scores from the mean is 1.55 for Class 1 and 1.79 for Class 2. This means that Class 2 has more variability in scores.

PRACTICE 16: APPLICATION

Two tutorial classes sat the same exam. Each tutorial class had eight students. Scores, out of 10, were as follows:

  Class 1: 9, 4, 8, 4, 5, 6, 5 and 7

  Class 2: 3, 7, 6, 7, 4, 8, 8 and 5

Which class had the greater variability in scores, as measured by the standard deviation?

When to use which measure of variability

The range, interquartile range (IQR) and standard deviation are all appropriate in different situations. Here is a general guide as to which measure to use and when:

- Range: simple to calculate and useful for evaluating a whole data set to show the variability within a data set and to compare spread between similar data sets.
- IQR: provides a clearer picture of the overall data set than the range because it removes extreme values, so it is less sensitive to extreme values. However, like the range, it is based on only two values from the data set (Q1, Q3). The IQR is useful for data sets with extreme values where the range will be skewed; common examples include house sale prices and household income levels.
- Standard deviation: a more powerful measure of variability since it considers every value in the data set. It allows us to compare different data sets, even when the means and standard deviations are different. However, the standard deviation is sensitive to extreme values, so take care when using it on such data sets. The standard deviation is most useful for normally distributed data, which we will discuss in Section 15.6 of this chapter.

15.5 REPRESENTING AND ANALYSING STATISTICAL INFORMATION USING DIAGRAMS

Statistical information can be visually presented in a variety of ways. Different representations are used to emphasise different aspects of data in different contexts. Most visual representations of statistical data are summative representations. In other words, just like the range, interquartile range and standard deviation, visual representations summarise the data, highlighting key values or features of the data set. This section describes three commonly used visual representations of data: box-and-whisker plots (or box plots for short), histograms, and stem-and-leaf plots.

Box-and-whisker plots

The three quartiles Q1, Q2 and Q3 used to calculate the interquartile range (IQR) are used in other forms of data representation. The following five values: the minimum value, Q1, Q2, Q3 and the maximum value are together called the *five-number summary*. The five-number summary is used to generate a visual representation of data called a box-and-whisker plot.

We will use the data set from Example 13 to illustrate how a box-and-whisker plot is created. In this example, a class of nine students completed a quiz and received the following scores: 63, 77, 52, 89, 90, 46, 30, 64 and 70.

For this set of data, Q1 = 49, Q2 = 64 and Q3 = 83.

The minimum value of this data set is 30 and the maximum value is 90.

So, the five-number summary for this data is:

minimum = 30, Q1 = 49, Q2 = 64, Q3 = 83, maximum = 90

These values are shown on a box-and-whisker plot below:

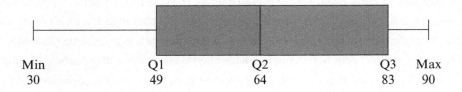

Min	Q1	Q2	Q3	Max
30	49	64	83	90

The box-and-whisker plot provides a visual picture of how the data is spread around the median (Q2). The span of the box shows the interquartile range, the middle half of the data. In this case, we can see that there is a bigger distance between Q3 and the median than between the median and Q1. This tells us that some students who did better than the median did quite a lot better than the median, whereas students who did worse than the median were relatively closer to the median. The upper quartile (Q3) is much closer to the maximum than the lower quartile (Q1) is to the minimum. This tells us that more students were closer to the maximum result than there were students close to the minimum result.

A box-and-whisker plot also provides a useful way of comparing two sets of data. For example, suppose we had another class of nine students who sat the same quiz and received the following scores: 32, 62, 88, 63, 63, 86, 68, 76 and 51. The five-number summary for this data is:

minimum = 32, Q1 = 56.5, Q2 = 63, Q3 = 81, maximum = 88

(Review Section 15.4 to see how to calculate Q1, Q2 and Q3.)

These values are shown on the second box-and-whisker plot below the box-and-whisker plot from the first class of nine students below:

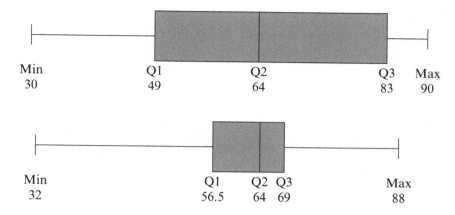

We can see that the second data set has a much smaller IQR than the first one. This means that the middle 50% of scores are bunched more closely than in the first class; more students scored close to the median in the second class compared to the first. Not as many students did exceptionally well in the second class but also given that Q1 is higher not as many people did as poorly. While one student in the second class did very well (the maximum value of 88), the middle 50% of students scored between 56.5 and 69, as compared to 49 and 83 for the first class. As to which class of students performed better, arguably it is the first class, given both distributions have the same median (Q2), the first class has the much larger IQR, and higher Q3 (83) compared to the second class (69). However, it must be noted that in the first class, Q1 falls just below the pass mark (presumably 50%), whereas in the second class, Q1 is well above the pass mark.

The advantages of a box-and-whisker plot include that it shows both the range and distribution of data, so any extreme values are part of the summary. Some sense of skewness in the data, that is many more points being on one side of the median relative to the other, is able to be seen according to where Q2 is positioned relative to Q1 and Q3. The disadvantages of the box-and-whisker plot include that it does not show individual data points and that it can only be used for numerical data.

EXAMPLE 17

A class of eight students completed a quiz and received the following scores:

63, 77, 52, 89, 46, 32, 64 and 70

Determine the five-number summary and use this to draw a box-and-whisker plot to represent the data.

Solution

The five-number summary is:

minimum $= 32$, Q1 $= 49$, Q2 $= 63.5$, Q3 $= 73.5$, maximum $= 89$

(See Example 14 for calculations of Q1, Q2 and Q3.)

Using the five-number summary, we can draw the box-and-whisker plot for the data:

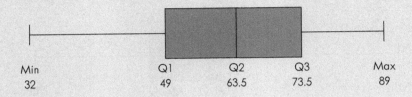

Min	Q1	Q2	Q3	Max
32	49	63.5	73.5	89

PRACTICE 17

A class of eleven students completed a quiz and received the following scores:

63, 77, 58, 52, 89, 70, 46, 32, 64, 69 and 70

Determine the five-number summary and use this to draw a box-and-whisker plot to represent the data.

EXAMPLE 18: APPLICATION

Ten houses have sold in Bedford Park in the last week for the following prices:

$410 000, $389 000, $476 000, $512 000, $350 000, $960 000 $432 000, $543 000, $529 000 and $499 000

Determine the five-number summary for this data and use it to draw a box-and-whisker plot of the data. What does the box-and-whisker plot tell you about the maximum value?

Solution

The five-number summary is:

minimum $= 350 000$, Q1 $= 410 000$, Q2 $= 487 500$, Q3 $= 529 000$, maximum $= 960 000$

(Q1, Q2 and Q3 were calculated in Practice 13 question)

We can draw the box-and-whisker plot using this information:

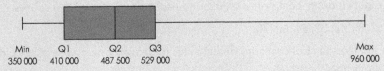

Min	Q1	Q2	Q3	Max
350 000	410 000	487 500	529 000	960 000

It is clear from the box-and-whisker plot that the maximum value of $960 000 is an outlier well above any other values.

PRACTICE 18: APPLICATION

The recorded daily maximum temperatures, in Celsius, for Adelaide over twelve days in January are as follows: 28, 31, 37, 35, 36, 30, 28, 32, 39, 41, 35 and 31. For Melbourne, the recorded daily temperatures for the same twelve days are as follows: 31, 26, 29, 35, 32, 36, 26, 34, 38, 41, 36 and 29.

Calculate the five-number summary for each set of data and use this to draw a box-and-whisker plot for each set of data one under the other. What do the box-and-whisker plots tell you about the relative temperature of the two cities over those twelve days?

Frequency table

A frequency table is a way of summarising data that shows how often a particular value occurs. It is often used to group data into categories and record a count for each category. For this reason, a frequency table is very useful for continuous data. For example, data such as a person's age, where it might be useful to group everyone who is between, say, 20 and 29 years into the same category and know how many people are in that category, rather than list each person's age separately.

Consider a survey of twenty people of the following ages: 25, 24, 27, 34, 41, 28, 44, 20, 22, 33, 44, 32, 34, 43, 35, 31, 37, 36, 29 and 30. A frequency table could record the number of people in their 20s, 30s and 40s as follows:

AGE	TALLY	COUNT
20–29	⦀⦀ II	7
30–39	⦀⦀ IIII	9
40–49	IIII	4

EXAMPLE 19: APPLICATIONS

Record the following sets of data in a frequency table and use the frequency table to answer the specified questions about the data.

a The neck measurements of 20 people in centimetres were recorded as follows:

39, 33, 37, 34, 39, 35, 37, 33, 35, 40, 39, 39, 37, 40, 35, 35, 35, 41, 37 and 32

What is one feature that the frequency table highlights about the data? How would the information in the frequency table be useful if you were manufacturing clothing?

Note: use groups of five, with the first group 30–34.

b Fifteen students gained the following percentages in an exam:

95, 45, 67, 55, 53, 77, 58, 81, 79, 88, 38, 82, 53, 60 and 69

Did the class as a whole do well on the exam? Why or why not?

Note: use groups of ten for your table, with the first group 30–39.

Tiffany Winn

Solutions

a

NECK MEASUREMENT	TALLY	COUNT
30–34	IIII	4
35–39	IИ IИ III	13
40–44	III	3

The frequency table highlights that the 35–39 category is more frequently occurring than the other categories. If this information reflects the neck sizes of the population as a whole (although a larger survey would be needed), then a clothing manufacturer would most likely make most of their collar sizes to fit the middle group of neck sizes, since this group is more common than the others.

b

EXAM PERCENTAGE	TALLY	COUNT
30–39	I	1
40–49	I	1
50–59	IIII	4
60–69	III	3
70–79	II	2
80–89	III	3
90–99	I	1

From the frequency table, we can quickly see that over half the students gained 60% or more in the exam, and only two students failed (assuming 50% or more is a pass), which one could argue is a good result.

PRACTICE 19: APPLICATION

Record the following sets of data in a frequency table and use the frequency table to answer the specified questions about the data.

a   Students in a class of 25 gained the following percentages in an exam:

95, 59, 45, 82, 67, 51, 55, 64, 53, 31, 52, 77, 58, 81, 49, 79, 57, 88, 38, 85, 82, 53, 90, 60 and 69.

Did the class as a whole do well on the exam? Why or why not?

Note: use groups of ten for you table, with the first group 30–39.

b   Eighteen people were surveyed as to how many hours they usually sleep each night. Their answers (in hours) were as follows:

10, 8, 7, 6, 7, 8, 5, 7, 10, 11, 9, 7, 8, 9, 8, 10, 11 and 9

What does the data tell you about how long most people sleep?

Note: choose an appropriate group size for your table. You may like to compare the difference between different group sizes.

Histograms

A histogram groups data into categories and shows the count per category, as a frequency table does, but a histogram shows this information visually in the form of bars. Like a frequency table, a histogram is very useful for continuous data. For example, data such as a person's height where it may be useful to group everyone who is between, say, 165 cm and 169 cm into the same category, rather than list each person's height separately. When drawing a histogram, the size of each data category is called the bin size, so, for example, the bin size for Example 19b is 10.

In order to draw a histogram correctly, it is important to remember the following points.

- There are no gaps between bars on a histogram, unlike for a bar or column chart, because a histogram represents continuous data.
- The bin sizes are marked on a histogram under the middle of the bar.
- If each bin size is the same, for example, all the groups are of size ten, then the height of the bars of the histogram shows the count per category.

The histogram below has been drawn based on the frequency table developed in the previous section on frequency tables and shown again below (left). Note that the bin size is 10 and the histogram highlights that the 30–39 year age group is the most frequent category, known as the mode or modal category.

AGE	TALLY	COUNT
20–29	卌 ||	7
30–39	卌 ||||	9
40–49	||||	4

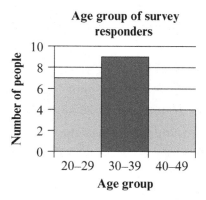

One advantage of a histogram is that it quickly highlights the overall spread of data. For example, in the histogram above we can quickly see that most of the people are between 20 and 39 years and less people are 40 to 49 years, and that the most people are in the 30 to 39 year category. The histogram is also particularly useful for large sets of data because individual data can be grouped in categories. The disadvantages of a histogram include that it does not show the individual data, so detailed information is lost, and that it can only be used for numerical data.

EXAMPLE 20

Use the frequency table from Example 19b, shown again below, to generate a histogram. Which is the modal category?

EXAM PERCENTAGE	TALLY	COUNT
30–39	I	1
40–49	I	1
50–59	IIII	4
60–69	III	3
70–79	II	2
80–89	III	3
90–99	I	1

Solution

From the histogram, it is easy to see that the modal category is 50–59%.

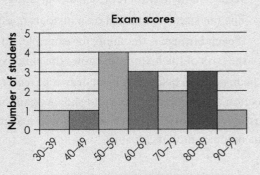

Exam scores

PRACTICE 20

Consider the following data:

    38, 27, 46, 54, 27, 53, 25, 43, 22, 31, 34, 36, 37, 36, 47, 22, 33, 51, 35 and 36.

Draw a histogram of the data, with data grouped in categories from 20–24, 25–29 and so on, and highlight the modal category.

EXAMPLE 21: APPLICATION

The heights of a group of thirty women are as follows:

    165, 163, 168, 166, 163, 168, 163, 169, 170, 177, 159, 156, 164, 166, 163, 168, 163, 169, 170, 180, 162, 165, 166, 171, 152, 172, 177, 159, 154 and 164

Draw a frequency table of the heights and use this to draw a histogram of the heights. Choose an appropriate bin size.

Solution

Because the spread of the data is from a minimum of 152 to a maximum of 180, a bin size of 5 would be appropriate. The frequency table for this data is shown on the left and the histogram is shown on the right.

HEIGHT	TALLY	COUNT
150–154	II	2
155–159	III	3
160–164	ⅢⅠ III	8
165–169	ⅢⅠ ⅢⅠ	10
170–174	IIII	4
175–179	II	2
180–184	I	1

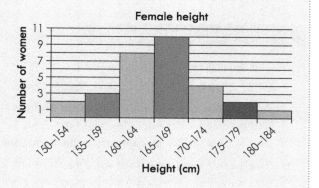

PRACTICE 21: APPLICATION

The lasting times (in hours) of 15 light bulbs are as follows:

506, 556, 561, 598, 630, 721, 734, 763, 765, 772, 778, 790, 882, 824 and 846

Draw a histogram of the data, choosing an appropriate bin size. Highlight the modal category.

Stem-and-leaf plots

A stem-and-leaf plot splits data items into two parts, one of which is called the stem and the other the leaf. Data items are ordered first by the stem and then by the leaf. For example, if we were looking at a temperature of 35 it would be split into a stem of 3 and leaf of 5, and a temperature of 41 would be split into a stem of 4 and leaf 1.

For example, if the temperatures in Darwin over two weeks were as follows:

32, 34, 31, 38, 29, 26, 42, 43, 33, 30, 28, 26, 40 and 38

A stem-and-leaf plot of the data would look like this:

Stem	Leaf
2	6 6 8 9
3	0 1 2 3 4 8 8
4	0 2 3

Note that all points in the data are ordered; for example, the points that are in the 20s are listed in order as 26, 26, 28 and 29.

One advantage of the stem-and-leaf plot is that it provides an overall picture of the spread of data while retaining individual data points. It shows the minimum and maximum

Tiffany Winn

values, and gaps and clusters in the data can be easily spotted. Large data sets can also be handled relatively easily with a stem-and-leaf plot. Disadvantages include that it is not easy to see the median, mean and mode from the stem-and-leaf plot, and perhaps that the plot does not give the visual picture that a histogram does; although it can easily be used to generate a histogram if necessary.

EXAMPLE 22: APPLICATIONS

a The maximum wind velocity (in km/h) in Bright Gully for the past twenty years is as follows:

84, 81, 88, 64, 72, 86, 81, 67, 70, 91, 82, 86, 92, 79, 77, 91, 68, 77, 82 and 83

Draw a stem-and-leaf plot of the data. Use the stem-and-leaf plot to find the minimum and maximum wind velocities over the twenty years. Using the categories designated by the stem values, which category is the modal one?

b Twelve swimming race times were measured as follows (in seconds):

13.2, 13.6, 12.5, 12.7, 14.0, 12.3, 12.5, 12.8, 13.6, 12.8, 14.5 and 15.0

Draw a stem-and-leaf plot of the data. Use the plot to identify the fastest and slowest race times. Which stem value designates the modal category?

Solutions

a The stem-and-leaf plot shows that the minimum wind velocity is 64 km/h and the maximum 92 km/h. The modal category is the one with the most items; this is the category including all data with a stem of 8, so between 80 and 89 (inclusive) km/h.

Stem	Leaf
6	478
7	02779
8	112234668
9	112

b In this stem-and-leaf plot, firstly note that the last digit is usually the leaf in a stem plot and any other digits make up the stem. Secondly, if the place value or meaning of the stem and the leaf is not clear, it must be specified in a key, usually located below the plot.

Stem	Leaf
12	355788
13	266
14	05
15	0

stem 12, leaf 3 means 12.3

The fastest race time is 12.3 seconds and the slowest 15.0 seconds. The modal category is the one with a stem of 12.

PRACTICE 22: APPLICATIONS

a   The heights of 15 students were measured, correct to the nearest centimetre:

161, 156, 143, 149, 156, 160, 142, 151, 166, 145, 167, 172, 143, 148 and 159

Draw a stem-and-leaf plot of the data. Use the plot to identify the minimum and maximum heights. Which stem value designates the modal category?

b   Consider the following data, representing the number of students in different tutorial classes for a particular university course:

22, 23, 44, 18, 26, 52, 22, 40, 21, 25, 29, 39, 31, 30, 41, 26, 27, 12, 37 and 33

Draw a stem-and-leaf plot of the data. Use the plot to identify the minimum and maximum class sizes. Which stem value designates the modal category?

Frequency and percentage polygons

Another useful representation for comparing different sets of data is a frequency polygon. A frequency polygon is a line graph created by plotting the frequency for each category or class of data. The horizontal axis of a frequency polygon is labelled with category midpoints and the frequencies of particular categories are plotted at their respective midpoints. A percentage polygon is identical to a frequency polygon except that it plots percentage values rather than frequencies. While the frequency polygon is often used because it is quick and easy to produce, the percentage polygon is a better fit for some statistical contexts, so we describe both graphs here.

Consider the data shown in the table below as the basis for both a frequency and a percentage polygon. This data was obtained by comparing the cost of eating out in Adelaide and Sydney across 200 restaurants. The restaurants are grouped into categories based on the average price of a main course. For example, in Adelaide there are 52 restaurants where the average price of a main course is between $20 and $25; in Sydney, there are 28 restaurants in that same price range.

	$10 TO LESS THAN $15	$15 TO LESS THAN $20	$20 TO LESS THAN $25	$25 TO LESS THAN $30	$30 TO LESS THAN $35
Adelaide — No. of restaurants	20	35	52	57	36
Sydney — No. of restaurants	25	38	28	64	45

This data is plotted onto a frequency polygon overpage. Class midpoints are shown as labels on the horizontal axis (for example, the midpoint of the $10 to $15 class is $12.5). The graph shows the number of restaurants in each class. It is common for frequency polygons to display a value of zero for the class below the lowest class and the one above the highest class.

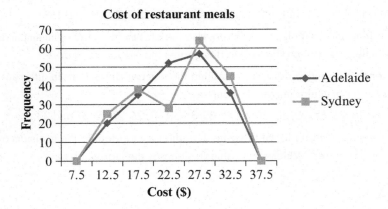

A percentage polygon is identical to a frequency polygon except that instead of the number of values in each category (the frequency) being plotted the percentage represented by each category (that is, the number of values in each category as a percentage of the whole) is plotted. For example, the data represented in the previous table is shown in table below with percentage values for each category calculated and included. The percentage is calculated by dividing the frequency value by the total number of values (200) and multiplying by 100 (since per cent means per 100). For example:

$$\frac{35}{100} \times 100 = \frac{3500}{200}$$
$$= \frac{35}{2}$$
$$= 17.5$$

Making Connections
To review how to calculate percentages revisit Chapter 6.

The frequency value of 35 in the $15 to $20 class for Adelaide restaurants is equivalent to 17.5% of restaurants.

	$10 TO LESS THAN $15	$15 TO LESS THAN $20	$20 TO LESS THAN $25	$25 TO LESS THAN $30	$30 TO LESS THAN $35
Adelaide — No. of restaurants (%)	20 (10%)	35 (17.5%)	52 (26%)	57 (28.5%)	36 (18%)
Sydney — No. of restaurants (%)	25 (12.5%)	38 (19%)	28 (14%)	64 (32%)	45 (22.5%)

The percentage polygon of this data is shown overpage. Class midpoints are shown as labels on the horizontal axis (for example, the midpoint of the $10 to $15 class is $12.5). The graph shows the percentage of restaurants in each class. It is common for percentage polygons to display a value of zero for the class below the lowest class and the one above the highest class. Of note is that this graph is similar to the frequency polygon (above) except that it shows percentage rather than frequency values.

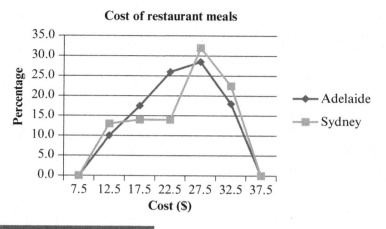

Cost of restaurant meals

EXAMPLE 23

Two schools of four hundred students each completed national testing and obtained the results shown in the table below. Draw both a frequency and a percentage polygon of the data. Which school is performing better? Comparing the shape of the two polygons, what do you notice?

	20 TO <30%	30 TO <40%	40 TO <50%	50 TO <60%	60 TO <70%	70 TO <80%	80 TO <90%	90 TO <100%
School A	20	24	12	108	116	72	32	16
School B	32	12	20	80	124	88	20	24

Solution

The frequency polygon is drawn using the data provided in the table. The graph compares the national testing results of two different schools by showing the number of students from each of the schools with a test score in a particular range. Class midpoints are shown as labels on the horizontal axis (for example, the midpoint of the 20% to <30% class is 25%). Entries from the table are shown

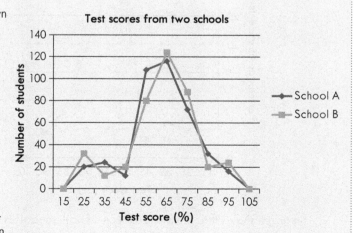

Test scores from two schools

as points on the graph. For example, for the 40 to < 50% category, the values of 12 for school A and 20 for school B are shown at the class midpoint (45%).

The percentages for each category are needed to draw the percentage polygon. These are calculated by dividing each frequency value by the total number of values (400) and multiplying by 100. For example, the 108 students from school A who scored from 50 to <60 on the test represent $\frac{108}{400} \times 100 = \frac{10\,800}{400} = \frac{108}{4} = 27\%$ of the total number of students who completed the test.

The table below shows the data from the original table together with percentage values for each category.

	20 TO <30%	30 TO <40%	40 TO <50%	50 TO <60%	60 TO <70%	70 TO <80%	80 TO <90%	90 TO <100%
School A	20 (5%)	24 (6%)	12 (3%)	108 (27%)	116 (29%)	72 (18%)	32 (8%)	16 (4%)
School B	32 (8%)	12 (3%)	20 (5%)	80 (20%)	124 (31%)	88 (22%)	20 (5%)	24 (6%)

A percentage polygon of the data from the table above is shown below. The graph compares the national testing results of two different schools by showing the percentage of students from each of the schools whose test score was in a particular range. Class midpoints are shown as labels on the horizontal axis (for example, the midpoint of the 20% to <30% class is 25%). Entries from the data table are shown as points on the graph. For example, for the 40 to <50% category, the values of 3% for school A and 5% for school B are shown at the class midpoint (45%).

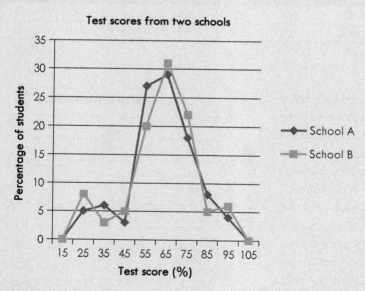

The graph shows that the performance of both schools is similar, although it could be argued that either school may be doing better depending on the criteria used to measure performance. The graphs show that school B has slightly more students achieving under

50% than school A, as indicated by the measurement points for school B being higher than the corresponding points for school A for most values below 50%. Therefore, it could be argued that school A is doing better from the point of view of the number of students that are passing. School B, however, has more students at 60% or higher than school A, as indicated by school B being slightly above school A in the rightmost section of the graph. So, it could be argued that school B is doing better from the point of view of the overall mastery of students of the material being examined. The frequency and percentage polygons are identical in shape, with the only difference between the graphs being that the first graph shows frequencies and the second percentages.

PRACTICE 23

Three hundred restaurants in Melbourne and Canberra were compared for their prices, with the results shown in the table below. Draw both a frequency polygon and a percentage polygon of the data. Which city boasts cheaper restaurants overall?

	$10 TO LESS THAN $15	$15 TO LESS THAN $20	$20 TO LESS THAN $25	$25 TO LESS THAN $30	$30 TO LESS THAN $35
Melbourne — No. of restaurants	60	69	66	57	48
Canberra — No. of restaurants	33	54	93	84	36

Cumulative frequency polygons and cumulative percentage polygons (ogives)

Sometimes it can be more useful to display data using a cumulative frequency or cumulative percentage polygon. These graphs are often called ogives.

A cumulative frequency polygon displays the cumulative frequency of each class. In other words, it displays the sum of the frequencies from all of the classes up to the upper boundary of each class in the distribution. Whereas the horizontal axis of a frequency polygon shows class midpoints, the horizontal access of a cumulative frequency polygon shows the upper boundaries of each class. The table below shows the cumulative frequencies for the data related to the average cost of restaurants in Adelaide and Sydney that we used in the previous section.

	$10 TO LESS THAN $15	$15 TO LESS THAN $20	$20 TO LESS THAN $25	$25 TO LESS THAN $30	$30 TO LESS THAN $35
Adelaide — No. of restaurants	20	55 (20 + 35)	107 (20 + 35 + 52)	164 (20 + 35 + 52 + 57)	200 (20 + 35 + 52 + 57 + 36)
Sydney — No. of restaurants	25	63 (25 + 38)	91 (25 + 38 + 28)	155 (25 + 38 + 28 + 64)	200 (25 + 38 + 28 + 64 + 45)

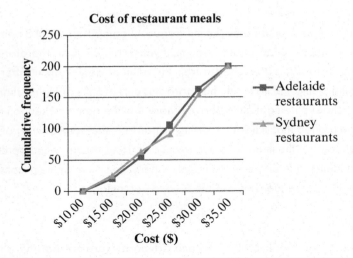

This graph shows the cumulative frequencies for the number of restaurants in a given price range in Adelaide and Sydney. For example, in Adelaide, there are 107 restaurants where the average price of a main course is anything up to $25; and in Sydney there are 91 such restaurants.

A cumulative percentage polygon is identical to a cumulative frequency polygon except that it displays percentages rather than frequencies. The table below shows both the cumulative frequencies and the cumulative percentages for the average cost of restaurants in Adelaide and Sydney.

	$10 TO LESS THAN $15	$15 TO LESS THAN $20	$20 TO LESS THAN $25	$25 TO LESS THAN $30	$30 TO LESS THAN $35
Adelaide – No. of restaurants (%)	20 (10%)	55 (27.5%)	107 (53.5%)	164 (82%)	200 (100%)
Sydney – No. of restaurants (%)	25 (12.5%)	63 (31.5%)	91 (45.5%)	155 (77.5%)	200 (100%)

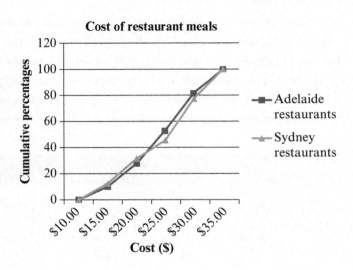

This graph shows the cumulative percentages from the table above for the number of restaurants in a given price range in Adelaide and Sydney. For example, in Adelaide, 53.5% of restaurants surveyed have an average price of anything up to $25; and in Sydney the same is true for 45.5% of restaurants.

If we compare the percentage polygon and the cumulative percentage polygon, we can note that each graph emphasises different aspects of the data. The first graph highlights the dip in available restaurants in Sydney in the $20 to $25 price range, and the jump in availability for Sydney in the $25 to $30 class. The ogive, on the other hand, shows that restaurants in the two cities are overall similarly priced but slightly cheaper in Adelaide, because the line graph for Adelaide is mainly slightly above that for Sydney.

EXAMPLE 24

Two schools of four hundred students each completed national testing and obtained the results shown in the table below. Draw both a cumulative frequency and a cumulative percentage polygon of the data. Which school is performing better? Compare the shape of the two polygons; what do you notice?

	20 TO <30%	30 TO <40%	40 TO <50%	50 TO <60%	60 TO <70%	70 TO <80%	80 TO <90%	90 TO <100%
School A	20	24	12	108	116	72	32	16
School B	32	12	20	80	124	88	20	24

Solution

First, the cumulative frequencies need to be calculated from the data in the table by adding up accumulated frequencies for each class. For example, in the <50% category, the cumulative frequency for school A is 20 + 24 + 12 = 56. The cumulative percentages are then calculated from the cumulative frequencies by multiplying by 100 and dividing by the total number of frequencies (400). For example, when the cumulative frequency is 56, the cumulative percentage is $\frac{56}{100} \times 100 = \frac{5600}{400} = \frac{56}{4} = 14\%$. The table below shows both the cumulative frequencies and the cumulative percentages for the given data.

	20 TO <30%	30 TO <40%	40 TO <50%	50 TO <60%	60 TO <70%	70 TO <80%	80 TO <90%	90 TO <100%
School A	20 (5%)	44 (11%)	56 (14%)	164 (41%)	280 (70%)	352 (88%)	384 (96%)	400 (100%)
School B	32 (8%)	44 (11%)	64 (16%)	144 (36%)	268 (67%)	356 (89%)	376 (94%)	400 (100%)

A cumulative frequency polygon of the data is shown below. This graph compares the national testing results of two different schools by showing the cumulative frequency of students from each of the schools whose test score was in a particular range. Entries from the cumulative frequency table are shown as points on the graph. For example, for the 40 to less than 50% category, the cumulative values are 56 for school A and 64 for school B.

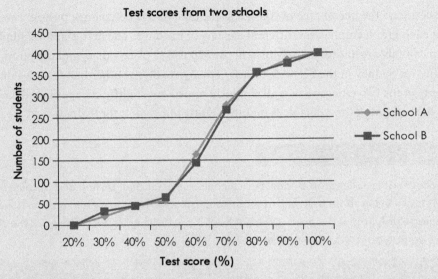

The graph below compares the national testing results of two different schools by showing the cumulative percentage of students from each of the schools whose test score was in a particular range. The cumulative percentage frequency data points are shown on the graph. For example, for the 40 to less than 50% category, the cumulative values of 14% for school A and 16% for school B are shown.

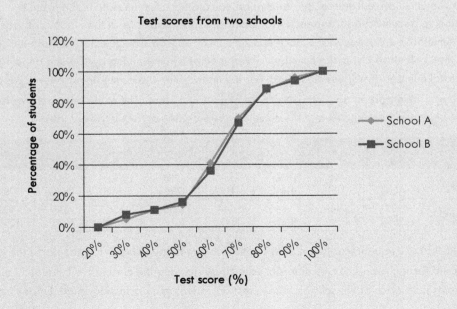

These graphs highlight the overall similarity between the results of school A and school B. The graphs show that school B has slightly more students achieving under 50% than school A, indicated by the measurement point for school B being noticeably higher than the corresponding point for school A on two of three points below 50%. So it could be argued that school A is doing better from the point of view of how many students are passing. School B, however, has more students at 60% or higher than school A, indicated by school B being slightly to the right of school A in the rightmost section of the graph. So it could be argued that school B is doing better from the point of view of the overall mastery of students of the material being examined. The two graphs are identical in shape, with the only difference between the graphs being that the first graph shows frequencies and the second percentages.

PRACTICE 24

Three hundred restaurants in Melbourne and Canberra were compared for their prices, with the results shown in the table below. Draw a cumulative frequency polygon and a cumulative percentage polygon of the data. Which city boasts cheaper restaurants overall?

	$10 TO LESS THAN $15	$15 TO LESS THAN $20	$20 TO LESS THAN $25	$25 TO LESS THAN $30	$30 TO LESS THAN $35
Melbourne — No. of restaurants	60	69	66	57	48
Canberra — No. of restaurants	33	54	93	84	36

15.6 THE NORMAL DISTRIBUTION

The **normal distribution** refers to a particular way in which data is spread, or distributed.

Data can be distributed in many ways. For example, in this histogram the data is located more to the right. This is known as being skewed right.

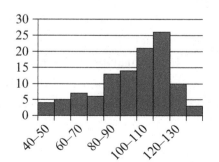

Normal distribution
A distribution of data where the data is evenly spread in a *bell-shaped* curve around the mean

Tiffany Winn

In this histogram the data is located more to the left. This is known as being skewed left.

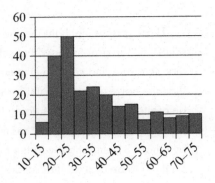

On occasions, such as in this histogram, the data is jumbled with no real pattern as to where it is located in the distribution.

In many situations, however, data tends to be spread around a central value with no bias left or right, and it gets close to what is called a *normal distribution*, which is illustrated below.

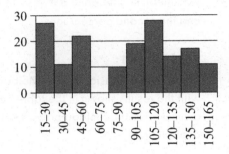

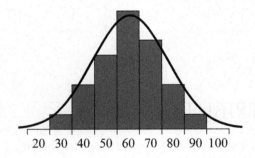

Often, the normal distribution is referred to as a 'bell curve' because it looks like the shape of a bell.

In a normal distribution:
- fifty per cent of data is above the mean, and fifty per cent of the data is below the mean
- the distribution of data is symmetric about the mean
- the mean, median and mode are all equal.

Examples of things that are approximately a normal distribution are:

- shoe size
- life spans of consumable items
- intelligence quotient (IQ)
- standardised test scores (such as NAPLAN)
- errors in measurement.

The normal distribution is a continuous distribution, meaning that it describes variables that are continuous. A continuous variable is a variable that can take on any value between two specified values. For example, the measurement of a group of people's heights is continuous because it can be any part of a whole unit, for example, 165.97 cm. On the other hand, counting the number of heads or tails in a collection of coin tosses is not continuous because the result can only be an integer number; it is not possible to have 3.5 heads.

The normal distribution and the standard deviation

The standard deviation, discussed in Section 15.4, is only regarded as an accurate, average measure for how far individual values are from the mean value if the distribution of data approximates a normal distribution. So it is useful before calculating the standard deviation to investigate the distribution of data.

The relationship between the standard deviation and the normal distribution is that the greater the standard deviation, the 'flatter' the distribution, as shown below. Note that the mean is denoted as μ and the standard deviation is σ. As σ decreases, values in the distribution are clustered more closely around the mean, so the distribution appears 'taller'.

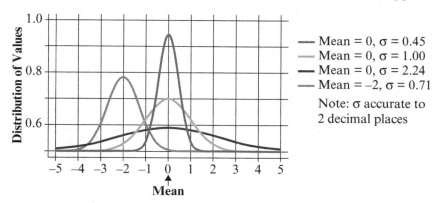

Mean = 0, σ = 0.45
Mean = 0, σ = 1.00
Mean = 0, σ = 2.24
Mean = −2, σ = 0.71

Note: σ accurate to 2 decimal places

A key property of the normal distribution is that a specific percentage of the data values are within 1, 2 and 3 standard deviations of the mean:

- 68% of data is within one standard deviation of the mean
- 95% of data is within two standard deviations of the mean
- 99.7% of data is within three standard deviations of the mean.

This is known as the 68–95–99.7 rule and is illustrated below.

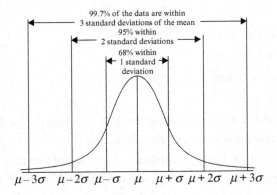

   This diagram represents a normal distribution and the percentages of data that fall within one, two and three standard deviations ($\pm\sigma, \pm2\sigma, \pm3\sigma$) from the mean μ.

EXAMPLE 25

Six hundred mathematics students sat for an exam. Their marks were normally distributed with a mean of 63 and standard deviation of 8. If a passing result is 50% or higher and a credit is between 65% and 74% (inclusive), will the proportion of students who failed be more or less than those who will receive a credit?

Solution

The first step is to draw roughly a normal distribution curve labelling each standard deviation with the appropriate percentage. Remember that the 68–95–99.7 rule tells us the percentages of data for a normal distribution lie within each standard deviation from the mean, and given $\mu = 63\%$ and $\sigma = 8$, you can determine the standard deviation boundaries on the normal distribution as follows:

BOUNDARY ON NORMAL DISTRIBUTION	GRADE (%)
$-3\sigma + \mu$	39
$-2\sigma + \mu$	47
$-\sigma + \mu$	55
μ	63
$\mu + \sigma$	71
$\mu + 2\sigma$	79
$\mu + 3\sigma$	87

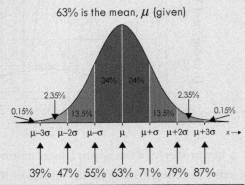

Standard deviation is 8 so each successive standard deviation is 8 above/below the previous one.

By looking at the graph and estimating, it is clear that there are a greater percentage of values from 65% to 74% than there are below 50%. Therefore, the proportion of students gaining a credit will be greater.

PRACTICE 25: GUIDED

Three hundred history students sat for an exam. Their marks were normally distributed with a mean of 60 and standard deviation of 5. A passing result is 50% or higher and a distinction is between 75% and 84% (inclusive). Draw roughly a normal distribution curve labelling each standard deviation with the appropriate percentage. Use this diagram to determine whether the proportion of students who fail will be more or less than those who get a distinction.

Given $\mu = 60\%$ and $\sigma = 5$ we need to first determine the standard deviation boundaries.

BOUNDARY ON NORMAL DISTRIBUTION	GRADE (%)
$-3\sigma + \mu$	45
$-2\sigma + \mu$	☐
$-\sigma + \mu$	55
μ	60
$\mu + \sigma$	☐
$\mu + 2\sigma$	70
$\mu + 3\sigma$	☐

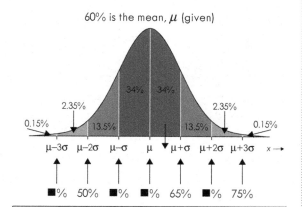

60% is the mean, μ (given)

34% 34%
2.35%   2.35%
0.15%   13.5%   13.5%   0.15%

$\mu-3\sigma$ $\mu-2\sigma$ $\mu-\sigma$ μ $\mu+\sigma$ $\mu+2\sigma$ $\mu+3\sigma$ $x \rightarrow$

■%   50%   ■%   ■%   65%   ■%   75%

Standard deviation is 5 so each successive standard deviation is 5 above/below the previous one.

By looking at the graph and estimating, it is clear that there are a greater percentage of values below 50% than there are over 75%. Therefore, the proportion of students failing will be greater than those getting a distinction.

Now try this one:

Five hundred biology students sat an exam. Their results were normally distributed with a mean of 57% and standard deviation of 8. Assume the following percentages for successive standard deviations: 68%, 95% and 99.7%. If a passing result is 50% or higher and a credit is between 65% and 74% (inclusive), will the proportion of students who fail be more or less than those who get a credit?

Tiffany Winn

EXAMPLE 26

Consider three normally distributed data sets, representing the exam results from three subjects each with an equal number of students. Data set A has a mean of 64% and standard deviation of 2. Data set B has a mean of 70% and standard deviation of 12. Data set C has a mean of 60% and standard deviation of 8.

a   In which subject did the greatest number of students pass, assuming the pass mark is 50% or more?

b   In which subject did the least number of students pass, assuming the pass mark is 50% or more?

c   Which subject had the most variance in results?

Solution

The standard deviations are 2 (data set A), 12 (data set B) and 8 (data set C), respectively, and the standard deviation boundaries for each data set will fall at the percentages shown in the table below.

	2σ BELOW MEAN	σ BELOW MEAN	MEAN	σ ABOVE MEAN	2σ ABOVE MEAN
Data set A	60	62	64	66	68
Data set B	46	58	70	82	94
Data set C	44	52	60	68	76

a   The pass mark of 50% is furthest below the mean, and closest to the tail of the distribution, in data set A. Therefore, the greatest number of students will have passed in data set A.

b   The pass mark of 50% is closest to the mean, so closest to the middle of the distribution, in data set C. Therefore, the least number of students will have passed in data set C.

c   The highest standard deviation is 12 for data set B so this topic had the most variance in results.

PRACTICE 26

Consider three normally distributed data sets, each with an equal number of data values, representing the height of males from three different countries. Data set A has a mean of 168 cm and standard deviation of 8. Data set B has a mean of 174 cm and standard deviation of 5. Data set C has a mean of 170 cm and standard deviation of 10.

a   In which country would you expect the greatest number of males to be between 170 and 180 cm (inclusive)?

b   In which country is the most number of males below 160 cm?

c   Which country has the greatest variance in height?

Standardising the normal distribution: the standard normal distribution

The number of standard deviations that a value is located away from the mean is known as the *standard score* or the *z-score*.

Suppose the mean (μ) in the graph below is 63% and the standard deviation (σ) is 4. The numbers in under the graph show the mean of the distribution and values at the first, second and third standard deviations above and below the mean. The *z*-score for 67% would be 1 (because it is one standard deviation above the mean), and for 55% would be −2 (because it is two standard deviations below the mean).

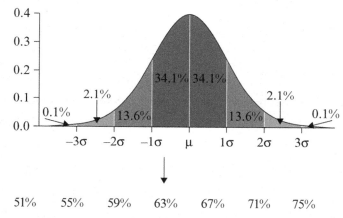

To convert a particular value to a standard score (*z*-score):

• first subtract the mean

• then divide by the standard deviation.

This can be written as a simple formula:

$$z = \frac{x - \mu}{\sigma}$$

where x = a particular data value, μ = the mean and σ = the standard deviation for the distribution.

For example, if the mean test score in a class is 62% with a standard distribution of 8, and Jen has 79%, then her *z*-score is calculated as follows:

$$z = \frac{x - \mu}{\sigma} = \frac{79 - 62}{8} = 2.125$$

This means Jen's score is roughly two standard deviations above the mean.

We can create a standard normal distribution from any distribution by shifting that distribution so that its mean is zero, as shown overpage. Standardising a normal distribution is useful because it enables values from different normal distributions to be compared easily.

Tiffany Winn

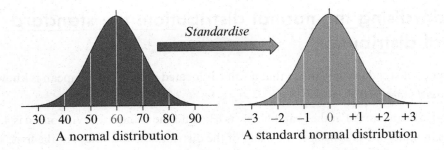

30 40 50 60 70 80 90	−3 −2 −1 0 +1 +2 +3
A normal distribution	A standard normal distribution

For example, comparing Anna to Jen (from the previous example), Anna got 75% in a class where the mean test score was 60% and the standard deviation 5. Did Anna do relatively better or worse than Jen?

Anna's z-score is calculated as follows:

$$z = \frac{x - \mu}{\sigma} = \frac{75 - 60}{5} = 3$$

This means Anna's score is three standard deviations above the mean. This is further above the mean than Jen's score (2.125) so even though Anna's actual percentage is lower than Jen's (75% as compared to 79%), Anna has done relatively better than Jen in the test.

EXAMPLE 27: APPLICATION

Six hundred mathematics students sat an exam. Their marks were normally distributed with a mean of 63 and standard deviation of 8. Assume the following percentages for successive standard deviations: 68%, 95% and 99.7%.

a   What would be the z-score of a student who got 78%?

b   In another mathematics exam, the mean was 67% and standard deviation was 4. Would a student who got 75% in this exam have done relatively better or worse than the student in part (a)?

Solution

a   The z-score of a student who got 78% is calculated as follows:

$$z = \frac{x - \mu}{\sigma} = \frac{78 - 63}{8} = \frac{15}{8} = 1.875$$

b   The z-score of a student who got 75% in an exam with a mean score of 67% and standard deviation of 4 would be:

$$z = \frac{x - \mu}{\sigma} = \frac{75 - 67}{4} = \frac{8}{4} = 2$$

So the student who got 75% (z-score of 2) did relatively better than the student who got 78% (z-score of 1.875).

PRACTICE 27: APPLICATION

Five hundred history students sat a history exam. Their results were normally distributed with a mean of 57% and standard deviation of 8. Assume the following percentages for successive standard deviations: 68%, 95% and 99.7%.

a What would be the z-score of a student who got 50%? Write your answer as a decimal.

b In another history exam, the mean was higher at 60% and the standard deviation was greater at 12. Would a student who got 50% in this exam have done the same, relatively better or relatively worse than the student in part (a)?

EXAMPLE 28: APPLICATION

Four hundred biology students sat an exam. Their results were normally distributed with a mean of 70% and standard deviation of 12. Assume the following percentages for successive standard deviations: 68%, 95% and 99.7%.

a What would be the z-score of a student who got 52%? Write your answer as a decimal.

b In another biology exam, the mean was lower at 55% and the standard deviation smaller at 6. Would a student who got 47% in this exam have done the same, relatively better or relatively worse than the student in part (a)?

Solution

a The z-score of a student who got 52% is calculated as follows:

$$z = \frac{x - \mu}{\sigma} = \frac{52 - 70}{12} = \frac{-18}{12} = -1.5$$

b The z-score of a student who got 55% in an exam with a mean score of 47% and standard deviation of 6 would be:

$$z = \frac{x - \mu}{\sigma} = \frac{47 - 55}{6} = \frac{-8}{6} = -1.33$$

So the student who got 47% (z-score of -1.33) did relatively better than the student who got 52% (z-score of -1.5), because the z-score of -1.33 is closer to the mean.

PRACTICE 28: APPLICATION

The heights were measured of three hundred males in country A. The mean height was 170 cm with a standard deviation of 6. Assume the following percentages for successive standard deviations: 68%, 95% and 99.7%.

a What would be the z-score of a person with a height of 165 cm?

b In country B, the mean was lower at 166 cm and the standard deviation greater than 10. Would a person who was 158 cm tall in country B be relatively shorter or taller than the person in part (a)?

15.7 CHAPTER SUMMARY

SKILL OR CONCEPT	DEFINITION OR DESCRIPTION	EXAMPLES
Population	A group of events, objects, results or individuals, all of which share some unifying characteristic.	• All of a person's red blood cells • All of the university graduates in Australia
Sample	A portion of a population that represents the whole population.	• Those of a person's red blood cells taken in a test tube for analysis • 1000 university graduates who completed a survey
Random sample	A sample where all members of the population have an equal chance of selection.	Use computerised random number generator for student sample based on student ID numbers
Sampling techniques	• Stratified random sampling: random sample divided into categories (strata) with strata sample size set according to population • Quota sampling: stratified but strata filled using any sampling method • Systematic sampling: sample selected in an organised way • Judgement sampling: based on the belief of the person doing sampling that the sample accurately represents the population • Convenience sampling: sample chosen in most convenient way possible	• Stratified random sample: specific sample must contain certain number of 1st, 2nd and 3rd year students • Quota sample: approach patrons in café until each stratum filled • Systematic sample: every thirtieth student in an alphabetic list • Judgement sample: the tutorial group you believe best represents the topic • Convenience sample: approach first forty people you see
Measure of central tendency	Indicates where the middle of a set of data lies.	Mean, median or mode
Mean	The arithmetic average of a set of numerical data calculated by summing the data and then dividing by the number of data values.	The mean of 3, 6, 7, and 12 $\dfrac{3 + 6 + 7 + 12}{4} = 7$
Median	The middle value in a data set. In a data set where there are two middle values, the median is the mean of those values.	The median of 3, 4, 6, 7 and 12 is 6. The median of 4, 5, 7 and 12 is the mean of 5 and 7, which is 6.
Mode	The most frequently occurring item often used with non-numerical data. A data set with more than one mode is called multimodal.	The mode of {red, black, brown, red} is red. The set {red, black, brown, red, green, black, purple} is bi-modal.
Range	The difference between the largest and smallest values.	For the data set 2, 5, 7, 10 and 22, the range is $22 - 2 = 20$.

SKILL OR CONCEPT	DEFINITION OR DESCRIPTION	EXAMPLES
Quartiles	Divide a set of data into four equal quarters.	For the data set 2, 5, 7, 10, 22, 23, 24, 24, 26 Middle quartile, Q2, is 22 (median) Lower quartile, Q1, is 6 (median of lower half of data) Upper quartile, Q3, is 24 (median of upper half of data)
Interquartile Range (IQR)	The difference between the upper and lower quartiles. $IQR = Q3 - Q1$	For the set of data 2, 5, 7, 10, 22, 23, 24, 24, 26 $IQR = 24 - 6 = 18$
Five-number summary	Five numbers that summarise a data set: the minimum value, Q1, Q2, Q3 and the maximum value.	For the set of data 2, 5, 7, 10, 22, 23, 24, 24, 26 The five-number summary is: • minimum = 2 • Q1 = 6 • Q2 = 22 • Q3 = 24 • maximum = 26
Standard deviation	Measure of the average distance of all data points from the mean.	For the set of data 2, 5, 11, the standard deviation is: $$\sqrt{\frac{(2-4)^2}{6} + \frac{(5-4)^2}{6} + \frac{(11-4)^2}{6}}$$ $$= \sqrt{\frac{4}{6} + \frac{1}{6} + \frac{7}{6}}$$ $$= \sqrt{\frac{12}{6}}$$ $$= \sqrt{2}$$ $$= 1.41 \text{ (2 d.p.)}$$
Box-and-whisker plot	Visual representation of the five-number summary.	For the set of data: 63, 77, 52, 89, 90, 46, 30, 64, 70 The box-and-whisker plot is shown below: 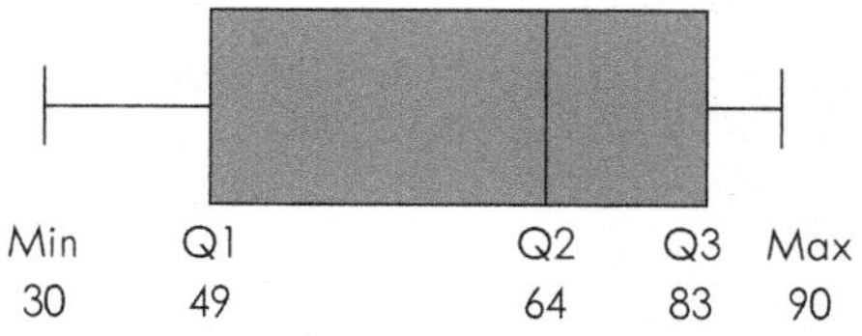 Min 30, Q1 49, Q2 64, Q3 83, Max 90
Frequency table	Shows how often a particular value occurs and is often used to group data into categories.	39, 33, 37, 34, 39, 35, 37, 33, 35, 40, 39, 39, 37, 40, 35, 35, 35, 41, 37, 32 A frequency table grouping data into categories of five is shown below: DATA / TALLY / COUNT 31–35 / 卌 IIII / 9 36–40 / 卌 卌 / 10 41–45 / I / 1

SKILL OR CONCEPT	DEFINITION OR DESCRIPTION	EXAMPLES
Histogram	Visually represents data from a frequency table.	**Neck measurements** A histogram of neck measurements (cm) on the horizontal axis (30–34, 35–39, 40–44) and Number of measurements on the vertical axis (0 to 15).
Stem-and-leaf plot	Splits data items into two parts, one called the stem and the other the leaf. Data items ordered by the stem and then the leaf.	For the data set: 32, 34, 31, 38, 29, 26, 42, 43, 33, 30, 28, 26, 40, 38, Stem | Leaf 2 | 6 6 8 9 3 | 0 1 2 3 4 8 8 4 | 0 2 3
Percentage polygon	Created by plotting the percentage value for each category above the midpoint of the category.	A line graph with Test score (%) on the horizontal axis (25 to 95) and Number of students (%) on the vertical axis (0.0 to 35.0), showing School A and School B.
Cumulative percentage polygon (ogive)	Displays at each class or category boundary point the total of the percentages of values up to that point. Values on the horizontal axis represent class boundaries.	**Test scores from two schools** A line graph with Test scores (%) on the horizontal axis (25 to 95) and Number of students (%) on the vertical axis (0 to 120), showing School A and School B.
Normal distribution	A particular shape or pattern of data arrangement, where it is spread around a central value with no bias left or right, in roughly a bell-shaped curve shape.	A bell-shaped curve with horizontal axis marked −3, −2, −1, +1, +2, +3 and values 4.6, 16.0, 27.4, 38.8, 50.2, 61.6, 73.0. Labels: 26 (−1.12), 33 (−0.51), 65 (+2.30).

SKILL OR CONCEPT	DEFINITION OR DESCRIPTION	EXAMPLES
68–95–99.7 rule	When data is normally distributed, 68% of data lies within one standard deviation of the mean, 95% of data within two standard deviations of the mean, and 99.7% of data within three standard deviations of the mean.	99.7% of the data are within 3 standard deviations of the mean; 95% within 2 standard deviations; 68% within 1 standard deviation. $\mu - 3\sigma \quad \mu - 2\sigma \quad \mu - \sigma \quad \mu \quad \mu + \sigma \quad \mu + 2\sigma \quad \mu + 3\sigma$
Standard normal distribution	Version of a normal distribution where the mean is shifted to zero for easy comparison with other normal distributions.	Standardise. 30 40 50 60 70 80 90 — A normal distribution. −3 −2 −1 0 +1 +2 +3 — A standard normal distribution
Standard score or z-score	The number of standard deviations that a value is located away from the mean.	For a distribution with mean 1010 and standard deviation of 20, the z-score of the value 970 would be: $$z\text{-score} = \frac{970 - 1010}{20} = \frac{-40}{20} = -2$$
Comparing normal distributions	Scores in two different normal distributions can be easily converted by using z-scores.	Quiz scores of two classes: class A has a mean of 75 and a standard deviation of 5, and class B has a mean of 70 and a standard deviation of 10. z-score for 72 in class A $$\frac{72 - 75}{5} = \frac{-3}{5} = -0.6$$ z-score for 65 in class B $$\frac{65 - 70}{10} = \frac{-5}{10} = \frac{-1}{2} = -0.5$$ So a z-score of 65 in class B is relatively slightly better than a z-score of 72 in class A.

15.8 REVIEW QUESTIONS

A SKILLS

1 Find the mean, median and mode of the following data sets.

a	6	7	2	10	7	8	7	1	6	5
b	34	45	68	91	19	56	45	34		
c	10	18	17	16	12	19	20	8	9	

2 Consider the following set of data.

6	7	5	7	7	8	7	5	9	7
4	10	6	8	8	9	5	6	4	8

   a   What is the range of the data?

   b   Work out the five-number summary.

   c   Calculate the interquartile range (IQR).

   d   Draw a box-and-whisker plot.

3 Consider the following set of data.

16	17	15	17	12	22	24	25	29
10	16	28	18	19	22	26	24	

   a   What is the range of the data?

   b   Work out the five-number summary.

   c   Calculate the interquartile range (IQR).

   d   Draw a box-and-whisker plot.

4 Calculate the standard deviation of the following set of numbers, to two decimal places.

12	12	8	5	13	19	16	11

5 Consider the following set of numbers.

10	11	8	5	13	16	14	3

   a   Calculate the standard deviation, accurate to two decimal places.

   b   Which set of numbers has greater variability — this set or the set in question 4? How do you know this?

6 Consider the following set of numbers.

6	9	15	17	17	8	7	5	19	17
4	10	16	18	3	9	15	16	14	8

   a   Draw a frequency table of the data, with data in groups of five and the first group being 1 to 5.

   b   Draw a histogram of the data, with a bin size of 5, highlighting the modal category.

c Draw a stem-and-leaf plot of the data, highlighting the median value.

d List one advantage and one disadvantage of the histogram in part (b) and the stem-and-leaf plot in part (c) relative to each other.

e Draw a percentage polygon of the data, with data grouped as in part (a).

f Draw an ogive of the data, with data grouped as in part (a).

7 Consider the following set of numbers.

| 16 | 9 | 25 | 17 | 27 | 8 | 3 | 2 | 19 | 27 |
| 4 | 20 | 26 | 18 | 20 | 4 | 20 | 16 | 14 | 22 |

a Draw a frequency table of the data, with data in groups of five and the first group 1 to 5.

b Draw a histogram of the data, with a bin size of 5, highlighting the modal category.

c Draw another histogram of the data, with a bin size of 10, highlighting the modal category.

d List one advantage and one disadvantage of the histograms in part (a) and (b) relative to the other one.

e Draw a stem-and-leaf plot of the data, highlighting the median value.

f Compare the histogram in part (c) and the stem-and-leaf plot. List one advantage and one disadvantage of each compared to the other one.

g Draw a percentage polygon of the data, with data in groups of 5 as per part (a).

h Draw an ogive of the data, with data in groups of 5 as per part (a).

8 Consider four normally distributed data sets, each with an equal number of data values. Data set A has a mean of 177 and standard deviation of 2. Data set B has a mean of 173 and standard deviation of 8. Data set C has a mean of 170 and standard deviation of 12. Data set D has a mean of 176 and standard deviation of 6.

a Which data set, if any, has the greatest number of values above the mean and why?

b Which data set has the most number of values above 178?

c Which data set has the least number of values above 180?

d Which data set has the most variance in values?

9 Consider a normal distribution of percentage values with a mean of 54% and standard deviation of 8. Assume the following percentages for successive standard deviations: 68%, 95% and 99.7%.

a Will the proportion of values less than 50% be more or less than the proportion of values that are 50% or more?

b Will the proportion of values less than 60% be more or less than the proportion of values that are 70% or more?

c What percentage of data values would you expect between 54% and 70%? 62% or more?

Tiffany Winn

10 Consider a normal distribution with a mean of 28 and standard deviation of 4.

    a What would be the z-score for a value of 30 in this distribution?

    b Compare the z-score in part (a) to the z-score of the value 32 in a distribution with a mean of 26 and standard deviation of 12. Which value is relatively lower?

    c Now compare the z-score in part (a) to a z-score of 28 in a distribution with a mean of 23 and standard deviation of 9. Which value is relatively lower?

B   APPLICATIONS

1   Sue spent the following amounts on her lunch each day over the course of two working weeks:

$12, $4, $10, $42.50, $7.50, $10.50, $6, $8, $6.50 and $7.50

Calculate the mean, median and mode of the data.

2   The following table shows the salaries of 21 employees in a small company:

JOB	NO. OF EMPLOYEES	SALARY
Manager	1	300000
Deputy manager	2	220000
Supervisor	2	115000
Sales representative	4	75000
Warehouse worker	8	56000
Clerical worker	4	52000

    a Find the five-number summary for the data.

    b Draw a box-and-whisker plot for the data.

3   The following table shows the number of different types of model kits sold by a small store.

JOB	NUMBER SOLD	COST ($)
Megaplane	1	300
Supersize racer	2	220
Grand prix race	2	115
Hero helicopter	4	75
Mini grand prix racer	3	56
Mini helicopter	4	52

    a Find the five-number summary for the data.

    b Draw a box-and-whisker plot for the data.

4   The maximum wind velocity (in km/h) in Breezy Gully for the past 14 years is as follows:

97   89   79   92   86   81   85

77   83   84   82   91   78   86

a   Draw a frequency table of the data, with data in groups of five and the first group 75 to 79.

b   Find the five-number summary.

c   Find the interquartile range (IQR).

d   Draw a box-and-whisker plot of the data.

5   The heights of a class of 19 twelve-year old school students were measured in centimetres.

132	144	147	146	150	152	145	138	137	146
144	145	147	157	147	143	153	150	148	

a   Arrange the scores in order and draw a histogram of the data, choosing an appropriate bin size.

b   Draw a percentage polygon for the data.

c   Draw an ogive for the data.

d   Highlight an advantage and a disadvantage of each of the data representations used in parts (a) to (d).

6   Twenty students took a maths test. Below are their results:

7	4	2	10	8	6	8	6	7	5
6	7	2	7	6	6	9	7	5	5

a   Prepare a frequency table of the scores.

b   Draw a box-and-whisker plot for the data.

7   Over a series of five AFL games, Sam had the following number of disposals per game: 18, 25, 41, 28 and 23. Over the same five games, Tom had the following number of disposals: 15, 39, 32, 28 and 21. On average, which player had more disposals per game? Which player had more variability in the number of disposals they had per game?

8   Find the interquartile range (IQR) for the Melbourne Cup winners' data shown in the frequency table below.

AGE OF HORSE	FREQUENCY
4-year-old	8
5-year-old	7
6-year-old	5
7-year-old	2

9   The lasting times (in hours) of 15 light bulbs are as follows:

506, 556, 561, 598, 630, 721, 734, 763, 765, 772, 778, 790, 882, 824 and 846

a   Draw a stem-and-leaf plot of the data.

b   Calculate the mean lasting time for a light bulb.

Tiffany Winn

10 Consider the following data of temperatures in Adelaide over 12 days.

| 31 | 21 | 41 | 40 | 27 | 39 | 38 | 31 | 31 | 25 | 36 | 35 |

 a Draw a histogram of the data with a bin size of ten and highlight which category is the mode.

 b Draw another histogram with a bin size of five and highlight the modal category.

 c Which of the histograms in (a) or (b) is most useful and why?

 d Draw a stem-and-leaf plot for the data, then highlight the median.

 e Draw a percentage polygon for the data.

 f Draw an ogive for the data.

 g Highlight an advantage and disadvantage of each of the data representations used in parts (c) to (f). Which histogram do you feel is most useful?

11 In order to gain entry to a particular university, students must sit a national test. The test scores are normally distributed with a mean of 600 and a standard deviation of 80. Students know that to get into the university, they must score better than at least 65% of the students who took the test. Jordan takes the test and scores 685. Jim gets 640, Viv gets 670 and Tom gets 590. Which students do well enough on the test to gain entry into university?

12 The battery life for a certain kind of computer is normally distributed with a mean of 20 hours and standard deviation of 5. Assume the following percentages for successive standard deviations: 68%, 95% and 99.7%.

 a For what percentage of computers does the battery last longer than 30 hours?

 b For what percentage of computers does the battery last less than five hours?

 c For what percentage of computers does the battery last between 15 and 25 hours?

13 The annual salaries of employees in a business employing 400 people are approximately normally distributed with mean $60000 and standard deviation $20000. Assume the following percentages for successive standard deviations: 68%, 95% and 99.7%.

 a How many employees earn more than $100000 annually?

 b How many employees earn between $40000 and $80000?

 c How many employees earn less than $80000?

14 In first semester, Amy sat a statistics exam in which scores were normally distributed with a mean of 69% and standard deviation of 6. Amy sat this exam and scored 72%. Assume the following percentages for successive standard deviations: 68%, 95% and 99.7%.

 a What is Amy's z-score?

 b In the second semester exam for the same topic, the mean was 60% and standard deviation 4 and Emma scored 63%. Did Emma do relatively better or worse than Amy?

 c In the summer semester exam for the same topic, the mean was 70%, the standard deviation 8 and Jack got 75%. How did Jack do compared to both Amy and Emma?

16

COUNTING AND PROBABILITY

CHAPTER OBJECTIVES

» Understand that probability involves assigning a numerical value to the likelihood of an event occurring

» Use counting techniques to determine the number of ways an event can occur

» Distinguish between experimental and theoretical probability

» Represent sample spaces using the most appropriate diagram

» Discern between mutually and non-mutually exclusive events, and independent and dependent events

» Calculate theoretical probabilities

THE MONTY HALL PROBLEM

The Monty Hall Problem is a probability conundrum that is loosely based on the US television game show *Let's Make a Deal* and named after its original host Monty Hall (source; montyhallproblem.com). The problem involves a game show situation where a contestant is shown three closed doors. Behind one door is a car, and behind the other two doors are goats.

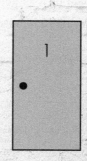
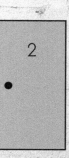

Contestants are asked to pick a door (say Door 1) behind which they think is the car. Once they have picked the first door, the host Monty opens another one of the doors (say Door 3) to reveal a goat. Monty always opens a door that reveals a goat. He then asks the contestant do you want to keep Door 1 as your selection or swap to Door 2? What would you do?

Did you say it does not matter, that there is a 50–50 chance of the car being behind either Door 1 or Door 2? This is what our intuition tells us about this situation, but this type of intuition is sometimes not reliable. In fact, in this case, you would have a better chance of winning the car if you were to swap doors. This is because Monty always picks a door that is incorrect. Look at this table, which involves what might happen after you pick Door 1 and Monty picks another incorrect door.

BEHIND DOOR 1	BEHIND DOOR 2	BEHIND DOOR 3	IF YOU STAY WITH DOOR 1	IF YOU SWAP DOORS
Car	Goat	Goat (Monty's pick)	**Win Car**	Goat
Goat	**Car**	Goat (Monty's pick)	Goat	**Win Car**
Goat	Goat (Monty's pick)	**Car**	Goat	**Win Car**

As this table shows the contestant that stays the with original door (Door 1 in this case) has a 1 out of 3 chance of winning the car, but if they swap doors they have a 2 out of 3 chance. The Monty Hall Problem is a good example of how we often cannot trust our intuition when it comes to determining the likelihood, or chance, that something may happen.

16.1 INTRODUCTION TO COUNTING AND PROBABILITY

Probability is the mathematics of chance and probability that tells us the likelihood of something, a particular event, occurring. For any event, we can assign a number between 0 and 1 to describe the likelihood that it will occur. An event that will never occur has a probability of 0, and is labelled an **impossible event**. An event that will always occur has a probability of 1, and is labelled a **certain event**. For all other events between these extremes a probability between 0 and 1 can be assigned.

We can use the following scale to represent differing probabilities:

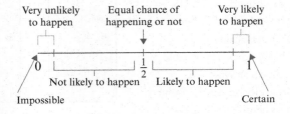

EXAMPLE 1

On a probability scale, mark the probability that:

a   it will snow in Adelaide on Christmas Day

b   if today is Thursday, tomorrow will be Friday

c   if you toss a 20 cent coin, you will get a tail

Solutions

a   We cannot say that it would be impossible for snow to fall in Adelaide on Christmas Day, but it would be exceedingly unlikely. So we would place this on a probability scale very close to 0.

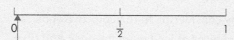

b   If today is Thursday, then it is certain that tomorrow will be Friday. So we place this on the probability scale at 1.

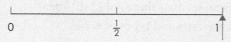

c   If we toss a coin, there is an equal chance of getting a tail (or a head) so this is placed on the probability scale at $\frac{1}{2}$.

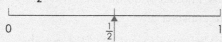

PRACTICE 1

On a probability scale, mark the probability that:

a   it will rain on every day in winter in Melbourne

b   Anzac Day will be on a Monday this year

c   you will win Cross Lotto at the next draw

A very informal way to determine the probability that an event might occur is using the formula:

$$\frac{\text{the number of ways a particular event can occur}}{\text{the total number of outcomes that could happen}}$$

We can see from this formula that to calculate probabilities we need to be able to count things, and when counting you are answering the question that begins with 'In how many ways?'

16.2 COUNTING PRINCIPLES

Suppose your local ice-cream bar sells blueberry, vanilla, strawberry or honey flavours, served in waffle, cup or plain cones. How many combinations of single-scoop ice-cream cones could you buy?

We can work this out by considering in turn our two choices. The first is that we have 4 choices of ice-cream flavours, and then for each flavour we have 3 choices of cones. We can represent all of our choices on what is called a tree diagram.

From this diagram, we can see that there are 12 different combinations of ice-cream flavours and types of cones.

Let's consider another situation. You are looking to buy a new car and have a choice of five colours, red, white, black, silver or green, and two transmissions, manual or automatic. How many combinations do you have to consider?

Again a tree diagram will help us sort out all of the possible combinations.

We can see from this diagram that there are 10 possible combinations of cars that you could buy.

In both examples, we can see a pattern developing. With the ice-cream choices there were 4 flavours and for each of those 3 different cones, which gave us 12 combinations ($4 \times 3 = 12$). In the car example, there were 5 different colours and for each of those 2 different transmissions, which gave us 10 different combinations ($5 \times 2 = 10$). These examples allow us to generalise what is known as the **Fundamental Principle of Counting**. This is where if an event A occurs in n ways and an event B occurs in m ways, then the events A and B can occur in succession in $n \times m$ ways.

> **Fundamental Principle of Counting**
> If an event A can occur in n ways and for each of these n ways, an event B can occur in m ways, then the events A and B can occur in succession in $n \times m$ ways

EXAMPLE 2: APPLICATION

You are making a trip from Adelaide to Sydney and then to Hong Kong. You can travel to Sydney by car, train, plane or bus, and from Sydney to Hong Kong by ship or plane. How many different routes are possible?

Solution

The first event to occur is the trip from Adelaide to Sydney and this can occur in 4 ways. For each of these ways, the second event, the trip to Hong Kong, can occur in 2 ways. Therefore, the number of different routes will be 4 × 2 = 8.

PRACTICE 2: APPLICATION

In a medical clinical study, participants are classified according to their blood type A, B, AB or O, and by their weight as underweight, normal slightly overweight, overweight or obese. Using these classifications, how many different types of participants are involved in the study?

The Fundamental Principle of Counting can be extended to more than two events occurring in succession. Let's look at how this works by considering car number plates.

In South Australia, a general-issue car number plate is in the form Snnn-LLL (an example is shown here) and no general issue number plate can contain the letter Q. How many general issue number plates do you think are possible?

We can diagrammatically represent the number plate using a set of boxes:

Each box represents a position on the number plate moving from left to right. We can insert into each box the number of possible digits or letters that can occupy each position.

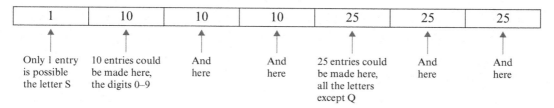

Moving from left to right we can successively apply the Fundamental Principle of Counting because for *each* of the ways that an entry can be made into one box the number of ways recorded in the next box will occur. Therefore, the total possible number plates will be:

$$1 \times 10 \times 10 \times 10 \times 25 \times 25 \times 25 = 15\ 625\ 000$$

Kathy Brady

In South Australia, car owners can also choose to have a personalised number plate. These can have the following combinations, and the letter Q is permitted in this case:

- 1 letter followed by 5 numbers (e.g. A12345)
- 2 letters followed by 4 numbers (e.g. AB1234)
- 3 letters followed by 3 numbers (e.g. ABC123)
- 4 letters followed by 2 numbers (e.g. ABCD12)
- 5 letters followed by 1 number (e.g. ABCDE1)

How many different personalised number plates would be possible? To answer this we need to consider each different combination of letters and numbers separately. Let's use sets of boxes to record the number of possibilities in each position and then apply the Fundamental Principle of Counting.

1 letter followed by 5 numbers

| 26 $*$ 10 $*$ 10 $*$ 10 $*$ 10 $*$ 10 | = 2 500 000

2 letters followed by 4 numbers

| 26 $*$ 26 $*$ 10 $*$ 10 $*$ 10 $*$ 10 | = 6 760 000

3 letters followed by 3 numbers

| 26 $*$ 26 $*$ 26 $*$ 10 $*$ 10 $*$ 10 | = 17 576 000

4 letters followed by 2 numbers

| 26 $*$ 26 $*$ 26 $*$ 26 $*$ 10 $*$ 10 | = 45 697 600

5 letters followed by 1 number

| 26 $*$ 26 $*$ 26 $*$ 26 $*$ 26 $*$ 10 | = 118 813 760

Addition
Principle
If the
possibilities
being counted
can be divided
into groups with
no possibilities
in common,
then the total
number of
possibilities
is the sum of
the number of
possibilities of
each group

To get the number of different personalised number plates we would then add the number of ways each of the combinations can occur—which will be a very large number indeed! This is an example of how the second counting principle, the **Addition Principle**, is applied. According to the Addition Principle, if the number of possibilities can be divided into groups with no possibilities in common, then the total number of possibilities is the sum of the number of possibilities of each group.

We can write the Addition Principle using the following mathematical notation:

$$n(A \cup B) = n(A) + n(B)$$

In this notation, the symbol $\cup$ means 'or', and we read the notation as 'the number of elements in A or B is equal to the number of elements in A plus the number of elements in B'.

EXAMPLE 3

When you draw a card from a pack of 52, in how many ways can you draw a king or a queen?

Solution

In a standard pack of 52 cards there are 4 kings and 4 queens. So using the notation given above, we know that:

$$n(K) = 4 \qquad n(Q) = 4$$

and therefore:

$$n(K \cup Q) = n(K) + n(Q)$$
$$= 4 + 4$$
$$= 8$$

Remember $\cup$ means 'or'.

PRACTICE 3

In how many ways can a number be selected from the set of numbers 1–35 that is either a multiple of 5 or a multiple of 8?

Let's now consider another situation that also involves drawing a card from a pack. In how many ways can you draw a black card or a king from a pack of 52?

Now there are 26 black cards in a pack, and we have already noted that there are 4 kings. So $n(B) = 26$ and $n(K) = 4$. However, there are 2 cards in the pack that are both black and king (king of clubs and king of spades). So we will use a new notation to describe these two cards: $n(B \cap K) = 2$. In this notation, the symbol $\cap$ means 'and'. We read this notation as 'the number of elements that are black and king'.

We have established that $n(B) = 26$ and $n(K) = 4$, but using these notations, the two cards (king of clubs and king of spades) have been double counted. So we need to adjust for the double counting to determine the number of ways that we can draw a king or a black card.

$$n(B) = 26 \qquad n(K) = 4 \qquad n(B \cap K) = 2$$
$$n(B \cup K) = n(B) + n(K) - n(B \cap K)$$
$$= 26 + 4 - 2$$
$$= 28$$

In subtracting the number of cards that are both black and king, we are making the adjustment for the double counting. This is an example of the application of the third principle of counting, the **Exclusion Principle**, in which there are two sets of possibilities that have elements in common, so the number of common elements must be subtracted from the sum of the two sets of possibilities.

Exclusion Principle
If set *A* and set *B* have some elements in common, then the number of common elements must be subtracted from the sum of set *A* and set *B*

Kathy Brady

We can write this using notation as:

$$n(A \cup B) = n(A) + n(B) - n(A \cap B)$$

Remember $\cup$ means 'or' and $\cap$ means 'and'.

The Exclusion Principle can be diagrammatically represented using what is known as a Venn diagram. A Venn diagram uses circles to show the relationships between different sets or groups of objects. Letters are used to label the circles according to what they represent and numbers are written in each circle to indicate how many objects are contained in the set or group. The Venn diagram that represents the example of drawing a card that is either black or a king would be:

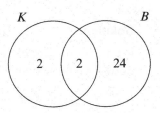

The information on this Venn diagram tells us that there are 2 cards that are king but not black, 24 cards that are black but not king and 2 cards, represented where the circles intersect, that are both black and king.

EXAMPLE 4

If $n(A) = 7$, $n(B) = 9$ and $n(A \cap B) = 2$, find $n(A \cup B)$.

Solution

This is an application of the Exclusion Principle and we can calculate $n(A \cup B)$ by substituting the given values into the rule.

$$
\begin{aligned}
n(A \cup B) &= n(A) + n(B) - n(A \cap B) \\
&= 7 + 9 - 2 \\
&= 14
\end{aligned}
$$

PRACTICE 4

If $n(A) = 23$, $n(B) = 16$ and $n(A \cup B) = 32$, find $n(A \cap B)$.

EXAMPLE 5

If $n(A \cap B) = 4$, $n(A \cup B) = 18$ and $n(B) = 7$, find $n(A)$.

Solution

This is an application of the Exclusion Principle and we begin by substituting the given values into the rule. Next we evaluate what we can, and then by observation determine the value that is required.

$$n(A \cup B) = n(A) + n(B) - n(A \cap B)$$

$$18 = n(A) + 7 - 4$$

$$18 = n(A) + 3$$

By observation, $n(A) = 15$.

PRACTICE 5

If $n(A \cap B) = 13$, $n(A \cup B) = 23$ and $n(A) = n(B)$, find $n(A)$.

EXAMPLE 6: APPLICATION

A university Drama class has 6 first year female students, 8 first year male students, 5 second year female students and 4 second year male students. In how many ways can the lecturer select a second year student or a female student to play the leading role? Draw a Venn diagram to represent this situation.

Solution

To use the language of the Exclusion Principle this example requires $n(S \cup F)$, where S represents second year students and F represents female students. We can now write the rule using these labels:

$$n(S \cup F) = n(S) + n(F) - n(S \cap F)$$

Reading through the information we find that $n(S) = 9$. This is the total number of second year students, both male and female. The $n(F) = 11$ because there are 6 first year female students and 5 second year female students. Finally, the number of students that are both in second year and female is 5, that is $n(S \cap F) = 5$. We can now substitute all of these values into the rule:

$$n(S \cup F) = n(S) + n(F) - n(S \cap F)$$

$$= 9 + 11 - 5$$

$$= 15$$

So the lecturer can select a student for the leading role in 15 ways.

As a Venn diagram:

The information on this Venn diagram shows that there are 4 second year students who are not female (the male students), 6 female students who are not in second year and 5 students who are both female and in second year.

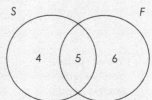

PRACTICE 6: APPLICATION

Two pain relief drugs, *A* and *B* were tested on 40 participants in a clinical trial.

• 23 reported relief from drug *A*

• 18 reported relief from drug *B*

• 11 reported relief from both *A* and *B*

a   Draw a Venn diagram to represent this information.

b   How many participants reported relief from with drug *A* or drug *B*?

c   How many did not get relief from either of the drugs?

d   How many participants reported relief from drug *A* only?

16.3 PERMUTATIONS AND COMBINATIONS

Permutations

Think about the letters MATH. How many different ways can we arrange these letters?

The one way to attempt to answer this question is to take a systematic approach to recording all of the different ways that these letters can be arranged, such as this:

MATH	AMTH	TMAH	HMAT
MAHT	AMHT	TMHA	HMTA
MTAH	ATMH	TAMH	HAMT
MTHA	ATHM	TAHM	HATM
MHAT	AHMT	THMA	HTMA
MHTA	AHTM	THAM	HTAM

Doing this we have found that there are 24 different ways we can arrange these letters. Each of these different arrangements is known as a **permutation** of the letters MATH.

However, listing arrangements as we did above is not the most efficient way of determining how many permutations a group of objects may have. A better way is to use a set of boxes as we did with the number plate example earlier in the chapter. For example, using the letters MATH:

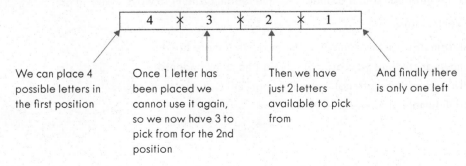

| 4 | ✳ | 3 | ✳ | 2 | ✳ | 1 |

We can place 4 possible letters in the first position

Once 1 letter has been placed we cannot use it again, so we now have 3 to pick from for the 2nd position

Then we have just 2 letters available to pick from

And finally there is only one left

> **Permutation**
> Any arrangement of a group of objects made in a definite order

Now we use the Fundamental Principle of Counting to multiply across the boxes to get, in this case, that the number of permutations is 24. As you can see, this is a more efficient way of counting permutations.

EXAMPLE 7

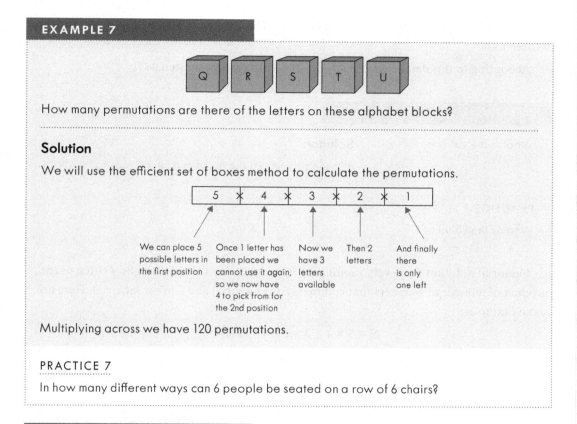

How many permutations are there of the letters on these alphabet blocks?

Solution

We will use the efficient set of boxes method to calculate the permutations.

| 5 | × | 4 | × | 3 | × | 2 | × | 1 |

We can place 5 possible letters in the first position

Once 1 letter has been placed we cannot use it again, so we now have 4 to pick from for the 2nd position

Now we have 3 letters available

Then 2 letters

And finally there is only one left

Multiplying across we have 120 permutations.

PRACTICE 7

In how many different ways can 6 people be seated on a row of 6 chairs?

EXAMPLE 8

In how many ways can 7 different paintings be hung in a row?

Solution

We have seen that drawing a set of boxes is an efficient way of finding the number of arrangements of objects. However, we do not need to draw the boxes. We can go straight into writing out the multiplication involved in calculating permutations. In this example, this will be:

$$7 \times 6 \times 5 \times 4 \times 3 \times 2 \times 1 = 5040 \text{ permutations}$$

PRACTICE 8

A band plays 8 songs on their set list. In how many different ways can they arrange the songs for their show?

Factorial notation

Factorial
notation
Factorial
notation is used
to represent
the product
of consecutive
numbers. The
product of n
consecutive
numbers are
written as n!

All of the examples of permutations that we have seen have involved finding the product of consecutive numbers such as $5 \times 4 \times 3 \times 2 \times 1$. There is an even shorter way of describing products such as this known as **factorial notation**. The product of n consecutive numbers are written as $n!$.

According to this definition, $5 \times 4 \times 3 \times 2 \times 1$ can then be written as $5!$.

EXAMPLE 9

What is the value of $4!$?

Solution

$4!$ is the product of the first 4 numbers $4 \times 3 \times 2 \times 1 = 24$.

PRACTICE 9

What is the value of $6!$?

Factorial notation has a very useful property in that any factorial can be written as the product of number or numbers that commence the factorial and a lower factorial. Here are some examples:

$$5! = 5 \times 4 \times 3 \times 2 \times 1 = 5 \times 4 \times 3!$$
$$6! = 6 \times 5 \times 4 \times 3 \times 2 \times 1 = 6 \times 5!$$

We can use this property to simplify and evaluate expressions using factorial notation.

CRITICAL THINKING

Using this factorial property, we can say that $1! = 1 \times 0!$
How can we use this relationship to determine a value for $0!$?

EXAMPLE 10

What is the value of:

a $\dfrac{5!}{3!}$

Solution

a $\dfrac{5!}{3!} = \dfrac{5 \times 4 \times 3!}{3!}$
Now the $3!$ in the numerator and the $3!$ in the denominator will cancel leaving us with $5 \times 4 = 20$.

Making
Connections
To review how to
cancel fractions
revisit Chapter 5.

b $\frac{6!}{4!}$

b $\frac{6!}{4!} = \frac{6 \times 5 \times \cancel{4!}}{\cancel{4!}}$

The 4! in both the numerator and the denominator will cancel leaving us with $6 \times 5 = 30$.

PRACTICE 10

Determine the value of:

a $\frac{4!}{6!}$

b $\frac{7!}{4!}$

Calculating permutations

In the examples we have looked at so far, we have been calculating the number of arrangements, or permutations, when all of the objects in a set have been used. However, there are other situations where not all of the objects in a set are being used when forming permutations.

EXAMPLE 11

How many 3 digit numbers can be formed from the set of numbers 1, 2, 3, 4, 5, 6 and 7 if each digit is used only once?

Solution

Let's use the set of boxes method for this example:

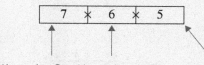

| 7 | ✳ | 6 | ✳ | 5 |

We can place 7 possible digits in the 1st position

Now we have 6 digits available to pick from

Then we have 5 digits available for the 3rd position

Multiplying across we get 210 permutations.

PRACTICE 11

In how many ways can 3 paintings from a collection of 8 be hung in a row?

In Example 11, we found that the permutation of 3 objects taken from a set of 7 was equal to $7 \times 6 \times 5$. Using the property of factorial notation, mentioned in the previous section, we see that:

$$7 \times 6 \times 5 = \frac{7 \times 6 \times 5 \times 4!}{4!}$$
$$= \frac{7!}{(7-3)!}$$

We can use this example as a model to create a generalised rule for calculating the number of permutations of r objects chosen from a set of n objects. This is written as P_r^n.

$$P_r^n = \frac{n!}{(n-r)!}$$

Where the n objects are all different, the r objects are used just once and they are arranged in a particular order.

EXAMPLE 12

Calculate the following permutation.

a P_2^7

Solution

a In this question, $n = 7$ and $r = 2$.

$$P_2^7 = \frac{7!}{(7-2)!}$$
$$= \frac{7!}{5!}$$
$$= \frac{7 \times 6 \times 5!}{5!}$$
$$= 7 \times 6$$
$$= 42$$

b P_3^6

b In this question, $n = 6$ and $r = 3$.

$$P_3^6 = \frac{6!}{(6-3)!}$$
$$= \frac{6!}{3!}$$
$$= \frac{6 \times 5 \times 4 \times 3!}{3!}$$
$$= 6 \times 5 \times 4$$
$$= 120$$

PRACTICE 12

Calculate the following permutations.

a P_1^8

b P_4^5

EXAMPLE 13: APPLICATION

There are 8 horses in a race field. In how many ways can they be placed 1st, 2nd and 3rd?

Solution

In this question, order is clearly important and we are finding how many sets of 3 can be formed from the field of 8. So this is a question that calculates a permutation when $n = 8$ and $r = 3$.

$$P_3^8 = \frac{8!}{(8-3)!}$$
$$= \frac{8!}{5!}$$
$$= \frac{8 \times 7 \times 6 \times 5!}{5!}$$
$$= 8 \times 7 \times 6$$
$$= 336$$

There are 336 ways the horses can be placed.

PRACTICE 13: APPLICATION

There are 18 teams in the Australian Football League. In how many ways can the teams be placed in the top 4 on the premiership table?

Combinations

So far we have considered choosing objects when their order in an arrangement is important. But what about when order is not important? To think about this, let's return to the letters MATH and start by considering the permutations of 3 letters chosen from these 4 letters.

MAT	MAH	THM	THA
MTA	MHA	TMH	TAH
AMT	AMH	HTM	HTA
ATM	AHM	HMT	HAT
TMA	HMA	MTH	ATH
TAM	HAM	MHT	AHT

Combination
A group of objects that disregards how they are ordered or arranged

In this listing, the top row of letters (in green) are the 4 different combinations of 3 letters taken from the letters MATH. Underneath are the different permutations of each of these combinations of letters. This listing highlights what happens when we do not concern ourselves with order. We can see that there are only 4 different combinations of letters if we are not concerned with order, and that there are 24 different permutations if order is important. When we are not concerned with order, we are forming what is known as a **combination**.

The difference between permutations and combinations is that we disregard the order of objects in combinations by eliminating all of the various permutations that can occur. The notation for combinations is C_r^n and the general formula for calculating combinations is:

$$C_r^n = \frac{n!}{r!(n-r)!}$$

Meaning r objects are chosen from a set of n objects, where repetition of objects is not possible and order is not important.

EXAMPLE 14

Calculate the following combinations.

a $\quad C_2^7$

b $\quad C_3^6$

Solution

a In this question, $n = 7$ and $r = 2$.

$$C_2^7 = \frac{7!}{2!(7-2)!}$$
$$= \frac{7!}{2!5!}$$
$$= \frac{7 \times 6 \times 5!}{2!5!}$$
$$= \frac{7 \times 6}{2!}$$
$$= \frac{7 \times 6}{2 \times 1}$$
$$= \frac{42}{2}$$
$$= 21$$

b In this question, $n = 6$ and $r = 3$.

$$C_3^6 = \frac{6!}{3!(6-3)!}$$
$$= \frac{6!}{3!3!}$$
$$\frac{6 \times 5 \times 4 \times 3!}{3!3!}$$
$$= \frac{6 \times 5 \times 4}{3!}$$
$$= \frac{6 \times 5 \times 4}{3 \times 2 \times 1}$$
$$= \frac{120}{6}$$
$$= 20$$

PRACTICE 14

Calculate the following combinations.

a $\quad C_5^8$ 　　　　　　　　　　b $\quad C_2^5$

EXAMPLE 15: APPLICATIONS

Solutions

a A club has 9 members. In how many ways can a committee of 3 be chosen from the members of the club?

a A committee is a group of people that are not arranged in any particular order. Therefore, this question involves the combinations when choosing 3 people from a group of 9.

$n = 9$ and $r = 3$

$$C_3^9 = \frac{9!}{3!(9-3)!}$$

$$= \frac{9!}{3!6!}$$

$$= \frac{9 \times 8 \times 7 \times 6}{3!6!}$$

$$= \frac{9 \times 8 \times 7}{3!}$$

$$= \frac{9 \times 8 \times 7}{3 \times 2 \times 1}$$

$$= \frac{504}{6}$$

$$= 84$$

There are 84 ways the committee can be selected.

b A pizza shop has 16 different toppings. In how many ways can a pizza be made that uses 4 toppings?

b The pizza toppings are not ordered in any particular way; rather they are randomly placed on the pizza. So this question involves the combinations when choosing 4 toppings from a group of 16.

$n = 16$ and $r = 4$

$$C_4^{16} = \frac{16!}{4!(16-4)!}$$

$$= \frac{16!}{4!12!}$$

$$= \frac{16 \times 15 \times 14 \times 13! \times 12!}{4!12!}$$

$$= \frac{16 \times 15 \times 14 \times 13}{4!}$$

$$= \frac{16 \times 15 \times 14 \times 13}{4 \times 3 \times 2 \times 1}$$

$$= \frac{43680}{24}$$

$$= 1820$$

There are 1820 different pizza combinations.

PRACTICE 15: APPLICATIONS

a How many different 5-card hands can be dealt from a pack of 52 cards?

b A student must answer 7 questions in an exam from a set of 10 questions. In how many ways can she choose the 7 questions?

Sometimes it may not be obvious whether we are required to use permutations or combinations. Some hints can usually be found in the nature of the situation being described in a question.

PERMUTATIONS	COMBINATIONS
The order of objects matters or is important.	Order of objects does not matter.
The question involves arranging letters, numbers, people or any other objects.	The question involves selecting committees, teams, or any other objects or items.
The key words are arrange or arrangement.	The key words are select or choose.

EXAMPLE 16: APPLICATION

You have one of each type of coin in your pocket: 5 cent, 10 cent, 20 cent, 50 cent, $1 and $2.

a   In how many ways can 4 of these coins be arranged in a row?

b   In how many ways can 4 of these coins be taken out of your pocket in one handful?

Solution

a   The key word in this question is 'arrange'. This tells us that order is important and, therefore, we need to use permutations.

$$n = 6 \text{ and } r = 4$$
$$P_4^6 = \frac{6!}{(6-4)!}$$
$$= \frac{6!}{2!}$$
$$= \frac{6 \times 5 \times 4 \times 3 \times \cancel{2!}}{\cancel{2!}}$$
$$= 4 \times 5 \times 4 \times 3$$
$$= 360$$

There are 360 ways the coins can be arranged in a row.

b   When you take 4 coins out of your pocket in one handful, there is no order involved in the selection. So this is a combination question.

$$n = 6 \text{ and } r = 4$$
$$C_4^6 = \frac{6!}{4!(6-4)!}$$
$$= \frac{6!}{4!2!}$$
$$= \frac{6 \times 5 \times 4 \times 3 \times \cancel{2!}}{4!\cancel{2!}}$$
$$= \frac{6 \times 5 \times \cancel{4 \times 3}}{\cancel{4 \times 3} \times 2 \times 1}$$
$$= \frac{6 \times 5}{2 \times 1}$$
$$= \frac{30}{2}$$
$$= 15$$

There are 15 ways that the coins can be taken out of your pocket.

16.4 WHAT IS PROBABILITY?

In the introduction to this chapter, we established that probability is the mathematics of chance and that probability tells us the likelihood of something, a particular event occurring.

For any event, we assign a number between 0 and 1 to describe the likelihood that it will occur. The notation that we use for the probability that an event will occur is $p(E)$. If it is impossible that an event will ever occur, it has a probability of 0. That is, if $p(E) = 0$, E will never occur. If an event is certain to occur it has a probability of 1. In other words, if $p(E) = 1$, E will always occur. All other events between these extremes can be assigned a probability between 0 and 1. The probability that an event will occur is written either as a fraction or as a decimal, but it will always be a number between 0 and 1.

The probability that an event will occur can be calculated in two possible ways: experimental probability and theoretical probability. We will commence this section by looking at experimental probability.

Experimental probability

When we think about tossing a coin, our intuition tells us that the chance of getting a 'head' will be $\frac{1}{2}$ or 0.5. However, if we were to toss a coin say 20 times it would be unusual to get exactly 10 'heads'. Rather, we might expect outcomes such as 9 'heads' and 11 'tails' or 12 'heads' and 8 'tails'. This is an example of what is known as experimental probability.

The experimental probability of an event occurring is based on the actual results of a probability activity, which is called an **experiment**.

Experimental probability describes what *actually* happens when an experiment is conducted. To calculate the experimental probability that an event would occur we need to use the following rule:

$$\text{Experimental probability} = \frac{\text{number of occurrences of the event}}{\text{total number of trials}}$$

So, the experimental probability of getting a 'head' in 20 **trials** that results in 9 'heads' and 11 'tails' is $\frac{9}{20}$. If we tossed 12 'heads' and 8 'tails' in 20 trails, then the experimental probability would be $\frac{12}{20}$.

Experiment
A probability activity that generates results known as outcomes

Trial
The performance of a probability experiment

Kathy Brady

EXAMPLE 17: APPLICATION

An ice-cream shop recorded the number of customers purchasing ice-creams in various flavours.

FLAVOUR	NO. CUSTOMERS
Vanilla	12
Chocolate	18
Strawberry	9
Peach	11

a   What is the experimental probability that a customer selected at random purchased a peach ice-cream?

b   What is the experimental probability that a customer selected at random did not purchase a chocolate ice-cream?

Solution

To answer both of these questions we first need to calculate the number of customers that the ice-cream shop had: $12 + 18 + 9 + 11 = 50$.

a   The number of customers that bought a peach ice-cream was 11.

Experimental probability of peach $= \frac{11}{50}$

b   The customers that did not buy a chocolate ice-cream would have purchased vanilla, strawberry or peach. Therefore, the number of customers that did not purchase a chocolate ice-cream was 39.

Experimental probability of not chocolate $= \frac{32}{50}$

PRACTICE 17: APPLICATION

A dice is rolled 60 times and the following results recorded.

Outcome	1	2	3	4	5	6
Frequency	10	9	10	12	8	11

Find the experimental probability of the following events.

a   Getting a 4

b   Getting an odd number

c   Getting a number greater than 3

Theoretical probability

Theoretical probability of an event occurring is what you *expect* to happen when a probability experiment is conducted (which we have discovered is not what might actually happen). It is important to note, however, that as the number of trials increases, the experimental probability of an event gets closer and closer to the theoretical probability.

To calculate the theoretical probability of a particular event occurring we need to think about the number of ways this event can occur when the experiment is conducted. We also need to determine the total possible number of **outcomes** associated with that experiment, known as the **sample space**. For example, when rolling a dice there are 6 possible outcomes: 1, 2, 3, 4, 5 and 6. This is the sample space. An **event** could be rolling an even number or rolling a number less than 3.

We calculate theoretical probability in much the same way as experimental probability by using the following rule:

$$\text{Theoretical probability} = \frac{\text{number of ways that an event can occur}}{\text{total possible number of outcomes}}$$

As an example, let's think about calculating the theoretical probability of getting a 4 when rolling a dice. There are 6 possible outcomes: 1, 2, 3, 4, 5, and 6. But there is just one way a 4 can occur.

$$\text{Theoretical probability of 4} = \frac{\text{number of ways a 4 can occur}}{\text{total possible number of outcomes}} = \frac{1}{6}$$

Experimental and theoretical probability

Experimental probability is frequently used in experimental and clinical contexts including in the social sciences, behavioural sciences, economics and medicine. These involve experimental circumstances where theoretical probability cannot be calculated. For example, to determine the effectiveness of a medical treatment, laboratory mice are injected with the drug and the numbers of mice that are cured are recorded. The experimental probability of a mouse being cured is calculated by dividing the number of cured mice by the total number of mice tested. This experimental probability of mice being cured can then be extended to the population of all mice.

In this example, it would not have been possible to calculate the theoretical probability of a cure. The experiment needed to be conducted to determine the probability of a cure. However, in order for experimental probability to be meaningful in research, the number of trials must be sufficiently large to draw valid conclusions. The principle disadvantage of using experimental probability is that it is not always possible to generalise using it. For this reason, theoretical probability is used in mathematical probability calculations, and we will be concentrating on using theoretical probability for the remainder of this chapter.

Kathy Brady

Outcome
The result of a single trial probability experiment

Sample space
All the possible outcomes that could occur in a probability experiment

Event
A subset of the outcomes from a specified sample space

16.5 REPRESENTING SAMPLE SPACES

We have seen that theoretical probability is calculated as:

$$\text{Theoretical probability} = \frac{\text{number of ways that an event can occur}}{\text{total possible number of outcomes}}$$

All of the outcomes that can occur in an experiment are called the sample space. So to calculate theoretical probability we need to have ways to count the outcomes in a particular sample space. A good way to do this is to use systematic or diagrammatic ways of representing sample spaces.

We can represent sample spaces by:

- listing all the possible outcomes
- creating an organised table
- using a 2D grid
- using a tree diagram.

Listing all the possible outcomes

For a straightforward probability experiment, the sample space can be written as a simple list. For example, when rolling a dice the sample space will be {1, 2, 3, 4, 5, 6} or when tossing a coin the sample space will be {H, T}. Notice the use of curly brackets that enclose the outcomes in a sample space. However, listing a sample space in this way is only practical and efficient if the sample space is small. For example, we would not list the full sample space if the experiment were to draw a card from a standard 52-card pack.

Creating an organised table

When considering the combination of two or more trials of an experiment, it is important to be able to identify all of the possible outcomes. An organised table provides a systematic way of recording outcomes to avoid omissions or duplications.

EXAMPLE 18

Use an organised table to represent the sample space when 10 cent, 20 cent and 50 cent coins are tossed simultaneously.

....

Solution

We need to set up a table to represent the outcomes for each of the different coins in separate columns.

10 CENT	20 CENT	50 CENT
H	H	H
H	H	T
H	T	H
H	T	T
T	H	H
T	H	T
T	T	H
T	T	T

All of the possible outcomes have been listed in a systematic way. Notice the patterns in this table: in the 50 cent column individual H and T alternate, in the 20 cent column pairs of H and T alternate, and in the 10 cent column sets of 4 H and T alternate. Using this pattern, we can create an organised table for tossing any number of coins or other cases where there are only two outcomes possible, for example, having a boy or a girl.

PRACTICE 18

Use an organised table to represent the sample space for 4 spins of this spinner.

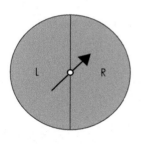

Using a 2D grid

A 2D grid is useful to represent the sample space for two trials of one particular experiment or when two different experiments are conducted simultaneously.

EXAMPLE 19

Use a 2D grid to represent the sample space for 2 spins of this spinner.

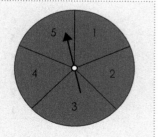

Solution

We set up a 2D grid much like the x and y axes of a graph. On one axes, we list the outcomes from one of the experiments, and on the other axis the outcomes from the other experiment. In this case, we are carrying out the same experiment twice. On the grid, we place a point to represent each member of the sample space.

	5	•	•	•	•	•
Spin 2	4	•	•	•	•	•
	3	•	•	•	•	•
	2	•	•	•	•	•
	1	•	•	•	•	•
		1	2	3	4	5
				Spin 1		

PRACTICE 19

Use a 2D grid to represent the sample space when simultaneously rolling an ordinary dice and spinning this spinner.

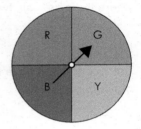

EXAMPLE 20

Two dice are rolled simultaneously and the sum of their faces is recorded. Use a 2D grid to represent this sample space for this experiment.

Solution

In this example, we set up the 2D grid in much the same way as the previous example. However, instead on plotting points to represent the members of the sample space we write the sums of the faces.

	6	7	8	9	10	11	12
	5	6	7	8	9	10	11
Dice 2	4	5	6	7	8	9	10
	3	4	5	6	7	8	9
	2	3	4	5	6	7	8
	1	2	3	4	5	6	7
		1	2	3	4	5	6
				Dice 1			

PRACTICE 20

Simultaneously, an ordinary dice is rolled and this spinner is spun. The sums are recorded. Use a 2D grid to represent the sample space for this experiment.

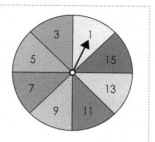

Using a tree diagram

We use a tree diagram as a systematic way of recording all the possible outcomes in an experiment that involves a sequence of trials.

Here is a tree diagram that represents the outcomes when a coin is tossed three times. The first two branches on the left of the tree diagram show the possible outcomes for the first toss of the coin.

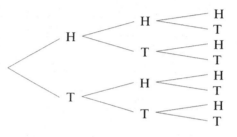

Then for each of those outcomes, the outcomes for the second toss are shown on the next 4 branches, and likewise for the outcomes for the third toss of the coin.

EXAMPLE 21: APPLICATION

A boutique sells handbags in two colours, black and white, and for each colour there are three different sizes: small, medium and large. Draw a tree diagram to represent all of the types of handbags that are possible.

Solution

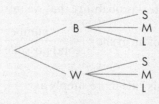

PRACTICE 21: APPLICATION

Draw a tree diagram to represent all of the outcomes when this spinner is spun followed by the toss of a coin.

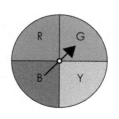

EXAMPLE 22: APPLICATION

A toy bag contains 2 blue toys, 3 green toys and one red toy. Ethan takes out three toys, one after the other. Draw a tree diagram to represent all of the possible sets of toys.

Solution

Because Ethan is taking toys out of the bag and not putting them back in the bag, the number of each colour toy will change depending upon which toys he has removed previously. This needs to be taken into account in the tree diagram.

Because there is only one red toy, once a R has been recorded on this tree diagram no further R branches can be drawn. Similarly, with the 2 blue toys, once two B have been recorded along branches of the tree diagram no further B branches can be drawn.

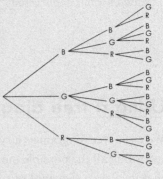

PRACTICE 22: APPLICATION

In Simone's sock drawer she has 2 pairs of white socks, 1 pair of pink socks and 3 pairs of blue socks. Simone randomly removes 3 pairs of socks from the drawer one after the other. Draw a tree diagram to represent all of the outcomes.

16.6 PROBABILITY OF A SINGLE EVENT

When calculating the probability of a single event when *all outcomes are equally likely*, we use the rule for theoretical probability that we introduced in Section 16.4.

$$\text{Probability of an event} = \frac{\text{number of ways that an event can occur}}{\text{total possible number of outcomes in sample space}}$$

We can write this rule more simply as:

$$p(E) = \frac{n(E)}{n(S)}$$

where $n(E)$ is the number of ways the event can occur and $n(S)$ is the total number of outcomes in the sample space.

EXAMPLE 23

A fair dice is rolled. What is probability of:

a a 5?

b an even number?

c a multiple of 3?

d a prime number?

e a number more than 6?

Solutions

When a fair dice is rolled, there are 6 possible outcomes in the sample space $\{1, 2, 3, 4, 5, 6\}$ so $n(S) = 6$. For each part, we need to decide how many ways the desired outcome can occur in order to determine the probability of each outcome.

a The number 5 can occur in one way, so $n(5) = 1$.

$$p(5) = \frac{n(5)}{n(S)} = \frac{1}{6}$$

b There are 3 ways that an even number can occur $\{2, 4, 6\}$, so $n(\text{even}) = 3$.

$$p(\text{even}) = \frac{n(\text{even})}{n(S)} = \frac{3}{6} = \frac{1}{2}$$

c There are 2 ways a multiple of 3 can occur $\{3, 6\}$, so $n(\text{multiple 3}) = 2$.

$$p(\text{multiple 3}) = \frac{n(\text{multiple 3})}{n(S)} = \frac{2}{6} = \frac{1}{3}$$

d There are 3 ways a prime number can occur $\{2, 3, 5\}$, so $n(\text{prime}) = 3$.

$$p(\text{prime}) = \frac{n(\text{prime})}{n(S)} = \frac{3}{6} = \frac{1}{2}$$

e When rolling a dice it is not possible to get a number greater than 6, therefore:

$$p(\text{more than 6}) = 0$$

PRACTICE 23

Each number on this spinner is equally likely to occur when it is spun.
What is the probability of:

a an 11?

b 7 or 9?

c a multiple of 3?

d an even number?

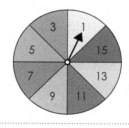

CRITICAL THINKING

In this question what does the term 'fair dice' mean? And why has it been included?

EXAMPLE 24

A card is drawn at random from a standard 52-card pack. What is the probability of:

a a 7?

b a black Queen?

c a Heart?

d not a Diamond?

Kathy Brady

Solutions

Since there are 52 different cards in a standard pack and each is equally likely to be drawn, the size of the sample space is 52, that is $n(S) = 52$. As in the previous example, for each part we need to determine how many ways the desired outcome can occur.

a There are four 7s in the pack {7 Hearts, 7 Diamonds, 7 Clubs, 7 Spades}, so $n(7) = 4$.

$$p(7) = \frac{n(7)}{n(S)} = \frac{4}{52} = \frac{1}{13}$$

b There are 2 black Queens in the pack {Q clubs, Q spades}, so $n(\text{black } Q) = 2$.

$$p(\text{black } Q) = \frac{n(\text{black } Q)}{n(S)} = \frac{2}{52} = \frac{1}{26}$$

c There are 13 Hearts in the pack {2H, 3H, 4H, 5H, 6H, 7H, 8H, 9H, 10H, Jack H, Queen H, King H, Ace H}, so $n(H) = 13$.

$$p(H) = \frac{n(H)}{n(S)} = \frac{13}{52} = \frac{1}{4}$$

d There are 4 suits in the pack (Hearts, Diamonds, Clubs, Spades) and 13 cards in each suit (as listed above). This means that there are 13 cards that are diamonds and therefore $52 - 13 = 39$ cards that are not. So $n(\text{not } D) = 39$.

$$p(\text{not } D) = \frac{n(\text{not } D)}{n(S)} = \frac{39}{52} = \frac{3}{4}$$

PRACTICE 24

A card is drawn at random from a standard 52-card pack. What is the probability of:

a a 5 or a Jack?   b Queen of Hearts?

c a black card?   d not a 10?

Sometimes it is useful to refer to one of the representations of the sample spaces that we have developed in order to determine the probability of a single event.

EXAMPLE 25

What is the probability of getting one 'head' and two 'tails' when 10 cent, 20 cent and 50 cent coins are tossed simultaneously?

Solution

In the previous section, we created an organised table to represent the sample space for this experiment.

We can see that there are 8 possible outcomes, so $n(S) = 8$. We can also use this table to identify how our desired outcome of one 'head' and two 'tails' can occur. The three ways that this outcome can occur are highlighted on the table, so $n(1H \text{ and } 2T) = 3$.

$$p(1H \text{ and } 2T) = \frac{n(1H \text{ and } 2T)}{n(S)} = \frac{3}{8}$$

10 cent	20 cent	50 cent
H	H	H
H	H	T
H	T	H
H	T	T
T	H	H
T	H	T
T	T	H
T	T	T

PRACTICE 25

A couple intend on having 3 children. Assuming that having a boy or a girl is equally likely, what is the probability they will have at least 2 girls?

EXAMPLE 26

Two fair dice are rolled simultaneously and the sum of their faces is recorded. What is the probability that the sum is:

a 2?　　　　　　**b** 5?　　　　　　**c** more than 10?

Solutions

The 2D grid that we generated in the previous section will assist with this example.

We can see on this 2D grid that there are $6 \times 6 = 36$ different ways we can sum the faces of the two dice. This is the size of the sample space, so $n(S) = 36$. This 2D grid is also useful for identifying the desired outcomes in each question.

	6	7	8	9	10	11	12
	5	6	7	8	9	10	11
Dice 2	**4**	5	6	7	8	9	10
	3	4	5	6	7	8	9
	2	3	4	5	6	7	8
	1	2	3	4	5	6	7
		1	**2**	**3**	**4**	**5**	**6**
				Dice 1			

a There is one way a sum of 2 can occur, which is highlighted in light green on the grid, so $n(\text{sum } 2) = 1$.

$$p(\text{sum } 2) = \frac{n(\text{sum } 2)}{n(S)} = \frac{1}{36}$$

b There are four ways a sum of 5 can occur, which are highlighted in grey on the grid, so $n(\text{sum } 5) = 4$.

$$p(\text{sum } 5) = \frac{n(\text{sum } 5)}{n(S)} = \frac{4}{36} = \frac{1}{9}$$

c Beware a sum of more than 10 does not include 10! If it did, then the wording would be 'a sum of 10 or more'. Therefore, there are 3 ways this outcome can occur, which are highlighted in dark green on the grid, so $n(\text{sum more than } 10) = 3$.

$$p(\text{sum more than } 10) = \frac{n(\text{sum more than } 10)}{n(S)} = \frac{3}{36} = \frac{1}{12}$$

PRACTICE 26

Simultaneously, a fair dice is rolled and this spinner is spun. The sums are recorded. Construct a 2D grid to represent the sample space for this experiment. What is the probability that the sum is:

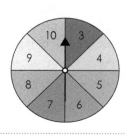

a even?　　　　　　**b** 3?

c 11 or 13?　　　　**d** less than 5?

Complement of an event

Complement of
an event
All outcomes
that are not
that event

The **complement of an event** is the set of outcomes in the sample space that are not included in the outcomes for that event. For example:

- If the event is {Heads}, the complement is {Tails}
- If the event is {Hearts}, the complement is {Diamonds, Clubs and Spades}
- When rolling a dice the event is {odd}, the complement is {2, 4, 6}.

We use the notation E' for the complement of the event E and an important, and very useful, relationship exists between E and E'. As an example of this relationship, let's think about rolling of a dice. In one roll of a dice, either a 5 will occur or it will not, with the complement of 5 being {1, 2, 3, 4, 6}.

Now $p(5) = \frac{1}{6}$ and $p(\text{not a } 5) = \frac{5}{6}$, and we can notice that:

$$p(5) + p(\text{not a } 5) = \frac{1}{6} + \frac{5}{6} = 1$$

This ought not be surprising because either getting a 5 or not getting a 5 *must occur,* and we have already established that the probability of something certainly occurring is equal to 1.

Therefore, in general terms we can write that:

$$p(E) + p(E') = 1$$

which can be re-arranged to:

$$p(E) = 1 - p(E')$$

We can use the probability of the complement of an event when it is easier to determine that probability and subtract from 1, rather than the probability of the required event.

EXAMPLE 27

If a number is selected at random from the set of numbers 1–50, what is the probability that the number is not a square number?

Solution

In the set of numbers 1–50, any particular number will either be a square number or it will not be a square number. These are complementary events. In this question, we are asked to find the probability of not a square number (E'). However, rather than listing and counting all of the not square numbers it is easier to think about the complement of that event: the square numbers $E' = \{1, 4, 9, 16, 25, 36, 49\}$.

So $n(E') = 7$ and since the $n(S) = 50$, we can calculate the $p(E')$.

$$p(E') = p(\text{square numbers}) = \frac{7}{50}$$

Now we can use the probability of the complementary event to find the probability of the event we require.

$$p(E) = p(\text{not square number})$$
$$= 1 - p(E')$$
$$= 1 - \frac{7}{50}$$
$$= \frac{43}{50}$$

PRACTICE 27: GUIDED

A set of marbles numbered 1–60 is placed in a bag. What is the probability that a marble drawn at random is not a multiple of 7?

The event E is drawing a marble that is not a multiple of 7.

$$E' = \{7, 14, \square, 35, \square, \square, \square\}$$
$$n(E') = 8$$
$$n(S) = \square$$
$$p(E') = p(\text{multiple } 7)$$
$$= \frac{\square}{60}$$
$$p(E) = p(\text{not multiple } 7)$$
$$= 1 - p(E')$$
$$= 1 - \frac{8}{\square}$$
$$= \frac{\square}{\square}$$

Now try this one:

What is the probability that a letter selected at random from the alphabet is not a consonant?

16.7 PROBABILITY OF TWO EVENTS

When two events take place at the same time, the probability of a particular combined outcome is calculated as for a single event:

$$p(E) = \frac{n(E)}{n(S)}$$

where $n(E)$ is the number of ways the particular combined outcome can occur and $n(S)$ is the number of outcomes in the sample space.

When calculating the probability of two events, 2D grids are particularly useful to both determine the size of the sample space and the number of desired outcomes.

EXAMPLE 28

An ordinary dice is rolled and this spinner is simultaneously spun.
Use a 2D grid to find the probability of:

a two 3's

b a 4 and a 2

c a 5 or a 6

d 4 or less on both

e a sum of 7

f a sum of 7 or 11

g a sum greater than 8

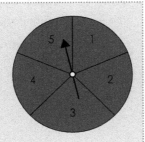

Solutions

The first thing we need to do is to construct a 2D grid to represent the sample space.

We can see on this 2D grid that there are $6 \times 5 = 30$ different combinations involving the spinner and the dice. This is the size of the sample space, so $n(S) = 30$. This 2D grid is also useful for identifying the desired outcomes in each question.

a There is one way that two 3s can occur, which is highlighted in grey on the grid above, so $n(\text{two 3s}) = 1$.
$$p(\text{two 3s}) = \frac{n(\text{two 3s})}{n(S)} = \frac{1}{30}$$

b There are two ways that a 4 and a 2 can occur, which are highlighted in dark green on the grid above, so $n(4 \text{ and } 2) = 2$.
$$p(4 \text{ and } 2) = \frac{n(4 \text{ and } 2)}{n(S)} = \frac{2}{30} = \frac{1}{15}$$

c There are 14 ways that a 5 or a 6 can occur (being careful not to double count), which are highlighted in light green on the grid about. So $n(5 \text{ or } 6) = 14$.
$$p(5 \text{ or } 6) = \frac{n(5 \text{ or } 6)}{n(S)} = \frac{14}{30} = \frac{7}{15}$$

d Getting a 4 or less on both is the complement of getting a 5 or a 6. That is the section of the grid above that is not shaded in light green. We can use the probability of complementary events here as we already know $p(5 \text{ or } 6)$.

$$p(4 \text{ or less on both}) = 1 - p(5 \text{ or } 6)$$
$$= 1 - \frac{7}{15}$$
$$= \frac{8}{15}$$

e To complete the next three questions we will start again with a fresh 2D grid for this experiment.

Dice	6	•	•	•	•	•
	5	•	•	•	•	•
	4	•	•	•	•	•
	3	•	•	•	•	•
	2	•	•	•	•	•
	1	•	•	•	•	•
		1	2	3	4	5
				Spinner		

There are 5 ways we can get a sum of 7 that are highlighted in dark green on this 2D grid so $n(\text{sum } 7) = 5$.

$$p(\text{sum } 7) = \frac{n(\text{sum } 7)}{n(S)} = \frac{5}{30} = \frac{1}{6}$$

f There is just one way we can get a sum of 11, which is highlighted in light green above. So altogether the total number of ways of getting 7 *or* 11 is 6, so $n(\text{sum } 7 \text{ or } 11) = 6$.

$$p(\text{sum } 7 \text{ or } 11) = \frac{n(\text{sum } 7 \text{ or } 11)}{n(S)} = \frac{6}{30} = \frac{1}{5}$$

g Beware here because a sum greater than 8 does not include 8! If it did, then the wording would be 'a sum of 8 or more'. Therefore, there are 6 ways this outcome can occur. These are highlighted in grey above and we also need to include the light green highlighting as this represents a sum of 11, so $n(\text{sum more than } 8) = 6$.

$$p(\text{sum more than } 8) = \frac{n(\text{sum more than } 8)}{n(S)} = \frac{6}{30} = \frac{1}{5}$$

PRACTICE 28

These two spinners are spun simultaneously. Use a 2D grid to find the probability of:

a two numbers that are the same

b a 3 and a 7

c a 7 or a 12

d a sum that is even

e a sum that is less than 10

f a sum that is no more than 10

g a sum of 32

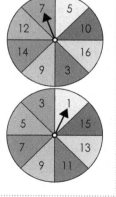

Mutually exclusive events

Let's think about two events that could occur when a card is drawn from a pack: drawing a 6 or drawing an ace. Since a card cannot be both a 6 and an ace, these events are known as **mutually exclusive events**.

Kathy Brady

Mutually exclusive events
Two or more events that have no outcomes in common

Recall that in Section 16.2 we considered the Addition Principle. This principle determined that if set A and set B had no elements in common then the total number of elements will be the sum of the elements in both set A and set B. We wrote the Addition Principle as:

$$n(A \cup B) = n(A) + n(B)$$

In this notation, the symbol $\cup$ means 'or', and we read the notation as 'the number of elements in A or B is equal to the number of elements in A plus the number of elements in B'.

We can apply the Addition Principle relationship to find the probability of mutually exclusive events. This is because mutually exclusive events have no outcomes in common.

If event A and event B have no outcomes in common, then the probability of either event A or event B is given by:

$$p(A \cup B) = p(A) + p(B)$$

EXAMPLE 29

When drawing a card from a pack of 52, what is the probability of drawing either a 6 or an ace?

Solution

As we have already established, drawing a 6 and drawing an ace are mutually exclusive because a card cannot be both a 6 and an ace. Let event A be drawing a 6 and event B be drawing an ace.

There are four 6s in the pack {6 Hearts, 6 Diamonds, 6 Clubs, 6 Spades}, so:

$$p(6) = \frac{n(6)}{n(S)} = \frac{4}{52} = \frac{1}{13}$$

There are four aces in the pack {ace Hearts, ace Diamonds, ace Clubs, ace Spades}, so:

$$p(ace) = \frac{n(ace)}{n(S)} = \frac{4}{52} = \frac{1}{13}$$

Therefore, the probability of drawing either a 6 or an ace is:

$$p(6 \cup ace) = p(6) + p(ace) = \frac{1}{13} + \frac{1}{13} = \frac{2}{13}$$

PRACTICE 29

When drawing a card from a pack of 52, what is the probability of drawing either a black card or a heart?

EXAMPLE 30

If the letters in the word PROBABILITY are written on individual cards and placed in a bag, what is the probability that a card drawn at random will be either an I or a consonant?

Solution

The events drawing an I and drawing a consonant are mutually exclusive because I is a vowel and no letter can be both a vowel and a consonant. Let event A be drawing an I and event B be drawing a consonant. The size of the sample space in this example is the number of letters in the word PROBABILITY, that is $n(S) = 11$.

There are two Is in the word PROBABILITY

$$p(I) = \frac{n(I)}{n(S)} = \frac{2}{11}$$

There are seven consonants in the word PROBABILITY

$$p(\text{consonant}) = \frac{n(\text{consonant})}{n(S)} = \frac{7}{11}$$

Therefore, the probability of drawing either an I or a consonant is:

$$p(I \cup \text{consonant}) = p(I) + p(\text{consonant}) = \frac{2}{11} + \frac{7}{11} = \frac{9}{11}$$

PRACTICE 30

When this spinner is spun, what is the probability that the arrow will land on a light green section or a multiple of 4?

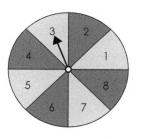

Non-mutually exclusive events

Now let's think about another two events that could occur when drawing a card from a pack: drawing a 4 or drawing a black card. These events are **non-mutually exclusive events** because it is possible for a card to be 4 and black (4 Clubs, 4 Spades).

In Section 16.2, we also considered the Exclusion Principle. This principle determined that if set A and set B have some elements in common the number of common elements must be subtracted from the sum of set A and set B so that we did not 'double count' the common elements. We write the Exclusion Principle as:

$$n(A \cup B) = n(A) + n(B) - n(A \cap B)$$

Remember $\cup$ means 'or' and $\cap$ means 'and'.

We can apply the same relationship to find the probability of non-mutually exclusive events.

If event A and event B have some outcomes in common, then the probability of either event A or event B is given by:

$$p(A \cup B) = p(A) + p(B) - p(A \cap B)$$

Non-mutually exclusive events
When two or more events have some outcomes in common

EXAMPLE 31

When drawing a card from a pack of 52, what is the probability of drawing either a 4 or a black card?

Solution

As we have already established, drawing a 4 and drawing a black card are non-mutually exclusive because it is possible for a card to be both a 4 and black. Let event A be drawing a 4 and event B be drawing a black card.

There are four 4s in the pack {4 Hearts, 4 Diamonds, 4 Clubs, 4 Spades}, so:

$$p(4) = \frac{n(4)}{n(S)} = \frac{4}{52}$$

There are 26 black cards in the pack {13 × Clubs, 13 × Spades}, so:

$$p(\text{black}) = \frac{n(\text{black})}{n(S)} = \frac{26}{52}$$

Now there are 2 black 4s in the pack {4 Clubs, 4 Spades}, so:

$$p(4 \cap \text{black}) = \frac{n(\text{black } 4)}{n(S)} = \frac{2}{52}$$

Therefore, the probability of drawing either a 4 or a black card is:

$$p(4 \cup \text{black}) = p(4) + p(\text{black}) - p(4 \cap \text{black})$$
$$= \frac{4}{52} + \frac{26}{52} - \frac{2}{52}$$
$$= \frac{28}{52}$$
$$= \frac{7}{13}$$

PRACTICE 31

What is the probability that a number picked at random from the set of numbers 1–100 is square or even?

EXAMPLE 32

When this spinner is spun, what is the probability that the arrow will land on a mid-green or a prime?

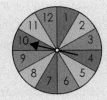

Solution

On this spinner are the numbers 1–12. Therefore, there are 5 prime numbers on the spinner {2, 3, 5, 7, 11}. One of the prime numbers {3} is on a mid-green sector; therefore, the events prime number and mid-green are non-mutually exclusive. The size of the sample space in this example are the number of sectors on the spinner, that is $n(S) = 12$.

There are five primes, so:

$$p(\text{prime}) = \frac{n(\text{prime})}{n(S)} = \frac{5}{12}$$

There are four mid-green sectors, so:

$$p(\text{mid-green}) = \frac{n(\text{mid-green})}{n(S)} = \frac{4}{12}$$

There is one prime on a mid-green, so:

$$p(\text{prime} \cap \text{mid-green}) = \frac{1}{12}$$

Therefore, the probability of the spinner landing on either a prime or a mid-green is:

$$p(\text{prime} \cup \text{mid-green}) = p(\text{prime}) + p(\text{mid-green}) - p(\text{prime} \cap \text{mid-green})$$
$$= \frac{5}{12} + \frac{4}{12} - \frac{1}{12}$$
$$= \frac{8}{12}$$
$$= \frac{2}{3}$$

PRACTICE 32

If the letters in the word PROBABILITY are written on individual cards and placed in a bag, what is the probability that a card drawn at random will be either a B or a consonant?

Independent events

When we are considering mutually exclusive events and non-mutually exclusive events, we use the word 'or' to describe the events. For example:

- Probability of drawing either a 4 *OR* a black card
- Probability that a card drawn at random will be either a B *OR* a consonant
- Probability that the arrow will land on a yellow *OR* a multiple of 4

In these situations, we apply the addition or exclusion rule to calculate the probability.

By contrast, there is a different type of probability event that requires 'and', rather than 'or'. These events are known as **independent events** and are where the outcomes from one event do not affect the outcomes from the other event.

To explore independent events let's think about tossing a coin *and* rolling a dice. These two events are considered to be independent because the outcomes from one event do not affect the outcomes from the other. We can use a 2D grid to represent the outcomes when these two independent events occur.

Independent events
When the outcomes from one event do not affect the outcomes from the other event

H	•	•	•	•	•	•
T	•	•	•	•	•	•
	1	2	3	4	5	6

Kathy Brady

Firstly, let's use this grid to find the probability of tossing a 'head'. The outcomes that result in a 'head' are shaded in green on the above diagram so the probability of a 'head' is:

$$p(H) = \frac{6}{12} = \frac{1}{2}$$

Now let's use the grid to find the probability of rolling a 4. The outcomes that result in a 4 are shaded in grey on the grid below.

H	•	•	•	•	•	•
T	•	•	•	•	•	•
	1	2	3	4	5	6

So the probability of a 4 is:

$$p(4) = \frac{2}{12} = \frac{1}{6}$$

What about the probability of getting a 'head' *and* a 4? We could write this as p(H ∩ 4), keeping in mind that ∩ means 'and'.

Notice that there are 12 outcomes in the sample space and only one of them is a 'head' *and* 4. Therefore:

$$p(H \cap 4) = \frac{1}{12}$$

We can also notice that:

$$p(H) \times p(4) = \frac{1}{2} \times \frac{1}{6} = \frac{1}{12}$$

It, therefore, looks like:

$$p(H \cap 4) = p(H) \times p(4)$$

We can generalise this relationship to what is known as the Multiplicative Rule.

If *A* and *B* are independent events, then the probability of both event *A* *and* event *B* occurring is:

$$p(A \text{ and } B) = p(A) \times p(B)$$

The Multiplicative Rule can also be written as $p(A \cap B) = p(A) \times p(B)$. The important thing to remember is that when we say 'and', we must use the Multiplicative Rule.

EXAMPLE 33: APPLICATION

Simon is in a Geography tutorial class of 20 students from which 4 will be randomly selected to present in class next week. He is also in an Education tutorial class of 16 students where 6 will be selected at random to present next week. What is the probability that Simon will be presenting in both classes next week?

Solution

What happens in Simon's Geography class does not have any affect on what happens in his Education class, therefore these events are mutually exclusive. The question is asking what is the probability that Simon will present in his Geography class *and* his Education class. Because we are saying 'and', we must multiply the probabilities. Let event A be being selected in the Geography class and event B be being selected in the Education. For event A, the $n(S) = 20$ and for event B, the $n(S) = 16$.

The $p(A) = p(\text{Geography}) = \frac{4}{20} = \frac{1}{5}$

The $p(B) = p(\text{Education}) = \frac{6}{16} = \frac{3}{8}$

Therefore:

$$p(A \cap B) = p(\text{Geography} \cap \text{Education})$$
$$= p(\text{Geography}) \times p(\text{Education})$$
$$= \frac{1}{5} \times \frac{3}{8}$$
$$= \frac{3}{40}$$

Simon has a $\frac{3}{40}$ chance of presenting in both classes next week.

PRACTICE 33: APPLICATION

The Faculty office has two photocopiers. Machine A breaks down on 3 days in every 40, and Machine B breaks down on 2 days in every 30. What is the probability that:

a machine A will break down on any particular day?

b machine B will break down on any particular day?

c both machines will break down on the same day?

d both will work on any particular day?

EXAMPLE 34: APPLICATION

Rosa is studying Mathematics, Economics and French. The probability that she will pass Mathematics is 0.9, pass Economics is 0.2 and pass French is 0.65. What is the probability that Rosa will pass:

a Mathematics and Economics?

b pass all 3 subjects?

Solution

In this example, we are calculating the probabilities as decimals rather than as fractions.

a This question is asking what is the probability that Rosa will pass Mathematics *and* Economics. Because we are saying 'and', we must multiply the probabilities.

$$p(M \text{ and } E) = p(M) \times p(E)$$
$$= 0.9 \times 0.2$$
$$= 0.18$$

Therefore, the probability that Rosa will pass both Mathematics and Economics is 0.18 (2 dec. pl.)

b   We can extend the multiplicative rule to more than two events as in this example. The probability that Rosa will pass all three subjects is:

$$p(M \text{ and } E \text{ and } F) = p(M) \times p(E) \times p(F)$$
$$= 0.9 \times 0.2 \times 0.65$$
$$= 0.117$$

So the probability that Rosa will pass all three subjects is 0.12 (2 dec. pl.)

PRACTICE 34: APPLICATION

Shane and Joshua take shots at the basket from the three-point line. From experience, Shane gets it in the basket 2 shots out of every 3; and Joshua makes the shot 4 out of 7 times. What is the probability (to 2 dec. pl.) that:

a   Shane makes the shot?

b   Joshua makes the shot?

c   they both get a basket?

d   Shane makes the shot, and Joshua does not?

e   they both miss?

Probability tree diagrams

Probability tree diagrams are an effective way to represent situations that involve independent events. The probabilities are written on the branches of the tree diagram and the outcomes are written at the end of each branch.

1st draw	2nd draw	Outcomes	Probability
	$\frac{2}{3}$ red	red, red	$\frac{2}{3} \times \frac{2}{3} = \frac{4}{9}$
$\frac{2}{3}$ red	$\frac{1}{3}$ blue	red, blue	$\frac{2}{3} \times \frac{1}{3} = \frac{2}{9}$
$\frac{1}{3}$ blue	$\frac{2}{3}$ red	blue, red	$\frac{1}{3} \times \frac{2}{3} = \frac{2}{9}$
	$\frac{1}{3}$ blue	blue, blue	$\frac{1}{3} \times \frac{1}{3} = \frac{1}{9}$

For example, this probability tree diagram represents the situation when a marble is drawn from a bag containing 2 red marbles and 1 blue marble. The colour of the marble is recorded and then it is return to the bag before a second marble is drawn.

We should notice on this probability tree diagram that we are using the word 'and' as we read along the branches. For example: 'red' and 'red', or 'blue' and 'red'. For this reason, the probabilities are multiplied at the ends of the branches.

EXAMPLE 35: APPLICATION

Raj is having technology problems. His computer boots up $\frac{4}{5}$ of the time, but his laptop will only boot up $\frac{2}{3}$ of the time. He turns on his computer first, and then his laptop.

a   Draw a probability tree diagram to represent this situation.

b   What is the probability that both devices will work?

c   What is the probability that the laptop is the only device that works?

Solution

When Raj turns on his computer it will either boot up or it will not. These are complementary events. So if the probability that the computer will boot up is $p(C)$, the probability that it will not is $p(C')$ and the $p(C) + p(C') = 1$. Therefore, $p(C') = \frac{1}{5}$, which is the probability that Raj's laptop will not work.

a   Using this notation, the following probability tree diagram represents this situation:

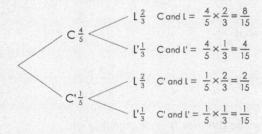

b   The probability that both devices will work is $p(C$ and $L)$. We can see on the probability tree diagram the $p(C$ and $L) = \frac{8}{15}$.

c   If the laptop is the only device that will work, then that means that the computer is not working, which is $p(C'$ and $L)$. From the probability tree diagram, we can see that $p(C'$ and $L) = \frac{2}{15}$.

PRACTICE 35: APPLICATION

The probability that Shane gets a basket from the three-point line is 0.8, and the probability that Joshua makes the shot is 0.65. They take turns to have a shot at the basket. Shane goes 1st and Joshua goes 2nd. Draw a probability tree diagram to represent this situation. What is the probability they both get their shots in? What is the probability that neither of them makes a shot? What is the probability that at least one of the shots goes in the basket?

EXAMPLE 36: APPLICATION

A toy bag contains 2 blue toys, 3 green toys and 1 red toy. Ethan takes out a toy, plays with it and puts it back. He then takes out a second toy to play with. Draw a probability tree diagram to represent this situation. What is the probability that:

a   the two toys that Ethan played with were the same colour?

b   Ethan did not play with a green toy?

c   Ethan played with at least one red toy?

Solution

There are 6 toys in the bag, so the size of the sample space is 6 at each pick because Ethan puts back the first toy before he takes out the second.

At each pick the probability of a blue is:
$$\frac{n(B)}{n(S)} = \frac{2}{6}$$

the probability of a green is:
$$\frac{n(G)}{n(S)} = \frac{3}{6}$$

and the probability of a red is:
$$\frac{n(R)}{n(S)} = \frac{1}{6}$$

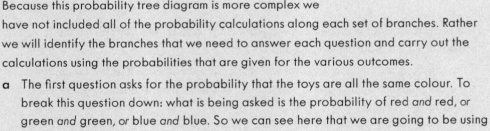

Because this probability tree diagram is more complex we have not included all of the probability calculations along each set of branches. Rather we will identify the branches that we need to answer each question and carry out the calculations using the probabilities that are given for the various outcomes.

a   The first question asks for the probability that the toys are all the same colour. To break this question down: what is being asked is the probability of red *and* red, *or* green *and* green, *or* blue *and* blue. So we can see here that we are going to be using the rules that apply to both mutually exclusive events (indicated by the word 'or') and independent events (indicated by the word 'and').

Using probability notation this will be written as:
$$p(\text{same colour}) = p(B \cap B) \cup p(G \cap G) \cup p(R \cap R)$$

Remember ∩ means 'and' and ∪ means 'or'. By using this notation, we can see where in the solution we need to multiply (∩) and where we need to add (∪).

On the probability tree diagram, we are following the BB branches, GG branches and RR branches to find the relevant probabilities (highlighted as * on the diagram). Now we can do the calculations:

$$p(\text{same colour}) = p(B \cap B) \cup p(G \cap G) \cup p(R \cap R)$$
$$= (\tfrac{2}{6} \times \tfrac{2}{6}) + (\tfrac{3}{6} \times \tfrac{3}{6}) + (\tfrac{1}{6} \times \tfrac{1}{6})$$
$$= \tfrac{4}{36} + \tfrac{9}{36} + \tfrac{1}{36}$$
$$= \tfrac{14}{36}$$
$$= \tfrac{7}{18}$$

b In the second part of this example, we need to determine the probability that Ethan did not play with a green toy. To do this we need to identify the branches that do not involve picking green (highlighted as ■ on the diagram). We can see that there are 4 ways that this can happen: B and B, or B and R, or R and B, or R and R. Using probability notation this will be written as:

$$p(\text{not green}) = p(B \cap B) \cup p(B \cap R) \cup p(R \cap B) \cup p(R \cap R)$$

Now we can do the calculations:

$$p(\text{not green}) = (\tfrac{2}{6} \times \tfrac{2}{6}) + (\tfrac{2}{6} \times \tfrac{1}{6}) + (\tfrac{1}{6} \times \tfrac{2}{6}) + (\tfrac{1}{6} \times \tfrac{1}{6})$$
$$= \tfrac{4}{36} + \tfrac{2}{36} + \tfrac{2}{36} + \tfrac{1}{36}$$
$$= \tfrac{9}{36}$$
$$= \tfrac{1}{4}$$

c In the final part of this example, we need to find the probability that Ethan played with at least one red toy. This is complementary to the event that Ethan played with no red toys and in this case it is more straightforward to use the complementary event.

The outcomes for no red toys are highlighted on the diagram (as ✖). We can see that there are 4 ways that this can happen: B and B, or B and G, or G and B, or G and G. Using probability notation this will be written as:

$$p(\text{no red}) = p(B \cap B) \cup p(B \cap G) \cup p(G \cap B) \cup p(G \cap G)$$

Now we can do the calculations for no red toys:

$$p(\text{no red}) = (\tfrac{2}{6} \times \tfrac{2}{6}) + (\tfrac{2}{6} \times \tfrac{3}{6}) + (\tfrac{3}{6} \times \tfrac{2}{6}) + (\tfrac{3}{6} \times \tfrac{3}{6})$$
$$= \tfrac{4}{36} + \tfrac{6}{36} + \tfrac{6}{36} + \tfrac{9}{36}$$
$$= \tfrac{25}{36}$$

Finally, we need to use the probability of this complementary event to calculate the probability that Ethan played with at least one red toy:

$$p(\text{at least 1 red}) = 1 - p(\text{no red})$$
$$= 1 - \tfrac{25}{36}$$
$$= \tfrac{11}{36}$$

PRACTICE 36: APPLICATION

In Simone's sock drawer she has 2 pairs of white socks, 1 pair of pink socks and 3 pairs of blue socks. Simone randomly removes a pair of socks, records the colour and then puts them back. She repeats this two more times. Draw a probability tree diagram to represent this situation. What is the probability that the three pairs of socks that Simone removed:

a   were all blue?   b   were all different colours?

c   did not include a pink pair?   d   included at least one white pair?

Dependent events

In some situations, the outcome of one event does have an affect on the outcomes of subsequent events. To illustrate this let's keep thinking about sock drawers. Imagine a sock drawer that contains 4 pairs of blue socks and 3 pairs of red socks. One pair of socks is removed from the drawer, but not put back, and then another pair is taken out. What is the probability that the second pair will be red?

If the first pair removed was red, p(second pair red) = $\frac{2}{6}$. This is because once the first pair has been removed there are 6 pairs left in the drawer and 2 of them are red. On the other hand, if the first pair removed was blue, p(second pair red) = $\frac{3}{6}$. This is because there are 6 pairs left in the drawer but now 3 of them are red. So, we can see that the probability that the second pair is red was dependent upon the colour that was removed the first time.

The probability of **dependent events** is calculated in a similar way to independent events using a multiplicative rule. However, if event A and event B are dependent events, then:

$$p(A \textit{ then } B) = p(A) \times p(B \text{ given that } A \text{ has occurred})$$

Often the probabilities of dependent events involve an object or item being selected from a set and not returned. This is known as drawing *without replacement*. Probability tree diagrams are very useful when calculating the probabilities of these and other types of dependent events.

> **Dependent events**
> For two events, when the outcome of the second event is affected by the outcome from the previous event

---

EXAMPLE 37: APPLICATION

Sophia has a box of identical chocolates containing 8 peppermint creams and 12 coffee creams. She selects a chocolate at random and eats it; she then selects another chocolate at random. Draw a probability tree diagram to represent this situation. What is the probability that:

a   both chocolates will be the same?

b   Sophia will eat at least one peppermint cream?

Solution

Because there are 20 chocolates in the box to begin with, the sample space for the first pick will be $n(S) = 20$. Once the first chocolate has been selected and consumed, the sample space for the second pick will be $n(S) = 19$. The numbers of peppermint and coffee creams remaining in the box after the first pick will be dependent on that pick. So the probability tree diagram will be:

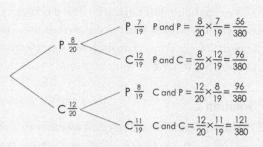

a The probability that both chocolates will be the same is the probability of *P and P*, or *C and C*. Using probability notation we would write this as:

$$p(\text{both same}) = p(P \cap P) \cup p(C \cap C)$$

To calculate this we can use the probabilities that have been given at the ends of the branches:

$$p(\text{both same}) = \frac{56}{380} + \frac{121}{380} = \frac{177}{380}$$

Alternatively, we could write the answer as 0.47 (2 dec. pl.).

b In the second part of this example, we need to find the probability that Sophia ate at least one peppermint cream. This is the complementary event to the event that Sophia ate no peppermint creams and in this case it is more straightforward to use the complementary event.

The only way that Sophia would have eaten no peppermint creams is if both chocolates that she ate were coffee creams. We can see from the probability tree diagram that:

$$p(\text{two C}) = \frac{121}{380}$$

Therefore:

$$p(\text{at least 1 peppermint}) = 1 - \frac{121}{380} = \frac{259}{380}$$

As a decimal, we can write this as a probability of 0.68 (2 dec. pl.).

PRACTICE 37: APPLICATION

Five balls numbered 3, 5, 6, 7 and 8 are placed in a bag. Two balls are selected at random without replacement. Draw a probability tree diagram to represent this situation. What is the probability that:

a both are even?

b both are odd?

c the first is odd and the second is even?

d one is odd and the other is even?

Kathy Brady

EXAMPLE 38: APPLICATION

The probability that Kerri will attend a Physics lecture is 0.1 if she has had a late night out with her friends and 0.7 if she does not. The probability that Kerri will go out with her friends is 0.66. Draw a probability tree diagram to represent this information. What is the probability that Kerri will not attend her lecture?

Solution

The first event that will occur is whether or not Kerri goes out with her friends. If the probability that she will go out is p(go out) = 0.66, then the probability that she will not go out is the complementary event, that is p(not go out) = 1 − 0.66 = 0.34. The probability that Kerri will attend her lecture is dependent on whether or not she goes out, and in

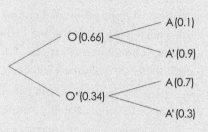

each case we need to use the complementary event to calculate the probability that she does not attend the lecture. The probability tree diagram will be as shown.

We can see from the tree diagram the two outcomes that will result in Kerri not attending her lecture are O and A', or O' and A'. We can write this using probability notation as:

p(not attend) = $p(O \cap A') \cup p(O' \cap A')$

And calculate the probability as:

p(not attend) = (0.66 × 0.9) + (0.34 × 0.3)

= 0.594 + 0.102

= 0.696

Therefore, the probability that Kerri will not attend her lecture is 0.7 (to 1 dec. pl.)

PRACTICE 38: APPLICATION

The probability that Jeffery will pass his Philosophy exam is 0.75 if he revises and 0.4 if he does not. The probability that he will revise is 0.33. Draw a probability tree diagram to represent this situation. What is the probability that Jeffery will not pass his exam?

EXAMPLE 39: APPLICATION

Marc has four 10-cent coins in his pocket and five 20-cent coins. He takes out two coins without replacement. What is the probability that he takes out:

a   40 cents?

b   at least 30 cents?

c   one 10-cent piece and one 20-cent piece?

Solution

It is not always necessary to draw a probability tree diagram. In this question, we can move straight to calculating the probabilities by considering the relevant outcomes.

a If Marc were to take two coins out of his pocket without replacement that totalled to 40 cents, then he has taken out two 20-cent pieces. On the first draw, the probability of a 20-cent would be $\frac{5}{9}$ and then the probability of a 20-cent when the second coin is taken out is $\frac{4}{8}$. This is because there are now 8 coins left in Marc's pocket, of which four are 20-cent pieces. Therefore, the probability of taking out two 20-cent pieces is:

$$p(\text{two } 20) = p(1\text{st } 20) \times p(2\text{nd } 20)$$
$$= \frac{5}{9} \times \frac{4}{8}$$
$$= \frac{20}{72}$$
$$= \frac{5}{18}$$

b The only way that Marc would not have at least 30 cents is if he took out a 10-cent coin each time. Therefore, we can use complementary events to find this probability by first finding the probability of two 10-cent coins. On the first draw, the probability of a 10-cent would be $\frac{4}{9}$ and then the probability of another 10-cent is $\frac{3}{8}$. This is because there are 8 coins left in Marc's pocket, of which three are 10-cent pieces. Therefore, the probability of taking out two 10-cent pieces is:

$$p(\text{two } 10) = p(1\text{st } 10) \times p(2\text{nd } 10)$$
$$= \frac{4}{9} \times \frac{3}{8}$$
$$= \frac{12}{72}$$
$$= \frac{1}{6}$$

Now using the probability of complementary events:

$$p(\text{at least 30 cents}) = 1 - \frac{1}{6} = \frac{5}{6}$$

c If Marc were to take out a 10-cent piece and a 20-cent piece, he could have taken the 10-cent first and the 20-cent second, or the reverse. So we need to consider both possibilities in finding this probability.

$$p(10 \text{ then } 20) = \frac{4}{9} \times \frac{5}{8} = \frac{20}{72} = \frac{5}{18}$$
$$p(20 \text{ then } 10) = \frac{5}{9} \times \frac{4}{8} = \frac{20}{72} = \frac{5}{18}$$

We now need to add these to find the probability of either outcome:

$$p(\text{one 10 and one 20}) = \frac{5}{18} + \frac{5}{18}$$
$$= \frac{10}{18}$$
$$= \frac{5}{9}$$

Kathy Brady

PRACTICE 39: APPLICATION

There are 4 male and 6 female students hoping to be selected for an overseas internship. Two students will be selected at random. What is the probability that:

a   both will be female?

b   at least one male is chosen?

c   one male and one female is chosen?

16.8 CHAPTER SUMMARY

SKILL OR CONCEPT	DEFINITION OR DESCRIPTION	EXAMPLES
Probability	Probability is the mathematics of chance.	
Describing a probability	For any event a number between 0 and 1 (written as a fraction or a decimal) is assigned to describe the likelihood that it will occur.	$p(\text{head}) = \frac{1}{2}$ $p(\text{rain tomorrow}) = 0.14$
Certain event	A *certain* event will have a probability of 1.	When tossing a coin $p(\text{head or tail}) = 1$
Impossible event	An *impossible* event will have a probability of 0.	When rolling a fair dice $p(\text{getting a 7}) = 0$
Probability scale	Used to record differing probabilities.	
Fundamental Principle of Counting	If event A occurs in *n* ways and for each of these *n* ways, event B occurs in *m* ways, then the events A and B can occur in succession in $n \times m$ ways.	3 different types of bread and 5 different fillings will result in $3 \times 5 = 15$ different sandwiches
Addition Principle	If the elements being counted can be divided into sets A and B with no elements in common, then the total number of elements is the sum of the elements in each set. Written as: $n(A \cup B) = n(A) + n(B)$ Where ∪ means 'or'	How many ways can you draw a Jack or an Ace from a pack? $n(J) = 4$ $n(A) = 4$ Therefore, $n(J \cup A) = n(J) + n(A)$ $= 4 + 4 = 8$
Exclusion Principle	If sets A and B have some elements in common, when finding the total number of elements the number of common elements must be subtracted from the sum of sets A and B. Written as: $n(A \cup B) = n(A) + n(B) - n(A \cap B)$ Where ∪ means 'or' and ∩ means 'and'	How many ways can you draw a Queen or a red card from a pack? $n(Q) = 4$ $n(\text{red}) = 26$ $n(Q \cap \text{red}) = 2$ Therefore, $n(Q \cap \text{red})$ $= n(Q) + n(\text{red}) - n(Q \cap \text{red})$ $= 4 + 26 - 2 = 28$
Venn diagram	Uses circles to show the relationships between different sets of objects. Letters are used to label the circles according to what they represent and numbers are written in each circle to indicate how many objects are contained in each set.	Venn diagram for drawing Queen or red from a pack:
Permutation	A permutation of a group of objects is any arrangement of the objects made in a definite order.	Permutations of letters A, B and C: ABC ACB BAC BCA CAB CBA

Kathy Brady

SKILL OR CONCEPT	DEFINITION OR DESCRIPTION	EXAMPLES
Factorial notation	Used to represent the product of consecutive numbers. The product of n consecutive numbers is written as $n!$	$6 \times 5 \times 4 \times 3 \times 2 \times 1 = 6!$
Calculating permutations	The number of permutations of r objects chosen from a set of n objects is written as P_r^n where $$P_r^n = \frac{n!}{(n-r)!}$$	$P_3^8 = \dfrac{8!}{(8-3)!} = \dfrac{8!}{5!}$ $= \dfrac{8 \times 7 \times 6 \times 5!}{5!}$ $= 8 \times 7 \times 6$ $= 336$
Combination	A combination is a group of objects that disregards how they are ordered or arranged.	Combinations 3 letters selected from the letters ABCDE: ABC ABD ABE ACE ACE ADE BCD BCE BDE CDE
Calculating combinations	The number of combinations of r objects that are chosen from a set of n objects is written as C_r^n where $$C_r^n = \frac{n!}{(n-r)!r!}$$	$C_4^7 = \dfrac{7!}{(7-4)!4!}$ $= \dfrac{7!}{3!4!}$ $= \dfrac{7 \times 6 \times 5 \times 4!}{3! \times 4!}$ $= \dfrac{7 \times 6 \times 5}{3 \times 2 \times 1}$ $= \dfrac{210}{6} = 35$
Experiment	A probability activity that generates results known as outcomes. The performance of an experiment is called a trial.	Examples of experiments: Tossing a coin Rolling a dice Drawing a card
Experimental probability	The probability of an event occurring based on the actual results of probability activity.	
Calculating experimental probability	Experimental probability = $\dfrac{\text{number of occurrences of the event}}{\text{total number of trials}}$	Experimental probability of getting a 'tail' in 25 trials that result in 11 'heads' and 14 'tails' is $\dfrac{14}{25}$.
Outcome	The result of a single trial probability experiment.	When rolling a dice, an outcome would be getting a 4.
Sample space	All the possible outcomes that could occur in a probability experiment.	When rolling a dice, the sample space is $\{1, 2, 3, 4, 5, 6\}$.
Event	A subset of the outcomes from a specified sample space.	When rolling a dice, an event would be getting an odd number.
Theoretical probability	The probability of an event occurring based on what you *expect* to happen when an experiment is conducted.	
Calculating theoretical probability	Theoretical probability = $\dfrac{\text{number of ways that an event can occur}}{\text{total possible number of outcomes}}$	When tossing a coin, tails can occur in one way and the total possible number of outcomes is two (H and T). So the theoretical probability of getting a tail is $\dfrac{1}{2}$.

SKILL OR CONCEPT	DEFINITION OR DESCRIPTION	EXAMPLES
Sample space list	For a straightforward probability experiment, the sample space can be written in a simple list.	 When spinning this spinner, the sample space is {1, 2, 3, 4, 5}.
Organised table	Provides a systematic way of recording all the possible outcomes for two or more trials of an experiment to avoid omissions or duplications.	Sample space for gender of 3 children in a family: <table><tr><th>1st child</th><th>2nd child</th><th>3rd child</th></tr><tr><td>G</td><td>G</td><td>G</td></tr><tr><td>G</td><td>G</td><td>B</td></tr><tr><td>G</td><td>B</td><td>G</td></tr><tr><td>G</td><td>B</td><td>B</td></tr><tr><td>B</td><td>G</td><td>G</td></tr><tr><td>B</td><td>G</td><td>B</td></tr><tr><td>B</td><td>B</td><td>G</td></tr><tr><td>B</td><td>B</td><td>B</td></tr></table>
2D grid	Useful to represent the sample space for two trials of one experiment or when two different experiments are conducted simultaneously.	2D grid for two tosses of a coin:
Tree diagram	A systematic way of recording all the possible outcomes in an experiment that involves a sequence of trials.	Tree diagram for drawing two balls from a bag with 2 red, 2 white and 2 blue balls.

SKILL OR CONCEPT	DEFINITION OR DESCRIPTION	EXAMPLES
Probability of a single event	$p(E) = \dfrac{n(E)}{n(S)}$ where $n(E)$ is the number of ways the event can occur, and $n(S)$, is the total number of outcomes in the sample space.	 When spinning this spinner to find probability of an odd number: $n(E) = n(\text{odd}) = 3$ $n(S) = 5$ $p(\text{odd}) = \dfrac{3}{5}$
Complementary events	The complement of an event, E, is all outcomes that are not part of that event. Notation for the complement is E'.	
Calculating probability of complementary events	The following relationship links the probability of E and E': $p(E) + p(E') = 1$ Which can be re-arranged to: $p(E) = 1 - p(E')$	A set of marbles numbered 1–60 is placed in a bag. What is the probability that a marble drawn at random is not a multiple of 9? There are 6 marbles that are multiple 9, therefore: $p(\text{multiple } 9) = \dfrac{6}{60} = \dfrac{1}{10}$ $p(\text{not multiple } 9) = 1 - \dfrac{1}{10} = \dfrac{9}{10}$
Mutually exclusive events	Two or more events that have no outcomes in common.	
Calculating probability of mutually exclusive events	If event A and event B have no outcomes in common, then the probability of either event A or event B is: $p(A \cup B) = p(A) + p(B)$	What is the probability of drawing a Jack or an Ace from a pack of cards? $p(J) = \dfrac{4}{52}$ $p(A) = \dfrac{4}{52}$ Therefore: $p(J \cup A) = p(J) + p(A)$ $= \dfrac{4}{52} + \dfrac{4}{52}$ $= \dfrac{8}{52} = \dfrac{2}{13}$
Non-mutually exclusive events	When two or more events have some outcomes in common.	
Calculating probability of non-mutually exclusive events	If event A and event B have some outcomes in common, then the probability of either event A or event B is given by: $p(A \cup B) = p(A) + p(B) - p(A \cap B)$	What is the probability of drawing a Queen or a red card from a pack of cards? $p(Q) = \dfrac{4}{52}$ $p(\text{red}) = \dfrac{26}{52}$ $p(Q \cap \text{red}) = \dfrac{2}{52}$ Therefore: $p(Q \cup \text{red})$ $= p(Q) + p(\text{red}) - p(Q \cap \text{red})$ $= \dfrac{4}{52} + \dfrac{26}{52} - \dfrac{2}{52}$ $= \dfrac{28}{52} = \dfrac{7}{13}$

SKILL OR CONCEPT	DEFINITION OR DESCRIPTION	EXAMPLES
Independent events	Two events are independent when the outcomes of one event do not affect the outcomes from the other event.	
Calculating the probability of independent events	If A and B are independent events, then the probability of both event A and event B occurring is: $$p(A \cap B) = p(A) \times p(B)$$ Where $\cap$ means 'and'	When tossing a coin and rolling a dice, what is the probability of a tail and an odd number? $p(T) = \dfrac{1}{2}$ $p(odd) = \dfrac{3}{6}$ $p(T \text{ and odd}) = \dfrac{1}{2} \times \dfrac{3}{6} = \dfrac{3}{12} = \dfrac{1}{4}$
Dependent events	When the outcomes of one event has an affect on the outcomes of subsequent events. Quite often involves situations where objects are selected without replacement.	
Calculating probability of dependent events	If event A and event B are dependent events, then: $p(A \text{ then } B) = p(A) \times p(B$ given that A has occurred$)$	Five balls numbered 2, 3, 5, 6 and 7 are placed in a bag. Two balls are selected at random without replacement. What is the probability that both are odd? $p(1st \text{ odd}) = \dfrac{3}{5}$ $p(2nd \text{ odd given 1st odd}) = \dfrac{2}{4}$ $p(\text{both odd}) = \dfrac{3}{5} \times \dfrac{2}{4} = \dfrac{6}{20} = \dfrac{3}{10}$
Probability tree diagram	Useful when calculating the probabilities of both independent and dependent events.	

16.9 REVIEW QUESTIONS

A SKILLS

1   On a probability scale, indicate the likelihood of the following events.

   a   You will win the lottery.

   b   You will meet someone today who has the same birthday as yours.

   c   It will rain in Melbourne in July.

   d   If you were to have a child it would be a girl.

2   A restaurant has on its menu 3 entrees, 6 main courses and 4 desserts.

   a   How many different three-course meals can be selected?

   b   The restaurant also has a special deal for a two-course meal comprising either an entrée and a main course, or a main course and a dessert. How many different two-course meals can be selected?

3   Calculate the following outcomes.

   a   If $n(A) = 12$, $n(B) = 9$ and $n(A \cup B) = 12$, find $n(A \cap B)$.

   b   If $n(A \cap B) = 5$, $n(A \cup B) = 28$ and $n(B) = 11$, find $n(A)$.

4   Evaluate:

   a $\dfrac{100!}{98!}$ b $\dfrac{48!}{50!}$

5   Evaluate:

   a P^7_3 b P^9_2 c C^8_3 d C^9_5

6

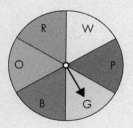

This spinner was spun 50 times and the following outcomes recorded:

W	7
P	5
G	10
B	8
O	11
R	9

What is the experimental probability that the spinner will land on B (to 2 dec. pl.)

7 In two different ways, list the sample space of the possible outcomes when this spinner is spun.

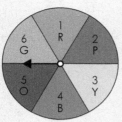

8 Use an organised table to represent the sample space of the possible orders that three students can make a presentation to the class.

9 Two spinners that are the same as in Question 7 are spun. For the first spinner the letter is recorded and for the second spinner the number is recorded. Use a 2D grid to represent the sample space for this experiment.

10 A new eatery on campus has an opening lunch special: a sandwich, a muffin and a drink for $10. Their menu is as follows:

Sandwiches: salad, chicken, ham

Muffins: blueberry, lemon

Drink: iced coffee, cola, water

Use a tree diagram to represent all of the possible lunch combinations available.

11 Eleven balls numbered 1–11 are placed in a bag and one ball is randomly drawn. What is the probability that the number drawn is:

a even?

b a multiple of 4?

c a 7?

d prime?

12 These two spinners are spun simultaneously. Construct a 2D grid to represent the sample space for this experiment. What is the probability of:

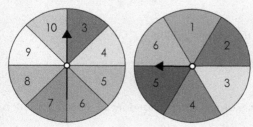

a two numbers that are the same?

b a 3 and a 7?

c a 7 or a 10?

d not a 9?

e a sum that is even?

f a sum that is less than 10?

g a sum that is no more than 10?

h a sum of 18?

13 If the letters in the word MATHEMATICS are written on individual cards and placed in a bag. What is the probability that a card drawn at random will be either an A or a consonant?

14 When this spinner is spun, what is the probability of getting a number less than 6 or an odd number?

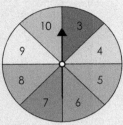

15 When a card is randomly drawn from a standard 52-card pack, what is the probability that the card will be a 7 or red?

16 These two spinners are spun simultaneously. What is the probability of getting:

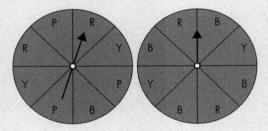

   a   a P and an R   b   two B's

17 A box contains 3 red, 2 yellow and 1 blue tickets. Two tickets are randomly selected without replacement. Draw a probability tree diagram to illustrate this experiment. Use the tree diagram to determine the probability that:

   a   both are red

   b   the first is red and the second is blue

   c   a red and a yellow are drawn

   d   a red is not drawn

18 Two cards are randomly drawn from a standard 52-card pack without replacement. What is the probability that:

   a   both cards are diamonds?

   b   both cards are the same colour?

   c   neither card is a spade?

B APPLICATIONS

1 A survey of customers at a supermarket found that 37% used the supermarket's loyalty card, 57% brought their own bags and 22% did both. Draw a Venn diagram to represent this information. Use the Venn diagram to determine the percentage of customers that:

   a use the loyalty card but do not bring their own bags

   b use the loyalty card given they bring their own bags

   c neither use the loyalty card nor bring their own bags

2 Stephanie has a lock on her locker that requires 3 numbers from 1–30 to be entered in a particular order. Unfortunately, Stephanie has forgotten the 3-number code required for her locker and so she decides to try all of the possibilities. What is the total number of possibilities that Stephanie could be required to try?

3 A pianist has prepared 10 well-practiced pieces. In how many ways can the pianist choose 6 of these pieces for a recital?

4 The following table show the marks and grade that first year students received for their end year Economics exam.

MARK (GRADE)	NO. OF STUDENTS
0–49 (Fail)	32
50–64 (Pass)	95
65–74 (Credit)	316
75–84 (Distinction)	213
85–100 (High Distinction)	144

What is the experimental probability (to 2 dec. pl.) that a student selected at random will have:

   a failed?

   b received between 50–74 marks?

   c received at least a Distinction?

5 Gregor Mendel (1822–1844) has been called the father of genetics. Through the experiments he conducted using a garden, that he was able to formulate laws that explain how we genetically inherit many of our characteristics. Through his experiments, Mendel concluded that any offspring inherits one gene from each parent to create to gene pair that will determine the inheritance of a particular characteristic (Source: eplantscience.com).

In his original experiment, Mendel identified the genes that control the height of the plant. These could be labelled as either T (for tall) or t (for short). The gene T is dominant, so a plant with the gene pair TT or Tt will be tall, whilst a plant with the gene pair tt will be short. The following table provides the possible gene pairs of an offspring of both parents having the gene pairs Tt:

		Parent 1	
		T	t
Parent 2	T	TT	Tt
	t	Tt	tt

a Use this table to find the probability that an offspring plant will be tall.

b Use this table to find the probability that an offspring plant will be short.

c Draw another table to represent the possible gene pairs in the offspring if the parents have gene pairs Tt and tt.

d Use this table to determine the probability that an offspring plant will be tall, and then short.

e Now draw another table to determine the probability that an offspring plant will be either tall or short if the parents have gene pairs TT and Tt.

6 The cook in the campus cafeteria decides at random what main course to serve for lunch. He chooses from pasta, schnitzel, burgers and quiche. Draw a 2D grid to represent the possible outcomes of what could be served on two consecutive days.

a What is the probability that he will serve:

i   quiche on both days?

ii the same meal on both days?

iii schnitzel or burgers on both days?

b Helen cannot eat pasta, what is the possibility that she will not be able to select a meal on at least one day?

7 A roulette wheel has 38 slots. Two of the slots are numbered 0 and 00 and the others are numbered 1–36. A player places their wager on a number between 1 and 36 and wins if the ball thrown on to the wheel lands on the selected number. What is the probability that:

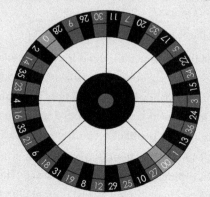

a a player wins in two consecutive spins of the wheel?

b the ball lands on 00 or a multiple of 3?

c the ball lands on an even number or a number greater than 25?

8 A football game, a soccer game and a netball game are being played in Melbourne on the weekend. There is an 80% probability that the football game will receive television coverage, a 50% probability that the soccer game will be televised, and a 30% probability that the netball game will be covered. A particular sports commentator covers 20% of football games, 40% of soccer games and 60% of netball games.

a What is the probability the commentator will be covering the netball?

b What is the probability the commentator will not be covering the football?

c Which sport is the commentator most likely to be covering?

d Which sport is the commentator least likely to be covering?

SOLUTIONS TO CHAPTER REVIEW QUESTIONS

CHAPTER 2

A   SKILLS

1 a   ten thousands   b   hundreds
   c   thousands   d   ones

2 a   135 419   b   3 411 728
   c   32 964 176

3 a   1222   b   977   c   19 601
   d   1842   e   17 481   f   252 427

4 a   5103   b   162   c   6237
   d   756 R4   e   153 576   f   62
   g   3792   h   193   i   168 630
   j   693 R1   k   3 842 818   l   87 R16

5 a   37   b   259
   c   45   d   15

6 a   19 860   b   5 823 800
   c   26 800 000   d   99 000

7 a   40 000   b   2 100 000
   c   125 000   d   9 010 000

B   APPLICATIONS

1   430 runs

2 a   spring 60 mm, summer 114 mm, autumn
       134 mm, winter 225 mm
   b   spring 60 mm, summer 110 mm, autumn
       130 mm, winter 230 mm
   c   533 mm
   d   spring 20 mm, winter 75 mm
   e   spring 20 mm, winter 80 mm

3   $69

4   $16

5   23 320

6   7140 m²

7   13 kilometres per litre

8   18 months

9   248 pieces

10 $130

11 $221

12 86 400 seconds

13 Yes, 882 pencils

14 230 boxes

15 12 buses

16 1100 students

17 6720 students

18 $494 000

CHAPTER 3

A   SKILLS

1 a   1, 2, 4, 7, 14, 28
   b   1, 2, 4, 23, 46, 92
   c   1, 3, 5, 9, 15, 25, 45, 75, 225
   d   1, 2, 4, 5, 8, 10, 16, 20, 25, 40, 50, 80, 100,
       200, 400
   e   1, 3, 9, 17, 51, 153

2   225 and 400 are perfect squares

3 a   11, 22, 33, 44, 55, 66, 77, 88, 99, 110
   b   7, 14, 21, 28, 35, 42, 49, 56, 63, 70
   c   9, 18, 27, 36, 45, 54, 63, 72, 81, 90
   d   22, 44, 66, 88, 110, 132, 154, 176, 198, 220
   e   13, 26, 39, 52, 65, 78, 91, 104, 117, 130

4 a   637 is a multiple of 13
   b   343 is a multiple of 7
   c   300 is not a multiple 9

d   508 is a multiple of 4

e   262 is not a multiple of 8

5 a prime   b neither

   c composite   d composite

   e neither

6 a Since the digits of 8208 add up to a
      multiple of 9, 8208 is divisible by 9.

   b Since 4020 ends in a zero, it is divisible
      by 10.

   c Since the last three digits of 6442 make
      a number that is not divisible by 8, 6442
      is not divisible by 8.

   d Since 2095 is not even, it is not divisible
      by 2.

   e Since 9876 is divisible by 2 (it is even)
      and divisible by 3 (digits add up to a
      multiple of 3), it is divisible by 6.

7 a $2 \times 5 \times 19$ b 17

   c $2 \times 3 \times 3 \times 47$ d $5 \times 7 \times 7$

   e 3×5

8 a 8

   b no common factors

   c 4

   d 15

   e 6

9 All the factors of 56 are: 1, 2, 4, 7, **8**, 14, 28
   and 56.

   All the factors of 120 are 1, 2, 3, 4, 5, 6, **8**,
   10, 12, 15, 20, 24, 30, 40, 60 and 120.

   Since 8 is the highest number that appears
   in each list, the HCF of both numbers is 8.
   This confirms 8(a).

   All the factors of 12 are: 1, 2, 3, 4, **6** and 12.

   All the factors of 36 are: 1, 2, 3, 4, **6**, 9, 12,
   18 and 36.

   All the factors of 90 are: 1, 2, 3, 5, **6**, 9, 10,
   15, 18, 30, 45 and 90.

Since 6 is the highest number that appears
in each list, the HCF of both numbers is 6.
This confirms 8(e).

10 a 840   b 9408   c 1320

    d 300   e 180

11 Multiples of 56 are 56, 112, 168, 224, 280,
    336, 392, 448, 504, 560, 616, 672, 728, 784,
    840, 896, …

    Multiples of 120 are 120, 240, 360, 480,
    600, 720, **840**, 960, …

    The lowest multiple that appears in both
    lists is 840, so the LCM of both numbers is
    840. This confirms 10(a).

    Multiples of 12 are 12, 24, 36, 48, 60, 72, 84,
    96, 108, 120, 132, 144, 156, 168, **180**, 192, …

    Multiples of 36 are 36, 72, 108, 144, **180**,
    216, …

    Multiples of 90 are 90, **180**, 270, …

    The lowest multiple that appears in all
    three lists is 180, so the LCM of the three
    numbers is 180. This confirms 10(e).

12 a 361   b 289   c 529

    d 400   e 225

13 a $24 = 2 \times 2 \times 2 \times 3$, so square root of
        24 cannot be found using this method

    b 30

    c 17

    d 21

    e $425 = 5 \times 5 \times 17$ so square root of 17
       cannot be found using this method

B   APPLICATIONS

1 Every three hours

2 6 bouquets (with 3 roses and 5 tulips each)

3 2 groups of 20, 4 groups of 10, 5 groups of 8

4 10 rows with 8 trees per row

5 No (there will be 4 left over)

6   150th caller will receive both prizes

7   In 30 days time

8   5 guards

9   48 hot dogs

10 17 trees

11 $30

12 400 cm^2

13 Board side length 24 cm. Square side length 3 cm and square area 3 cm^2

14 2025 cm^2

15 2000 cm or 20 m

CHAPTER 4

A   SKILLS

1   a   −7   b   4
    c   11   d   76
    e   80   f   −27
    g   0   h   9
    i   −17   j   90

2   a   −18   b   16
    c   11   d   9
    e   −2   f   −7
    g   −20   h   −2
    i   8   j   19

3   a   −4   b   −6
    c   −30   d   56
    e   42   f   33
    g   −6   h   −16
    i   12   j   320

4   a   −40   b   −24
    c   −196   d   18
    e   −4   f   78
    g   −2304   h   40

i   10   j   120
k   −24   l   −16

5   a   −61   b   −917
    c   −40   d   15
    e   −12   f   2
    g   −8   h   49
    i   −91   j   36
    k   20   l   −8

B   APPLICATIONS

1   No (gained 9 yards)

2   3 kg above (+3 kg)

3   15th floor

4   Not quite; they were $200 under or −$200

5   −$8

6   9 m below sea level (−9 metres)

7   266 less than original population

8   52.5 degrees Celsius

9   Profitable by $7624 (+$7624)

10 75 degrees Celsius (+75 degrees Celsius)

11 Fell 14 points (−14 points)

12 $1 (+1)

CHAPTER 5

A   SKILLS

1   a $\frac{11}{4}$ b $\frac{32}{3}$

2   a $3\frac{3}{8}$ b $8\frac{2}{9}$

3   a   20   b   48

4   a $\frac{2}{5} < \frac{3}{7}$ b $\frac{7}{10} > \frac{5}{9}$

5   a $\frac{7}{10}, \frac{4}{5}, \frac{17}{20}$ b $\frac{7}{12}, \frac{11}{16}, \frac{5}{6}, \frac{7}{8}$

6   a $\frac{11}{3}$ b $\frac{5}{7}$

7   a $\frac{5}{6}$ b $\frac{87}{56}$
    c $4\frac{3}{8}$ d $5\frac{29}{35}$

8 a $\dfrac{11}{40}$ b $\dfrac{61}{60}$

   c $1\dfrac{14}{15}$ d $1\dfrac{17}{30}$

9 a $\dfrac{2}{3}$ b $\dfrac{7}{96}$

   c $7\dfrac{8}{11}$ d 4

10 a $4\dfrac{1}{2}$ b $\dfrac{3}{4}$

   c $1\dfrac{1}{2}$ d $5\dfrac{2}{5}$

B   APPLICATIONS

1 $\dfrac{11}{441}$

2 G

3 $\dfrac{53}{100}$

4 $3\dfrac{1}{3}$ cm

5 $\dfrac{9}{14}$

6 a $1\dfrac{1}{5}$ m   b $2\dfrac{22}{25}$ m^2

7 20°C

8 30 grams

CHAPTER 6

A   SKILLS

1 38.63, 38.36, 36.83, 36.38, 33.86, 33.68

2 a 17.84   b 53.956

   c 6.28   d 22.708

3 a 34.02   b 13.6773

   c 143.5   d 346

4 a 85.84   b 143.55   c 12.69

5 a 19.9   b 581.37   c 67.675

6 a 4000   b 0.025   c 115

7 a 4   b 87

8 a 12%   b 75%

9 a 1700   b 1400

10 a 62.8%   b 141%

   c 44%   d 12.5%

11 a $\dfrac{9}{20}$ b $\dfrac{11}{40}$

   c $\dfrac{2}{125}$ d $\dfrac{57}{125}$

12 a 0.208   b 1.46

   c 2.6   d $0.8\overline{1}$

B   APPLICATIONS

1 a Border, Hayden, Waugh, Ponting

   b 3rd

2 a 383.7 mm   b 380 mm

3 15.85 cm

4 11.25 cm

5 95.9 mm

6 a $1040   b $832

   c $32 profit

7 26.67% decrease

8 181.82% increase

9 $4080

10 a $6615   b $15

11 7.8 hours

CHAPTER 7

A   SKILLS

1 a 16 : 3   b 1 : 5

   c 164 : 101   d 30 : 1

   e 2 : 9   f 1 : 9

   g 1 : 12   h 7 : 1

   i. 2 : 13   j 2 : 27

   k 19 : 51   l 7 : 1

   m 67 : 114   n 435 : 2

   o 1 : 157

2 a $\dfrac{7}{8}$, 87.5%, 0.875

   b $\dfrac{95}{11}$, 863.6%, 8.636

   c 5, 500%, 5

   d $\dfrac{1}{3}$, 33.3%, 0.333

   e $\dfrac{2}{5}$, 40%, 0.4

   f $\dfrac{4}{11}$, 36.4%, 0.364

3   a   20 emails per 3 days

    b   3 square metres of carpet per $275

    c   0.25 L per serve

    d   1 nurse per 5 patients

    e   1 million hits per 3 months

4   a   53.6 heart beats per minute

    b   5.67 shots per basket

    c $3.23 per kilogram

    d   302 kilojoules per cookie

    e $3.20 per day

5   a   No   b   Yes   c   No

    d   Yes   e   No   f   Yes

6   a $x = 5$ b $x = 52.5$

    c $x = 20$ d $x = 32$

    e $x = 38$ f $x = 2.1$

B   APPLICATIONS

1   a   9 males : 4 females

    b   2 oranges : 1 apple

    c   125 handmade chocolates : 1 substandard chocolate

    d   Price : earnings = 14 : 1

    e   15 minutes : 14 minutes

2   a   8 kilometres : 15 minutes

    b   11 kilometres : 2 days

    c   11 hits : 30 appearances at the plate

    d $4 : 1 hour

    e $403 : 1 night

3   a   13.1 kilometres per minute

    b $1.06 per litre

    c   13.5 points per unit

    d   6.67 milligrams of medication per minute (2 d.p.)

    e   96 minutes per day

4   3 : 1

5   $0.06 per gram

6   1 : 5500

7   $1.60 per notebook (2 d.p.)

8   $\frac{1}{6}$ litres of oil

9   In the first class (females : males = 15 : 17)

10   18 cm

11   Incumbent : total = 248 : 471, opponent : total = 1007 : 2355

12   38 people per ride (rounded to nearest whole)

13   8.57 litres per minute (2 d.p.)

14   250 g packet: 1.04 cents per gram, 425 g packet: 1.08 cents per gram, 700 g packet: 0.99 cents per gram, so the 700 g packet is the best value

15   67.2 cm long

16   8 milligrams codeine and 192 milligrams ibuprofen

17   Packet of 4 is $0.67 per battery, packet of 8 is $0.53 per battery, packet of 10 is $0.59 per battery, so the packet of 8 is the cheapest per battery

18   7.65 m

CHAPTER 8

A   SKILLS

1   a 6^6

    b $\left(\frac{1}{3}\right)^4$

    c $(-0.75)^3$

2   a   256   b $\frac{8}{27}$ c -3.375

3   a 4^9 b $(-7)^6$ c $(-3)^3$

    d $\frac{11^2}{7^2}$ e 8^6 f   1

4   a $\frac{1}{6^2}$ b $\frac{1}{7^2}$

    c $\frac{1}{5^4}$ d $\frac{1}{4^8}$

5   a   7   b   10

    c $\frac{1}{12}$ d $\frac{1}{4}$

6 a $\dfrac{1}{6^4}$ b $\dfrac{1}{7^{\frac{1}{2}}}$ c $\dfrac{1}{3}$

7 a 3.71×10^8 b 8.54×10^{-6}

8 a $481\,000$ b 0.000001095

9 a 4.234×10^{-2} b 2.349×10^{-2}

 c 5.25×10^3 d 3.75×10^5

10 a 4   b 0   c -6

12 a $\log 21$ b $\log 4000$

 c $\log 1 = 0$ d $\log 9$

 e $\log 25$ f $\log 3$

 g $\log 20$ h $\log 9$

13 a $\log 8$ b $\log 72$

 c $\log 45$ d $\log 6$

 e $\log 20$

B   APPLICATIONS

1 a 10^8 b 10^{24}

2 c 8000 e 25

3 a 1.392×10^6

 b 4.76×10^9

 c 3.42×10^3, 9.4 years

 d May 2015

4 a 27×10^{-9}

5 a $10 \times \log 10^6$

 b 60

 c 110

 d The vacuum is 10^6 more intense than the whisper.

6 a pH $= 4$, acidic

 b pH $= 12$, alkaline

CHAPTER 9

A   SKILLS

1 a $2x - 3y + 2$ b $-2e + 3h + 5$

 c $\dfrac{5b}{a}$ d $3x - \dfrac{y}{4}$

2 a $p + 8$ b $4xyz$

 c $b - \dfrac{a}{4}$ d $\dfrac{d+3}{c}$ or $\dfrac{c}{d+3}$

3 a 10   b 9

 c 4   d 100

4 a 4   b 12

 c -1 d -1

5

x	-2	-1	0	1	2
$-4x + 3$	11	7	3	-1	-5

6 a $2d + 7$

 b $-2x + 5$

 c $7a^2 + 2a$

 d $-3xy + x^2y + xy^2$

7 a $5t - 10$ b $-3p + 12$

 c $2ab + 2ac$ d $3y^2 + 2yz$

B   APPLICATIONS

1

m	5	10	15	20	25
$25 + 40m$	225	425	625	825	1025

2 a $x, x + 5, 2x - 3$ b $4x + 2$

3 a $400t$

 b $t + 50$

 c $200(t + 50)$

 d $400t + 200(t + 50)$

 e $600t + 1000$

4 $S = 2\pi rh + 2\pi r^2$

5 a $7m + 4f$ b $14m + 8f$

 c $4m + 4f$ d $18m + 12f$

 e $\$324$

CHAPTER 10

A   SKILLS

1 a chart   b Quadrant 2

 c line of best fit; scatter plot

 d categorical   e segmented

 f vertical   g pictograph

h   Cartesian plane

i   length; proportional

j   Quadrant 4

k   graphs

l   percentage

m   origin

n   numerical

o   line graph

p   horizontal

q   continuous

2

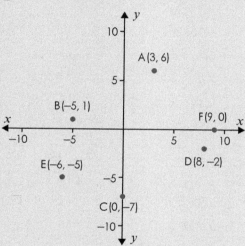

3   a   Quadrant 3   b   Quadrant 1
     c   Quadrant 4   d   Quadrant 2

4

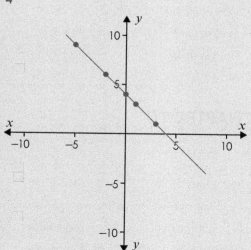

B   APPLICATIONS

1   a   50

     b   7

     c   58; 2000

     d   43

2

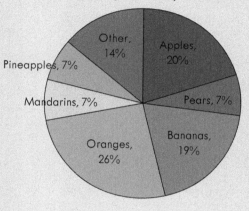

Australian fruit production (30 June 2014)

a   266, 000kg

3   a   Perth: 32 degrees; Hobart: 22 degrees

     b   September, 5 degrees

     c   February, 10 degrees

4   a   Qld, approx. 79%

     b   NSW, approx. 21%

     c   NSW, approx. 25%

     d   approx. 11%

     e   approx. 82%

5

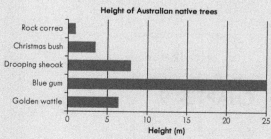

Height of Australian native trees

6

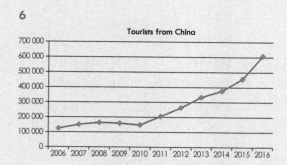

Tourists from China

7

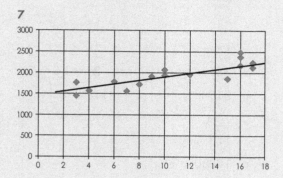

CHAPTER 11

A   SKILLS

1 a

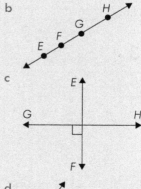

b

c

d

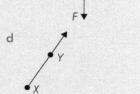

e

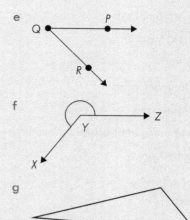

f

g

h

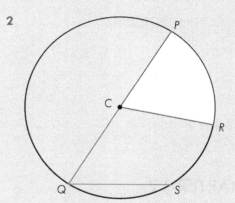

2

3 complement $= 31°$; supplement $= 121°$

   b supplement $= 54°$

4 a $y = 40°$

   b $p = 145°; q = 145°; r = 35°$

   c $a = 125°; b = 55°$

   d $x = 58°$

   e $y = 108°$

5 a $z = 103°$

   b $e = 43°$

   c $w = 34°$

d $a = 61°; b = 133°$

e $y = 110°; x = 75°$

f $d = 108°$

6 a rectangular prism

   b oblique cone

   c right pentagonal-based pyramid

   d octahedron

7 10

8 a 15 cm

   b 10.4 mm

   c 2.8 cm

9 a 4.6 m

   b $x = 14.3$ cm; $y = 17.4$ cm

   c 2.2 mm

B   APPLICATIONS

1 a 3 m   b 3.9 m

2 a 11.4 m   b 11.5 m

3 180 m

4 26.5 km

5 a 1.25 m   b 3.7 m

CHAPTER 12

A   SKILLS

1 a 0.074 km   b 0.061 μg

   c 4900 kL   d 1 300 000 cm^2

   e 0.67 cm^3 f 5.5 kL

2 a 25.7 mm

   b 21.9 cm

   c 52.28 m

3 a 1.52 hectares

   b 272 mm^2

   c 19.2 m^2

d 942 cm^2

e 76.56 m^2

4 a 144 mm^3 b 800 cm^3

   c 28.26 m^3 d 480 cm^3

   e 4.19 m^3 f 785 mm^3

5 1356.48 cm^3

B   APPLICATIONS

1 a 5   b 1.5 m^2

   c 1.8 m   d 2.1 × 1.8

   e 7.56 m^2 f 2.1 × 1.5

   g 3.15 m^2 h 12.21 m^2

2 a 12 m   b 10.38 m^2

   c 14.35 m^3 d 14.53 kL

3 a 1 000 000 m^2

   b 5 831 000 m^2

   c 583.1 hectares

4 a 376.8 m   b 11 304 m^2

5 b 23.73 cm

   c 11.87 cm

   d 6993 cm^3

   e 7 L

   f The size of the ball indicates its capacity in litres

CHAPTER 13

A   SKILLS

1 a $m = 7$

   b $k = -2$

2 a

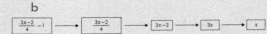

   b

3 a $x = 4$ b $c = -5$

  c $d = -7$ d $y = -21$

4 a $x - 14 = 3; x = 17$

  b $x + 8 = 12; x = 4$

  c $\frac{x}{6} = 7; x = 42$

  d $7x = 84; x = 12$

5 a $x = 2$ b $x = -4$

  c $x = -17$ d $x = -9$

6 a $x = 11$ b $x = -2$

  c $x = \frac{18}{7}$ d $x = 3$

  e $x = 3$ f $x = \frac{8}{7}$

7 $V = 628; A = 121$

8 a $h = \frac{3V}{\pi r^2}$ b $r = \sqrt{\frac{3V}{\pi h}}$

9 $I = \sqrt{\frac{V}{R}}; I = 5$

  $P = \frac{A}{1 + rt}; P = \$166.67$

B   APPLICATIONS

1 a $240 + 2w$ b $w = 90$ metres

2 a 168

  b $\frac{m + 168}{3}$

  c $\frac{m + 168}{3} = 89; m = 99$

3 a $2h + 74$ b $10h - 247$

  c $13h - 173$

  d $13h - 173 = 1062$

  e $h = 95$

  f liver $= 264$; kidney $= 7$

4 a $f + 240$

  b $\frac{f}{5} + 40$

  c $\frac{11}{5}f + 280$

  d $\frac{11}{5}f + 280 = 4680$

  e 2000

  f carbohydrates $= 2240$; protein $= 440$

5 $a = 6$

6 a $10 - x$ b $0.3(10 - x)$

  c $0.1x + 0.3(10 - x); 3 - 0.2x$

  d 1.5 L   e $3 - 0.2x = 1.5$

  f 7.5 L   g 2.5 L

7 29.3

8 1200

CHAPTER 14

A   SKILLS

1

x	-2	-1	0	1	2
y	9	7	5	3	1

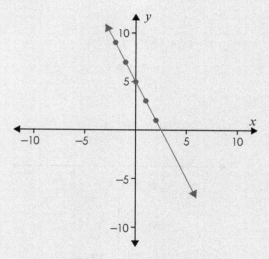

2 x-intercept $= 5$; y-intercept $= 4$

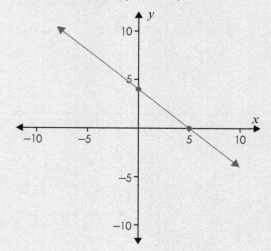

3   a   negative   b   undefined

    c   positive   d   zero

4   a $-\dfrac{2}{3}$ b -1

5   (a) and (c) are parallel; (b) and (d) are parallel

    Both (a) and (c) are perpendicular to (b) and (d)

6

SLOPE	y-AXIS INTERCEPT	EQUATION
-0.8	$(0, 2)$	$y = -0.8x + 2$
5	$(0, 2.5)$	$y = 5x + 2.5$
$\dfrac{5}{6}$	$\left(0, \dfrac{3}{2}\right)$	$y = \dfrac{5}{6}x + \dfrac{3}{2}$
undefined	$(0, 0)$	$x = 0$

7   a $y = -5x + 1$

    b $y = -4x + 13$

8   a

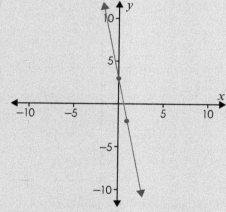

    b

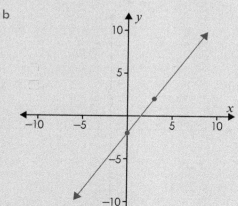

c

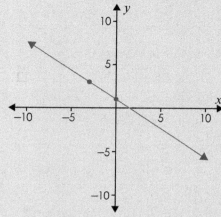

B   APPLICATIONS

1   a

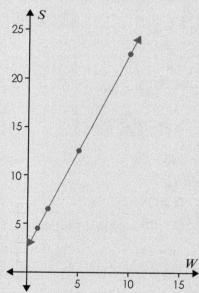

The graph cannot extend into the negative quadrants because neither the weight nor the length of the spring can be negative values.

    b   Slope $= 2$

    c   Using $(1, 4.5)$: $S - 4.5 = 2(W - 1)$

    d $S = 2W + 2.5$

    e 2.5

    f   Yes

    g   This value is the length of the spring before weight was added

h $S = 2 \times 15 + 2.5$
    $S = 32.5\,\text{cm}$

i $10 = 2W + 2.5$
    $W = 3.75\,\text{kg}$

2 a   Degrees F is the dependent variable; degrees C is the independent variable

b   Degrees F on the y-axis; degrees C on the x-axis

c   Slope $= \frac{9}{5}$; vertical axis intercept $= 32$

d

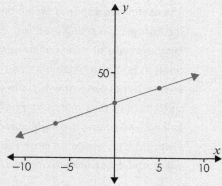

e   The graph can be extended into the negative quadrants because both degrees C and degrees F can be recorded in negative values.

f   The graph crosses the vertical axis when degrees $C = 0$.

g   The graph crosses the horizontal axis at -17.8. This is value of degrees C that equal to degrees $F = 0$.

h $F = \frac{9}{5} \times 10 + 32; F = 50°$

i $20 = \frac{9}{5}C + 32; C = \left(\frac{20}{3}\right)°$

CHAPTER 15

A   SKILLS

1 a   Mean 5.9, Median 6.5, Mode 7

b   Mean 49, Median 45, Mode 34 and 45

c   Mean 14.3 (1 d.p.), Median 16, since every value occurs only once the set

can be described either as having no mode or as every number being the mode

2 a   Range $= 6$

b   Minimum 4, Q1 5.5, Q2 7, Q3 8, Maximum 10

c   IQR $= 2.5$

d

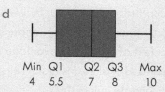

Min	Q1	Q2	Q3	Max
4	5.5	7	8	10

3 a   Range $= 19$

b   Min 10, Q1 16, Q2 19, Q3 24.5, Max 29

c   IQR $= 8.5$

d

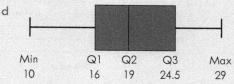

Min		Q1	Q2	Q3	Max
10		16	19	24.5	29

4 4.06

5 a   4.18

b   This set has slightly greater variability because its standard deviation is slightly bigger.

6 a

DATA VALUE	TALLY	FREQUENCY
1–5	III	3
6–10	NHII	7
11–15	III	3
16–20	NHII	7

b

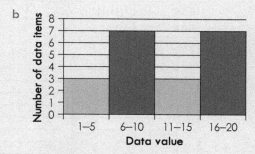

There are two modal categories, shown in the darker shade of blue: 6–10 and 16–20.

c

Stem	Leaf
0	3 4 5 6 7 8 8 9 9
1	0 4 5 5 6 6 7 7 7 8 9

The median value is $\dfrac{(10 + 14)}{2} = 12$.

d In this case, the histogram is probably a better representation because it provides more information about the overall pattern of the data. However, which diagram is better may depend on the context of use. One advantage of the stem-and-leaf plot is that it retains individual data values.

e

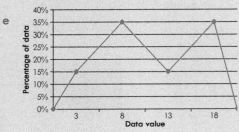

f

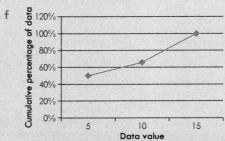

7 a

DATA VALUE	TALLY	FREQUENCY
1–5	IIII	4
6–10	II	2
11–15	I	1
16–20	IIIIIIII	8
21–15	III	3
26–30	II	2

b

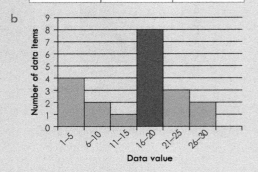

The modal category, as highlighted on the histogram, is 16–20.

c

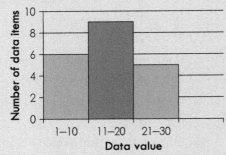

d The usefulness depends on the analysis the histogram is being used for. Histogram (b) provides a finer-grained analysis because of the bin size of 5, so trends that are lost with a bin size of 10, such as only one data item being in the 11–15 range, are highlighted. Histogram (c) would be more useful for a broader, less detailed analysis.

e

Stem	Leaf
0	2 3 4 8 9
1	4 6 6 7 8 9
2	0 0 0 2 5 6 7 7

The median value, as highlighted on the stem-and-leaf plot, is 18.

f The histogram provides a visual picture of the data (an advantage) that allows trends to be easily visible. On the other hand, individual data points are lost (a disadvantage). The stem-and-leaf plot retains individual data points (an advantage) but offers only one 'bin size' (groups of 10) so it can be harder to highlight interesting anomalies in the data (a disadvantage).

g

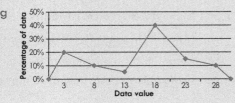

h

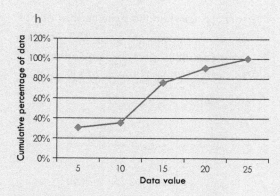

8 a Each data set will have the same number of data values above the mean because each data set is normally distributed and has an equal number of data values. In a normal distribution, 50% of values occur above the mean and 50% below.

b A

c A

d C

9 a The proportion of values less than 50% will be *less* than the proportion of values that are 50% or more.

b The proportion of values less than 60% will be *more* than the proportion of values that are 70% or more.

c i 47.5%   ii 16%

10 a 0.5

b The scores are the same (in both cases $z = 0.5$)

c The z-score in part (a) is relatively lower.

B APPLICATIONS

1 Mean $11.45; Median $7.75; Mode $7.50

2 a Minimum = 52000, Q1 = 56000, Q2 = 75000, Q3 = 95000, Maximum = 300000

b

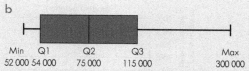

3 a Minimum = 52, Q1 = 54, Q2 = 75, Q3 = 115, Maximum = 300

b

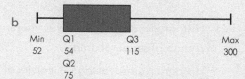

4 a

DATA VALUE	TALLY	FREQUENCY
75–79	|||	4
80–84	||||	4
85–89	|||	3
90–94	||	2
95–99			1

b Minimum = 77, Q1 = 81, Q2 = 84.5, Q3 = 89, Maximum = 97

c IQR = 8

d

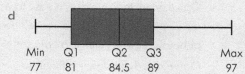

5 a This histogram uses a bin size of 5 but a bin size of 10 would also have been appropriate.

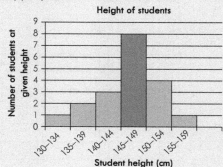

b

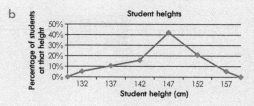

c
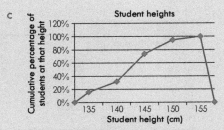

d   Histogram advantage – gives visual picture of shape of data

Histogram disadvantage – individual data values lost

Percentage polygon advantage – displays percentage of data values in each group rather than a count of data values, which can be more useful for understanding the size of one data group relative to another

Percentage polygon disadvantage – individual data values lost

Ogive advantage – makes it easy to work out what percentage of values lie above or below a given value in a data set

Ogive disadvantage – difficult to compare frequencies between each data group

6 a
DATA VALUE	TALLY	FREQUENCY
1		0
2	II	2
3		0
4	I	1
5	III	3
6	THL	5
7	THL	5
8	II	2
9	I	1
10	I	1

b
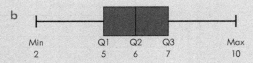

	Min	Q1	Q2	Q3	Max
	2	5	6	7	10

7   Tom and Sam had the same number of average disposals per game (27). Tom had more variability (standard deviation for Tom = 8.37 (2 d.p.), Sam = 7.72 (2 d.p.)).

8   IQR = 2

9 a
Stem	Leaf
5	06 56 61 98
6	30
7	21 34 63 65 72 78 90
8	24 46 82

b   Mean lasting time = 715 hours (rounded to nearest hour)

10 a
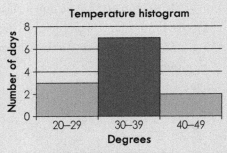

The modal category, as shown on the graph, is 30–39.

b
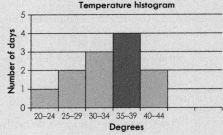

The modal category, as shown on the graph, is 35–39.

c   Histogram (b) provides more information about the breakdown of temperatures and highlights the relative frequency of days of 35–39 degrees compared to other days. Histogram (a) provides less detail but a clear overall picture of the data. Either could be more useful depending on context.

d

Stem	Leaf
2	1 5 7
3	1 1 1 5 6 8 9
4	0 1

The median value is 33, as shown on the graph (midpoint between 31 and 35).

e

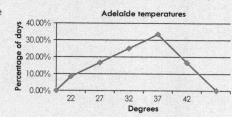

f

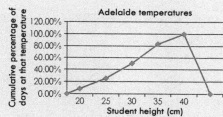

g Histogram bin size 10 advantage — provides excellent visual representation of overall data trends

Histogram bin size 10 disadvantage — individual data values are lost

Histogram bin size 5 advantage — provides more detail while still giving good overall picture of data

Histogram bin size 5 disadvantage — individual data values are lost

Stem-and-leaf plot advantage — individual data values are retained

Stem-and-leaf plot disadvantage — fixed bin size

Percentage polygon advantage —more useful for understanding the size of one data group relative to another

Percentage polygon disadvantage — individual data values lost

Ogive advantage — makes it easy to work out what percentage of values lie above or below a given value in a data set

Ogive disadvantage — difficult to compare frequencies between each data group

11 Jim, Viv and Jordan will all gain entry to the University. Tom will not.

12 a 2.5%   b 0.15%   c 68%

13 a 10   b 272   c 336

14 a Amy's z-score is 0.5

b. Emma did better than Amy (z-score 0.75)

c. Jack did better than Amy but worse than Emma (z-score 0.625)

CHAPTER 16

A   SKILLS

1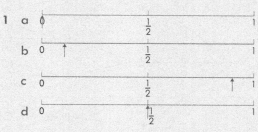

2 a 72   b 42

3 a 7   b 22

4 a 9900   b $\frac{1}{2450}$

5 a 210   b 72

c 56   d 126

6 0.16

7 {1, 2, 3, 4, 5, 6}

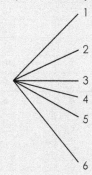

8 ABC

ACB

BAC

BCA

CAB

CBA

9

10

11 a $\dfrac{5}{11}$ b $\dfrac{2}{11}$

c $\dfrac{1}{11}$ d $\dfrac{5}{11}$

12

10	11	12	13	14	15	16
9	10	11	12	13	14	15
8	9	10	11	12	13	14
7	8	9	10	11	12	13
6	7	8	9	10	11	12
5	6	7	8	9	10	11
4	5	6	7	8	9	10
3	4	5	6	7	8	9
	1	2	3	4	5	6

a $\dfrac{1}{12}$ b $\dfrac{1}{24}$

c $\dfrac{1}{4}$ d $\dfrac{7}{8}$

e $\dfrac{1}{2}$ f $\dfrac{7}{16}$

g $\dfrac{9}{16}$ h 0

13 $\dfrac{9}{11}$

14 $\dfrac{5}{8}$

15 $\dfrac{7}{13}$

16 a $\dfrac{3}{32}$ b $\dfrac{1}{16}$

17 a $\dfrac{1}{5}$ b $\dfrac{1}{10}$

c $\dfrac{2}{5}$ d $\dfrac{1}{5}$

18 a $\dfrac{1}{17}$ b $\dfrac{4}{17}$ c $\dfrac{19}{34}$

B   APPLICATIONS

1 a 15%   b 39%   c 28%

2 24 360

3 210

4 a 0.04   b 0.51   c 0.45

5 a $\dfrac{3}{4}$ b $\dfrac{1}{4}$

c

		PARENT 1	
PARENT 2		T	t
	t	Tt	tt
	t	Tt	tt

$\dfrac{1}{2}; \dfrac{1}{2}$

d

		PARENT 1	
PARENT 2		T	T
	T	TT	TT
	t	Tt	Tt

1; 0

6

Q	QP	QS	QB	QQ
B	BP	BS	BB	BQ
S	SP	SS	SB	SQ
P	PP	PS	PB	PQ
	P	S	B	Q

Day 1 / Day 2

a $\dfrac{1}{16}$ b $\dfrac{1}{4}$ c $\dfrac{3}{4}$ d $\dfrac{7}{16}$

7 a $\dfrac{1}{1444}$ b $\dfrac{13}{38}$

c $\dfrac{23}{38}$

8 a 0.18   b 0.38

c soccer   d football

GLOSSARY

Addition Principle

If the possibilities being counted can be divided into groups with no possibilities in common, then the total number of possibilities is the sum of the number of possibilities of each group

Algebraic expression

A mathematical sentence that combines variables, constants and arithmetic operations

Angle

Formed when two rays meet at a common endpoint, known as the vertex

Area

The measurement that describes the size of a two-dimensional figure in terms of square units

Base

The value to be multiplied in an exponential

Capacity

The quantity that a three-dimensional container can hold

Categorical data

Observations that are classified or sorted into groups or categories

Certain event

An event that will have a probability of 1

Circle

A set of points that are all the same distance from a fixed centre point

Circumference (measurement)

The distance, or perimeter, around the boundary, or edge, of a circle

Coefficient

A number that is used to multiply a variable or variables in an algebraic term

Collecting like terms

The process of simplifying an algebraic expression by adding, or subtracting, the coefficients of like terms

Collinear points

Three, or more, points that lie on the same line

Combination

A group of objects that disregards how they are ordered or arranged

Complement of an event

All outcomes that are not that event

Complementary angles

Two angles that add up to 90°

Composite number

A whole number greater than 1 that can be divided exactly by whole numbers other than 1 and itself

Concurrent lines

Three, or more, lines that all intersect, or cross, at the same point

Constant

A value that is known and does not change

Continuous data

Numerical data that can occur in an infinite range

Degree

The unit of measure for the size of an angle

Dependent events

For two events, when the outcome of the second event is affected by the outcome from the previous event

Dependent variable

A variable in a linear relationship that has its value determined by the value of the other variable (the independent variable)

Discrete data

Numerical data that involves exact number values

Distributive law

In general terms, for any three values a, b and c:

$a(b + c) = a \times b + a \times c$ or

$a(b - c) = a \times b - a \times c$

Dividend

The number being divided into

Divisor

The number of groups that the dividend is being split into

Equation

A mathematical sentence that describes the equality of two algebraic expressions, and comprises left and right-hand sides that are separated by an equal sign

Event

A subset of the outcomes from a specified sample space

Exclusion Principle

If set A and set B have some elements in common, then the number of common elements must be subtracted from the sum of set A and set B

Experimental

A probability activity that generates results known as outcomes

Exponent

A value that indicates how many times to carry out a multiplication

Exterior angle of a polygon

An angle that lies on the outside of the polygon, which is formed by any side of the polygon and

the extension of its adjacent side

Factor

A whole number that multiplies with another whole number with the result being the product. A factor can also be defined as a whole number that divides into another whole number without a remainder.

Factorial notation

Factorial notation is used to represent the product of consecutive numbers. The product of n consecutive numbers are written as $n!$

Fraction

Any number that can be written in the form $\frac{a}{b}$, where a and b are whole numbers and b is not zero

Fractional exponent $\frac{1}{3}$

The fractional exponent $\frac{1}{3}$ represents the cube root of the base. In general terms: $a^{\frac{1}{3}} = \sqrt[3]{a}$

Fractional exponent $\frac{1}{2}$

The fractional exponent $\frac{1}{2}$ represents the square root of the base. In general terms: $a^{\frac{1}{2}} = \sqrt{a}$

Fundamental Principle of Counting

If an event A can occur in n ways and for each of these n ways, an event B can occur in m ways, then the events A and B can occur in succession in $n \times m$ ways

General form of a linear equation

A linear equation is written in the form $Ax + By + C = 0$, where A, B and C are constants

Highest Common Factor (HCF)

The largest number that is a factor of two or more numbers

Horizontal line

A straight-line graph that has a slope of 0 and the y values of each point are equal

Impossible event

An event that will have a probability of 0

Independent events

When the outcomes from one event do not affect the outcomes from the other event

Independent sample

A sample where the choice of one member does not influence the choice of another

Independent variable

A variable in a linear relationship that can be any chosen value

Integers

The set of numbers that include the counting numbers (1, 2, 3, 4, …), zero (0), and the negative of the counting numbers (−1, −2, −3, −4, …)

Interior angle of a polygon

An angle that is formed within a polygon where two adjacent sides meet

Interquartile range

The difference between the upper (Q3) and the lower quartile (Q1) of a dataset

Kite

A quadrilateral with pairs of adjacent sides equal in length

Like terms

When two or more algebraic terms contain exactly the same variable or variables, which are raised to exactly the same powers

Line

A collection of points in a straight path that extend infinitely

Line of best fit

A straight line that represents the data in a scatter plot

Line segment

A part of a line that is enclosed by two distinct endpoints

Linear equation

An equation involving two variables that produce a straight line when graphed

Logarithm of a positive number

The power to which a base of 10 must be raised to produce a positive number

Lowest common multiple

The smallest number that is a multiple of two or more numbers

Mean

The arithmetic average of a collection of numbers, which is often referred to as the average

Measure of central tendency

Indicates the middle data point of a set of data

Median

The middle value in a data set

Mode

The most frequently occurring value in a data set

Multiple

A whole number that can be divided by another number without leaving a remainder

Mutually exclusive events

Two or more events that have no outcomes in common

Negative exponent

A negative exponent in an exponential expression is equal to the reciprocal of the same expression with a positive exponent. In general terms:
$$a^{-x} = \frac{1}{a^x}$$

Negative slope

Occurs when a line goes downward from left to right

Non-mutually exclusive events

When two or more events have some outcomes in common

Normal distribution

A distribution of data where the data is evenly spread in a *bell-shaped* curve around the mean

Numerical data

Values or observations that can be counted or measured and recorded as numbers

Ordered pair

A pair of numbers that describes the position of any point on the coordinate plane, written in the form (x, y)

Outcome

The result of a single trial probability experiment

Parallel lines

Two, or more, lines in the same plane that never meet

Parallel lines (linear equations)

Two straight lines that have slopes that are equal

Parallelogram

A quadrilateral that has opposite sides parallel and equal in length, and opposite angles equal in size

Perfect square

The product of a whole number multiplied by itself

Perimeter

The length, or distance, around the boundary or edge of a two-dimensional closed figure

Permutation

Any arrangement of a group of objects made in a definite order

Perpendicular lines

Lines are at right angles to each other

Perpendicular lines (linear equations)

Two straight lines that have the product of their slopes equal to −1

Plane

A collection of points that form a two-dimensional surface that extends infinitely in all directions

Point

An exact location in space

Point–intercept form of a linear equation

A linear equation that is written in the form $y - y_1 = m(x - x_1)$, where m is the slope of the line and (x_1, y_1) is a point on the line

Polygon

A simple closed figure made up of straight, line segments

Polyhedron

A solid three-dimensional figure with faces that are all polygons

Population

A group of events, objects, results or individuals, all of which share some unifying characteristic

Positive slope

Occurs when a line goes upwards from left to right

Power rule for exponents

When a value in exponential format is raised to another power the exponent in the resultant expression is found by finding the product of the two original exponents. In general terms: $(a^x)^y = a^{xy}$

Power rule for logarithms

When the logarithm of a value is multiplied by another value, the result is the logarithms of the first value raised to the power of the second value. In general terms: $n \log a = \log a^n$

Prime number

A whole number greater than 1 with only two whole number factors: 1 and itself

Prism

A polyhedron with two opposite end faces that are identical polygons

Product rule for exponents

When two numbers with the same base are multiplied, the resultant product is obtained by adding their exponents. In general terms: $a^x \times a^y = a^{x+y}$

Product rule for logarithms

When the logarithms of two numbers are added, the resultant value is the logarithm of the product of these

numbers. In general terms: $\log a + \log b = \log (ab)$

Product

The result of the multiplication of two or more numbers

Proportion

Two or more relationships can be expressed as ratios that are equal

Pyramid

A polyhedron with a base that is a polygon and a single point located above the base called the apex

Quotient

The result of a division

Quotient rule for exponents

When two numbers with the *same base* are divided the resultant quotient is obtained by subtracting the exponent in the denominator from the exponent in the numerator. In general terms: $\frac{a^x}{a^y} = a^{x-y}$

Quotient rule for logarithms

When the logarithms of two numbers are subtracted, the resultant value is the logarithm of the quotient of the first number divided by the second number. In general terms: $\log a - \log b = \log \left(\frac{a}{b}\right)$

Random sample

A sample where all members of the population have an equal chance of selection

Range

The difference between largest and smallest values in a set of data

Rates

A special ratio that expresses a comparison between two quantities of different units

Ratio

A comparison between two or more quantities that have the same units

Ray

A part of a line that has one endpoint and continues infinitely in the other direction

Regular polyhedron

A polyhedron with faces that are all identical regular polygons

Rounding

The process of approximating a number to a given place value

Rule in a formula

Indicates how the subject should be calculated

Sample

A portion of a population that represents the whole population

Sample space

All the possible outcomes that could occur in a probability experiment

Significant figure

An approximation to a specified number of digits, starting from the highest place value digits in the number

Skew lines

Two lines in space that do not intersect and are not parallel

Slope

A number that describes the steepness and direction of a straight-line graph

Slope–intercept form of a linear equation

A linear equation that is written as $y = mx + c$, where m is the slope and c is the y-axis intercept at the point $(0, c)$

Solution of an equation

The value of the variable required to make the LHS of the equation equal to the RHS of the equation

Space

An infinite three-dimensional region in which objects can have both position and direction

Square root

A value that, when multiplied by itself, gives the number

Standard deviation

A measure of the average distance of all data points from the mean and is represented by the Greek letter σ

Subject of a formula

The value that is being determined

Supplementary angles

Two angles that add up to 180°

Term

A component of an algebraic expression that is a constant, a variable or the product of constants and variables

Trapezium

A quadrilateral that has only one pair of parallel sides

Trial

The performance of a probability experiment

Uniform cross-section

Cross-sections that are always identical to each other and the base or ends of the figure

Unit rate

A rate in which the second named value is one

Variable

A letter that is used to represent an unknown value or a value that can change

Vertical line

A straight line that has a slope that is undefined and the x values of each point are equal

Vertically opposite angles

Angles that are opposite each other, which are formed when two lines intersect

Volume

The amount of space that a three-dimensional figure occupies

Whole number

A positive number with no fraction or decimal part that is comprised of at least one of the digits from 0 to 9

x-intercept

The point where a graph crosses the x-axis, when $y = 0$

y-intercept

The point where a graph crosses the y-axis, when $x = 0$

INDEX